Mao-Hong Yu

Generalized Plasticity

Mao-Hong Yu
Guo-Wei Ma
Hong-Fu Qiang
Yong-Qiang Zhang

Generalized Plasticity

With 315 Figures

Springer

Professor Mao-Hong Yu
Xian Jiaotong University
School of Civil Engineering and Mechanics
710049 Xian, People's Republic of China
E-mail: mhyu@mail.xjtu.edu.cn

Co-Authors

Guo-Wei Ma
Nanyang Technological University, Singapore

Hong-Fu Qiang
Yi'an Hi-Tech Research Institute, China

Yong-Qiang Zhang
National University of Singapore, Singapore

Library of Congress Control Number: 2005932191

ISBN-10 3-540-25127-8 Springer Berlin Heidelberg New York
ISBN-13 978-3-540-25127-0 Springer Berlin Heidelberg New York

Springer is a part of Springer Science+Business Media.

springeronline.com

Printed in Germany

Camera-ready by the Author and SPI, India
Cover design: *design & production* GmbH, Heidelberg

Printed on acid-free paper SPIN 11009092 62/3141/SPI 5 4 3 2 1 0

Appreciate the beauty of the universe; discover the truth of the universe.

Zhuang Zi (369–286 B.C. China)

I have loved the principle of beauty in all things, and if I had had time I would have made myself remembered.

Keats, John (1795–1821, UK)

A thing of beauty is a joy for ever.

Keats, John (1795–1821, UK)

"Beauty is truth, truth is beauty," — that is all ye know on earth, and all ye need to know.

Keats, John (1795–1821, UK)

Preface

Generalized plasticity is a generalization of the unified strength theory to the theory of plasticity. It is the unification of metal plasticity for Tresca materials, Huber-von Mises materials, and twin-shear materials. It is also the unification of geomaterial plasticity for Mohr-Coulomb materials and generalized twin-shear materials. Moreover, it leads to unification of metal plasticity and plasticity of geomaterials, in general. It is a companion volume to *Unified Strength Theory and Its Applications* published by Springer in 2004.

Generalized Plasticity is based on the lectures on the unified theory of materials and structures given by the author at the School of Civil Engineering and Mechanics, Xi'an Jiaotong University in Xi'an, China and at the Nanyang Technological University of Singapore in 1996. It is a course entitled "Generalized Plasticity" for Ph.D. students at Xi'an Jiaotong University since 1993. The main contents are the unified yield function (unified strength theory) of material, the unified slip line field theory for plane strain problem, unified characteristics field theory for plane stress problem, unified characteristics field theory for spatial axisymmetric problem, limit pressure and shakedown pressure of a pressure vessel, the plastic zone analysis at a crack tip under small-scale yielding and the unified fracture criterion.

Several chapters in this book have been presented in conferences and published in various journals. They are: Unified Strength Theory (Yu, 1991, 1992, 1994, 2002, 2004); Unified Slip Line Field Theory for Plane Strain Problem (Yu, Yang, et al., 1997, 1999); Unified Characteristics Field Theory for Plane Stress Problem (Yu and Zhang, 1998, 1999; Zhang, Hao and Yu, et al., 2003); Unified Characteristics Field Theory for Spatial Axisymmetric Problem (Yu and Li, 2001); Unified Solution for Limit Pressure of a Pressure Vessel (Wang and Fan, 1998; Zhao et al., 1999); Unified Solution for Shakedown Pressure of a Pressure Vessel (Xu and Yu, 2004, 2005); Analysis of Plastic Zone at Crack Tip (Qiang et al., 1998, 2004); Unified Fracture Criterion (Yu, Fan, Che, Yoshimine, et al., 2003, 2004; Qiang and Yu, 2004). The beauty of the unified strength theory discussed in Chap. 5 is a part of a closing lecture delivered at the International Symposium on Developments in Plasticity and Fracture: Centenary of M.T. Huber Criterion, held at Cracow, Poland in 2004. The garden of the flowers of strength theory, the beauty of the Huber-von Mises criterion, and the beauty of the unified strength theory were discussed at the lecture.

The analytical results of the generalized plasticity are a series of results. It is different from the conventional plasticity. As an example, a unified solution of bearing capacity of a plane strain structure by using the unified slip line field theory is shown in Fig. 1. The conventional solution of bearing capacity of a structure is adapted only for one kind of material. It is shown in Fig. 1 at $b = 0$. This result is obtained by using the Mohr-Coulomb strength theory (inner bound or lower bound as shown in Fig. 2a), Fig. 2b is a special case for $\alpha = 1$ materials. It can also be obtained by using the unified strength theory with $b = 0$. The unified solution includes a series of solutions and encompasses the solution of the Mohr-Coulomb strength theory as a special case. It is also possible to obtain a series of new solutions for different values of parameter b and different ratios of tension and compression strength of material, i.e., $\alpha = \sigma_t/\sigma_c \neq 1$.

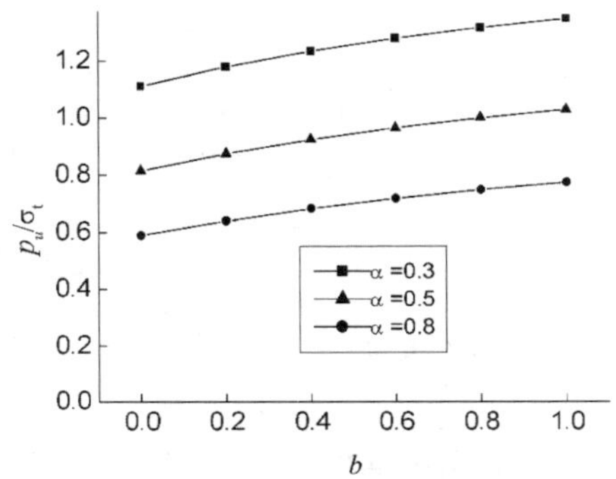

Fig. 1 Limit loads of a plane strain structure.

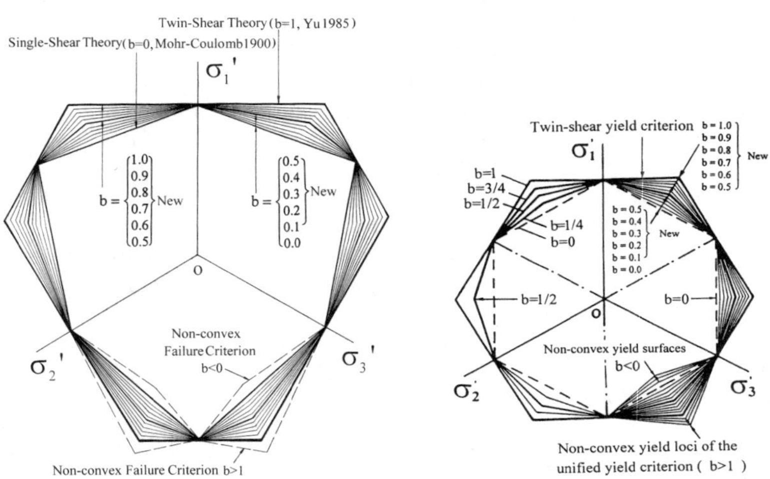

(a) $\alpha = \sigma_t/\sigma_c \neq 1$ materials (b) $\alpha = \sigma_t/\sigma_c = 1$ materials

Fig. 2 Yield loci of the unified strength theory on the deviatoric plane.

The unified strength theory and the unified slip line field theory for plane strain problem can be expressed in terms of another material parameter, such as friction angle φ, it is widely used in geomechanics and geotechnical engineering. The unified solutions for a plane strain problem in terms of the friction φ are shown in Fig. 3. The description of the unified solutions of plane strain problems can be seen in Chap. 9.

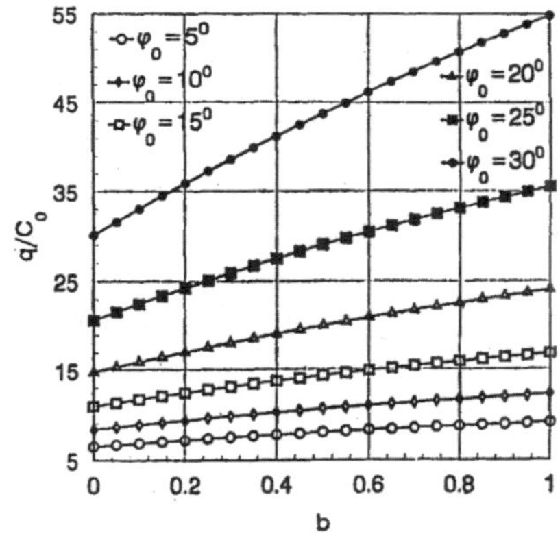

Fig. 3 Unified solutions of a plane strain problem in terms of the friction φ.

For the plane stress problems, the yield loci of the unified strength theory in plane stress state is shown in Fig. 4, and the unified solution of bearing capacity of a plane stress structure by using the unified characteristics line field theory is shown in Fig. 5. It can be seen that a series of new results are given whereas the Mohr-Coulomb theory and the Tresca criterion can give only one result. It is a special case of the unified solution by using the unified strength theory and the unified characteristics line theory with $b = 0$ and $b = 0$, $\alpha = 1$.

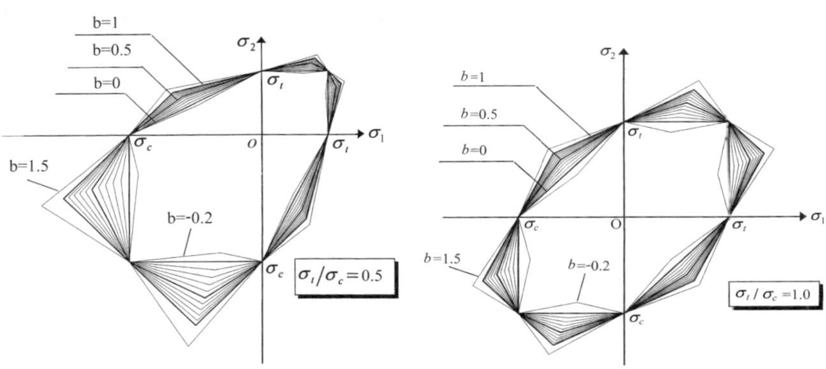

(a) $\alpha = \sigma_t/\sigma_c \neq 1$ materials (b) $\alpha = \sigma_t/\sigma_c = 1$ materials

Fig. 4 Yield loci of the unified strength theory in plane stress state.

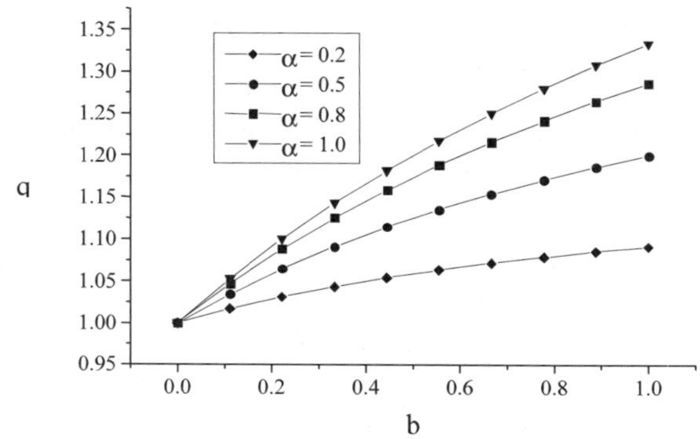

Fig. 5 Limit loads of a plane stress structure.

The analytical results are clearly illustrated to show the effects of yield criterion on plastic limit behaviors for plane strain problems, plane stress problems, axisymmetric problems, other engineering structures, the shape and size of plastic zone at crack tip, discontinuous bifurcation, and angle of shear band. Generalized plasticity gives us a series of results, which can be adapted for different materials and structures.

The contents of the book can be divided into five parts as follows:

Part One. The unified strength theory, material parameters in the unified strength theory, yield surfaces, yield loci, reasonable choice of the yield criterion, and the beauty of the unified strength theory are described in Chaps. 4 and 5.

Part Two. Plastic stress–strain relation and concrete plasticity, discussed in Chaps. 6 and 7.

Part Three. Twin-shear slip field and the unified slip-line field theory for plane strain problems, twin-shear characteristics field and the unified characteristics line theory for plane stress problems, and unified characteristics line field theory for axisymmetric problems and high velocity penetration problem are explained in Chaps. 8–12.

Part Four. The unified solution of plastic zone at crack tip under small-scale yielding is given. Based on the unified strength theory, a unified fracture criterion, a new closed form of plastic core region model, and variation for the angle of initial crack growth versus crack inclination under different loading conditions are obtained. They are described in Chaps. 13 and 14.

Part Five. Chapter 15 is devoted to the unified solutions of limit loads and shakedown loads for pressure vessels.

Stress state analysis and basic behaviors of materials under complex stress are discussed in Chaps. 2 and 3. The description of the stress state may be found in a number of books covering mechanics of materials, solid mechanics, and elasticity and plasticity. Only some basic formulae and figures as well as some new ideas are given here. Brief summaries, problems, and references and bibliography are given at the end of the chapters.

I would like to express my gratitude for the support of the National Natural Science Foundation of China (Grants nos. 59779028, 59924033, and 50078046), the Ministry of Education of China, the China Academy of Launch Vehicle Technology, the Aircraft Strength Research Institute of China, the Department of Science and Technology of Xi'an Jiaotong University, as well as the School of Civil and Environmental Engineering, Nanyang Technological University, Singapore.

I am indebted to many authors and colleagues, especially Dr Ma at Nangyang Technological University, Singapore, Prof. Qiang at Xi'an Hi-Tech Research Institute, and Dr Zhang at National University of Singapore for going through the manuscript of this book. I am also indebted to Dr Li YM, Dr Yang SY, Dr Zhao JH, Dr Wang F, Dr Li JC, Dr Song L, Dr Liu YH, Dr Fan W, Dr Zan YW, Dr Wei XY, Dr Xu SQ, Dr Yoshimine M, Dr Yang JH, Dr Zheng H, Dr Gao JP, Dr Liu FY, and Mr Liu JY and Ms Zeng WB for their research work at Xi'an Jiaotong University. Thanks are due to Ms He, Ms Lei, and others for their support during the course of writing this book. I would like to thank many professors from other universities and many research scientists and engineers from various institutions for their work in the research and application of the unified strength theory. I would also like to acknowledge the support from all other individuals and universities, research organizations, journals, and publishers.

I would also like to express my sincere thanks to Dr Dieter Merkle, Editorial Director, Engineering and Editorial Department and International Engineering Department, Springer-Verlag, Germany, and his team for their excellent editorial work on this manuscript.

The study of generalized plasticity based on the unified strength theory is just the beginning. A lot of research in generalized plasticity is still to be done.

Mao-Hong Yu
Spring 2005

Contents

Notations

Stresses and Invariants

σ_1, σ_2, σ_3	major principal stress, intermediate principal stress and minor principal stress
σ_{13}, σ_{12}, σ_{23}	normal stresses acted on the orthogonal octahedron element
m	intermediate principal stress state parameter
σ_r, σ_θ, σ_z	radial stress, circumferential stress and axial stress in polar coordinates
$\sigma_r^{\ r}$, $\sigma_z^{\ r}$, $\sigma_\theta^{\ r}$	residual stresses of the cylinder in polar coordinates
σ_{ij}	stress tensor
σ	normal stress
τ_{13}	maximum principal shear stress
τ_{12}, τ_{23}	intermediate shear stress or minimum shear stress
μ_σ	Lode stress parameter
μ_τ μ_τ'	twin-shear parameter for stress state $\mu_\tau = \tau_{12}/\tau_{23}$, $\mu_\tau' = \tau_{23}/\tau_{13}$
θ	stress angle corresponding to the twin-shear parameter
τ_8 or τ_{oct}	octahedral shear stress
σ_8 or σ_{oct}	octahedral normal stress
I_1, I_2, I_3	invariants of the stress tensor σ_{ij}
σ_m	hydrostatic stress or mean stress
S_1, S_2, S_3	deviatoric stresses
J_1, J_2, J_3	invariants of the deviatoric stress tensor
K_I, K_{II}, K_{III}	stress-intensity factors for mode-I, mode-II and mode-III

Strain and Flow Vector

$_1$, $_2$, $_3$	principal strain
^{e}	elastic strain
^{p}	plastic strain
d ^{e}	elastic strain increment
d ^{p}	plastic strain increment
$d\varepsilon_{ij}^{e}$	elastic strain increment tensor

$d\varepsilon_{ij}^{p}$	plastic strain increment tensor
ij	strain tensor
ε_r , ε_θ	radial and circumferential strain
ε_z	longitudinal strain
ε_r , ε_θ	radial and circumferential strain
C_{ijkl}	stiffness tensor
$d\lambda$	proportional positive scalar factor
$\{a_1\}$, $\{a_2\}$, $\{a_3\}$	flow vectors
C_1, C_2, C_3 or C_1', C_2', C_3'	parameters defining the yield surface

Material Parameters

σ_y	yield stress
σ_t	uniaxial tensile strength
σ_c	uniaxial compressive strength
α	ratio of tensile strength to compressive strength
m	compressive-tensile strength ratio of materials
τ_0 or τ_y	pure shear strength or shear yield strength of materials
b	failure criterion parameter in the unified strength theory
β	coefficient in the unified strength theory that represents the effect of the normal stress on failure
C_0	cohesive strength
C_t	twin-shear cohesion
C_{uni}	unified cohesion
φ_0	friction angle
φ_t	twin-shear friction angle
φ_{uni}	unified friction angle
E	elastic modulus (Young's modulus)
G	shear modulus
λ	Lame constant
v	Poisson's ratio
κ	hardening parameter of material
K_{Ic} , K_{IIc} , K_{IIIc}	stress-intensity factors of materials under mode-I, mode -II and mode-III

Miscellaneous

f, f'	yield criterion function
F, F'	yield criterion function or fracture criterion function
ρ	density of the material
u, v, w	displacements
$[\sigma]$	allowable tensile stress, $[\sigma] = \sigma_t / n$
p	internal pressure subjected to the cylinder

p_e	elastic limit pressure
p_p	plastic limit pressure
$p_{\max}$	shakedown pressure of the cylinder
r	radial variable
r_i , r_e	the internal and external radii of the cylinder
r_p	plastic zone radius of the cylinder subjected to internal pressure p

1 Introduction

1.1 Linear Elasticity

A structure deforms when it is subjected to external forces. The deformation is elastic if it is reversible, that is, if the deformation vanishes instantaneously as the external forces are removed. It is also assumed that the relationships of force-displacement and stress-strain are linear. Most engineering materials possess to a certain extent the property of linear-elasticity.

The stress states of materials in structures are usually biaxial and triaxial, or in general multiaxial or polyaxial. It is very important to see how the stress combinations affect the strength of material and structure. Serious errors may be resulted if the uniaxial stress assumptions are extended directly to the multiaxial stress state.

It will be assumed that the material of an elastic structure is homogeneous and continuously distributed over its volume so that the smallest element cut from the structure possesses the same specific physical properties as the structure. It will also be considered that the material of a structure is isotropic, i.e., the properties are the same in all directions.

On the other hand, engineering materials are generally very complicated in compositions. The important material steel, for instance, consists of various kinds of crystals oriented in different directions. The material is very far from being homogeneous in microscope scale. The solutions in view of the theory of elasticity, however, based on the assumptions of homogeneity and isotropy can be applied to steel structures with very great accuracy on the macro-scale level.

The explanation is that the crystals are very small; there are millions of them in one cubic inch of space. While the elastic properties of a single crystal may be very different in different directions, the crystals are ordinarily distributed at random and the elastic properties of larger pieces of metal represent averages of the properties of the crystals. Therefore, as long as the geometrical dimensions of a structure are very huge in comparison with the dimensions of a single crystal the assumption of homogeneity can be used with very high accuracy. And if the crystals are orientated randomly the material can be treated as isotropic (Timoshenko and Goodier 1970; Fung 1977).

The assumptions of isotropy and homogeneity for materials are also used in plasticity. Discussions of the plasticity for heterogeneous and anisotropic materials are not in the scope of this book.

1.2 Classical Plasticity

The deformation of a material is plastic if it is irreversible. A brittle material such as glass, concrete, or rock under tension has only elastic deformation before it fails. On the other hand, metals, rocks and concrete under high confining pressure can undergo substantial plastic deformation before failure and therefore show ductile material properties.

The "Theory of Plasticity" or "Plasticity" is a branch of solid mechanics regarding the plastic deformation and the limit load carrying capacity of materials and structures. It deals with the theories and methods of yield initiation of materials under complex stress state and calculation of stresses and strains in a deformed structure after part or the whole structure has yielded. It is necessary, as for elasticity, to establish equations of equilibrium and compatibility and to determine the experimental relations between stress and strain, besides, a condition of yield initiation is needed. It is called yield criterion or yield function.

In 1864, Tresca presented two notes dealing with the experimental investigations on plastic flow of metals under great pressure to the French Academy. He first postulated a yield condition for the continuum problem. It is the well-known Tresca yield criterion.

Saint-Venant (Barre de Saint-Venant 1797–1886) was the first to set up the fundamental equations of plasticity and to use them to solve several practical problems. The fundamental equations of plasticity are based on the following assumptions.

1. The volume of materials does not change during plastic deformation;
2. The directions of principal strains coincide with the directions of principal stresses;
3. The maximum shearing stress at each point is equal to a specific constant. It means that the Tresca criterion was used as a yield condition. Sometimes, it is referred as the maximum shear stress criterion or single-shear criterion. The yield surface of the Tresca criterion in stress space and its yield locus in deviatoric plane are shown in Fig. 1.1. The yield loci and the yield surfaces will be described in detail in Chapter 6.

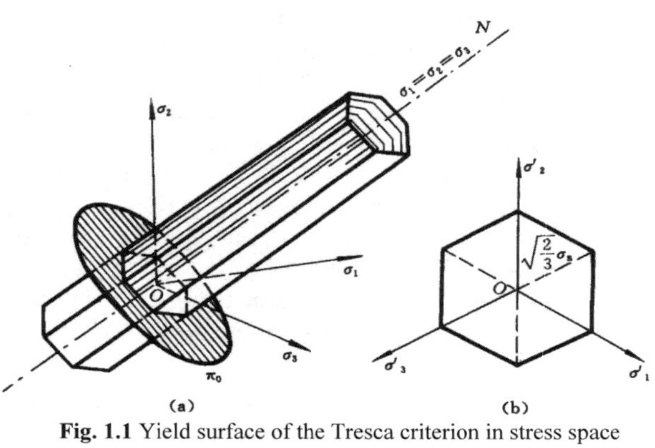

Fig. 1.1 Yield surface of the Tresca criterion in stress space

Using these hypotheses, Saint-Venant solved several problems such as torsion of circular shafts, pure bending of rectangular prismatic bars, and the plastic deformation of hollow circular cylinders under the internal pressure. In fact, Saint-Venant initiated the study of a complete new field of solid mechanics. Saint-Venant called the new subject "Plasticodynamics". Now plasticity has been the object of considerable research.

Prandtl (Ludwig Prandtl 1875–1953) made further progress in the field of plasticity. He solved the more complicated two-dimensional problem of a semi-infinite body under a uniform pressure p distributed over a strip of width. He obtained the limit pressure p_{limit} of strip as follows

$$p_{\text{limit}} = \sigma_s \left(1 + \frac{\pi}{2}\right)$$

The first systematic treatment of plasticity was given by Nadai in 1931. Rapid progress in plasticity began with the appearance of the paper of Prandtl and the book of Nadai. After the Second World War, the subject of plasticity constitutes an important branch in the solid mechanics. Some books relating the plasticity were published near 1950 by Sokolovsky (1946), Freudental (1950), Hill (1950), Nadai (1950), Prager and Hodge (1951) et. al.

After the Tresca yield criterion, the Huber-von Mises yield criterion was proposed by Huber in 1904 and von Mises in 1913. The yield surface of the Huber-Mises yield criterion in stress space and its yield locus in deviatoric plane are shown in Fig.1.2.

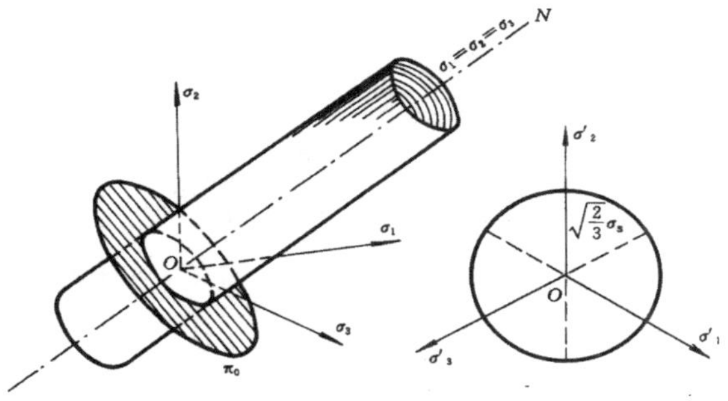

Fig. 1.2 Yield surface of the Huber-von Mises criterion in stress space

The classical plasticity is widely applied in mechanical engineering and metal forming. It is assumed by the Tresca criterion and the Huber-von Mises criterion that these materials have the same strength in both tension and compression and the ratio of shear yield stress τ_y to the tensile yield stress σ_y is τ_y/σ_y=0.5 based on the Tresca criterion or 0.577 on the Huber-von Mises criterion.

In 1961, a new yield criterion was proposed by Haythorthwaite and Yu. It is introduced from the maximum deviatoric stress by Haythorthwaite and the twin-shear stresses by Yu, and referred to respectively as the maximum deviatoric stress yield criterion and the twin-shear yield criterion. The yield surface of the twin-shear yield criterion in stress space and its yield locus in π plane are shown in Fig.1.3. The twin-shear yield criterion is adapted for those material have the same strength in tension and compression and the ratio of the shear yield stress τ_y to the tensile yield stress σ_y is $\tau_y/\sigma_y = 0.667$. The mathematical expression of the twin-shear yield criterion is linear. It is convenient for analytical solution of limit analysis of structure.

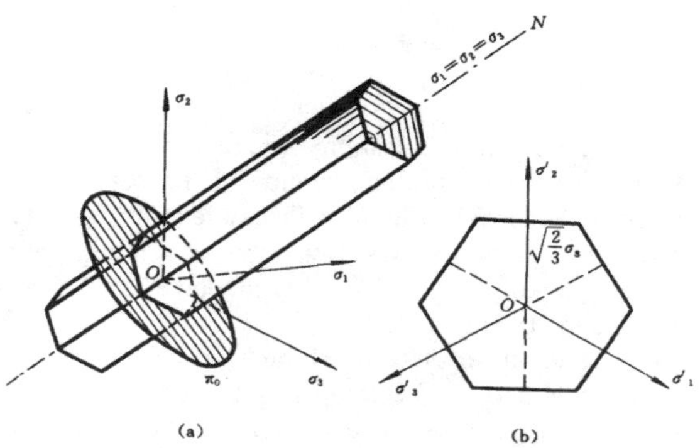

Fig. 1.3 Yield surface of the Twin-shear criterion in stress space

The yield loci in deviatoric plane and meridian plane of the three yield criteria are shown in Fig.1.4.

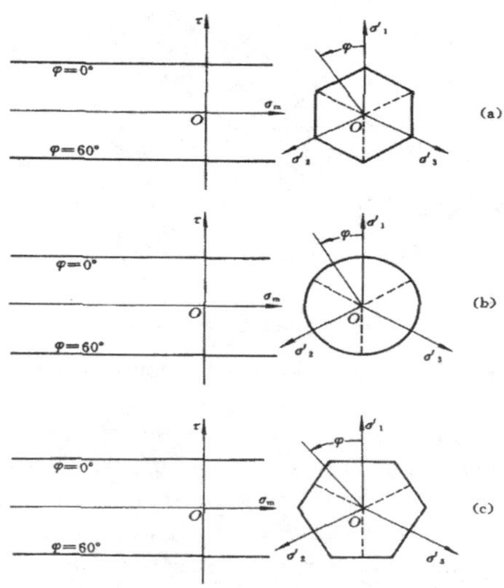

Fig. 1.4 Yield loci of the three yield criteria

Classical plasticity is widely used for metallic materials. However, classical plasticity models are not able to predict basic failure phenomena of rocks, soils, concrete, polymers and other materials. Applications in concrete structural engineering, geotechnical engineering, mining and petroleum engineering, nuclear power plant, underground excavation, structure safety problems under earthquake, and solid mechanics problems at meso/micro scales call for more realistic and more accurate solutions.

Now developments in plasticity theory is an active field of mechanics. The concrete plasticity, soil plasticity, rock plasticity and computational plasticity etc. are developed.

1.3 Concrete Plasticity

Concrete as a structural material has been used widely in many major constructions such as tall buildings, bridge, dam, offshore platforms, reactor vessels, nuclear containment structures, etc. Plastic analysis of these structures has become increasingly important.

Concrete is a composite material, consisting of coarse aggregate and a continuous matrix, which itself comprises a mixture of cement paste and smaller sand particles. Its physical behavior is quite complex. In engineering application, only the mechanical behavior instead of the composition of the material is concerned, which is developed on the basis of a homogeneous continuum. Also, the material is assumed to be initially isotropic.

In recent years, a good progress has been achieved in the area of constitutive modeling of concrete materials. Various predictive models with two-parameter, three-parameter and multi-parameter have been proposed and used for analysis of concrete structures. The models are based on the principles of continuum mechanics, neglecting the microstructure effect of the concrete materials. To this end, nonlinear elasticity and plasticity are found adequate in characterizing the macroscopic stress-strain behavior of concrete.

The theory of concrete plasticity, i.e. the basic concepts and applications of plasticity modeling to concrete materials were presented by Chen (1982) and Nielsen (1984, 1999). Systematical descriptions on the "Plasticity in Reinforced Concrete" and "Limit Analysis and Concrete Plasticity" were given.

The Mohr-Coulomb strength theory and some multi-parameter criteria are widely used in concrete plasticity. The yield locus of the Mohr-Coulomb strength theory in deviatoric plane is shown in Fig.1.5. The solid line is the yield locus of the generalized twin-shear criterion (Yu et al. 1985).

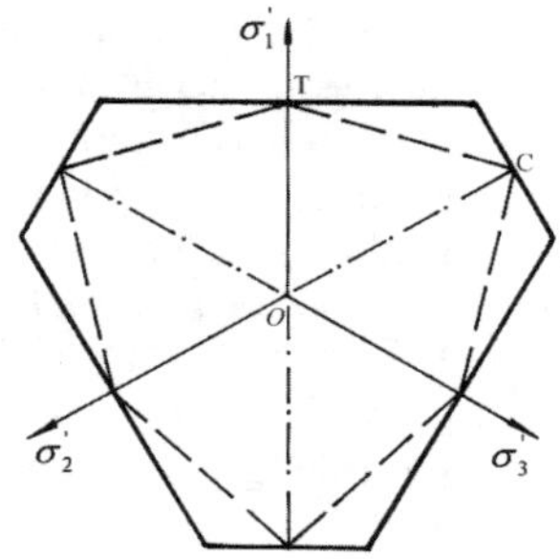

Fig. 1.5 Two bounds of the convex yield criteria

1.4 Soil Plasticity

Soils are complex materials consisting of solid, air and water. It exhibits a wide range of behavior depending on classification, stress history, void ratio, density and characteristics of the disturbing force.

The plastic behavior of soils was studied by Roscoe, Schofield and Thurairajah (1963), Roscoe (1968), Roscoe and Burland (1968), Schofield and Wroth (1968) et al in Cambridge University. Soil Plasticity was presented by Roscoe (1968), Schofield and Wroth (1968), Chen (1975) and Salencon (1977) in 60's and 70's of the last century and continued by Chen and Baladi (1985). The soil plasticity was also implemented into computational code by Zienkiewicz and Pande (1977), Zienkiewicz and Humpheson (1977), Zienkiewicz and Mroz (1984), Desai (1984), Vermeer and de Borst (1984), de Boer R (1988) and others. Over the last 30 years, significant advances in this area have led to the development of several constitutive models, which may efficiently represent the behavior of soils subjected to complex stress. Now, the soil plasticity is still one of the most active research areas.

The Mohr-Coulomb strength theory (Mohr 1900) and the Drucker-Prager (1952) criterion are widely used in soil plasticity. The yield loci in deviatoric plane and meridian plane of the Drucker-Prager criteria and the Mohr-Coulomb criterion are shown in Figs.1.6 and 1.7. Two bounds of failure criteria are shown in Fig.1.8.

Systematical descriptions on the soil mechanics and applications were given by Chen (1975), Salencon (1977) and Chen and Baladi (1985), Zeng, Shen and Gong (2002).

Soil plasticity based on the Critical State Soil Mechanics were described by Schofield and Wroth (1968), Atkingson & Bransby (1978), Atkingson (1981) and Wood (1990).

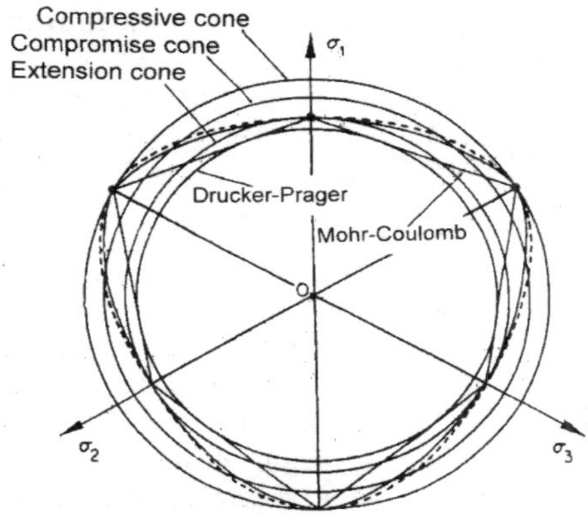

Fig. 1.6 Some varieties of the Drucker-Prager criteria

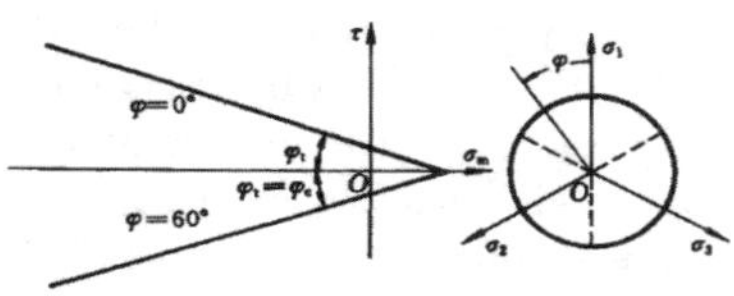

Fig. 1.7 Yield loci of the Drucker-Prager criteria

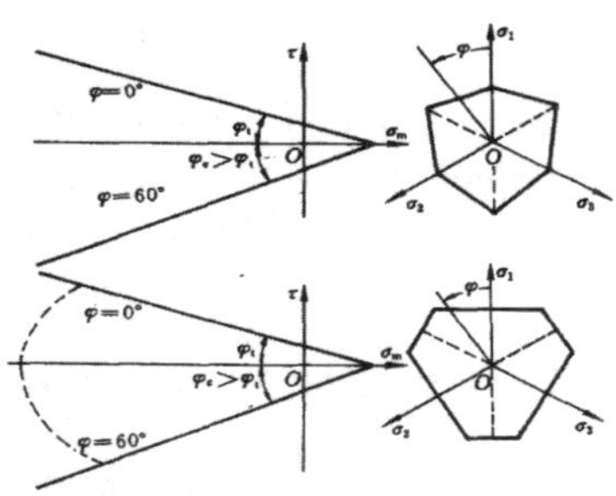

Fig. 1.8 Yield loci of the two bounds (single-shear and twin-shear criteria)

1.5 Rock Plasticity

The plastic behavior of rock was not generally known until 60's of the last century when the rigid test equipment was produced. The first congress of international society of rock mechanics was hold in Lisbon in 1966. The flow behavior of rocks were studied by Broms (1966), Goodman, Taylor and Brekke (1968) and others. The bearing capacity of rock and concrete block was studied by Chen and Drucker (1969). Non-associated plasticity for soils, concrete and rock were studied by Vermeer and de Borst (1984) et al.

The Mohr-Coulomb strength theory and the Hoek-Browm (1980) criterion are widely used in rock plasticity.

"Finite Element Code for Soil and Rock Plasticity" was presented by Vermeer (1991, 1998). The concept of plastic zone of rock is widely used for rock mechanics and engineering.

1.6 Generalized Plasticity

The theory of plasticity was originally developed for the description of the behavior of metallic materials. From the microscopic point of view, the deformational mechanisms of metal are quite different from those of concrete, rock and soils. The former is due to the arrangement of dislocations of polycrystals, while the later is due to the initiation and nucleation of microcracks at the aggregate-mortar interface as well as through the mortar, air or cracks. However, if we do not limit our interpretations of the behavior "plastic" and "yielding" in the usual sense, the classical theory of plasticity can be extended to approximate the concrete, rock and soils behavior under various circumstances (Chen 1978). In other words, the theory of plasticity, when not interpreted too narrowly, provides a very flexible mathematical model that can be used to describe a wide variety of behaviors of concrete, rock and soils. The continuum theory of plasticity was given by Khan et al. (1995).

The generalized stress-strain behaviour of wet clay was studied by Roscoe and Burland (1968). The generalized cap model for geological materials was studied by Sandler, DiMaggio and Baladi (1976). The term of Generalized Plasticity was introduced by Zienkiewicz and Mroz (1984), Pastor and Zienkiewicz (1986), and by Pastor, Zienkiewicz and Chan (1990). The word of 'generalized' means the extension of plastic conception for metallic materials to geomaterials. Now, generalized plasticity is widely studied and applied in literature and engineering.

1.7 Generalized Plasticity Based on the Unified Strength Theory

Great effort has been devoted to the formulation of yield criteria for various materials during the past 100 years (Yu 2002). Most yield criteria are suitable for only a certain type of materials. The plasticity theory for metallic materials is mainly based on the Tresca yield criterion and the Huber-Mises yield criterion. The plasticity theory of soils is mainly based on the Drucker-Prager criterion; and the plasticity theory for concrete and rock is mainly based on the Mohr-Coulomb criterion. What is the relationship among various yield criteria and different "Plasticity"? Many scientists have devoted considerable effort to this topic.

Based on a unified mechanical model that takes into account the effects of all the stress components on the failure of materials, a new unified strength theory that has a unified mathematical expression was proposed by Yu in 1991. The Mohr–Coulomb strength theory, the twin-shear stress theory, and many other new criteria can be deduced from the unified strength theory. A relationship among most of the prevailing failure criteria and yield criteria is established

The unified strength theory contains two families of yield criteria for metal materials and geomaterials. The first family is the convex yield criteria. A series of convex yield criteria can be deduced from the unified strength theory by giving a certain value to material parameter b ($0<b<1$) and the tension–compressive strength ratio $\alpha=\sigma_t/\sigma_c$, where parameter b reflects the influence of intermediate principal shear stress and the choose of yield criteria.

Fig. 1.9 shows the limit loci on deviatoric plane of the unified strength theory for $\alpha=1/2$ material and $\alpha=1$ materials. Detailed description is given in Chapter 4 and 5.

The unified strength theory embraces all the criteria from the lower bound to the upper bound. The unified strength theory is very simple, while it can be used widely.

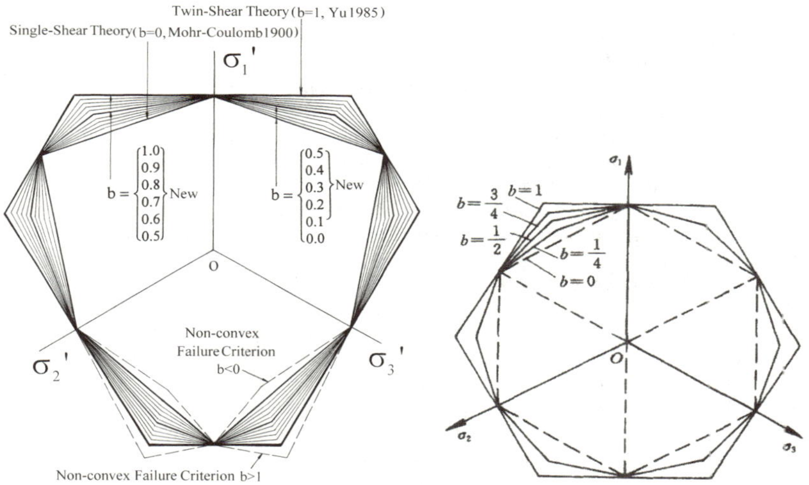

(a) $\alpha\neq 1$ (b) $\alpha=1$

Fig. 1.9 Yield loci of the unified strength theory on the deviatoric plane

The unified yield criterion can be deduced from the unified strength theory when $\alpha=1$. It contains a series of convex yield criteria, as shown in Fig. 1.9 (b). The single-shear yield criterion and the twin-shear yield criterion can be derived with b=0 and b=1, respectively. These two criteria can also be obtained from the single-shear strength theory (the Mohr–Coulomb strength theory) and the twin-shear strength theory when $\alpha=1$.

The unified strength theory is convenient to derive analytic solution for elastic limit design, elasto-plastic analysis, plastic limit analysis of structures and other engineering problems. Research results show that the yield criterion has significant influence on the elasto-plastic behavior and load-carrying capacities of structures.

A series of new results for generalized plasticity by using the unified strength theory were obtained. They will be described in Chapter 7 to 15. They are: Twin-shear slip line theory (Yu, Liu and Ma 1994), the unified slip line theory for plane strain problem (Yu et al. 1997,1999); Twin-shear characteristics field for plane stress problem (Yan and Bu 1993), the unified characteristics field theory for plane stress problem (Yu and Zhang 1998,1999; Zhang, Hao and Yu 2003); the unified characteristics field theory for spatial axisymmetric problem (Yu, Li and Zhang 2001); Plastic zone analysis at crack tip (Qiang et al. 1998), The unified fracture criterion (Yu and Fan 2003, 2004; Qiang 2003, Qiang and Yu 2004), the unified solution of limit load for pressure vessel (Wang and Fan 1998; Zhao et al. 1999); the unified solution of shakedown load for pressure vessel (Xu and Yu 2004, 2005).

References and Bibliography

Atkingson JH & Bransby PL (1978), *The Mechanics of Soils: An Introduction to Critical State Soil Mechanics*. McGraw-Hill, Maidenhead.

Atkingson JH (1981) *Foundation and Slops: An Introduction to Application Critical State Soil Mechanics*. McGraw-Hill, Maidenhead.

Bishop AW (1971) Shear strength parameters for undisturbed and remoulded soil specimens. *Stress-Strain Behaviour of Soils Proc.* (Roscoe Memorial Symposium, Cambridge University, Cambridge, England 1971), Parry RHG ed. Foulis, pp 1–59.

de Boer R (1988) On plastic deformation of soils. *Int. J. Plasticity* **4,** 371–391.

Broms BB (1966) A note of strength properties of rock. *Proc. First Congress Int. Society of Rock Mechanics*, Lisbon, Vol.2, 69–70.

Butterfield R and Harkness RM (1971) The kinematics of Mohr–Coulomb materials. *Stress-Strain Behaviour of Soils.* Parry RHG ed. Foulis, pp 220–233.

Buyukozturk O, Nilson AH and Slate FO (1971) Stress-strain response and fracture of a concrete model in biaxial loading. *ACI Journal* **68**(8), 590–599.

Chakrabarty J (1987) *Theory of Plasticity*. McGraw-Hill, New York.

Chen WF (1975) *Limit Analysis and Soil Plasticity*. Elsevier, Amsterdam.

Chen WF (1982) *Plasticity in Reinforced Concrete.* McGraw–Hill, New York, pp 190–252.

Chen WF and Baladi GY (1985) *Soil Plasticity: Theory and Implementation.* Elservier, Amsterdam.

Dimaggio FL and Sandler IS (1971) Material model for granular soils. *J. Engrg. Mechanics*, **97**, 935–950.

Fan W (2003) Study on Structural Strength Theory. Dissertation of Doctor at Xi'an Jiaotong University, Xi'an, China (in Chinese)

Franklin JA (1971) Triaxial strength of rock materials. *Rock Mech.* **3**, 86–98.

Freudental AM (1950) *Inelastic Behavior of Engineering Materials and Structures*. Wiley, New York.

Fung YC (1977) *A First Course In Continuum Mechanics* (2nd ed.). Prentice-Hall, Inc. Englewood Cliffs.

Goodman RE, Taylor RL and Brekke TL (1968) A model for the mechanics of jointed rock. *J. Soil Mechanics Foundation* **94**, 637–659.

Guowei M., Iwasaki S. , Miyamoto Y. and Deto H.(1998) Plastic Limit Analysis of Circular Plates with Respect to the Unified Yield Criterion, *Int. J. of Mech. Science*, 40(10): 963–976.

Hill R (1950) *The Mathematical Theory of Plasticity.* Clarendon, Oxford.

Johnson W and Mellor PB (1962) *Plasticity for Mechanical Engineers*. Van Nostrand, Lonton and New York.

Khan AS (1995) *Continuum Theory of Plasticity*. John Wiley & Sons, New York.

Liu XQ, Ni XH, Yan S *et al.* (1998) Application of the Twin Shear Strength Theory in Strength Calculation of Gun Barrels. In : *Strength Theory: Application, Development and Prospect for the 21st Cencury*, Yu MH and Fan SC eds. Beijing and New York: Science Press, 1039-1042.

Martin J.B (1975) Plasticity: Fundamentals and General Results. The MIT Press.

Mendelson A (1968) *Plasticity: Theory and Application*. MaCmillan, New York.

von Mises RV (1928) Mechanik der plastischen Formanderung von Kristallen. *Z. Agnew. Math. Mech.*, **8**, 161–185.

Mogi K (1967a) Effect of the intermediate principal stress on rock failure. *J. Geophysics Res.* **72**, 5117–5131.

Mogi K (1967b) Effect of the triaxial stress system on fracture and flow of rock. *Phys. Earth Planet Inter.* **5**, 318–324.

Nadai A (1950) *Theory of Flow and Fracture of Solids. Vol. 1*. McGraw-Hill, New York.

Ni XH, Liu XQ, Liu YT *et al.*(1998) Calculation of stable loads of strength differential thick cylinders and spheres by the twin shear strength theory. In : *Strength Theory: Application, Development and Prospect for the 21st Cencury*, Yu MH and Fan SC eds. Beijing and New York: Science Press, 1043-1046.

Chen WF (1982) *Plasticity in Reinforced Concrete*. McGraw–Hill, New York, pp 190–252.

Nielsen M.P (1984, 1999) *Limit Analysis and Concrete Plasticity*. CRC Press, London.

Parry RHG (1971) ed. *Stress-Strain Behaviour of Soils Proc.* (Roscoe Memorial Symposium, Cambridge University, Cambridge, England 1971),. Foulis.

Pastor M. and Zienkiewicz O.C. (1986), A generalized plasticity hierarchical model for sand under monotonic and cyclic loading. In: Pande G.N. et al. (eds), Numerical Models in Geomechanics, Jackson and Son, London, 131-150.

Pastor M., Zienkiewicz O.C. and Chan A.H.C. (1990), Generalized plasticity and the modeling of soil behaviour. Int. J. Numerical and Analysis Methods Geomech., 14, 151-190.

Pastor M., Zienkiewicz O.C. and Chan A.H.C. (1990) A generalized plasticity model for the dynamic behavior of sand, including liquefaction. In: Handbook of Materials behavior Models. Lemaitre J. ed. Academic Press.

Prandtl L (1925) Spannungsverteilung in plastischen Koerpern, *Proc.of the First Int.Congress on Applied Mechanics*, Delft Technische Boekhandel en Druckerij, J.Waltman, Jr., 43–54.

Prager W (1947) An introduction to the mathematical theory of plasticity. *J. Appl. Phys.* **18**, 375–383.

Prager W (1949) Recent developments in the mathematical theory of plasticity. *J. Appl. Phys.*, **20**, 235–241.

Prager W (1959) *An Introduction to Plasticity*. Addison-Wesley, Reading, Mass.

Prager W and Hodge PG Jr (1951) *Theory of Perfectly Plastic Solids*. Wiley, New York.

Qiang HF, Xu YH and Zhu JH (1998) unified solutions of crack tip plastic zone under small scale yielding. In: *Strength Theory: Applications, Developments and Prospects for the 21st Century*. Yu MH and Fan SC eds. Science Press, New York, Beijing, 823–829.

Qiang HF, Han X and Liu GR (2003) A unified criterion for mixed mode crack initation. Presented at the 5th Euromech Solid Mechanics Conference, Elias C. Aifantis Editor, Greece.

Qiang HF, Yu MH and Yang YC (2004) Unified fracture criterion: A generalization of the T-criterion and the unified strength theory. Presented at the International Symposium on Developments in Plasticity and Fracture Centenary of M.T. HUBER Criterion. August 12–14, 2004, Cracow, PolanRoscoe KH, Schofield AN and Thurairajah A (1963) Yielding of clays in states wetter than critical. *Géotechnique* **13**, 211–240.

Roscoe KH ed. (1968) *Engineering Plasticity*. Cambridge University Press, Cambridge.

Roscoe KH and Burland JB (1968) On the generalized stress-strain behaviour of wet clay. *Engineering Plasticity*. Cambridge University Press, Cambridge, pp 535–609.

Schofield AN and Wroth CP (1968) *Critical State Soil Mechanics*. McGrw-Hill, London.

de Saint-Venant B (1870) Memoire sur l'establissement des equations differentielles des mouvements interieurs operes dans les corps solides ductiles au dela des limites ou l' elasticite pourrait les ramener a leur premier etat. *Comptes Rendus hebdomadaire s des Seances de l'A cademie de s Sciences*, **70**, 473–480.

Salencon J (1977) *Applications of the Theory of Plasticity in Soil Mechanics* (Translation from France by Lewis RW and Virlogeux H). Wiley, Chichester.

Timoshenko S.P. and Goodier J.N., Theory of Elasticity. Third edn. Mcgraaw-Hill, New York, 1970.

Tresca H (1864) Sur I'e coulement des corps solids soumis a de fortes pression, *Comptes Rendus hebdomadaires des Seances de l' Academie des Sciences*, Rend **59**, 754–758.

Vermeer PA and de Borst R (1984) Non-associated plasticity for soils, concrete and rock. *HERON* Vol. 29, No.3.

Vermeer PA (1991) *PLAXIS—Finite Element Code for Soil and Rock Plasticity. Version 4. 0.* Balkema, Rotterdam.

Vermeer PA (1998) *PLAXIS—Finite Element Code for Soil and Rock Plasticity.* Balkema, Rotterdam.

Wang F. and Fan S.C. (1998) Limit pressures of thick-walled tubes using different yield criteria. In: *Strength Theory: Application, Development and Prospect for the 21st Cencury*, Yu MH and Fan SC eds. Beijing and New York: Science Press, 1047–1052.

Westergaard HM (1952) *Theory of Elasticity and Plasticity.* Harvard University Press, Cambridge.

Wood DM (1990) *Soil Behaviour and Critical State Soil Mechanics.* Cambridge University Press, Cambridge.

Xu QQ and Yu MH (2004) Shakedown analysis of thick cylinder. *J of Mechanical Engineering*, (in Chinese).

Xu QQ and Yu MH (2005) Shakedown analysis of thick cylinder based on unified strength theory. *J of Pressure Vessel and Pipes*, to be published.

Yu Mao-Hong (1961) General behaviour of isotropic yield function (in Chinese), *Res. Report of Xi'an Jiaotong University.* Xi'an.

Yu Mao-Hong (1961) Plastic potential and flow rules associated singular yield criterion (in Chinese), *Res. Report of Xi'an Jiaotong University.* Xi'an.

Yu Mao-Hong (1983) Twin shear stress yield criterion. *Int. J. Mech.* Sci. 25 (1) 71–74.

Yu Mao-Hong, He Li-nan and Song Ling-yu (1985) Twin shear stress theory and its generalization, *Scientia Sinica (Sciences in China)*, English Edition, Series A, **28**(11), 1174–1183.

Yu MH and He LN (1991) A new model and theory on yield and failure of materials under the complex stress state, in: *Mechanical Behaviour of Materials—6,* (ICM-6). Jono, M. and Inoue, T. Ed, Pergamon Press, Oxford, Vol. 3, 841–846.

Yu Mao-Hong (1994) Unified strength theory for geomaterials and its application. *Chinese J. of Geotech. Eng.,* **16**(2), 1–9.(in Chinese, English Abstract)·

Yu Mao-Hong, Zeng Wen-Bing, (1994) New theory of engineering structural analysis and its application. *J. of Engineering Mechanics,* **11**(1), 9–20. (in Chinese, English abstract)

Yu Mao-Hong, Liu Jian-Yu and Ma Guo-wei (1994), Twin-shear slip line theory: orthogonal and non-orthogonal slip line fields. *Journal of Xi'an Jiaotong University*, 28(2):122–126.

Yu MH, Yang SY, Liu CY and Liu JY (1997) Unified plane-strain slip line theory. *China Civil Engrg. J.* **30**(2), 14–26 (in Chinese, English abstract).

Yu Mao-hong (1998) *Twin-Shear Theory and Its Applications* (in Chinese). Science Press, Beijing.

Yu MH and Fan SC eds (1998) *Strength Theory: Applications, Developments and Prospects for the 21st Century.* New York, Beijing: Science Press.

Yu MH, Zhang YQ and Yang SY (1998) Another important generalization of the unified strength theory. *Journal of Xi'an Jiaotong University* **32**(12) : 108–110 (in Chinese, English abstract).

Yu MH and Zhang YQ (1998) Unified characteristics field theory for plastic plane stress problem. *Journal of Xi'an Jiaotong University* **33**(4):1–4 (in Chinese, English abstract).

Yu MH, Li JC and Zhang YQ (2001) Unified characteristics line theory of spatial axisymmetric plastic problem. *Science in China* (Series E), English edn. **44**(2), 207–215; Chinese edn. **44**(4), 323–331.

Yu MH (2002a) Advances in strength theories for materials under complex stress state in the 20th Century. *Applied Mechanics Reviews*, **55**(3), 169–218.
Yu MH (2002b) *Concrete Strength Theory and Its Applications*. Higher Education Press, Beijing (in Chinese).
Yu MH, Fan W and Che AL, Yoshimine M and Iwatate T (2003) Application of the Unified Strength theory in analyzing fracture strength. Presented at the Fifth Int. Conf. on Fracture and Strength of Solids and the Second Int. Conf. On Physics & Chemistry of Fracture and Failure Prevention, , October 20–22, 2003, Sendai, Japan.
Yu Mao-hong (2004) *Unified Strength theory and Its Applications*. Springer, Berlin.
Yu MH, Fan W, Che AL, Yoshimine M and Iwatate T (2004) Application of the Unified Strength theory in analyzing fracture strength. *Key Engineering Materials*, Vols. 261–263: 111–116.
Yu Mao-hong, Yoshimine M, Oda Y et al. (2004) The Beauty of Strength Theories— Closing Lecture at the *International Symposium on the Developments in Plasticity and Fracture*, Centenary of M.T.Huber, Cracow, Poland.
Zhang YQ, Liu YH and Yu MH (1999) The unified solution of some plane stress problems. *Journal of Xi'an Jiaotong University* **33**(6):60–63. (in Chinese, English abstract).
Zhang YQ (2000) Analysis of damage and Discontinuous bifurcations for Elarto-Plastic Geomaterials. Dissertation of Doctor at Xi'an Jiaotong University, Xi'an, China (in Chinese)
Zhang YQ, Hao H. and Yu Mao-Hong., A unified characteristic theory for plastic plane stress and strain problems. *Journal of Applied Mechanics, 2*003, **70** : 649–655.
Zhang YQ and Yu MH (2001) Discontinuous bifurcations of metallic materials for plane stress. Chinese *J. of Mechanical Engineering*, 37(4): 62–85 (in Chinese, English Abstract).
Zhang YQ and Yu MH (2001) Discontinuous bifurcations of elasto-plastic materials for plane stress. *Acta Mechanica Sinica*, 33(5): 706–713 (in Chinese, English Abstract).
Zhao JH,Zhang YQ, Li JC (1999) The solutions of some plastic problems in plane strain by using the unified strength theory and the unified slip field theory. *Chinese J of Mechanical Engineering* **35**(6), 61–66.
Zheng YR, Shen ZJ and Gong XN (2002) The principles of Geotechnical Plasticity Mechanics. China Architecture and Building Press, Beijing (in Chinese).
Zhuang JH (1998) The analysis of limit inner pressure of thick-walled tube and sphere shell with different strength in tension and compression by theory of generalized twin-shear strength theory. *Mechanics and Practices,* 29(4): 39–41.
Zienkiewicz OC,Valliappan S and King IP (1968) Stress analysis of rock as a 'non-tension' material. *Geotechnique* **18**, 56–66.
Zienkiewicz OC and Pande GN (1977) Some useful forms of isotropic yield surfaces for soil and rock mechanics. *Finite Elements in Geomechanics*. Gudehus G ed. Wiley, London, pp 179–190.
Zienkiewicz OC and Humpheson C (1977) Viscoplasticity: A generalized model for description of soil behavior. *Numerical Methods in Geotechnical Engineering*. Desai CS and Christian JT eds. McGraw-Hill, New York.
Zienkiewicz OC (1982) Generalized plasticity and some models for geomechanics. *Appl. Math. and Mechanics* **3**, 267–280 (in Chinese).
Zienkiewicz OC and Mroz Z (1984) Generalized plasticity formulation and applications to geomechanics. In: *Mechanics of Engineering Materials*. Desai CS and Gallagher RH eds. Wiley, New York, pp 655–679.

2 Stress Space and Stress State

2.1 Elements

In applied mechanics and engineering, materials and structures are generally regarded as continua. This permits us to describe the behaviour and consequences of materials and structures by means of continuous functions. A material is a point (element), and a structure is a body. The structure may be considered as a partly ordered set of material elements (points) filling a structure (body). The cube is often used as an element. An element that can fill a space without gaps and overlapping is called the spatial equipartition.

Various polyhedra used in continuum mechanics are spatial equipartitions. They are the cubic element, regular hexagonal element, isoclinal octahedron element, dodecahedron element, orthogonal octahedron element and pentahedron element (Yu 1998, 2004).

2.2 Stress at a Point, Stress Invariants

A general state of stress at a point can be determined by a stress tensor σ_{ij}, which stands for nine components:

$$\sigma_{ij} = \begin{bmatrix} \sigma_x & \tau_{xy} & \tau_{xz} \\ \tau_{yx} & \sigma_y & \tau_{yz} \\ \tau_{zx} & \tau_{zy} & \sigma_z \end{bmatrix} \tag{2-1}$$

It can be seen in any course of mechanics of materials, elasticity, mechanics of solids or plasticity, by three-dimensional transformations, that there exists a coordinate system σ_1, σ_2, σ_3 where the state of stress at the same point can be described by the following:

$$\sigma_i = \begin{bmatrix} \sigma_1 & 0 & 0 \\ 0 & \sigma_2 & 0 \\ 0 & 0 & \sigma_3 \end{bmatrix} \tag{2-2}$$

The stresses σ_1, σ_2, σ_3 are referred to as the principal stresses.

An element of material subjected to principal stresses σ_1, σ_2 and σ_3 acting in mutually perpendicular directions (Fig. 2.2) is said to be in a state of triaxial stress or three-dimensional stress. If one of the principal stresses equals zero, this is referred to as the plane stress state or biaxial stress state. The triaxial stress and biaxial stress are called the polyaxial stresses, multiaxial stresses or complex stress. The principal planes are the planes on which the principal stresses occur on mutually perpendicular planes.

The principal stresses are the three roots of the equation:

$$\begin{aligned}&\sigma^3-(\sigma_x+\sigma_y+\sigma_z)\sigma^2+(\sigma_x\sigma_y+\sigma_y\sigma_z+\sigma_z\sigma_x-\tau_{xy}^2+\tau_{yz}^2+\tau_{zx}^2)\sigma\\&-(\sigma_x\sigma_y\sigma_z+2\tau_{xy}\tau_{yz}\tau_{zx}-\sigma_x\tau_{yz}^2-\sigma_y\tau_{zx}^2-\sigma_z\tau_{xy}^2)=0\end{aligned} \tag{2-3}$$

which can be rewritten as

$$\sigma^3-I_1\sigma^2+I_2\sigma-I_3=0 \tag{2-4}$$

where I_1, I_2, I_3 are

$$\begin{aligned}&I_1=\sigma_x+\sigma_y+\sigma_z=\sigma_1+\sigma_2+\sigma_3\\&I_2=\sigma_x\sigma_y+\sigma_y\sigma_z+\sigma_z\sigma_x-\tau_{xy}^2-\tau_{yz}^2-\tau_{zx}^2=\sigma_1\sigma_2+\sigma_2\sigma_3+\sigma_3\sigma_1\\&I_3=\sigma_1\sigma_2\sigma_3\end{aligned} \tag{2-5}$$

The quantities I_1, I_2 and I_3 are independent of the direction of the axes chosen; they are called the three invariants of the stress at a point (or invariant quantities).

2.3 Deviatoric Stress Tensor, Deviatoric Tensor Invariants

It is convenient in the study of strength theory and plasticity to split the stress tensor into two parts, one called the deviatoric stress tensor S_{ij} and the other the spherical stress tensor p_{ij}. The relation is

$$\sigma_{ij}=S_{ij}+p_{ij}=S_{ij}+\sigma_m\delta_{ij} \tag{2-6}$$

The spherical stress tensor is the tensor whose components are $\sigma_m\delta_{ij}$, where σ_m is the mean stress, i.e.,

$$p_{ij}=\sigma_m\delta_{ij}=\sigma_m\begin{bmatrix}1&0&0\\0&1&0\\0&0&1\end{bmatrix}=\begin{bmatrix}\sigma_m&0&0\\0&\sigma_m&0\\0&0&\sigma_m\end{bmatrix} \tag{2-7}$$

where

$$\sigma_m=(\sigma_x+\sigma_y+\sigma_z)/3=(\sigma_1+\sigma_2+\sigma_3)/3=I_1/3 \tag{2-8}$$

It is apparent that σ_m is the same for all possible orientations of the axes, hence it is named the spherical stress. Also, since σ_m is the same in all directions, it can be considered to act as a hydrostatic stress.

The deviatoric stress tensor S_{ij} can be determined as follows

$$S_{ij} = \sigma_{ij} - p_{ij} = \sigma_{ij} - \sigma_m \delta_{ij}$$

$$= \begin{bmatrix} \sigma_x - \sigma_m & \tau_{xy} & \tau_{xz} \\ \tau_{yx} & \sigma_y - \sigma_m & \tau_{yz} \\ \tau_{zx} & \tau_{zy} & \sigma_z - \sigma_m \end{bmatrix} \tag{2-9}$$

The invariants of the deviatoric stress tensor are denoted by J_1, J_2, J_3 and can be obtained as follows

$$J_1 = S_1 + S_2 + S_3 = 0$$

$$J_2 = \frac{1}{2} S_{ij} S_{ij} = \frac{2}{3}\left(\tau_{13}^2 + \tau_{12}^2 + \tau_{23}^2\right)$$

$$= \frac{1}{6}\left[(\sigma_1 - \sigma_2)^2 + (\sigma_2 - \sigma_3)^2 + (\sigma_3 - \sigma_1)^2\right]$$

$$J_3 = \left|S_{ij}\right| = S_1 S_2 S_3 = \frac{1}{27}(\tau_{13} + \tau_{12})(\tau_{21} + \tau_{23})(\tau_{31} + \tau_{32})$$

2.4 Stresses on the Oblique Plane

If the three principal stresses σ_1, σ_2, σ_3 acting on three principal planes, respectively, at a point are given, we can determine the stresses acting on any plane through this point. This can be done by consideration of the static equilibrium of an infinitesimal tetrahedron formed by this plane and the principal planes, as shown in Fig. 2.1. In this figure we have shown the principal stresses acting on the three principal planes. These stresses are assumed to be known. We wish to find the stresses σ_α, τ_α acting on the oblique plane whose normal has direction cosines l, m and n.

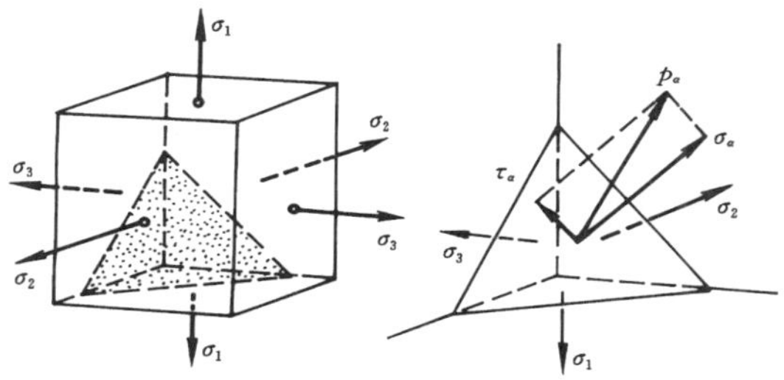

Fig. 2.1 Stress on an infinitesimal tetrahedron

2.4.1 Stresses on the Oblique Plane

The normal stress σ_α and shear stress τ_α acting on this plane can be determined as follows:

$$\sigma_\alpha = \sigma_1 l^2 + \sigma_2 m^2 + \sigma_3 n^2$$

$$\tau_\alpha = \sigma_1^2 l^2 + \sigma_2^2 m^2 + \sigma_3^2 n^2 - (\sigma_1 l^2 + \sigma_2 m^2 + \sigma_3 n^2)$$

$$\vec{p}_\alpha = \vec{\sigma}_\alpha + \vec{\tau}_\alpha$$

2.4.2 Principal Shear Stresses

The three principal shear stresses τ_{13}, τ_{12} and τ_{23} can be obtained as follows:

$$\tau_{13} = \frac{1}{2}(\sigma_1 - \sigma_3)$$

$$\tau_{12} = \frac{1}{2}(\sigma_1 - \sigma_2) \quad (2\text{-}10)$$

$$\tau_{23} = \frac{1}{2}(\sigma_2 - \sigma_3)$$

The maximum shear stress acts on the plane bisecting the angle between the largest and smallest principal stresses and is equal to half of the difference between these principal stresses

$$\tau_{max} = \tau_{13} = \frac{1}{2}(\sigma_1 - \sigma_3) \quad (2\text{-}11)$$

The corresponding normal stresses σ_{13}, σ_{12} and σ_{23} acting on the sections where τ_{13}, τ_{12} and τ_{23} are acting, respectively, are

$$\sigma_{13} = \frac{1}{2}(\sigma_1 + \sigma_3)$$

$$\sigma_{12} = \frac{1}{2}(\sigma_1 + \sigma_2) \quad (2\text{-}12)$$

$$\sigma_{23} = \frac{1}{2}(\sigma_2 + \sigma_3)$$

The directions of the principal stresses and the principal shear stresses are shown in Fig. 2.3.

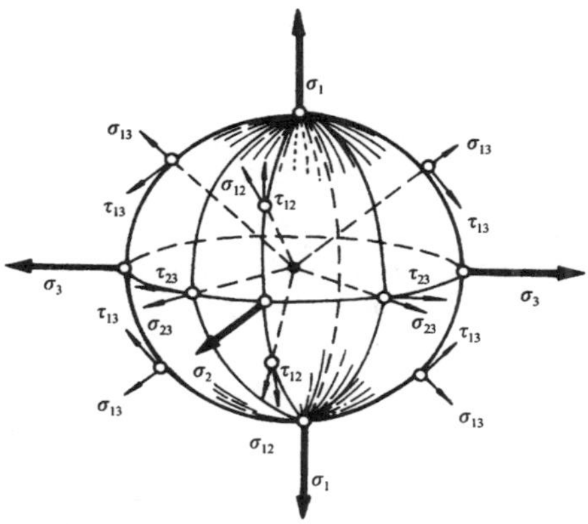

Fig. 2.2 Directions of the principal stresses and the principal shear stresses

The three principal stresses, three principal shear stresses and the stress on oblique plane can be illustrated by three stress circles, it is referred to as the Mohr circle, as shown in Fig. 2.3 (refer to Kussmaul 1981). The magnitude of the normal stress and shear stress of any plane are equal to the distance of the corresponding stress point on the stress circle The three principal shear stresses are evidently equal to the radius of the three Mohr circle. A detail description of the stress circle can be found in Johnson and Mellor (1962), Kussmaul (1981), Chakrabarty (1987) and others.

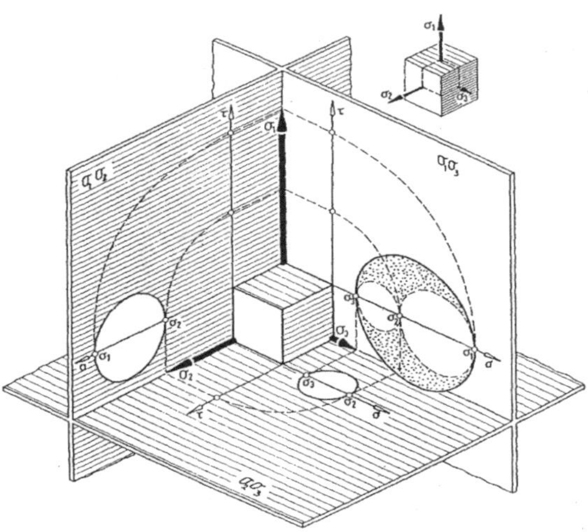

Fig. 2.3 The principal stresses, the principal shear stresses and stress circles

2.4.3
Octahedral Shear Stress

If the normal direction of the oblique plane makes equal angles with all the principal axes, and

$$l = m = n = \pm\frac{1}{\sqrt{3}} \tag{2-13}$$

These planes are called the octahedral plane and the shear stresses acting on it are called the octahedral shear stresses. The normal stress, called the octahedral normal stress σ_8 (or σ_{oct}), acting on this plane equals the mean stress

$$\sigma_8 = \frac{1}{3}(\sigma_1 + \sigma_2 + \sigma_3) = \sigma_m \tag{2-14}$$

A tetrahedron similar to this one can be constructed in each of the four quadrants above the x–y plane and in each of the four quadrants below the x–y plane. On the oblique face of each of these eight tetrahedra the condition $l^2 = m^2 = n^2 = 1/3$ will apply. The difference between the tetrahedra will be in the signs attached to l, m and n. The eight tetrahedra together form an octahedra as shown in Fig. 2.4e, and the eight planes form the faces of this octahedron. The normal stress is given by Eq. (2-14) and the octahedral shear stress τ_8 (sometimes denoted as τ_{oct}) acting on the octahedral plane is

$$\begin{aligned} \tau_8 &= \frac{1}{3}[(\sigma_1 - \sigma_2)^2 + (\sigma_2 - \sigma_3)^2 + (\sigma_3 - \sigma_1)^2]^{1/2} \\ &= \frac{1}{\sqrt{3}}[(\sigma_1 - \sigma_m)^2 + (\sigma_2 - \sigma_m)^2 + (\sigma_3 - \sigma_m)^2]^{1/2} \end{aligned} \tag{2-15}$$

The direction cosines l, m and n of principal planes, principal shear stress planes and the octahedral plane, as well as the normal stresses and shear stresses are listed in Table 2.1.

Table 2.1. Direction cosines of the principal planes, the principal shear stress planes and the octahedral plane

Principal plane				Principal shear stress planes			Octa. plane
l =	± 1	0	0	$\pm\frac{1}{\sqrt{2}}$	$\pm\frac{1}{\sqrt{2}}$	0	$\frac{1}{\sqrt{3}}$
m =	0	± 1	0	$\pm\frac{1}{\sqrt{2}}$	0	$\pm\frac{1}{\sqrt{2}}$	$\frac{1}{\sqrt{3}}$
n =	0	0	± 1	0	$\pm\frac{1}{\sqrt{2}}$	$\pm\frac{1}{\sqrt{2}}$	$\frac{1}{\sqrt{3}}$
$\sigma =$	σ_1	σ_2	σ_3	$\sigma_{12} = \frac{\sigma_1 + \sigma_2}{2}$	$\sigma_{13} = \frac{\sigma_1 + \sigma_3}{2}$	$\sigma_{23} = \frac{\sigma_2 + \sigma_3}{2}$	σ_8
$\tau =$	0	0	0	$\tau_{12} = \frac{\sigma_1 - \sigma_2}{2}$	$\tau_{13} = \frac{\sigma_1 - \sigma_3}{2}$	$\tau_{23} = \frac{\sigma_2 - \sigma_3}{2}$	τ_8

2.5 Hexahedron, Octahedron, Dodecahedron

According to the stress state, various polyhedral elements can be drawn as shown in Fig. 2.4. They are:

a. Cubic element (σ_1, σ_2, σ_3), the principal stress element: three principal stresses σ_1, σ_2, σ_3 act on this element.
b. Quadrangular prism element (τ_{13}, σ_{13}, σ_2), the maximum shear stress element; the maximum shear stress τ_{13} and respective normal stress σ_{13}, as well as the intermediate principal stress σ_2 act on this element.
c. Quadrangular prism element (τ_{12}, σ_{12}, σ_3), the intermediate principal shear stress element (when $\tau_{12} \geq \tau_{23}$), the intermediate principal shear stress τ_{12} and the respective normal stress σ_{12}, as well as the minimum principal stress σ_3 act on this element.
d. Quadrangular prism element (τ_{23}, σ_{23}, σ_1), the minimum principal shear stress element (when $\tau_{12} \leq \tau_{23}$), the minimum principal shear stress τ_{23} and the respective normal stress σ_{23}, as well as the maximum principal stress σ_1 act on this element.
e. Isoclinal octahedron element (τ_8, σ_8), the isoclinal octahedron element, the octahedral normal stresses σ_8 and octahedral shear stresses τ_8 act on this element.
f. Dodecahedron element (τ_{13}, τ_{12}, τ_{23}; σ_{13}, σ_{12}, σ_{23}), the dodecahedron element, the principal shear stresses τ_{13}, τ_{12}, τ_{23} and the respective normal stresses σ_{13}, σ_{12}, σ_{23} act on this element.
g. Orthogonal octahedron element (τ_{13}, τ_{12}; σ_{13}, σ_{12}), the orthogonal octahedron element, the principal shear stresses τ_{13}, τ_{12} and the respective normal stresses σ_{13}, σ_{12} act on this element. This element can also be referred to as the twin-shear element.

h. Orthogonal octahedron element (τ_{13}, τ_{23}; σ_{13}, σ_{23}), the orthogonal octahedron element, the principal shear stresses τ_{13}, τ_{23} and the respective normal stresses σ_{13}, σ_{23} act on this element. This element can also be referred to as the twin-shear element.

All the polyhedral elements shown in Fig. 2.4 are spatial equipartitions.

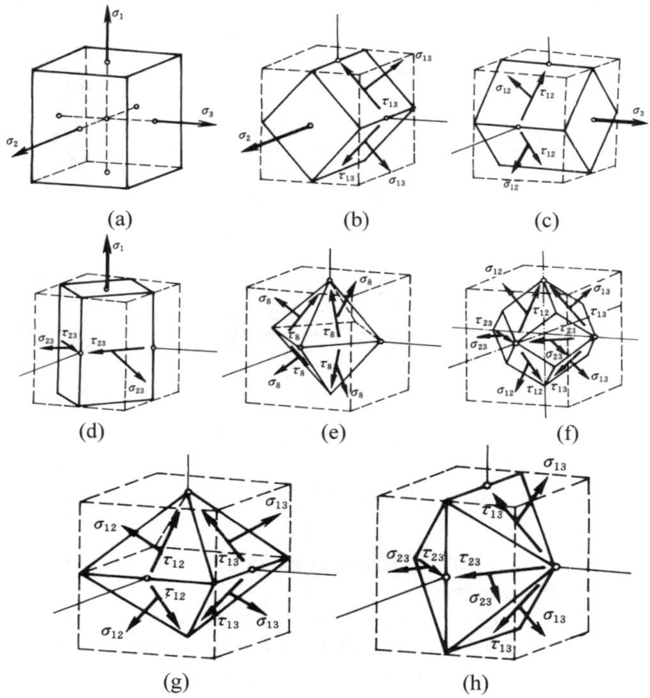

Fig. 2.4a–h Various polyhedral elements

2.6 Stress Space

The stress point P (σ_1, σ_2, σ_3) in stress space can be expressed by other forms, such as $P(x,y,z)$, $P(r,\ \theta,\ \xi,)$ or $P(J_2\ \ \theta,\ \xi,)$. The geometrical representation of these transfers can be seen in Fig. 2.5 and Fig. 2.6.

For the straight line OZ passing through the origin and making the same angle with each of the coordinate axes, the equation is

$$\sigma_1 = \sigma_2 = \sigma_3 \tag{2-16}$$

The equation of the π_0-plane is

$$\sigma_1 + \sigma_2 + \sigma_3 = 0 \tag{2-17}$$

The stress tensor σ_{ij} can be divided into the spherical stress tensor and deviatoric stress tensor. The stress vector σ can also be divided into two parts: the hydrostatic stress vector σ_m and the mean shear stress vector τ_m.

$$\sigma = \sigma_m + \tau_m \tag{2-18}$$

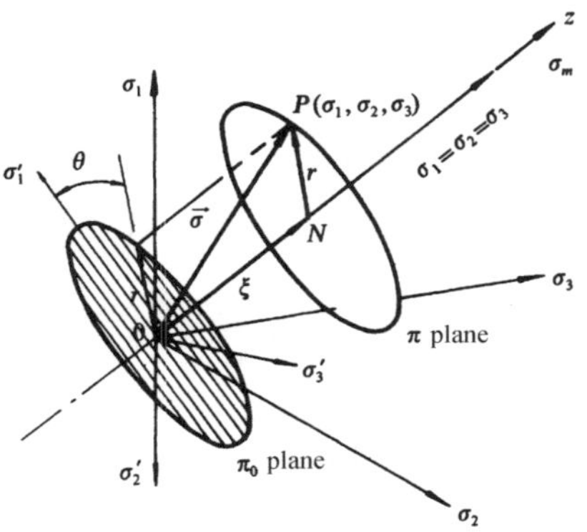

Fig. 2.5 Cylindrical coordinates

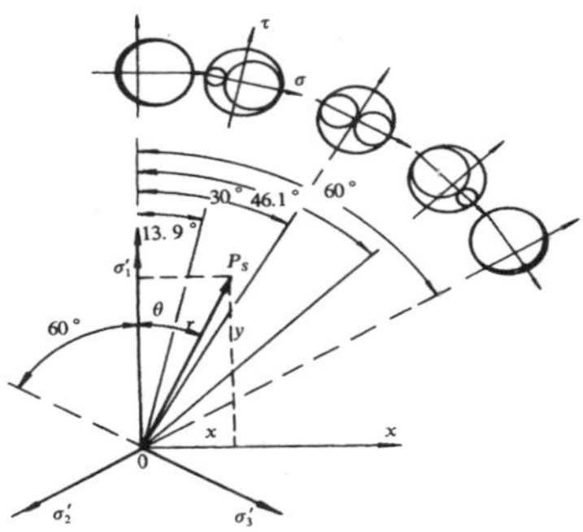

Fig. 2.6 Stress state in the π-plane

Their magnitudes are given by

$$\xi = \frac{1}{\sqrt{3}}(\sigma_1 + \sigma_2 + \sigma_3) \tag{2-19}$$

$$\begin{aligned} r &= \sqrt{\frac{1}{3}\left[(\sigma_1 - \sigma_2)^2 + (\sigma_2 - \sigma_3)^2 + (\sigma_3 - \sigma_1)^2\right]} \\ &= \sqrt{3}\tau_8 = \sqrt{2J_2} = 2\tau_m \end{aligned} \tag{2-20}$$

in which σ_8 is the octahedral normal stress and τ_8 is the octahedral shear stresses.

$$\begin{aligned} \tau_m &= \sqrt{\frac{\tau_{13}^2 + \tau_{12}^2 + \tau_{23}^2}{3}} \\ &= \sqrt{\frac{1}{12}\left[(\sigma_1 - \sigma_2)^2 + (\sigma_2 - \sigma_3)^2 + (\sigma_3 - \sigma_1)^2\right]} \end{aligned} \tag{2-21}$$

The π-plane is parallel to the π_0-plane and is given by

$$\sigma_1 + \sigma_2 + \sigma_3 = C \tag{2-22}$$

in which C is a constant. The spherical stress tensor σ_m is the same for all points in the π-plane of stress space and

$$\sigma_m = \frac{C}{3} \tag{2-23}$$

The projections of the three principal stress axes in stress space σ_1, σ_2, σ_3 are σ_1', σ_2', σ_3'. The relationship between them is

$$\begin{aligned} \sigma_1' &= \sigma_1 \cos\beta = \sqrt{\frac{2}{3}}\sigma_1 \\ \sigma_2' &= \sigma_2 \cos\beta = \sqrt{\frac{2}{3}}\sigma_2 \\ \sigma_3' &= \sigma_3 \cos\beta = \sqrt{\frac{2}{3}}\sigma_3 \end{aligned} \tag{2-24}$$

in which β is the angle between $O'A, O'B, O'C$ and the three coordinates as shown in Fig. 2.7.

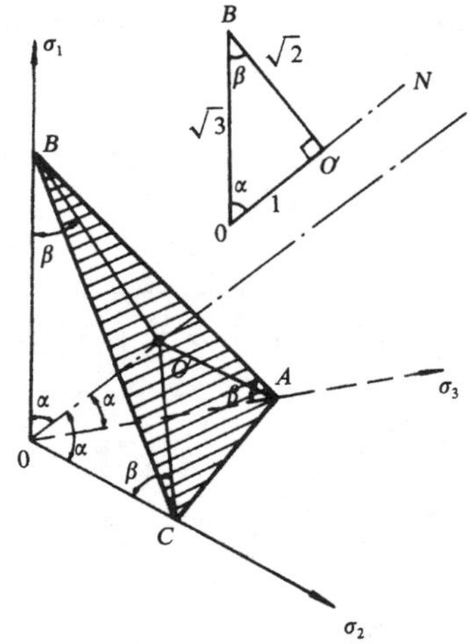

Fig. 2.7 Deviatoric plane

2.6.1 Relationship between (σ_1,σ_2,σ_3) and (x,y,z)

The relationships between the coordinates of the deviatoric plane and the principal stresses are:

$$x = \frac{1}{\sqrt{2}}(\sigma_3 - \sigma_2)$$

$$y = \frac{1}{\sqrt{6}}(2\sigma_1 - \sigma_2 - \sigma_3) \tag{2-25}$$

$$z = \frac{1}{\sqrt{3}}(\sigma_1 + \sigma_2 + \sigma_3)$$

$$\sigma_1 = \frac{1}{3}\left(\sqrt{6}y + \sqrt{3}z\right)$$

$$\sigma_2 = \frac{1}{6}\left(2\sqrt{3}z - \sqrt{6}y - 3\sqrt{2}x\right) \tag{2-26}$$

$$\sigma_3 = \frac{1}{6}\left(3\sqrt{2}x - \sqrt{6}y + 2\sqrt{3}z\right)$$

2.6.2 Relationship between ($\sigma_1, \sigma_2, \sigma_3$) and (ξ , r, θ) or (J_2, τ_m, θ)

The relationship between the cylindrical coordinates (ξ, r, θ) and the principal stresses ($\sigma_1, \sigma_2, \sigma_3$) are

$$\xi = |ON| = \frac{1}{\sqrt{3}}(\sigma_1 + \sigma_2 + \sigma_3) = \frac{I_1}{3} = \sqrt{3}\sigma_m \tag{2-27}$$

$$\begin{aligned} r = |NP| &= \frac{1}{\sqrt{3}}\left[(\sigma_1 - \sigma_2)^2 + (\sigma_2 - \sigma_3)^2 + (\sigma_3 - \sigma_1)^2\right]^{\frac{1}{2}} \\ &= \left(S_1^2 + S_2^2 + S_3^2\right)^{\frac{1}{2}} = \sqrt{2J_2} \\ &= \sqrt{3}\tau_8 = 2\tau_m \end{aligned} \tag{2-28}$$

$$\theta = \tan^{-1}\left(\frac{x}{y}\right) \tag{2-29}$$

From Eq. (2.25) and Eq. (2.28) we can obtain

$$\cos\theta = \frac{y}{r} = \frac{\sqrt{6}S_1}{\sqrt{2J_2}} = \frac{\sqrt{3}}{2}\frac{S_1}{\sqrt{J_2}} = \frac{2\sigma_1 - \sigma_2 - \sigma_3}{2\sqrt{3}\sqrt{J_2}} \tag{2-30}$$

The second and third invariants of the deviatoric stress tensor are

$$J_2 = -(S_1S_2 + S_2S_3 + S_3S_1) \tag{2-31}$$

$$J_3 = S_1S_2S_3 \tag{2-32}$$

Three principal deviatoric stresses can be deduced

$$\begin{aligned} S_1 &= \frac{2}{\sqrt{3}}\sqrt{J_2}\cos\theta \\ S_2 &= \frac{2}{\sqrt{3}}\sqrt{J_2}\cos\left(\frac{2\pi}{3} - \theta\right) \\ S_3 &= \frac{2}{\sqrt{3}}\sqrt{J_2}\cos\left(\frac{2\pi}{3} + \theta\right) \end{aligned} \tag{2-33}$$

These relationships are suitable to the conditions $\sigma_1 \geq \sigma_2 \geq \sigma_3$ and $0 \leq \theta \leq \pi/3$. The limit loci in the π-plane has threefold symmetry, so if the limit loci in the range of 60° are given, then the limit loci in π-plane can be obtained.

The three principal stresses can be expressed as follows:

$$\begin{aligned} \sigma_1 &= \frac{1}{\sqrt{3}}\xi + \sqrt{\frac{2}{3}}r\cos\theta \\ \sigma_2 &= \frac{1}{\sqrt{3}}\xi + \sqrt{\frac{2}{3}}r\cos(\theta - 2\pi/3) \\ \sigma_3 &= \frac{1}{\sqrt{3}}\xi + \sqrt{\frac{2}{3}}r\cos(\theta + 2\pi/3) \end{aligned} \tag{2-34}$$

The principal stresses can also be expressed in terms of the first invariant I_1 of the stress tensor and the second invariant of the deviatoric stress J_2 as follows:

$$\sigma_1 = \frac{I_1}{3} + \frac{2}{\sqrt{3}}\sqrt{J_2}\cos\theta$$

$$\sigma_2 = \frac{I_1}{3} + \frac{2}{\sqrt{3}}\sqrt{J_2}\cos(\theta - \frac{2\pi}{3}) \qquad (2\text{-}35)$$

$$\sigma_3 = \frac{I_1}{3} + \frac{2}{\sqrt{3}}\sqrt{J_2}\cos(\theta + \frac{2\pi}{3})$$

The principal shear stresses can also be obtained

$$\tau_{13} = \sqrt{J_2}\sin\left(\theta + \frac{\pi}{3}\right) = \sqrt{2}\tau_{\mathrm{m}}\sin\left(\theta + \frac{\pi}{3}\right)$$

$$\tau_{12} = \sqrt{J_2}\sin\left(\frac{\pi}{3} - \theta\right) \qquad (2\text{-}36)$$

$$\tau_{23} = \sqrt{J_2}\sin(\theta)$$

2.7 Stress State Parameters

The stress state at a point (element) is determined by the combination of the three principal stresses (σ_1, σ_2, σ_3). Based on the characteristics of the stress state and by introducing a certain parameter, it can be divided into several types. Lode (1926) introduced a stress parameter μ_σ as follows:

$$\mu_\sigma = (2\sigma_2 - \sigma_1 - \sigma_3)/(\sigma_1 - \sigma_3) \qquad (2\text{-}37)$$

which is referred to as the Lode stress parameter. The Lode parameter can be expressed in terms of principal shear stress as follows

$$\mu_\sigma = \frac{2\sigma_2 - \sigma_1 - \sigma_3}{\sigma_1 - \sigma_3} = \frac{\tau_{23} - \tau_{12}}{\tau_{13}} \qquad (2\text{-}38)$$

In fact, there are three principal shear stresses τ_{13}, τ_{12} and τ_{23} in the three-dimensional principal stress state. However, the three principal shear stresses τ_{13}, τ_{12} and τ_{23} are not independent and only two principal shear stresses are dependent variables, because the maximum principal shear stress τ_{13} equals the sum of the other two shear stresses. This relationship is expressed as follows:

$$\tau_{13} \equiv \tau_{12} + \tau_{23} \tag{2-39}$$

Subsquently, Yu introduced the "twin shear stress" concept into the analysis of the stress state and offered two twin–shear stress parameters (Yu 1991, 1992):

$$\mu_\tau = \frac{\tau_{12}}{\tau_{13}} = \frac{\sigma_1 - \sigma_2}{\sigma_1 - \sigma_3} = \frac{S_1 - S_2}{S_1 - S_3} \tag{2-40}$$

$$\mu_\tau' = \frac{\tau_{23}}{\tau_{13}} = \frac{\sigma_2 - \sigma_3}{\sigma_1 - \sigma_3} = \frac{S_2 - S_3}{S_1 - S_3} \tag{2-41}$$

$$\mu_\tau + \mu_\tau' = 1, \quad 0 \le \mu_\tau \le 1, 0 \le \mu_\tau' \le 1 \tag{2-42}$$

The twin-shear stress parameters are simpler and have an explicit physical meaning. They can reflect the state of the intermediate principal stress and can represent the status of stress state.

The twin-shear stress parameters have nothing to do with the hydrostatic stress. They instead represent the status of the deviatoric stress state and the stress angle on the deviatoric plane in stress space, as shown in Fig. 2.6. Five different stress states are shown in Fig. 2.6. They are $\theta = 0^\circ$ ($\mu_\tau = 1$), $\theta = 13.9^\circ$ ($\mu_\tau = 3/4$, $\mu_\tau' = 1/4$), $\theta = 30^\circ$ ($\mu_\tau = \mu_\tau' = 0.5$), $\theta = 46.1^\circ$ ($\mu_\tau = 1/4$, $\mu_\tau' = 3/4$) and $\theta = 60^\circ$ ($\mu_\tau = 0$, $\mu_\tau' = 1$). According to the meaning of the twin–shear stress parameters, we know that:

If $\mu_\tau = 1$ ($\mu_\tau' = 0$, stress angle equals $\theta = 0^\circ$), the stress states include three following cases:

1. $\sigma_1 > 0, \quad \sigma_2 = \sigma_3 = 0,$ uniaxial tension stress state;
2. $\sigma_1 = 0, \quad \sigma_2 = \sigma_3 < 0,$ equal biaxial compression stress state;
3. $\sigma_1 > 0, \quad \sigma_2 = \sigma_3 < 0,$ uniaxial tension, equal biaxial compression stress state.

If $\mu_\tau = \mu_\tau' = 0.5$ (stress angle equals $\theta = 30^\circ$), the corresponding stress states are as follows:

1. $\sigma_2 = \frac{1}{2}(\sigma_1 + \sigma_3) = 0$, pure shear stress state;
2. $\sigma_2 = (\sigma_1 + \sigma_3)/2 > 0,$ biaxial tension and uniaxial compression stress state;
3. $\sigma_2 = (\sigma_1 + \sigma_3)/2 < 0,$ uniaxial tension and biaxial compression stress state.

If $\mu_\tau = 0$ ($\mu_\tau' = 1$, stress angle equals $\theta = 60^\circ$), then the corresponding stress states are as follows:

1. $\sigma_1 = \sigma_2 = 0, \sigma_3 < 0,$ uniaxial compression stress state;
2. $\sigma_1 = \sigma_2 > 0, \sigma_3 = 0,$ equal biaxial tension stress state;
3. $\sigma_1 = \sigma_2 > 0, \sigma_3 < 0,$ equal biaxial tension and uniaxial compression stress state.

According to the twin-shear stress parameters and the magnitude of the two smaller principal shear stresses, the stress state can be divided into three kinds of conditions as follows:

1. Extended tension stress state, that is, $\tau_{12} > \tau_{23}$, $0 \le \mu_\tau' < 0.5 < \mu_\tau \le 1$. The stress state (uniaxial tension and biaxial compression) can be expressed by deviatoric stress, and the absolute magnitude of the tensile stress is a maximum, so it can be called the extended tension stress state. When the intermediate principal stress σ_2 equals the minimum principal stress σ_3, then $\mu_\tau = 1$ ($\mu_\tau' = 0$). If $\sigma_2 = \sigma_3 = 0$, the extended tension stress state becomes the uniaxial tension stress state.
2. Extended shear stress state, that is, $\tau_{12} = \tau_{23}$, $\sigma_2 = (\sigma_1 + \sigma_3)/2$. The two smaller stress circulars are equal, the second deviatoric stress $S_2 = 0$ and the magnitude of the other two deviatoric stresses are identical, but one is tensile and the other is compressive. The two twin-shear stress parameters are identical, that is, $\mu_\tau = \mu_\tau' = 0.5$. If $\sigma_2 = (\sigma_1 + \sigma_3)/2 = 0$, the extended shear stress state becomes the pure shear stress state.
3. Extended compression stress state, that is, $\tau_{12} < \tau_{23}$, $0 \le \mu_\tau < 0.5 < \mu_\tau' \le 1$. If $\sigma_1 = \sigma_2 = 0$, $\sigma_3 < 0$, this stress state becomes the uniaxial compression stress state.

The twin-shear parameters simplify the Lode parameter and have a clear physical meaning. Their relationships are:

$$\mu_\tau = \frac{1 - \mu_\sigma}{2} = 1 - \mu_\tau' \tag{2-43}$$

$$\mu_\tau' = \frac{1 + \mu_\sigma}{2} = 1 - \mu_\tau \tag{2-44}$$

Some types of stress states and stress state parameters including the Lode parameter and the twin-shear stress parameters are summarized in Table 2.2.

The relationships among various shear stresses are listed in Table 2.3. It is convienent for comparing the definitions of the stress parameters. Different sympose or expressions may be used in different applications.

Table 2.2. Principal stresses, shear stresses and stress state parameters

Stress state		Principal stress	Principal shear stress	Deviatoric stress	Stress angle	Parameter of stress state		
						μ_τ	μ_τ'	μ_σ
Extended tension	Pure tension, Equal Biaxial compression	$\sigma_2=\sigma_3$	$\tau_{12}=\tau_{13}$ $\tau_{23}=0$	$S_2=S_3$ $S_1=S_2+S_3$	0°	1	0	−1
	$\tau_{23}=\frac{\tau_{12}}{3}$, $\tau_{13}=4\tau_{23}$	$\sigma_2<\frac{\sigma_1+\sigma_3}{2}$	$\tau_{12}>\tau_{23}$	$S_1=S_2+S_3$	13.9°	$\frac{3}{4}$	$\frac{1}{4}$	$\frac{-1}{2}$
Pure shear		$\sigma_2=\frac{\sigma_1+\sigma_3}{2}$	$\tau_{12}=\tau_{23}$	$S_1=\lvert S_3\rvert$ $S_2=0$	30°	0.5	0.5	0
Extended compression	$\tau_{12}=\frac{\tau_{23}}{3}$, $\tau_{13}=4\tau_{12}$	$\sigma_2>\frac{\sigma_1+\sigma_3}{2}$	$\tau_{12}<\tau_{23}$	$\lvert S_3\rvert=S_1+S_2$	46.1°	$\frac{1}{4}$	$\frac{3}{4}$	$\frac{1}{2}$
	Pure compression equal biaxial compression	$\sigma_2=\sigma_1$	$\tau_{12}=0$ $\tau_{23}=\tau_{13}$	$S_1=S_2$ $\lvert S_3\rvert=S_1+S_2$	60°	0	1	+1

Table 2.3. Relationships among various shear stresses and J_2

	q_τ	τ_8	τ_s	$\tau_\pi=r$	J_2	S_{ij}
Generalized shear stress q_τ	q_τ	$\frac{3}{\sqrt{2}}\tau_8$	$\sqrt{3}\tau_s$	$\sqrt{\frac{3}{2}}\tau_\pi$	$\sqrt{3J_2}$	$\sqrt{\frac{3}{2}S_{ij}S_{ij}}$
Octhedral shear stress τ_8	$\frac{\sqrt{2}}{3}q_\tau$	τ_8	$\sqrt{\frac{2}{3}}\tau_s$	$\frac{1}{\sqrt{3}}\tau_\pi$	$\sqrt{\frac{2}{3}J_2}$	$\sqrt{\frac{1}{3}S_{ij}S_{ij}}$
Pure shear stress τ_s	$\frac{1}{\sqrt{3}}q_\tau$	$\sqrt{\frac{3}{2}}\tau_8$	τ_s	$\frac{1}{\sqrt{2}}\tau_\pi$	$\sqrt{J_2}$	$\sqrt{\frac{1}{2}S_{ij}S_{ij}}$
shear stress on deviatoric plane $\tau_\pi=r$	$\sqrt{\frac{2}{3}}q_\tau$	$\sqrt{3}\tau_8$	$\sqrt{2}\tau_s$	τ_π	$\sqrt{2J_2}$	$\sqrt{S_{ij}S_{ij}}$
Second invariant J_2 of deviatoric stress	$\frac{1}{3}q_\tau^2$	$\frac{3}{2}\tau_8^2$	τ_s^2	$\frac{1}{2}\tau_\pi^2$	J_2	$\frac{1}{2}S_{ij}S_{ij}$

Summary

Elements and stress states are described briefly in this chapter. Stress states can be studied in many courses, such as elasticity, plasticity, mechanics of solids, rock mechanics, soil mechanics. The basic formulas are given here only.

The twin-shear stresses, the twin-shear element and the twin-shear stress parameter are new concepts. They are used in following chapters.

The relationships among various shear stresses and J_2 are listed in Table 2.3. Various different notations may be used at different textbook.

References

Chakrabarty J (1987) *Theory of Plasticity*. McGraw-Hill, New York.
Johnson W. and Mellor P.B (1962), *Plasticity for Mechanical Engineers*. D.Van Nostrand Co. London.
Kussmaul Karl (1981) *Festigkeitslehre* I. MPA Stuttgart, University Stuttgart, Stuttgart.
Yu Mao-hong (1998) *Twin-Shear Theory and Its Applications*. Science Press, Beijing. (in Chinese)
Yu Mao-hong (2004) *Unified Strength Theory and Its Applications*. Springer, Berlin.

3 Basic Characteristics of Yield of Materials under Complex Stress

3.1 Introduction

The stress-strain curve of a mild steel under uniaxial stress shows that there exist a tensile yield point and a compressive yield point at which the material will begin to deform plastically. In this case the stress is uniaxial and the magnitudes of the tensile yield point σ_{yt} and the compressive yield point σ_{yc} are identical,

$$\sigma_{yt} = \sigma_{yc} = \sigma_y \tag{3-1}$$

Material behaviour is elastic if stress $\sigma < \sigma_y$. This yield stress can readily be determined based on the uniaxial test. What if, however, there are several stresses acting on an element in multiaxial stress state, i.e., biaxial stress (σ_1,σ_2) or triaxial stress (σ_1,σ_2,σ_3)? What combination of these stresses will cause yielding? We will now extend the definition for yielding from the uniaxial concept of a yield stress σ_y to a general three-dimensional state of stress or multiaxial stresses.

A law defining the limit of elastic behavior (or deciding what combination of multiaxial stresses will cause yielding) under any possible combination of stresses is called yield criterion. The mathematical expression of yield criterion is

$$F = (\sigma_1, \sigma_2, \sigma_3) = 0 \quad \text{or} \quad F(I_1, J_2, J_3) = 0 \tag{3-2}$$

The yield criterion is a function of the stress state and the material parameters. The suitability of any proposed yield criterion must be verified by experiment.

It is necessary to study some basic characteristics of yield for engineering materials under complex stress for research into a general yield function. A large number of experiments have laid the groundwork for the theoretical research of yield function. Some basic characteristics of yield behavior of materials under complex stress are summarized in this chapter.

3.2 Strength Difference Effect (SD Effect)

The strength of most brittle materials is greater under compression than that under tension. Fig. 3.1 shows the stress-strain curve of cast iron subjected to a uniaxial load. It is seen that the compressive strength of cast iron (σ_c) is about 3 to 5 times greater than its tensile strength (σ_t). Concrete and rock also have this kind of characteristics with compressive strength 10 times greater than tensile strength. The general stress-strain relation of rock under uniaxial stress is shown in Fig. 3.2.

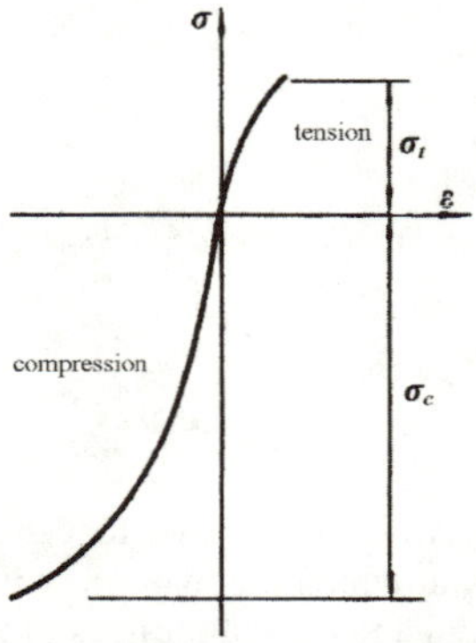

Fig. 3.1 Stress-strain curve of cast iron

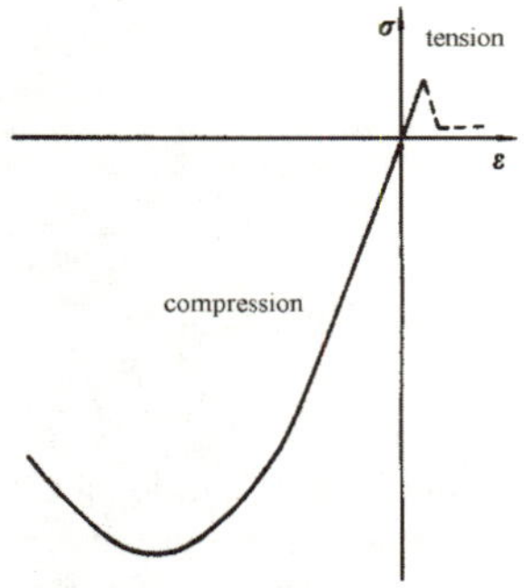

Fig. 3.2 Stress-strain curve of rock

For metals, we traditionally think that they have the uniform strength whether they are subjected to compressive or tensile load. However, there were a number of reports (Chait 1972; Rauch and Leslie 1972; Drucker 1973; Spitzig et al. 1975, 1976; Richmond and Spitzig 1980; Lewandowski and Lowhaphandu 1998) suggesting that there is a significant difference for high-strength steels between the compressive strength and tensile yield strength; this is called the strength difference effect or the SD effect. Some of these materials are high-strength stainless steels and high-strength aluminum alloys widely used in the aviation and automobile, electric and chemistry industries. Fig. 3.3 shows the stress-strain curve of stainless steel subjected to tension and compression at normal temperature (Spitzig et al. 1975, 1976, Richmond and Spitzig 1980). It is seen that both curves show an obvious SD effect. Generally, for metallic materials, the more obvious the strength difference effect, the higher the strength of the material.

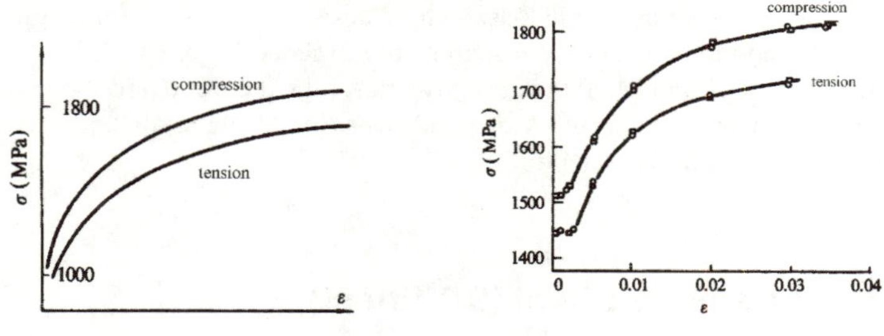

Fig. 3.3 Stress-strain curves in tension and in compression (Richmond and Spitzig 1980; Spitzig et al. 1975)

It is shown that the tensile strength σ_t and the compressive strength σ_c of most materials are different, i.e. $\sigma_t \neq \sigma_c$. To ensure this condition, it is necessary to impose the further restriction in the yield function that

$$F(-\sigma_i) \neq F(\sigma_i), \quad \text{or} \quad F(-\sigma_1, -\sigma_2, -\sigma_3) \neq F(\sigma_1, \sigma_2, \sigma_3) \qquad (3\text{-}3)$$

3.3 Effect of Hydrostatic Stress

Hydrostatic stress, or mean stress $\sigma_m = (\sigma_1 + \sigma_2 + \sigma_3)/3$, has a great influence on material strength for brittle materials. Many studies have been devoted to the effect of hydrostatic stress. In an early work, von Karman experimented on rock strength. He applied certain confining pressures on test rocks, then gradually increased the axial pressure while the confining pressures remained unchanged. The experimental results of his tests are shown in Fig. 3.4. The strength of rocks is increased with the increasing of the confining pressure. The relationship between the limit stress circle and the confining pressure can also be obtained. A systematical study was done by Bridgman (1964).

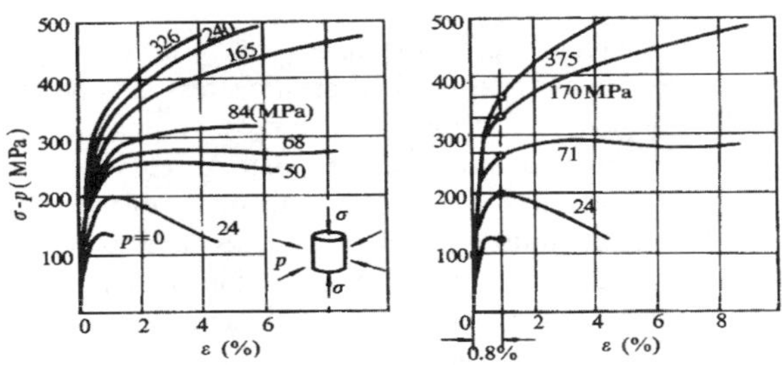

(a) (b)

Fig. 3.4 Stress-strain relation of marble under different σ_m (von Karman 1911)

Compressive stress-strain curves for an aged nickel alloy at three different hydrostatic pressures were given by Spitzig et al. (1975, 1976) and Richmond and Spitzig (1980). These curves are shown in Fig. 3.5. It is shown that the high-strength alloys have both the SD effect and the effect of hydrostatic stress. The linear dependence of yield stress on hydrostatic stress for nickel alloys is extended to tensile mean stress.

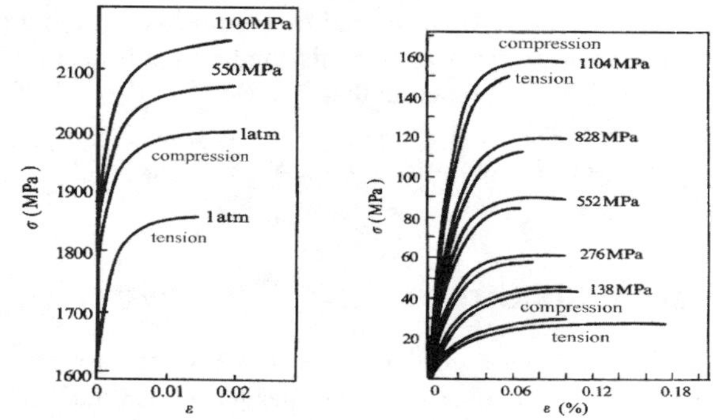

Fig. 3.5 SD effect and σ_m effect of maraging steel and polyethylene (Spitzig et al. 1976)

This behaviour implies that the general yield function must satisfy the condition as follows:

$$F(\sigma_i) \neq F(\sigma_i \pm \sigma_m) \neq F(S_i) \quad \text{or} \quad \frac{\partial F}{\partial \sigma_i} \neq 0 \tag{3-4}$$

The effects of hydrostatic stress can also be observed in some other materials subjected to high confining pressure loads.

The effect of σ_m on the yield stress for stainless steel was given by Richmond and Spitzig (1980). A marked linear effect of hydrostatic stress is indicated. Spitzig and Richmond (1979) also gave the stress-strain curves of polyethylene at various hydrostatic pressures. Four compressive stress-strain curves and two tensile stress–strain curves are shown in Fig. 3.6. The dependence of the proportional limit and yield stress at 1% offset strain on hydrostatic stress are shown in Fig. 3.6a for polyethylene and in Fig. 3.6b for polycarbonate. The rectangular experimental points in Fig. 3.6 represent tensile test results and the circles represent compressive test results. From these curves we can find that the effects of hydrostatic stress for both polyethylene and polycarbonate are linear.

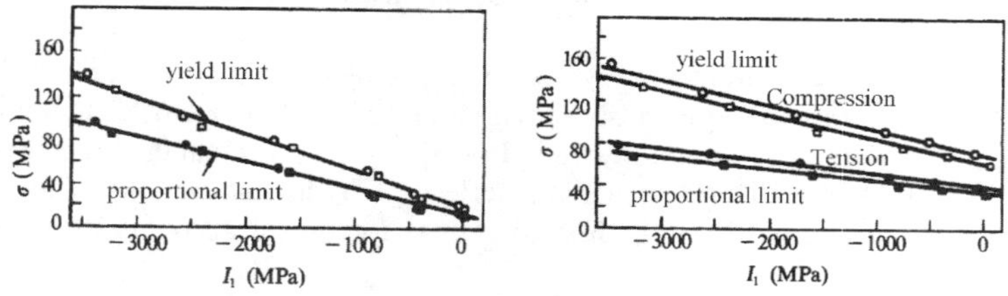

(a) polyethylene (b) polycarbonate

Fig. 3.6 Relation of yield stress on hydrostatic stress for two polymers

The combined effect of the SD effect and the effect of hydrostatic stress has been found in the tests. This phenomenon has also been observed in other hydrostatic pressure tests for other materials.

In the triaxial test, the axial stress σ_1 minus lateral compressive stress (confining pressure or hydrostatic pressure σ_3) yields the maximum shear stress, which is τ_{max}= $(\sigma_1 - \sigma_3)/2$. Therefore the result is shown as the relationship between shear strength and hydrostatic pressure. The general result from a great number of lateral confining compressive stress tests for rock is shown in Fig. 3.7.

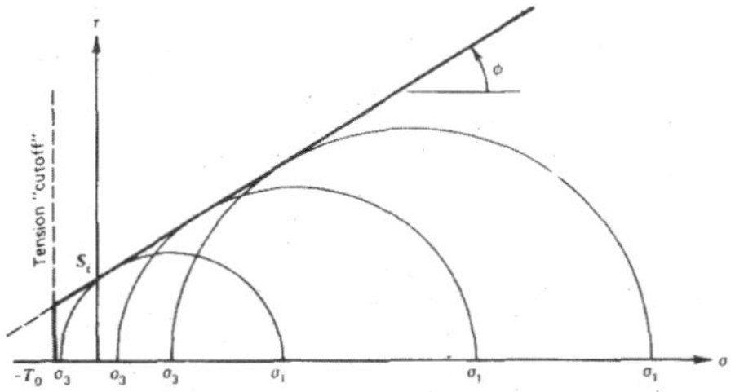

Fig. 3.7 Shear strength changes with hydrostatic pressure of rock (Goodman 1980)

The shear strength of rock increases with the development of hydrostatic pressure ($\sigma_2 = \sigma_3 = p$). The linear dependence of shear strength on hydrostatic stress in the low-pressure region is apparent.

3.4 Effect of Normal Stress

It is worth noting that the yield strength of a material usually depends on the difference of the principal stresses, which is the magnitude of the shear stress. Fig. 3.8 shows the values of the minimum shear stresses τ plotted against the normal stress σ for various rock materials obtained by Jaeger and Cook (1979). In Fig. 3.8, A represents marble, B is for Trachyte, C is for Trachyte with a smoother surface, D is for sandstone and μ is the coefficient of friction. Similar results were also observed for granite, basalt and losse. Fig. 3.9 was obtained for granite rock from the Laxiwa Hydraulic Power Station on the Yellow River in China. The similar results were also obtained from three different experiments of loess at Xi'an area.

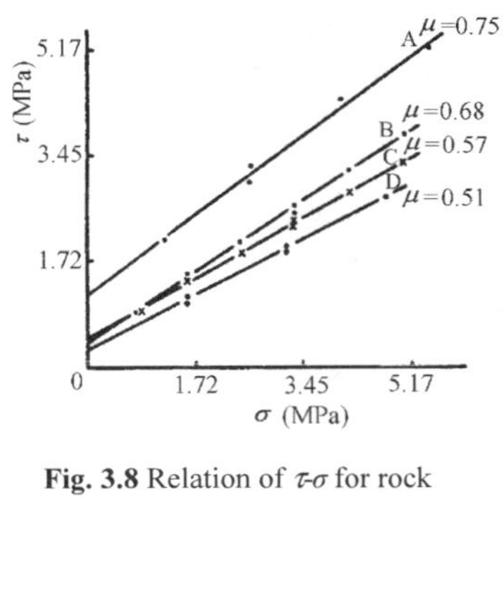

Fig. 3.8 Relation of τ-σ for rock

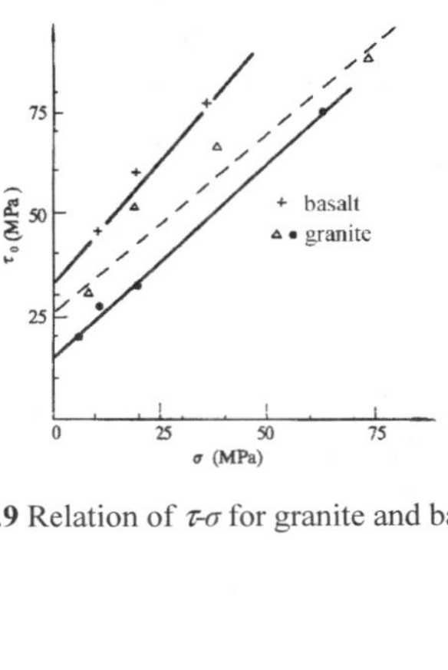

Fig. 3.9 Relation of τ-σ for granite and basalt

The dependence of strength on normal stress is the most marked characteristic of geomaterials. The general relationship between the shear strength τ and the normal stress σ can be considered linear and can be expressed as follows:

$$F = F(\tau_{ij} + \beta\sigma_{ij}) \tag{3-5}$$

or

$$\tau = C + \sigma \tan\varphi \tag{3-6}$$

3.5 Effect of Intermediate Principal Stress

In order to investigate the effect of the intermediate principal stress, Lode performed a lot of experiments on this problem. Lode introduced a stress parameter μ_{Lode} to represent the status of the intermediate principal stress:

$$\mu_{Lode} = \frac{2\sigma_2 - \sigma_1 - \sigma_3}{\sigma_1 - \sigma_3}, \quad (-1 \leq \mu_\sigma \leq 1) \tag{3-7}$$

which is called the Lode parameter. The Lode parameter, however, could not be expressed explicitly. The Lode parameter can be simplified by introducing a new concept of the twin-shear stress state parameter as follows (Yu 1990b, 1992).

$$\mu_\tau = \frac{\tau_{12}}{\tau_{13}} = \frac{\sigma_1 - \sigma_2}{\sigma_1 - \sigma_3}, \quad (0 \leq \mu_\tau \leq 1) \tag{3-8}$$

$$\mu_\tau' = \frac{\tau_{23}}{\tau_{13}} = \frac{\sigma_2 - \sigma_3}{\sigma_1 - \sigma_3}, \quad (0 \leq \mu_\tau' \leq 1) \tag{3-9}$$

These stress parameters are equivalent. They reflect the changes in the states of three principal stresses when increasing the intermediate principal stress σ_2 from the value of the minimum principal stress to the maximum principal stress, i. e., from $\sigma_2 = \sigma_3$ to $\sigma_2 = \sigma_1$. Increasing the intermediate principal stress means decreasing the intermediate principal shear-stress τ_{12}, while means increasing the intermediate principal shear-stress τ_{23}.

Their relationships are:

$$\mu_\tau = \frac{1-\mu_\sigma}{2} = 1 - \mu_\tau' \; ; \quad \mu_\tau' = \frac{1+\mu_\sigma}{2} = 1 - \mu_\tau \tag{3-10}$$

The twin-shear stress parameters are simpler and more straightforward than the Lode parameter.

3.5.1 Metals

Lode (1926) performed the experiments on thin-walled tubes made of iron, copper and nickel. The results are shown in Fig. 3.10. All materials show the effect of the intermediate principal stress. Taylor and Quinney (1931) did experiments on mild steel, copper and aluminum thin-walled tubes that were subjected to tension and torsion. Their results are shown in Fig. 3.11.

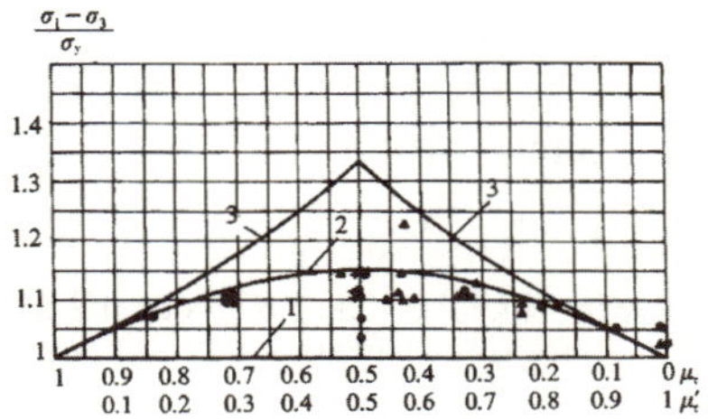

Fig. 3.10 The σ_2 effect (Lode 1926)

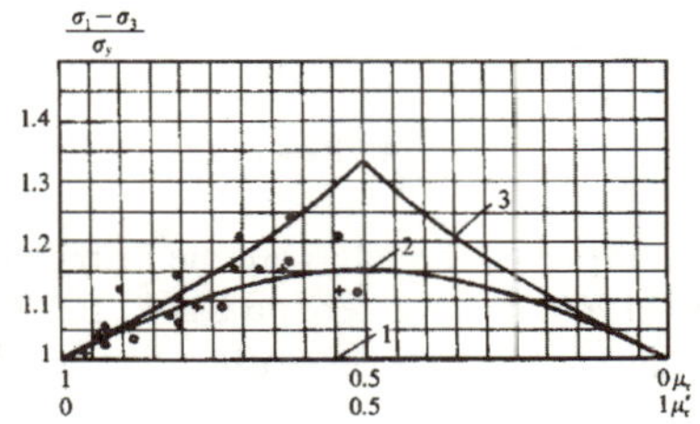

Fig. 3.11 The σ_2 effect (Taylor, Quinney 1931)

Fig. 3.12 shows the results for aluminum alloy thin-walled tubes that were subjected to combined stresses by Ivey (1961). The experimental results prove the existence of the effect of intermediate principal stress in aluminum alloys.

In 1964, Mair et al. experimented on pure copper thin-walled tubes subjected to combined tension and torsion stresses. The results are shown in Fig. 3.13.

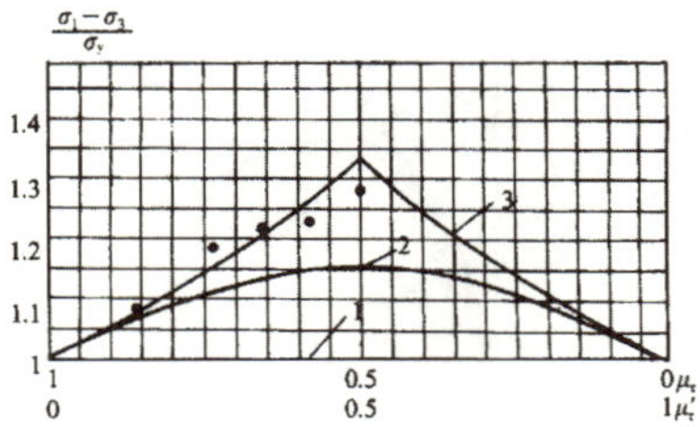

Fig. 3.12 Aluminum alloy (Ivey 1961)

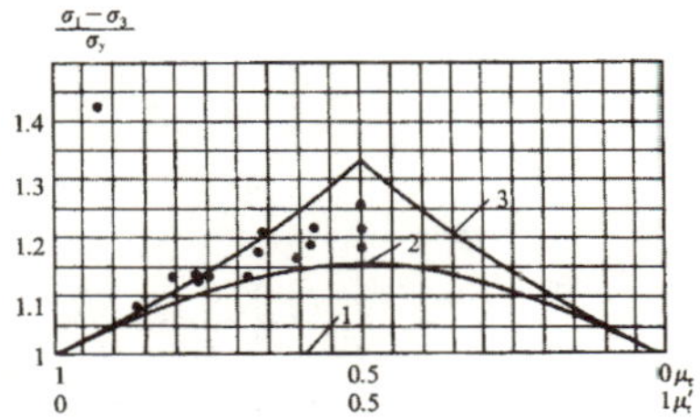

Fig. 3.13 Copper (Mair 1964)

From the above experiments, we can conclude that the effect of intermediate principal stress exists in all kinds of materials in different degrees. The difference may reach 10%~33%.

3.5.2 Rock

Fig. 3.14 shows the results of experiments on different coals (Hobbs 1962). It indicates that the strength of coal increases quickly as the intermediate principal stress increases. However, when the intermediate principal stress reaches a certain value, the strength of coal decreases gradually. Fig. 3.15 shows the results of experiments on granite (Mazanti and Sowers 1965).

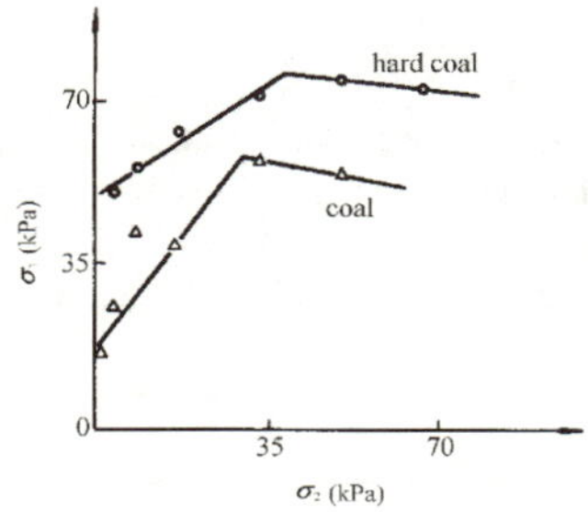

Fig. 3.14 Effect of σ_2 in coal

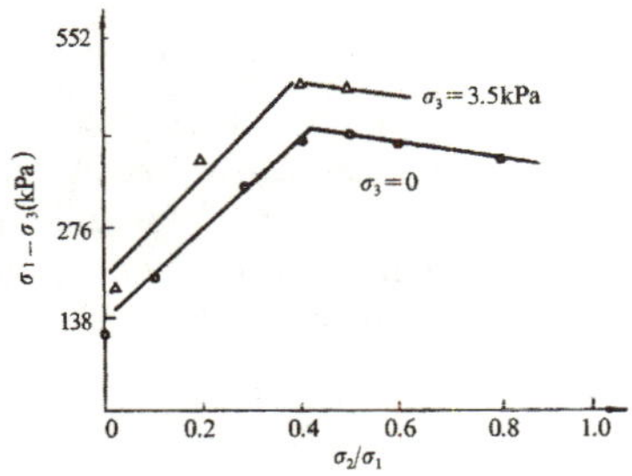

Fig. 3.15 Effect of σ_2 in granite

The effect of the intermediate principal stress for rock was also found by Hoskins et al. (1969, see: Jaeger and Cook 1979) as show in Fig.3.16. The results of the true triaxial experiments on marble obtained by Michelis (1985, 1987) are shown in Fig. 3.17.

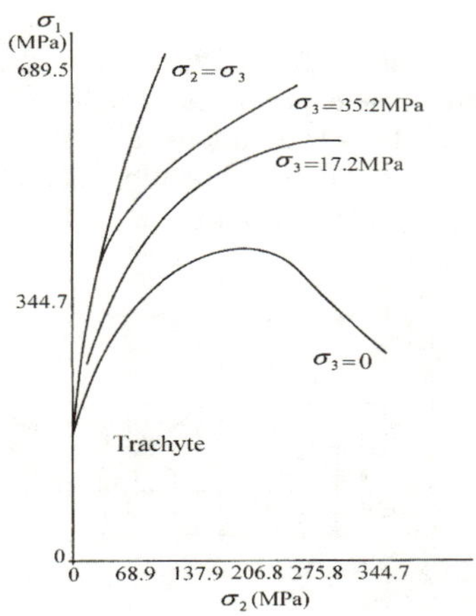

Fig. 3.16 Effect of σ_2 for rock

(Hoskins 1969, see: Jaeger and Cook 1979)

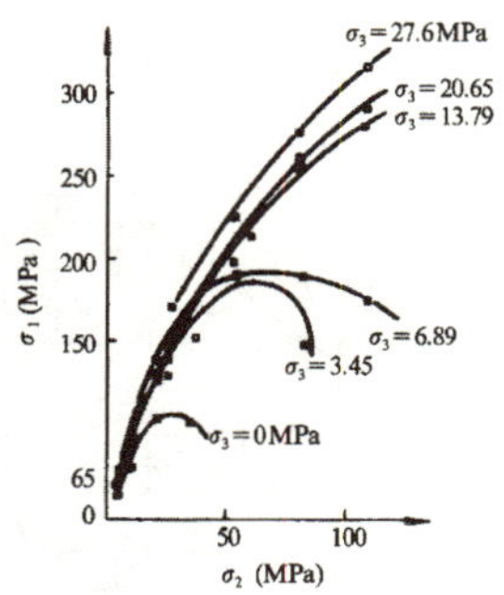

Fig. 3.17 Effect of σ_2 for rock

(Michelis 1985, 1987)

It is obvious that when σ_2 increases from the minimum value of $\sigma_2 = \sigma_3$ to the maximum value of $\sigma_2 = \sigma_1$ under constant σ_3, the strength of the rock gradually increases and reaches a peak. The strength then gradually decreases to the ending strength, which is ($\sigma_2 = \sigma_1 > \sigma_3$) slightly higher than the beginning strength ($\sigma_1 > \sigma_2 = \sigma_3$).

To test the effect of the intermediate principal stress on scarlet sandstone, Gao and Tao (1993) performed three groups of experiments with a rigid true triaxial machine. The minimum principal stresses were 0.4 and 8 MPa, respectively. The relation curves between the maximum principal stress strength limit σ_1 and the intermediate principal stress σ_2 are shown in Fig. 3.18. The rule is the same as the experimental results of Hoskins and Michelis for rock (Li and Xu 1990). Some true triaxial experimental results from various rocks under different conditions were summarized by Gao and Tao (1993). These results are shown in Fig. 3.18.

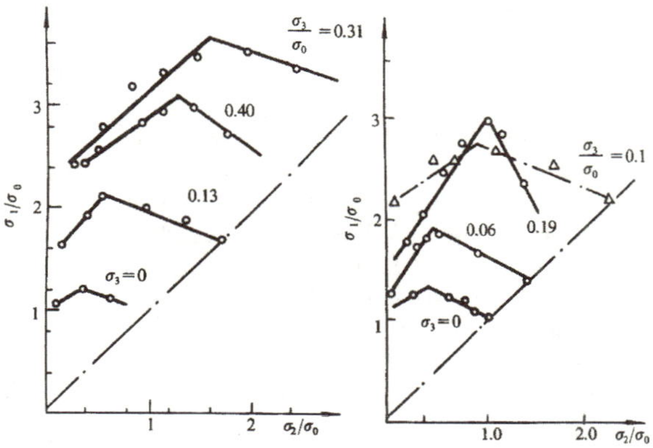

Fig. 3.18 The σ_2 effect curves of the true triaxial experiments (Gao and Tao 1993)

It can be seen from these results that all experimental results indicate remarkable effects of the intermediate principal stress. The effect of the intermediate principal stress, which ranges from the minimum value of 18% to the maximum value of 75%. Normally, it is about 25%~40%. This rule is similar to the rule for concrete that is discussed in next section.

Research on the intermediate principal stress for rock began with Foppl and Böker and von Karman at the beginning of the twentieth century. However, it has been discussed with various opinions until the late 1960s. Jaeger and Cook felt that the effect of the intermediate principal stress should be solved, since it is a problem of great significance in theory and practical matters. They also pointed out, "it can be surmised that the effect of increasing the intermediate principal stress is to increase the strength from that obtained in triaxial stress conditions to a higher value. An analytical formulation of this transition is so complex that its meaning is not obvious." (Jaeger and Cook 1979).

The effect of the intermediate principal stress has been confirmed and is considered a significant feature of the yield of rock.

1. The intermediate principal stress has an obvious influence on the strength of rock. Let the minimum principal stress σ_3 equal a certain value, then the strength of rock under various increasing σ_2 ($\sigma_3 < \sigma_2 \leq \sigma_1$) is greater than the strength under confining compression ($\sigma_3 = \sigma_2 < \sigma_1$). Therefore the value of strength of rock from confining triaxial tests is lower, and the strength of rock is increased by 20–30 % when the effect of the intermediate principal stress is taken into account. The

higher the minimum principal stress is, the greater the effect of the intermediate principal stress.

2. When the intermediate principal stress σ_2 increases from the lowest limit $\sigma_2=\sigma_3$ to $\sigma_2=\sigma_1$, the strength of rock will first increase to a certain peak value, and then decrease after that value. The yield strength of rock under $\sigma_2=\sigma_1>\sigma_3$ is somewhat greater than that with $\sigma_3=\sigma_2<\sigma_1$.

3.5.3 Concrete

True triaxial testing of concrete developed the same as that for rock in the 1960s because their test facilities and results are very similar. For the true triaxial test, in which the three loads can be separately controlled, Michelis began his experiment with σ_3 unchanged, and then increased σ_1 with σ_2 equal to 3.15 MPa, 6.89 MPa, 13.8 MPa and 27.58 MPa, respectively. He then plotted four stress–strain curves as shown in Fig. 3.19 with his test results. Fig. 3.19 also gives the relationship between ε_2 and $(\sigma_1-\sigma_3)$. It can be seen that σ_2 obviously increases the strength and deformation of concrete.

Michelis obtained various stress–strain curves at different intermediate principal stresses when σ_3 equals 3.45 MPa, 3.89 MPa and 13.79 MPa, respectively. Figure 3.19 shows the σ_2 effect when σ_3 =3.15 MPa (Michelis 1985).

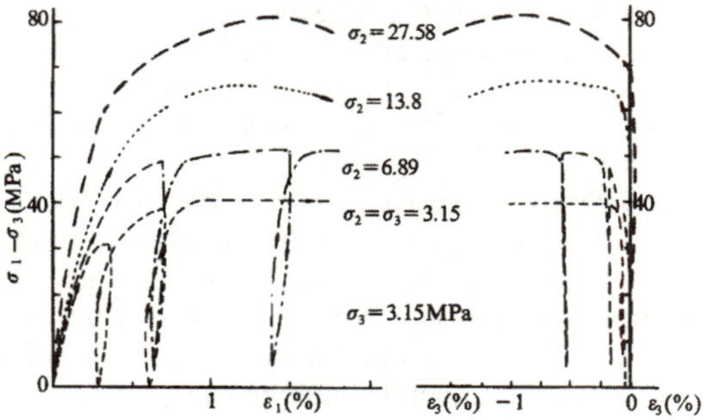

Fig. 3.19 The σ_2 effect when σ_3 =3.15 MPa (Michelis 1985)

Glomb's results showed that the intermediate principal stress does affect the yield strength of concrete, and the magnitude is risen up to 23-26%. Mier (1986) and Wang et al. (1987) determined the effect of biaxial strength they got an increase of magnitude of about 1.385–1.622 times, while other researchers obtained values of 1.2–1.6 times.

All of the above tests for the effect of biaxial strength were made in the plane stress state. In this case, because one of principal stresses equals to zero, i.e., the minimum principal stress σ_3=0, the effect of the two–dimensional stress is also the effect of the intermediate principal stress. Many true triaxial tests have been done over the past three decades. Fig. 3.20 is a strength curve made by Launay and

Gachon (1972). The changes in the strength of concrete under different intermediate principal stresses when the minimum principal stress σ_3 is $\sigma_3=0$, $\sigma_3=0.2\sigma_0$, $\sigma_3=0.4\sigma_0$, $\sigma_3=0.6\sigma_0$, $\sigma_3=0.8\sigma_0$ and $\sigma_3=\sigma_0$, respectively were given. All of these data show that the intermediate principal stress has a marked effect on the strength of materials.

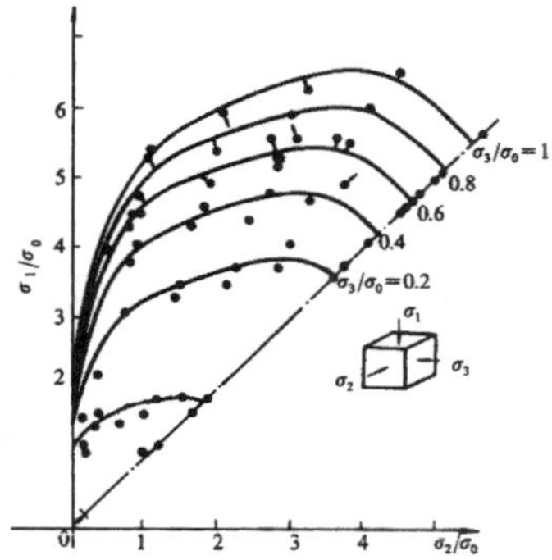

Fig. 3.20 The σ_2 effect curve for concrete under different σ_3 (Launay and Gachon 1972)

The effect of intermediate principal stress on soil was also observed by Shibata and Karube (1965), Ko and Scott (1968), Sutherland and Mesdary (1969), Green and Bishop (1969), Bishop (1971), Butterfield and Harkness (1971), Ergun (1981) and Matsuoka and Nakai (1974, 1985).

The effect of intermediate principal stress is an important characteristic of materials (Michelis 1985).

3.6 Effect of Intermediate Principal Shear-Stress

The effect of intermediate principal shear-stress τ_{12} or τ_{23} on sand was studied by Bishop (1966) and Green and Bishop (1969). The variation of friction angle of sand with intermediate principal shear stress $2\tau_{23}=(\sigma_2-\sigma_3)$ is shown in Fig.3.21. The friction angles of shearing resistance φ' with $(\sigma_2-\sigma_3)$ at porosities of n=0.37,0.38, 0.39, etc., to n=0.42 are plotted. The results of tests show that the peak strength of dense Ham River sand increases as the intermediate principal shear-stress $2\tau_{23}=(\sigma_2-\sigma_3)$ increases.

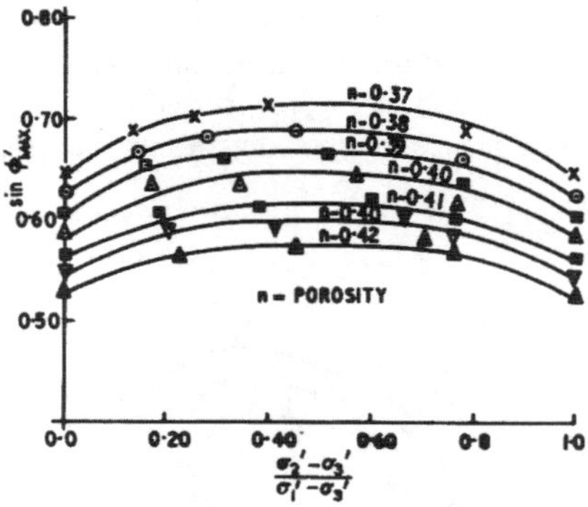

Fig. 3.21 Variation of friction angle with intermediate principal shear-stress (Bishop 1966)

Fig. 3.22 shows the same result given by Green and Bishop (1969). They indicated that: "The commonly used Mohr-Coulomb theory will tend to underestimate the strength by about 5° over most of the range. This would be a significant error in many analyses of field problems" (Green and Bishop 1969).

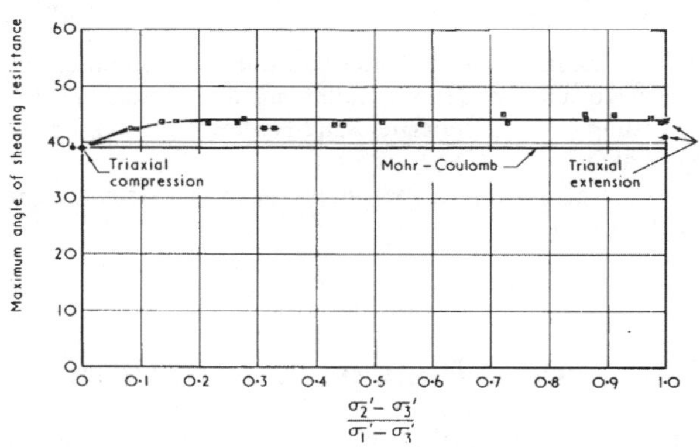

Fig. 3.22 Peak strengths of Dense Ham River sand (Green and Bishop 1969)

The similar results are summarized by Ergun (1981) and Ramamurthy-Tokhi (1981), as shown in Fig. 3.23.

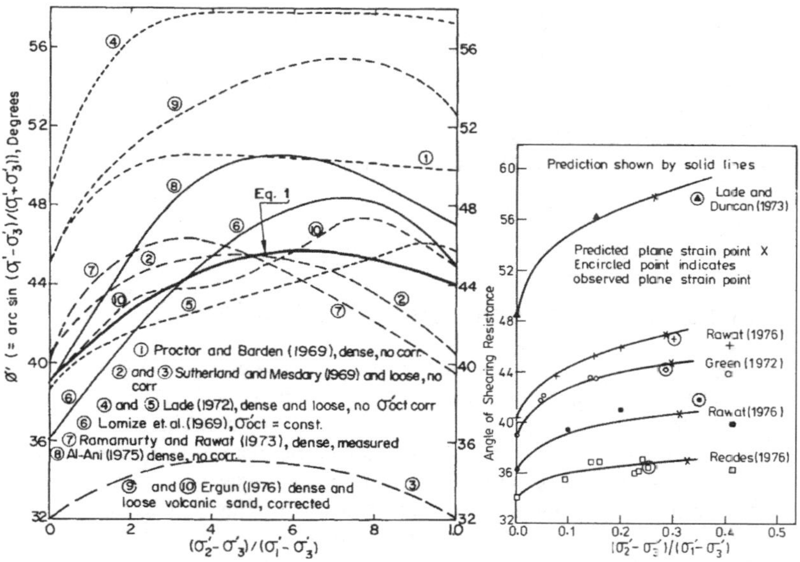

(a) Ergun (1981) (b) Ramamurthy-Tokhi (1981)

Fig. 3.23 Variation of friction angle with intermediate principal shear stress

The effect of intermediate principal shear-stress on rock was found experimentally by Kwasniewski, Takahashi and Li (2003). The fine- to medium-grained Slask sandstone was taken from the Upper Silesian Coal Basin, Poland. The results of true triaxial compression tests show that the ultimate strength of sandstone increases as the intermediate principal shear stress $2\tau_{23}=(\sigma_2-\sigma_3)$ increases, as shown in Fig. 3.24.

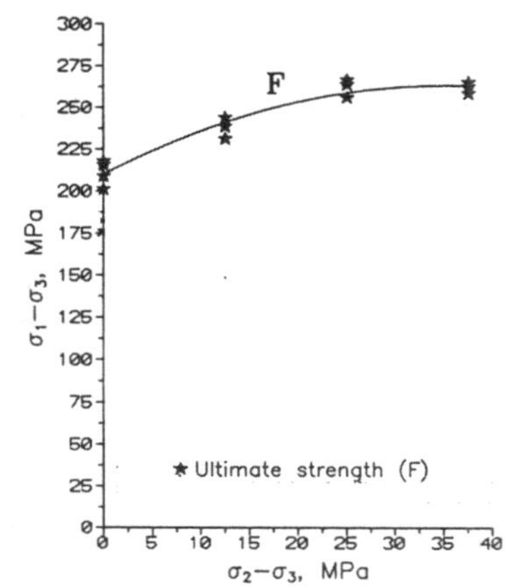

Fig. 3.24 Effect of intermediate principal shear stress on the strength of sandstone

The effect of intermediate principal shear-stress is similar to the effect of intermediate principal stress; however, it is more complex than the effect of intermediate principal shear stress. In point of fact the effect of intermediate principal shear-stress can be obtained from the true triaxial test; unfortunately, the relation of ultimate strength of rock with intermediate principal shear-stress $\tau_{23}=(\sigma_2-\sigma_3)/2$ or $\tau_{12}=(\sigma_1-\sigma_2)/2$ has not been given before Kwasniewski, Takahashi and Li (2003).

3.7 Bounds of the Convex Strength Theories

The function of strength theory for an isotropic material $F(\sigma_i)=0$ is a symmetric function of the principal stresses σ_i $(i=1, 2, 3)$. The yield loci must have three-fold symmetry for the three stresses axes σ_1, σ_2, σ_3, as shown in Fig. 3.25. It represents a cross section of the yield loci in the deviatoric plane. The projections of the coordinate axes σ_1, σ_2, σ_3 on the deviatoric plane are designated by σ_1', σ_2', σ_3'. It should be noted that interchanging the arbitrarily chosen indices 1, 2, 3 on the coordinate axes will not alter the physical conditions under which yield occurs. When performing experiments, it is only necessary to explore one of the six 60°-sections shown in the figure; the other sections are then given by three-fold symmetry.

The effect of the strength difference (the SD effect) does not require the limit stress in tension to be the same as the limit stress in compression. This means that the distances *OT* and *OC* would not have to be equal, as shown in Fig. 3.25(a). If the yield stresses in tension and compression are identical, the yield loci must have six-fold symmetry. In this case, in order to determine the yield locus experimentally, it is only necessary to explore any one of the typical 30°-sectors shown in Fig. 3.25(b).

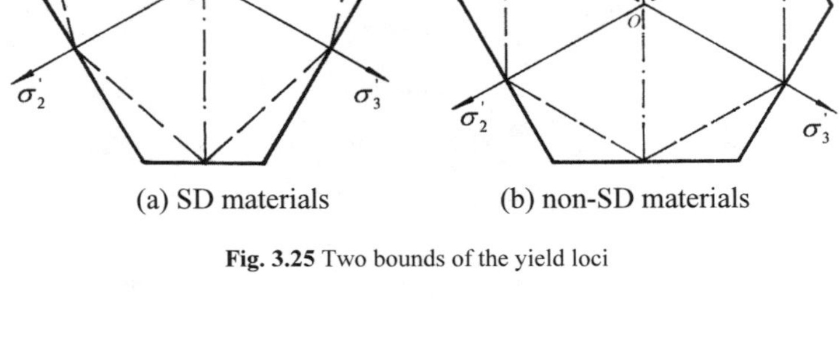

(a) SD materials (b) non-SD materials

Fig. 3.25 Two bounds of the yield loci

A fundamental postulate concerning the convex of yield surfaces was proposed by Drucker (1951) with the convexity of the yield surface determined.

According to convexity, the yield surface can be convex and satisfy the other requirements only if it lies between the two bounds as illustrated in Fig. 3.25. The lower bound is the yield locus of the Mohr-Coulomb strength theory; it can be thought of as the inner limit of convexity. The upper bound is the yield locus of the twin-shear strength theory proposed by Yu et al. in 1985; it can be thought of as the outer limit of convexity and will be described in Chap. 4. Most of the experimental results are situated between these two limit loci.

Two bounds of the yield criteria for isotropic materials with same yield stresses in tension and compression (non-SD material) are shown in Fig. 3.25(b). The inner (lower) bound is the yield locus of the single-shear yield criterion (Tresca 1864). The outer (upper) bound is the yield locus of the twin-shear yield criterion (Yu 1961a). The Huber-von Mises circle circumscribes the inner bound and inscribes the outer bound.

Summary

Some characteristics of yield of materials under complex stresses are discussed in this chapter. They are the SD effect, the effect of hydrostatic stress, the effect of normal stress, the effect of intermediate principal stress, the effect of intermediate principal shear-stress, symmetry and the convexity of yield surfaces. The research on the SD effect, the effect of hydrostatic stress and the effect of normal stress have developed rapidly because they can easily be carried out with relatively ordinary experimental facilities, and can be explained by theory on hand. However, research on the effect of intermediate principal stress has been more difficult and time-consuming. This is because its experiments are difficult, and it requires more accurate as well as more expensive facilities. Research on the effect of the intermediate principal stress and the effect of intermediate principal shear-stress will have to continue, and has become an interesting as well as significant topic for scholars.

Strength theory deals with the yield of materials under the complex stress state. It is difficult to find a general law for the varieties of yield of materials under the complex stress. However, considerable experimental and theoretical studies have provided us with valuable data for comparison, verification and study of the avaiable criteria.

References

Bishop AW (1966) The strength of soils as engineering materials. Sixth Rankine Lecture, Geotechnique, 16(2), 89–128.

Bishop AW (1971) Shear strength parameters for undisturbed and remoulded soil specimens. *Stress-Strain Behaviour of Soils Proc.* (Roscoe Memorial Symposium, Cambridge University, Cambridge, England 1971), Parry RHG ed. Foulis, pp 1–59.

Bridgman PW (1964) *Studies in Large Plastic Flow and Fracture , with Special Emphasis on the effects of Hydrostatic Pressure.* Harvard University Press, Cambridge.

Byerlee JD (1967a) Theory of friction based on brittle fracture. *J. Appl. Phys*. 38, 2928–34.

Byerlee JD (1967b) Frictional characteristics of grahite under high confining pressure. *J. Geophys. Res*. 72, 3639–48.

Byerlee JD (1970) Static and kinetic friction of granite at high normal stress. *Int. J. Rock Mech. Min. Sci. and Geomech. Abstrs* 7, 577–582.

Butterfield R and Harkness RM (1971) The kinematics of Mohr–Coulomb materials. *Stress-Strain Behaviour of Soils*. Parry RHG ed. Foulis, pp 220–233.

Chait R (1972) Factors influencing the strength differential of high strength steels. *Metallurgical Transactions* 3, 365–371.

Cornforth DH (1964) Some experiments on influence of strain conditions on the strength of sand. Geotechnique 143 - 167.

Drucker DC (1973) Plasticity theory, strength differential (SD) phenemenon, and volume expansion in metals and plastics. *Metall. Trans*. 4, 667–673.

Ergun MU (1981) Evaluation of three-dimensional shear testing. *Proc. of 10th Int. Conf. on Soil Mechanics and Foundation Engng*, Stockholm, 593–596.

Finn WD and Mittal HK (1963) Shear strength of soil in a general stress space. ASTM STP no. 361. Laboratory shear testing of soils. 42–48.

Goodman R.E (1980) Introduction to Rock Mechanics. John Wiley and Sons,New York.

Green GE (1972) Strength and deformation of sand measured in an independent stress control Cell. Roscoe memorial Symposium 'Stress-strain behaviour of soils'. G.T. Foulis and Co. 285-323.

Green GE and Bishop AW (1969) A note on the drained strength of sand under generalized strain conditions. Geotechnique, 19(1), 144–149.

Hanbly, E.C. and Roscoe, K.H. (1969) Observations and predictions of stresses and strains during plane strain of wet clays. Proc. 7th Int. Conf, Soil Mech. Found. Engg.(1), 173−181.

Handin J, Heard HC and Magouirk JN (1967) Effect of the intermediate principal stress on the failure of limestone, dolomite and glass at different temperatures and strain rates. *J. Geophys. Res*. 72, 611–640.

Hobbs DW (1966) A study of the behaviour of a broken rock under triaxial compression, and its application to mine roadways. *Int. J. Rock Mech. Min. Sci.* 3, 11–43.

Hoskins ER, Jaeger JC, and Rosengren KJ (1968) A medium scale direct friction experiment, *Int. J. Rock Mech. Min. Sci*., **5**, 143−154.

Hoskins ER (1969) The failure of thick-walled hollow cylinders of isotropic rock. *Int. J. Rock Mech. Min. Sci.,* **6**, 99−125.

Ichibara M. and Matsuzawa H (1973) Application of plane strain tests to earth pressure. Proc. 8th Int. conf. Soil Mech. Found. Engg.(1.1), 185 − 190. Moscow.

Jaeger JC (1963) Extension failures in rocks subject to fluid pressure, *J. Geophys. Res.,***68**, 6066-6067.

Jaeger JC and Hoskins ER (1966), Rock failure under the confined Brazilian test, *J. Geophys. Res.,* 71- 2651-2659.

Jaeger JC and Cook NGW (1979) Fundamentals of Rock Mechanics. 3rd. Chapman and Hall, London.

Kwagniewski. M., Takahashi, M. & Li. X (2003) Volume changes in sandstone under true triaxial compression conditions. In: ISRM 2003-Technology roadmap for rock mechanics, South African Institute of Mining and Metallurgy. 683-688

Ko HY and Scott RF (1967) Deformation of sand in shear. *J. of Soil Mechanics and Foundations* 93, 283–310.

Ko HY and Scott RF (1968) Deformation of sand at failure. *J. of Soil Mechanics and Foundations* 94, 883–898.

Lade, P.V. and Duncan J.M. (1973). Cubical triaxial tests on cohesionless soil. ASCE J.Soil Mech. and Found. Div. No.lO, 793–812.

Launay P and Gachon H (1971) Strain and ultimate strength of concrete under triaxial stress. *Proc. 1st Int. Conf. Struc. Mech. Reactor Technol.,* Belin, paper H1/3.

Lee KL (1970) Comparison of plane strain and triaxial tests on sand. ASCE J. Soil Mech. and Found. Div. No. 3, 901–923.

Leussink H and Wittke w (1963). Difference in triaxial and plane strain shear strength. ASTM STP No. 361 Laboratory shear testing of soils. 77–89.

Lewandowski JJ and Lowhaphandu P (1998) Effects of hydrostatic pressure on mechanical behaviour and deformation processing of materials. *Int. Materials Reviews* 43(4), 145–187.

Mansfield EH (1971) Biaxial yield criteria. *J. of the Royal Aeronautical Society* 75 (732), 849–850.

Matsuoka H and Nakai T (1974) Stress-deformation and strength characteristics of soil under three different principal stresses. *Proc. of Japan Society of Civil Engineers* 232, 59–70.

Matsuoka H and Nakai T (1985) Relationship among Tresca, Mises, Mohr–Coulomb and Matsuoka–Nakai failure criteria. *Soils and Foundations* (*Japan*) 25(4), 123–128.

Michelis P (1985) Polyaxial yielding of granular rock. *J. Eng. Mech.* 111, 1049–1066.

Michelis P (1985) True triaxial cycle behavior of concrete and rock in compression. *Int. J. of Plasticity,* 3(2), 249–270.

NagaraJ, T.S. and Somashekar, B.V. (1979) stress deformation and strength of soils in plane strain. VI Asian Reg. Conf. on Soll Mech. and Found. Engg. (I), 43–46·

Parry RHG (1971) A study of influence of intermediate principal stress on φ values using a critical state theory.IV ARC on Soll Mech.Found.Enggo(1),159–165, Bangkok.

Ramamurthy T and Tokhi VK (1981) Relation of triaxial and plane strain strength. *Proceedings of 10th Int. Conf. on Soil Mech. and Fund. Engrg., Stockholm,* pp 755–758.

Rauch GC and Leslie WC (1972) The extent and nature of the strength-differential effect in steels. *Metallurgical Transactions* 3, 373–381.

Rawat PC (1976) Shear behaviour of cohesionless materials under generalized conditions of stress and strain. Ph.D. thesis, IoloT. DeLhi, 415 pp.

Reades DW (1972) Stress-strain characteristics of a sand under three dimensional oading. Ph.D. thesis, Uni of London.

Richmond O and Spitzig WA (1980) Pressure dependence and dilatancy of plastic flow. *Theoretical and Applied Mechanics,* 15th ICTAM, pp 377–386.

Rowe PW(1962) The stress-dilatancy relationship for static equilibrium of an asse~bly of particles in contact. Proc.Royal Society. Vol. 269t London, 500 –527.

Spitzig WA, Sober RJ and Richmond O (1975) Pressure dependence of yielding and associated volume expansion in tempered martensite. *Acta Met.* 23, 885–893.

Spitzig WA, Sober RJ and Richmond O (1976) The effect of hydrostatic pressure on the deformation behavior of maraging and HY-80 steels and its implications for plasticity theory. *Metall. Trans.* 7A, 1703–1710.

Spitzig WA (1979) Effect of hydrostatic pressure on plastic flow properties of iron single crystals. *Acta Met.* 27, 523–534.

Spitzig WA and Richmond O (1979) Effect of hydrostatic pressure on the deformation behavior of polyethylene and polycarbonate in tension and compresion. *Polymer Engng.* 19, 1129–1139.

Shibata T and Karube D (1965) Influence of the variation of the intermediate principal stress on the mechanical properties of normally consolidated clays. *Proc. Sixth Int. Conf. on Soil Mechanics and Found Engrg. Vol. 1.* pp 359–363.

Sutherland HB and Mesdary MS (1969) The influence of the intermediate principal stress on the strength of sand. *Proc. of 7th Int. Conf. on Soil Mechanics and Foundation Engineering,* Vol. 1. Mexico City, pp 391–399.

Yu Mao-Hong (1961) General behaviour of isotropic yield function (in Chinese), *Res. Report of Xi'an Jiaotong University.*Xi'an.

Yu Mao-Hong (1961) Plastic potential and flow rules associated singular yield criterion (in Chinese), *Res. Report of Xi'an Jiaotong University.*Xi'an.

Yu Mao-Hong (1983) Twin shear stress yield criterion. *Int. J. Mech.* Sci. 25 (1) 71–74.

Yu Mao-Hong, He Li-nan and Song Ling-yu (1985) Twin shear stress theory and its generalization, *Scientia Sinica* (*Sciences in China*), English Edition, Series A, **28**(11), 1174–1183.

4 Unified Strength Theory and Its Material Parameters

4.1 Introduction

Great effort has been devoted to the formulation of strength theories, failure criteria and yield criteria. Many versions of these were presented during the past 100 years. The single-shear criterion (Tresca 1864), the Huber-von Mises criterion (1904, 1913) and the twin-shear criterion (Yu 1961a) can be suitable for those materials that have the identical strength both in tension and compression. For these materials the shear yield stresses are τ_y=0.5σ_y, τ_y=0.577σ_y and τ_y=0.667σ_y, respectively, where τ_y is the shear yield strength and σ_y is the uniaxial yield strength of materials. The Drucker–Prager criterion contradicts the experimental results of geomaterials. The Mohr–Coulomb strength theory (1900) and the twin-shear strength theory (Yu 1985) are two bounds of the convex strength theory. Each one mentioned above is suitable for only a certain type of materials.

What is the relationship among various strength theories? Can we propose a unified strength theory that adapted to more kinds of materials?

Before the study, we should to discuss the general behavior of yield functions of materials under complex stress state.

For an isotropic material, the yield function can generally be expressed in terms of the three principal stresses or stress invariants as

$$F(\sigma_1, \sigma_2, \sigma_3) = 0, \quad \text{or} \quad F(I_1, J_2, J_3) = 0 \tag{4-1}$$

The general yield function can also be expressed in terms of cylindrical coordinates (Haigh–Westgaard coordinates) as

$$F(\xi, \rho, \theta) = 0 \tag{4-2}$$

It is evident that all the effects of the three components σ_1, σ_2, σ_3 must be included in the general yield function. It means that the three stress invariants I_1, J_2, J_3 have to be incorporated into the expressions of the general yield function. In other words, the general mathematical expression of yield function must include all the three components of σ_1, σ_2, σ_3 or I_1, J_2, J_3.

The basic characteristics of material under the complex stress have been summarized in Chap.3. The following general behaviors must be considered in yield function.

1. SD effect;
2. Hydrostatic stress effect;

3. Normal stress effect;
4. Effect of the intermediate principal stress;
5. Effect of intermediate principal shear stress;
6. Three-fold symmetry of the yield surface.

The mathematical expression of a yield function for isotropic materials $F(\sigma_1, \sigma_2, \sigma_3)=0$ is a symmetric function of the principal stresses σ_i (i=1, 2, 3). So, the limit surface of yield function is three-fold symmetry.

The yield function may also be expressed in the form of the principal shear stresses as follows:

$$f(\tau_{13}, \tau_{12}, \tau_{23})=0 \quad \text{or} \quad f(\tau_{13}, \tau_{12}, \tau_{23}; \sigma_{13}, \sigma_{12}, \sigma_{23})=0 \tag{4-3}$$

4.2 Mechanical Model of the Unified Strength Theory

Mechanical and mathematical modelling are powerful means for establishing and understanding the development of a new theory. Mechanical modelling is an abstraction, a formation of an idea or ideas that may involve the subject with special configurations. Mathematical modelling may involve relationships between continuous functions of space, time and other variations (Tayler 1986; Meyer 1985; Besseling and van der Liesen 1994).

To express the general nature of the strength theory, the cubic element is often used. It is clear that there are three principal stresses σ_1, σ_2 and σ_3 acting on the cubic element as shown in Fig. 4.1a. Fig. 4.1b shows the single-shear element. The Tresca-Mohr–Coulomb strength theory can be introduced from this model. The effect of the intermediate principal stress σ_2 and the effect of intermediate principal shear stress (τ_{12} or τ_{23}), however, was not taken into account in the Tresca-Mohr–Coulomb strength theory.

A new twin-shear stress element and multi-shear element were proposed by Yu (see: Yu 1983, 1985, 1992). It is shown in Fig. 4.2a. The principal stress state (σ_1, σ_2, σ_3) can be converted into the principal shear stress state (τ_{13}, τ_{12}, τ_{23}) as shown in Fig.4.3(a) and (b). Since there are only two independent principal shear stresses, the shear stress state can also be converted into the twin-shear stress state (τ_{13}, τ_{12}, σ_{13}, σ_{12}) or (τ_{13}, τ_{23}, σ_{13}, σ_{23}). This stress state corresponds to the model shown in Fig. 4.2a. The eight sections that two groups of shear stress act on consist of the orthogonal octahedral elements, so the twin-shear mechanical model can be obtained as shown in Fig. 4.2a.

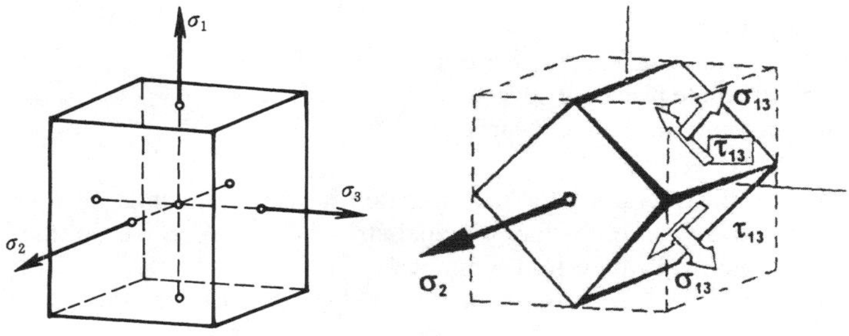

Fig. 4.1 Cubic element and single-shear element

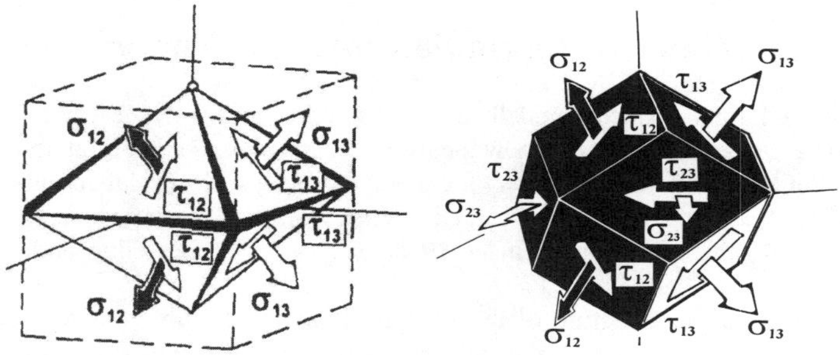

Fig. 4.2 Twin-shear element and multi-shear element

The twin-shear stress model is different from the regular octahedral model proposed by Ros and Eichinger (1926) and Nadai (1947). The orthogonal octahedral model consists of two groups of four sections that are perpendicular to each other and are acted on by the maximum shear stress τ_{13} and the intermediate principal stress τ_{12} or τ_{23}.

The multi-shear element is shown in Fig.4.2 (b). It is clear that there are three principal shear stresses τ_{13}, τ_{12} and τ_{23} in the three-dimensional principal stress state σ_1, σ_2 and σ_3. However, only two principal shear stresses are independent variables among τ_{13}, τ_{12}, τ_{23} because the maximum principal shear stress equals the sum of the other two, that is,

$$\tau_{13} = \tau_{12} + \tau_{23} \,. \tag{4-4}$$

The effect of intermediate principal shear-stress (τ_{12} or τ_{23}) can be taken into account naturally in the twin-shear element and the multi-shear element.

4.3 Unified Strength Theory

Considering all the stress components acting on the element and the different effects of various stresses on the failure of materials, the unified strength theory assumes that the yielding of materials begins when the sum of the two larger principal shear stresses and the corresponding normal stress function reaches a magnitude C. The mathematical modelling is given as follows

$$F = \tau_{13} + b\tau_{12} + \beta(\sigma_{13} + b\sigma_{12}) = C, \quad \text{when} \quad \tau_{12} + \beta\sigma_{12} \geq \tau_{23} + \beta\sigma_{23}$$

(Extended tension stress state) (4-5a)

$$F' = \tau_{13} + b\tau_{23} + \beta(\sigma_{13} + b\sigma_{23}) = C, \quad \text{when} \quad \tau_{12} + \beta\sigma_{12} \leq \tau_{23} + \beta\sigma_{23}$$

(Extended compression stress state) (4-5b)

where b is a parameter that reflects the influence of the intermediate principal shear stress τ_{12} or τ_{23} on the failure of material; β is the coefficient that represents the effect of the normal stress on failure; C is a strength parameter of material; τ_{13}, τ_{12} and τ_{23} are principal shear stresses and σ_{13}, σ_{12} and σ_{23} are the corresponding normal stresses acting on the sections where τ_{13}, τ_{12} and τ_{23} act. They are defined as

$$\tau_{13} = \frac{1}{2}(\sigma_1 - \sigma_3), \quad \tau_{12} = \frac{1}{2}(\sigma_1 - \sigma_2), \quad \tau_{23} = \frac{1}{2}(\sigma_2 - \sigma_3)$$

$$\sigma_{13} = \frac{1}{2}(\sigma_1 + \sigma_3), \quad \sigma_{12} = \frac{1}{2}(\sigma_1 + \sigma_2), \quad \sigma_{23} = \frac{1}{2}(\sigma_2 + \sigma_3) \tag{4-6}$$

The magnitude of β and C can be determined by experimental results of uniaxial tension strength σ_t and uniaxial compression strength σ_c, the experimental conditions are:

$$\sigma_1 = \sigma_t, \quad \sigma_2 = \sigma_3 = 0$$

$$\sigma_1 = \sigma_2 = 0, \quad \sigma_3 = -\sigma_c \tag{4-7}$$

So the material constants β and C can be determined:

$$\beta = \frac{\sigma_c - \sigma_t}{\sigma_c + \sigma_t} = \frac{1-\alpha}{1+\alpha}, \quad C = \frac{2\sigma_c\sigma_t}{\sigma_c + \sigma_t} = \frac{2}{1+\alpha}\sigma_t \tag{4-8}$$

Substituting β and C into the Eq. (4-5a) and (4-5b), the unified strength theory is now obtained. It can be expressed in terms of principal stresses as follows:

$$F = \sigma_1 - \frac{\alpha}{1+b}(b\sigma_2 + \sigma_3) = \sigma_t, \quad \text{when} \quad \sigma_2 \leq \frac{\sigma_1 + \alpha\sigma_3}{1+\alpha},$$

(Extended tension stress state) (4-9a)

$$F' = \frac{1}{1+b}(\sigma_1 + b\sigma_2) - \alpha\sigma_3 = \sigma_t, \text{ when } \sigma_2 \geq \frac{\sigma_1 + \alpha\sigma_3}{1+\alpha},$$

(Extended compression stress state) (4-9b)

$$F'' = \sigma_1 = \sigma_t, \text{ when } \sigma_1 > \sigma_2 > \sigma_3 > 0 \quad (4\text{-}9c)$$

The unified strength theory with the tension cutoff (similar to the Mohr–Coulomb theory with tension cutoff suggested by Paul in 1961) has to be supplemented in the state of three tensile stresses. It is expressed as Eq. (4-9c).

The relationship among shear strength τ_0, the uniaxial tensile strength σ_t and uniaxial compressive strength σ_c can be determined as follows:

$$b = \frac{(1+\alpha)\tau_0 - \sigma_t}{\sigma_t - \tau_0} = \frac{1+\alpha-B}{B-1}, \ \alpha = \frac{\sigma_t}{\sigma_c}, \ B = \frac{\sigma_t}{\tau_0} = \frac{1+b+\alpha}{1+b} \quad (4\text{-}10)$$

The ratio of shear strength to tensile strength of materials can be introduced from the unified strength theory as follows:

$$\alpha_\tau = \frac{\tau_0}{\sigma_t} = \frac{1+b}{1+b+\alpha} \quad (4\text{-}11)$$

It is shown that:

1. The ratio of shear strength to tensile strength $\alpha_\tau = \tau_0 / \sigma_t$ of brittle materials ($\alpha_\tau < 1$) is lower than that of ductile materials ($\alpha_\tau = 1$). This agrees with the experimental data.
2. The limit surface may be non-convex when the ratio of shear strength to tensile strength $\alpha_\tau < 1/(1+\alpha)$ or $\alpha_\tau > 2/(2+\alpha)$.
3. The shear strength of the material is lower than the tensile strength of material. This is true for non-SD materials. It needs, however, further study for SD materials.

4.4 Special Cases of the Unified Strength Theory

4.4.1 Special Cases of the Unified Strength Theory (Varying *b*)

The unified strength theory contains a series of yield criteria for metal materials ($\alpha=1$) and for other materials ($\alpha \neq 1$).

It is worthy to point out that the parameter b is an important parameter in the unified strength theory.

The b is a parameter of intermediate principal shear stress τ_{12} or τ_{23} in Eq. (4-8). It reflects the influence of the intermediate principal shear stress on the failure of a material.

The b is also a parameter of intermediate principal stress σ_2 in Eq. (4-12). It also reflects the influence of the intermediate principal stress σ_2 on the failure of a material.

We can see below that b is also the parameter that determines the formulation of a failure criterion. A series of convex failure criteria can be obtained when the parameter varies in the range of $0 \le b \le 1$. The parameter b has the clear physical meaning. The unified strength theory give us a possibility to choose a reasonable yield criterion for research and applications.

The five types of failure criteria with the values of b=0, b=1/4, b=1/2, b=3/4 and b=1 are introduced from the unified strength theory. In addition, the unified strength theory can also introduce a family of non-convex failure criterion when $b<0$ or $b>1$.

1. *b*=0

The Mohr–Coulomb strength theory can be derived from the unified strength theory with b=0 as follows:

$$F = F' = \sigma_1 - \alpha\sigma_3 = \sigma_t \tag{4-12}$$

2. *b* =1/4

A new failure criterion is obtained from the unified strength theory with b=1/4 as follows:

$$F = \sigma_1 - \frac{\alpha}{5}(\sigma_2 + 4\sigma_3) = \sigma_t\,,\ \sigma_2 \le \frac{\sigma_1 + \alpha\sigma_3}{1+\alpha}$$

(Extended tension stress state) (4-13a)

$$F' = \frac{1}{5}(4\sigma_1 + \sigma_2) - \alpha\sigma_3 = \sigma_t,\ \sigma_2 \ge \frac{\sigma_1 + \alpha\sigma_3}{1+\alpha}$$

(Extended compression stress state) (4-13b)

3. *b* =1/2

A new failure criterion is derived from the unified strength theory with b=1/2 as follows:

$$F = \sigma_1 - \frac{\alpha}{3}(\sigma_2 + 2\sigma_3) = \sigma_t\,,\ \sigma_2 \le \frac{\sigma_1 + \alpha\sigma_3}{1+\alpha}$$

(Extended tension stress state) (4-14a)

$$F' = \frac{1}{3}(2\sigma_1 + \sigma_2) - \alpha\sigma_3 = \sigma_t,\ \sigma_2 \ge \frac{\sigma_1 + \alpha\sigma_3}{1+\alpha}$$

(Extended compression stress state) (4-14b)

Since the Drucker–Prager criterion cannot match with the practice for geomaterials, this criterion is more reasonable and can be substituted for the Drucker–Prager criterion.

4. b=3/4

A new failure criterion is deduced from the unified strength theory with b=3/4 as follows

$$F=\sigma_1-\frac{\alpha}{7}(3\sigma_2+4\sigma_3)=\sigma_t,\quad \sigma_2\le\frac{\sigma_1+\alpha\sigma_3}{1+\alpha}$$

(Extended tension stress state) (4-15a)

$$F=\frac{1}{7}(4\sigma_1+3\sigma_2)-\alpha\sigma_3=\sigma_t,\quad \sigma_2\ge\frac{\sigma_1+\alpha\sigma_3}{1+\alpha}$$

(Extended compression stress state) (4-15b)

5. b=1

A new failure criterion is deduced from the unified strength theory with b=1. The mathematical expression is

$$F=\sigma_1-\frac{\alpha}{2}(\sigma_2+\sigma_3)=\sigma_t,\quad \text{when } \sigma_2\le\frac{\sigma_1+\alpha\sigma_3}{1+\alpha}$$

(Extended tension stress state) (4-16a)

$$F'=\frac{1}{2}(\sigma_1+\sigma_2)-\alpha\sigma_3=\sigma_t,\quad \text{when } \sigma_2\ge\frac{\sigma_1+\alpha\sigma_3}{1+\alpha}$$

(Extended compression stress state) (4-16b)

This is the generalized twin-shear strength model proposed by Yu in 1983 (Yu 1983; Yu et al. 1985).

4.4.2 Special Cases of the unified strength theory (Varying α)

1. α=1, The Unified Yield Criterion for non-SD materials

When the tensile strength and the compressive strength are identical, the tension–compressive strength ratio $\alpha=\sigma_t/\sigma_c$ equals 1. A unified yield criterion can be deduced from the Yu unified strength theory. The mathematical expression of the unified yield criterion is expressed as follows. It also contains a series of yield criteria.

$$F=\sigma_1-\frac{1}{1+b}(b\sigma_2+\sigma_3)=\sigma_s,\qquad \sigma_2\le\frac{\sigma_1+\sigma_3}{2}$$

(Extended tension stress state) (4-17a)

$$F'=\frac{1}{1+b}(\sigma_1+b\sigma_2)-\sigma_3=\sigma_s,\qquad \sigma_2\ge\frac{\sigma_1+\sigma_3}{2}$$

(Extended compression stress state) (4-17b)

in which b is a parameter that reflects the influence of the intermediate principal shear stress τ_{12} or τ_{23} on material strength. It can be determined from the shear yield strength τ_y and the tensile strength σ_y of the materials:

In general cases, the unified yield criterion can be expressed by 12 equations as follows

$$f=\sigma_1-\frac{1}{1+b}(b\sigma_2+\sigma_3)=\pm\sigma_y\,;\quad f=\sigma_1-\frac{1}{1+b}(\sigma_2+b\sigma_3)=\pm\sigma_y$$

$$f=\sigma_2-\frac{1}{1+b}(b\sigma_1+\sigma_3)=\pm\sigma_y\,;\quad f=\sigma_2-\frac{1}{1+b}(\sigma_1+b\sigma_3)=\pm\sigma_y \tag{4-18}$$

$$f=\sigma_3-\frac{1}{1+b}(b\sigma_2+\sigma_1)=\pm\sigma_y\,;\quad f=\sigma_3-\frac{1}{1+b}(\sigma_2+b\sigma_1)=\pm\sigma_y$$

The Tresca yield criterion and the twin-shear yield criterion are special cases of the unified yield criterion when b=0 and b=1, respectively. The Huber-von Mises criterion can be approximated by the unified yield criterion by letting b=0.5. In fact, the unified yield criterion contains a series of yield criteria that are varying the parameter b. The unified yield criterion with b=0, b=1/4, b=1/2, b=3/4 and b=1 can be adapted to most kinds of metallic materials whose tensile strength is the same as its compressive strength.

The relations among the tensile yield stress σ_y, shear yield stress τ_y and the parameter b in the unified yield criterion (α=1) can be determined from the ratio of shear yield stress to tensile yield stress

$$b=\frac{2\tau_y-\sigma_y}{\sigma_y-\tau_y}, \tag{4-19}$$

or

$$\tau_y=\frac{b+1}{b+2}\sigma_y$$

Inversely, the ratio of shear yield stress to tensile yield stress can be given as

$$\alpha_\tau=\frac{\tau_y}{\sigma_y}=\frac{b+1}{b+2} \tag{4-20}$$

Some conclusions for non-SD materials can be made from this condition:

1. The shear yield stress is lower than tensile yield stress for metallic materials.
2. Yield surfaces are convex when $0\le b\le 1$ or $1/2\le\alpha_\tau\le 2/3$.
3. Yield surfaces are nonconvex when $b<0$ and $b>1$, or the ratio of shear yield stress to tensile yield stress $\alpha_\tau<1/2$ and $\alpha_\tau>2/3$.

For example, if the ratio of the shear yield stress to the tensile yield stress of material is $\alpha_\tau = \tau_y / \sigma_y = 0.45$, it can be determined from Eq. (4-19) that the parameter $b = -1/6$. This means that the yield criterion is nonconvex.

2. α =1/2, new series of failure criteria

The main disadvantage of the unified yield criterion is that it is only suitable to metallic materials having the same strengths both in tension and in compression. It cannot adapt to those materials that have different strength in tension and compression, or to the cases where the strength is pressure dependent, such as iron, high-strength steels, polymers and geomaterials. It can be solved by using the unified strength theory with $0<\alpha<1$. A series of failure criteria can be obtained from the unified strength theory with $0<\alpha<1$. For an example, we take the α =1/2. A new series of failure criteria can be obtained as follows

$$F = \sigma_1 - \frac{1}{2(1+b)}(b\sigma_2 + \sigma_3) = \sigma_t, \quad \text{when } \sigma_2 \le \frac{\sigma_1 + \alpha\sigma_3}{1+\alpha}$$

(Extended tension stress state) (4-21a)

$$F' = \frac{1}{1+b}(\sigma_1 + b\sigma_2) - \frac{1}{2}\sigma_3 = \sigma_t, \quad \text{when } \sigma_2 \ge \frac{\sigma_1 + \alpha\sigma_3}{1+\alpha}$$

(Extended compression stress state) (4-21b)

Fig. 4.3 shows the relationship among the unified yield criterion, the unified strength theory, the single-shear yield criterion (the Tresca yield criterion), the single-shear strength theory (the Mohr–Coulomb strength theory), the twin-shear yield criterion (Yu 1961), The twin-shear strength theory (Yu 1985) as well as some new failure criteria. A great number of new failure criteria can be introduced from the Yu unified strength theory. Three new failure criteria are introduced from the Yu unified strength theory when b=1/4, b=1/2 and b=3/4 as shown in the third row in Fig.4.3.

Example 4.1

Consider a metallic material with the same yield stress in tension and compression. If its shear yield stress is $\tau_y = 0.63\sigma_y$ and Poisson's ratio is ν =0.3, find an available yield criterion

Solution

1. Find an available yield criterion

The Tresca yield criterion (single-shear yield criterion) predicts the shear yield stress is $\tau_y = 0.5\sigma_y$, and the Huber-von Mises yield criterion (octahetral shear stress yield criterion) predicts the shear yield stress is $\tau_y = 0.677\sigma_y$. Obviously, these two yield criteria do not fit this kind of material with $\tau_y = 0.63\sigma_y$.

The parameter b in the unified yield criterion can be determined by using Eq. (4-19)

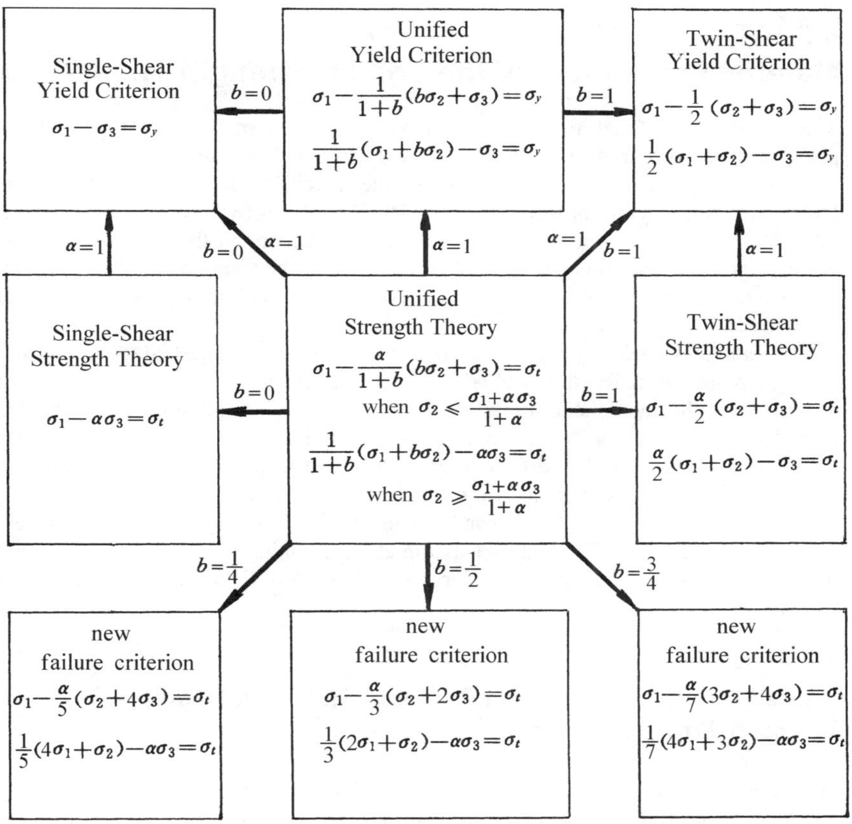

Fig. 4.3 Unified strength theory and its special cases

$$b = \frac{2\tau_y - \sigma_y}{\sigma_y - \tau_y} = \frac{2\bar{\alpha} - 1}{1 - \bar{\alpha}} = \frac{2\times 0.63 - 1}{1 - 0.63} = 0.7 \tag{4-22}$$

substitution of b=0.7 into Eqs. (4-19a) and (4-19b), a new available yield criterion is obtained as follows

$$f=\sigma_1-\frac{1}{1.7}(0.7\sigma_2+\sigma_3)=\sigma_{y,}, \quad \text{if } \sigma_2\le\frac{1}{2}(\sigma_1+\sigma_3) \tag{4-23a}$$

$$f'=\frac{1}{1.7}(\sigma_1+0.7\sigma_2)-\sigma_3=\sigma_y, \quad \text{if } \sigma_2\ge\frac{1}{2}(\sigma_1+\sigma_3) \tag{4-23b}$$

4.5 Material Parameters of the Unified Strength Theory

The use of a yield function is always connected with the material parameter. These parameters are required to be simple and easy to get. The uniaxial tensile strength σ_y is used for the material parameter in the Tresca yield criterion (1864), the Huber-von Mises yield criterion (1904, 1913) and the twin-shear yield criterion (Yu 1961). It is the same for the unified strength theory in the case of α=1. The Tresca criterion, the Huber-von Mises criterion and the twin-shear stress criterion can be suitable for those materials that have the identical strength in tension and compression.

The uniaxial tensile strength σ_t and the ratio of tension strength and compression strength α are used for the Mohr-Coulomb strength theory, the Drucker-Prager criterion and the twin-shear strength theory which are two-parameter criteria. The two parameters in the unified strength theory remain the same. Some other material parameters are also used in different applications. It needs some transfermation of the mathematical expressions of yield function.

The unified strength theory expressed in terms of principal stresses has been described in Eq. (4-9). The materials parameters are uniaxial tensile strength σ_t and the ratio of tension strength and compression strength α. Other material parameters can also be used.

4.5.1 Unified strength theory with Principal Stress and Compressive Strength $F(\sigma_1,\sigma_2,\sigma_3,\alpha,\sigma_c)$

In soil and rock mechanics and engineering, the compressive strength σ_c is often adopted. Rewriting Eq. (4-9a), (4-9b) in terms of the principal stress and compressive strength σ_c, we have

$$F=\frac{1}{\alpha}\sigma_1-\frac{1}{1+b}(b\sigma_2+\sigma_3)=\sigma_c, \quad \text{when } \sigma_2\le\frac{\sigma_1+\alpha\sigma_3}{1+\alpha} \tag{4-24a}$$

$$F'=\frac{1}{\alpha(1+b)}(\sigma_1+b\sigma_2)-\sigma_3=\sigma_c, \quad \text{when } \sigma_2\ge\frac{\sigma_1+\alpha\sigma_3}{1+\alpha} \tag{4-24b}$$

The unified strength theory can be expressed in terms of another material parameter m as follows:

$$\sigma_1 - \frac{1}{m(1+b)}(b\sigma_2 + \sigma_3) = \sigma_t \qquad (\sigma_2 \le \frac{m\sigma_1 + \sigma_3}{m+1}) \tag{4-25a}$$

$$\frac{1}{1+b}(\sigma_1 + b\sigma_2) - \frac{\sigma_3}{m} = \sigma_t \qquad (\sigma_2 \ge \frac{m\sigma_1 + \sigma_3}{m+1}) \tag{4-25b}$$

where $m=\sigma_c/\sigma_t$ is the compressive-tensile strength ratio of the material. The ratio m is an index of the material strength difference effect and $m \geqslant 1$ in general. The unified strength theory can also be expressed by other terms.

4.5.2
Unified Strength Theory with Stress Invariant and Tensile Strength $F(I_1, J_2, \theta; \sigma_t, \alpha)$

The principal stress state (σ_1, σ_2, σ_3) can be converted into the principal shear stress state (τ_{13}, τ_{12}, τ_{23}), invariates of stress tensor (I_1, I_2, I_3) or invariates of stress tensor (I_1,I_2,I_3). The principal shear stress state can be described in terms of stress invariant. The unified strength theory can also be expressed in terms of stress invariant $F(I_1, J_2, \theta)$ and material constants σ_t, α as follows:

$$F = (1-\alpha)\frac{I_1}{3} + \frac{\alpha(1-b)}{1+b}\sqrt{J_2}\sin\theta + (2+\alpha)\sqrt{\frac{J_2}{3}}\cos\theta = \sigma_t, \quad 0^\circ \le \theta \le \theta_b \tag{4-26a}$$

$$F' = (1-\alpha)\frac{I_1}{3} + \left(\alpha + \frac{b}{1+b}\right)\sqrt{J_2}\sin\theta + \left(\frac{2-b}{1+b} + \alpha\right)\sqrt{\frac{J_2}{3}}\cos\theta = \sigma_t, \quad \theta_b \le \theta \le 60^\circ \tag{4-26b}$$

where I_1 is the first stress invariant, J_2 is the second deviatoric stress invariant and θ is the stress angle corresponding to the twin–shear parameter $\mu_\tau=\tau_{12}/\tau_{23}$ or $\mu'_\tau=\tau_{23}/\tau_{13}$. The stress angle at the corner θ_b can be determined by the condition $F=F'$.

$$\theta_b = \text{arctg}\frac{\sqrt{3}(1+\beta)}{3-\beta}, \quad \beta = \frac{1-\alpha}{1+\alpha} \tag{4-27}$$

4.5.3
Unified Strength Theory with Stress Invariant and Compressive Strength $F(I_1, J_2, \theta, \alpha, \sigma_c)$

The unified strength theory can also be expressed in terms of stress invariant $F(I_1, J_2, \theta)$ and material constant α, σ_c as follows:

$$F=\frac{1-\alpha}{3\alpha}I_1+\frac{1-b}{1+b}\sqrt{J_2}\sin\theta+\frac{2+\alpha}{\alpha\sqrt{3}}\sqrt{J_2}\cos\theta=\sigma_c \quad 0^\circ\le\theta\le\theta_b \tag{4-28a}$$

$$F'=\frac{1-\alpha}{3\alpha}I_1+\frac{\alpha+\alpha b+b}{\alpha(1+b)}\sqrt{J_2}\sin\theta+\frac{2+\alpha+\alpha b-b}{\alpha\sqrt{3}(1+b)}\sqrt{J_2}\cos\theta=\sigma_c\theta_b\le\theta\le 60^\circ \tag{4-28b}$$

4.5.4 Unified Strength Theory with Principal Stress and Cohesive Parameter $F(\sigma_1,\sigma_2,\sigma_3,C_0,\varphi)$

In Eq. (4-9a) (4-9b), we adopt the material constants σ_t and the tension–compression ratio α. In geotechnical engineering the cohesion C_0 and the friction angle coefficient φ reflecting the material properties are used. The relationships among the tensile strength σ_t, the tension-compression ratio α, the material parameter C_0 and φ can be obtained as follows:

$$\sigma_t=\frac{2C_0\cdot\cos\varphi}{1+\sin\varphi}, \qquad \alpha=\frac{1-\sin\varphi}{1+\sin\varphi} \tag{4-29}$$

By substituting Eq. (4-29) into Eqs. (4.9a) and (5.9b), the Yu unified strength theory can be expressed in terms of C_0 and φ as

$$F=\sigma_1-\frac{1-\sin\varphi}{(1+b)(1+\sin\varphi)}(b\sigma_2+\sigma_3)=\frac{2C_0\cos\varphi}{1+\sin\varphi},$$

$$\text{when}\quad \sigma_2\le\frac{1}{2}(\sigma_1+\sigma_3)-\frac{\sin\varphi}{2}(\sigma_1-\sigma_3) \tag{4-30a}$$

$$F'=\frac{1}{1+b}(\sigma_1+b\sigma_2)-\frac{1-\sin\varphi}{1+\sin\varphi}\sigma_3=\frac{2C_0\cos\varphi}{1+\sin\varphi},$$

$$\text{when}\ \sigma_2\ge\frac{1}{2}(\sigma_1+\sigma_3)-\frac{\sin\varphi}{2}(\sigma_1-\sigma_3) \tag{4-30b}$$

4.5.5 Unified Strength Theory with Stress Invariant and Cohesive Parameter $F(I_1,J_2,\theta,C_0,\varphi)$

The unified strength theory can be also expressed by the stress invariant, stress angle and material parameters cohesion C_0 and friction angle φ.

$$F=\frac{2I_1}{3}\sin\varphi+\frac{2\sqrt{J_2}}{1+b}\left[\sin\left(\theta+\frac{\pi}{3}\right)-b\sin\left(\theta-\frac{\pi}{3}\right)\right]+\frac{2\sqrt{J_2}}{(1+b)\sqrt{3}}\cdot$$
$$\left[\sin\varphi\cos\left(\theta+\frac{\pi}{3}\right)+b\sin\varphi\cos\left(\theta-\frac{\pi}{3}\right)\right]=2C_0\cos\varphi,\quad 0^\circ\leq\theta\leq\theta_b \qquad (4\text{-}31a)$$

$$F'=\frac{2I_1}{3}\sin\varphi+\frac{2\sqrt{J_2}}{1+b}\left[\sin\left(\theta+\frac{\pi}{3}\right)-b\sin\theta\right]$$
$$+\frac{2\sqrt{J_2}}{(1+b)\sqrt{3}}\left[\sin\varphi\cos(\theta+\frac{\pi}{3})+b\sin\varphi\cos\theta\right]=2C_0\cos\varphi,\quad \theta_b\leq\theta\leq 60$$
$$(4\text{-}31b)$$

4.6 Other Material Parameters of the Unified Strength Theory

The unified strength theory in terms of three principal stresses Eq.(4.9) is introduced from the mathematical modeling equation (4.5) as follows

$$F=\tau_{13}+b\tau_{12}+\beta(\sigma_{13}+b\sigma_{12})=C,\quad \text{when}\quad \tau_{12}+\beta\sigma_{12}\geq\tau_{23}+\beta\sigma_{23}$$
(Extended tension stress state) (4-5a)

$$F'=\tau_{13}+b\tau_{23}+\beta(\sigma_{13}+b\sigma_{23})=C,\quad \text{when}\quad \tau_{12}+\beta\sigma_{12}\leq\tau_{23}+\beta\sigma_{23}$$
(Extended compression stress state) (4-5b)

The material parameter β and C are determined by experimental results of uniaxial tension strength σ_t and uniaxial compression strength σ_c, the experimental conditions are:

$$\sigma_1=\sigma_t,\ \sigma_2=\sigma_3=0 \quad \text{(uniaxial tension)}$$
$$\sigma_1=\sigma_2=0,\ \sigma_3=-\sigma_c \quad \text{(uniaxial compression)} \qquad (4\text{-}32)$$

So the material constants β and C can be determined.

The material parameters β and C can also be determined by other experimental results. A lot of experimental results may be used for the determination of the material parameter in the unified strength theory.

4.6.1
Material parameter β and C are determined by experimental results of uniaxial tension strength σ_t and shear strength τ_0

The material parameter β and C of the unified strength theory can be determined by experimental results of uniaxial tension strength σ_t and pure shear strength τ_0, the experimental conditions are:

$$\begin{aligned} &\sigma_1 = \sigma_t, \ \sigma_2 = \sigma_3 = 0 \quad \text{(uniaxial tension)} \\ &\sigma_1 = -\sigma_3 = 0, \ \sigma_2 = \tau_0 \quad \text{(pure shear)} \end{aligned} \tag{4-33}$$

4.6.2
Material parameter β and C are determined by experimental results of uniaxial compressive strength σ_c and shear strength τ_0

The material parameter β and C of the unified strength theory can be determined by experimental results of uniaxial compressive strength σ_c and pure shear strength τ_0, the experimental conditions are:

$$\begin{aligned} &\sigma_1 = \sigma_2 = 0, \ \sigma_3 = -\sigma_c \quad \text{(uniaxial compression)} \\ &\sigma_1 = -\sigma_3 = 0, \ \sigma_2 = \tau_0 \quad \text{(pure shear)} \end{aligned} \tag{4-34}$$

4.6.3
Material parameter β and C are determined by experimental results of uniaxial compressive strength σ_c and biaxial compressive strength σ_{cc}

The material parameter β and C of the unified strength theory can be determined by experimental results of uniaxial tension strength σ_t and biaxial compressive strength σ_{cc}, the experimental conditions are:

$$\begin{aligned} &\sigma_1 = \sigma_2 = 0, \ \sigma_3 = -\sigma_c \quad \text{(uniaxial compression)} \\ &\sigma_1 = 0, \ \sigma_2 = \sigma_3 = \sigma_{cc} \quad \text{(biaxial compression)} \end{aligned} \tag{4-35}$$

4.6.4
Material parameter β and C are determined by experimental results of uniaxial compressive strength σ_c and biaxial compressive strength σ_{cc}

The material parameter β and C of the unified strength theory can be determined by experimental results of uniaxial tension strength σ_t and biaxial compressive strength σ_{cc}, the experimental conditions are:

$$\sigma_1 = \sigma_t, \ \sigma_2 = \sigma_3 = 0 \ \text{(uniaxial tension)}$$
$$\sigma_1 = 0, \ \sigma_2 = \sigma_3 = \sigma_{cc} \quad \text{(biaxial compression)} \qquad (4\text{-}36)$$

4.6.5 Material parameter β and C are determined by experimental results of uniaxial compressive strength σ_c and biaxial compressive strength σ_{cc}

The material parameter β and C of the unified strength theory can be determined by experimental results of pure shear strength τ_0 and biaxial compressive strength σ_{cc}, the experimental conditions are:

$$\sigma_1 = -\sigma_3 = 0, \ \sigma_2 = \tau_0 \quad \text{(pure shear)} \qquad (4\text{-}37)$$
$$\sigma_1 = 0, \ \sigma_2 = \sigma_3 = \sigma_{cc} \quad \text{(biaxial compression)} \qquad (4\text{-}38)$$

4.6.6 Three-parameter unified strength theory

The unified strength theory can be generalized to as a three-parameter unified strength criterion. The mathematical modeling equation of the three-parameter unified strength criterion is:

$$F = \tau_{13} + b\tau_{12} + \beta(\sigma_{13} + b\sigma_{12}) + A\sigma_m = C,$$
$$\text{when} \quad \tau_{12} + \beta\sigma_{12} \ge \tau_{23} + \beta\sigma_{23} \qquad (4\text{-}39a)$$
$$F' = \tau_{13} + b\tau_{23} + \beta(\sigma_{13} + b\sigma_{23}) + A\sigma_m = C,$$
$$\text{when} \quad \tau_{12} + \beta\sigma_{12} \le \tau_{23} + \beta\sigma_{23} \qquad (4\text{-}39b)$$

where b is again the parameter that reflects the influence of the intermediate principal shear stress τ_{12} or τ_{23} on the failure of material; β is the coefficient that represents the effect of the normal stress on failure; σ_m is average stress; A is the coefficient that represents the effect of the average stress on failure; C is a strength parameter of material; τ_{13}, τ_{12} and τ_{23} are principal shear stresses and σ_{13}, σ_{12} and σ_{23} are the corresponding normal stresses acting on the sections where τ_{13}, τ_{12} and τ_{23} act.

Another kind of the three parameters criterion of the unified strength theory can be obtained by using the different parameters β_a and β_b. The mathematical modeling equation of this kind of three-parameter unified strength criterion is:

$$F = \tau_{13} + b\tau_{12} + \beta_a(\sigma_{13} + b\sigma_{12}) = C,$$
$$\text{when} \quad \tau_{12} + \beta\sigma_{12} \ge \tau_{23} + \beta\sigma_{23} \qquad (4\text{-}40a)$$
$$F' = \tau_{13} + b\tau_{23} + \beta_b(\sigma_{13} + b\sigma_{23}) = C,$$

$$\text{when} \quad \tau_{12} + \beta\sigma_{12} \le \tau_{23} + \beta\sigma_{23} \qquad \text{(4-40b)}$$

The three material parameters β, C and A or β_a, β_b and C can be determined by three experimental conditions. Interesting readers may be referred to the recent book (Yu 2004).

4.7 Yield Surfaces and Yield Loci

Yield criterion is a function of three principal stresses σ_1, σ_2, σ_3 as follows:

$$F = F(\sigma_1, \sigma_2, \sigma_3, K_1, K_2) = 0 \qquad \text{(4-41)}$$

It can be interpreted for an isotropic material in terms of a geometrical representation of the stress state obtained by taking the principal stresses as coordinates, as shown in Fig. 4.4. The yield surface in a three-dimensional principal stress space was introduced by Haigh and Westergaard in 1920. Sometimes, it is called the Haigh–Westergaard space. The advantage of such a space lies in its simplicity and visual presentation.

Every point in this principal stress space corresponds to a state of stress (σ_1, σ_2, σ_3), as shows in Fig.4.4.

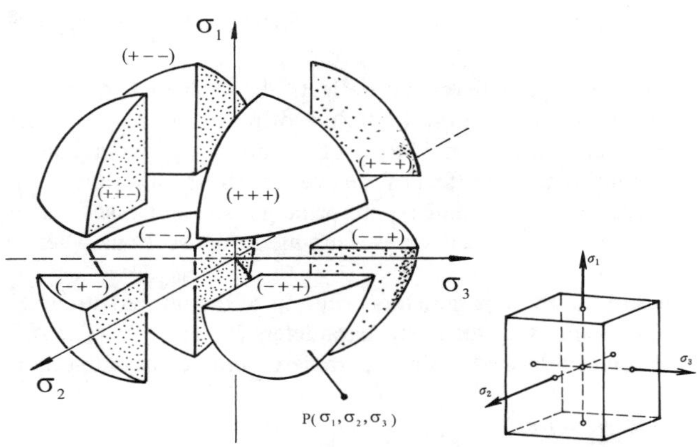

Fig. 4.4 Eight quadrants in principal stress space

The three-dimensional principal stresses (σ_1, σ_2, σ_3) can be regarded as a three-dimensional space of principal stresses. If we take the tensile stress as positive while taking the compressive stress as negative, the stress state may combine the space stresses into various magnitudes and signs of stress combinations. The stress point P(σ_1, σ_2, σ_3) of different signs could combine up to eight quadrants of (+ + +), (+ + −), (+ − +), (+ − −), (− + +), (− + −), (− − +) and (− − −). A stress point could be situated anywhere within the three-dimensional space of the principal stresses.

The uniaxial tensile and compressive tests can give us two limit points in the three-dimensional stress space, that is, A ($\sigma_1 = \sigma_t, \sigma_2 = 0, \sigma_3 = 0$) and B ($\sigma_1 = 0, \sigma_2 = 0, \sigma_3 = -\sigma_c$), as shown in Fig. 4.5.

In stress space the yield criterion defines a surface that is generally referred to as the yield surface. If a stress point is situated in the yield surface, it means the yield function $f < 0$, and the material will be elastic; if a stress point reaches the yield surface, it means the yield function $f = 0$, and yield of the material will occur.

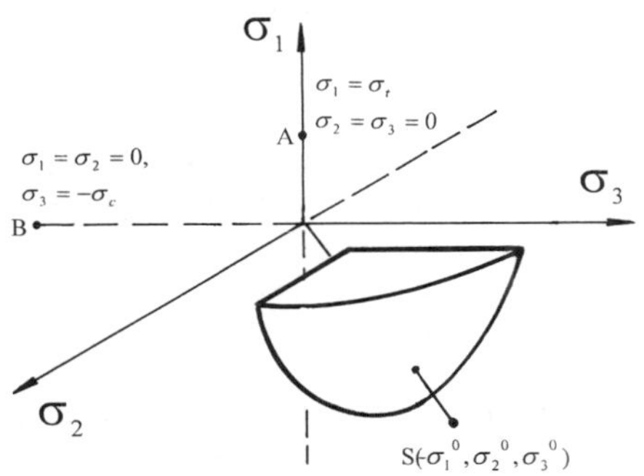

Fig. 4.5 Tensile limit point and compressive limit point

The state of stress at any point in a body or a structure may be represented by a vector emanating from the origin O (Fig. 4.6). The isoclinic axis ***ON*** is equally inclined to the three axes, its direction cosines are ($1/\sqrt{3}$, $1/\sqrt{3}$, $1/\sqrt{3}$). The stress vector ***Or***, whose stress components are (σ_1, σ_2, σ_3), may be resolved into a vector ***OO′*** along isoclinic asis ***ON*** and a vector ***Or***$_0$ in a plane that is perpendicular to ***ON*** and passes through the origin. The vector ***OO′*** is of magnitude $\sqrt{3}\sigma_m$ and represents the hydrostatic stress with components (σ_m, σ_m, σ_m). The vector ***Or***$_0$ represents the deviatoric stress with components $f(S_1, S_2, S_3)$ and magnitude $\sqrt{2J_2}$. For any given state of stress , the deviatoric stress vector will lie in the plane passing through O and perpendicular to ***ON***. This plane is known as the deviatoric plane in stress space or the π_0-plane. Its equation is $\sigma_1 + \sigma_2 + \sigma_3 = 0$ in the principal stress space. The planes that are parallel to the π_0-plane are called the π-planes and are given by $\sigma_1 + \sigma_2 + \sigma_3 = C$, where C is a constant.

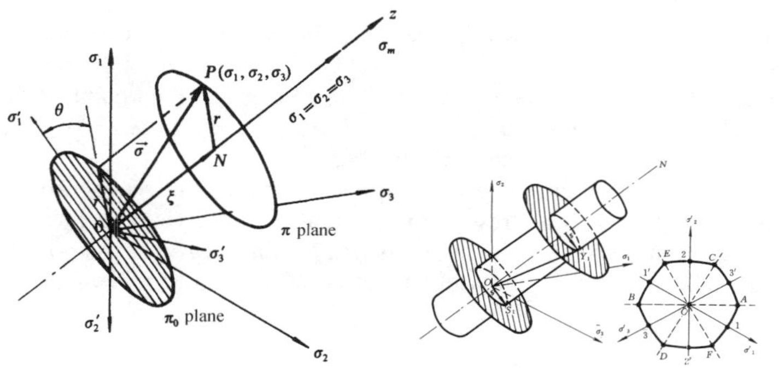

Fig. 4.6 Cylindrical coordinates and a yield surface for metal in principal stress space

If a hydrostatic stress has no effect on yielding, it follows that yielding can depend only on the magnitude and the direction of the deviatoric stress vector $\boldsymbol{Or}_0$ in the π_0-plane or the deviatoric stress vector $\boldsymbol{O'r}$ in the π-plane. The yield surfaces are therefore regarded as a prismatic surface whose generators are perpendicular to the deviatoric plane. Any stress state in which the stress point lies on the prismatic surface corresponds to a state of yielding. Any point inside the prismatic surface represents an elastic state of stress.

The general shape of a yield surface in a three-dimensional stress space for metallic materials with same strength in tension and compression can be determined by its cross-sectional shapes in the deviatoric planes, because the shapes on any π-plane are identical. The cross sections of the yield surface are the intersection of the yield surface with the deviatoric plane, called the yield locus.

The shape of yield loci on the π-plane are similar for linear pressure sensitive material, as show in Fig.4.7(a). The shape and size of yield loci on the π-plane are different for non-linear pressure sensitive material, as show in Fig.4.7(b).

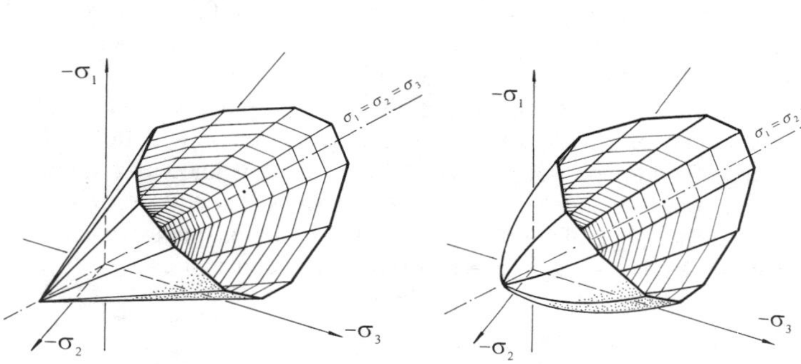

(a) Linear pressure sensitive (b) Nonlinear pressure sensitive

Fig. 4.7 The yield surface of the linear and non-linear pressure sensitive material

Consider the yield locus together with the orthogonal projections of the stress axes on the deviatoric plane, the yield locus is symmetrical with respect to the projections of the σ_1, σ_2, σ_3 axes. The yield locus, therefore, is threefold symmtric. The yield loci on the deviatoric plane,the axes x,y and projections of the stress axex σ_1', σ_2', σ_3' are taken in the plane of the paper (Fig. 4.6).

4.8
Yield Loci of the Unified Strength Theory in the π-Plane

The relationships between the coordinates of the deviatoric plane and hydrostatic stress axis z with the principal stresses are:

$$x=\frac{1}{\sqrt{2}}(\sigma_3-\sigma_2),$$
$$y=\frac{1}{\sqrt{6}}(2\sigma_1-\sigma_2-\sigma_3), \qquad (4\text{-}42)$$
$$z=\frac{1}{\sqrt{3}}(\sigma_1+\sigma_2+\sigma_3)$$

$$\sigma_1=\frac{1}{3}(\sqrt{6}y+\sqrt{3}z)$$

$$\sigma_2=\frac{1}{6}(2\sqrt{3}z-\sqrt{6}y-3\sqrt{2}x) \qquad (4\text{-}43)$$

$$\sigma_3=\frac{1}{6}(3\sqrt{2}x-\sqrt{6}y+2\sqrt{3}z)$$

The ratio between the tensile radius and the compressive radius in π plane is given by

$$K=\frac{r_t}{r_c}=\frac{1+2\alpha}{2+\alpha}=\frac{3-\sin\phi}{3+\sin\phi} \qquad (4\text{-}44)$$

By substituting Eq. (4-42) into the unified strength theory Eq. (4-25a) and (4-25b), the equations of the unified strength theory in the deviatoric plane can be obtained:

$$F=-\frac{\sqrt{2}(1-b)}{2(1+b)}\alpha x+\frac{\sqrt{6}(2+\alpha)}{6}y+\frac{\sqrt{3}(1-\alpha)}{3}z=\sigma_t \qquad (4\text{-}45a)$$

$$F' = -(\frac{b}{1+b}+\alpha)\frac{\sqrt{2}}{2}x + (\frac{2-b}{1+b}+\alpha)\frac{\sqrt{6}}{6}y + \frac{\sqrt{3}(1-\alpha)}{3}z = \sigma_t \quad \text{(4-45b)}$$

A great number of new failure criteria can be generated from the unified strength theory by changing α and b. The shape and size of yield loci of the unified strength theory are changed with α ,b and hydrostatic stress axis z. The shape of the yield loci is similar for a certain values of α and b, but the size of the yield loci are changed with different hydrostatic stress σ_m.

The variation of the unified strength theory with b is shown in Fig.4.7. Ten special cases with values of b=0, b=0.1, b=0.2, b=0.3, b=0.4, b=0.5, b=0.6, b=0.7, b=0.8, b=0.9 and b=1 are given.

The two bounds of convex yield loci are the Mohr-Coulomb theory and the twin-shear strength theory proposed by Yu in 1985. The yield locus of the twin-shear strength theory is the upper bound of the convex yield loci, as shown in Fig. 4.7.

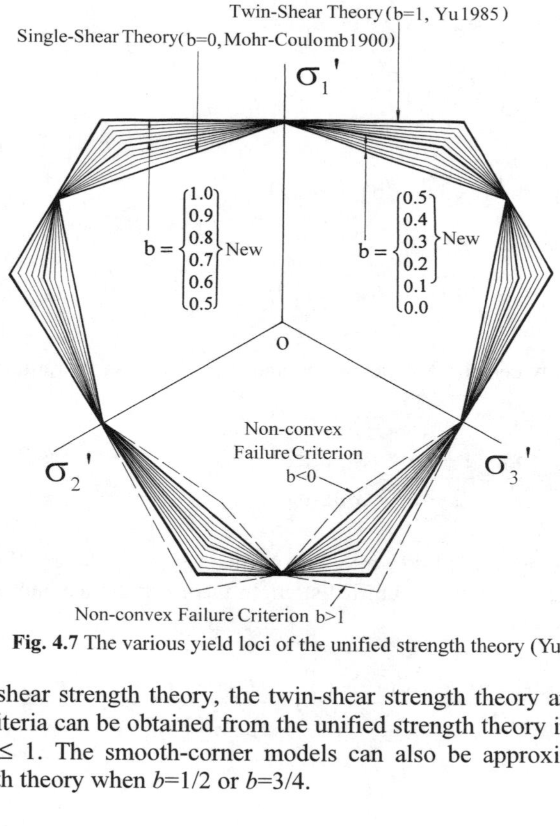

Fig. 4.7 The various yield loci of the unified strength theory (Yu 1992)

The single-shear strength theory, the twin-shear strength theory and a series of new failure criteria can be obtained from the unified strength theory in the range of $0 \le b \le 1$, $0 \le \alpha \le 1$. The smooth-corner models can also be approximated by the unified strength theory when b=1/2 or b=3/4.

The convex failure criteria can be obtained by varying the value of α (α<1) and b ($0 \le b \le 1$).

If $\alpha = \sigma_t/\sigma_c = 1$, that is, the tensile strength equals the compression strength, then the radii r of the yield locus of the unified strength theory on the axes σ_1, σ_2, σ_3 and $-\sigma_1$, $-\sigma_2$, $-\sigma_3$ are identical. The ratio between the tensile radius and the compressive radius is given by

$$K = \frac{1+2\alpha}{2+\alpha} = \frac{3-\sin\phi}{3+\sin\phi} = 1$$

which means that the irregular dodecahedron is converted to regular dodecahedron, and the yield surfaces of the unified yield criterion for $\alpha = \sigma_t/\sigma_c = 1$ materials change to a series of infinite prisms.

A new unified yield criterion can be deduced from Eqs. (4.45a) (4-45b). The equations of the unified yield criterion for $\alpha = \sigma_t/\sigma_c = 1$ materials on the deviatoric plane can be obtained as follows.

$$F = -\frac{\sqrt{2}(1-b)}{2(1+b)}x + \frac{\sqrt{6}}{2}y = \sigma_t \tag{4-46a}$$

$$F' = -\frac{\sqrt{2}(1+2b)}{2(1+b)}x + \frac{\sqrt{6}}{2(1+b)}y = \sigma_t \tag{4-46b}$$

A series of yield loci for $\alpha = \sigma_t/\sigma_c = 1$ materials with b=0, b=0.1, b=0.2, b=0.3, b=0.4, b=0.5, b=0.6, b=0.7, b=0.8, b=0.9 and b=1 on the deviatoric plane can be obtained as shown in Fig.4.8.

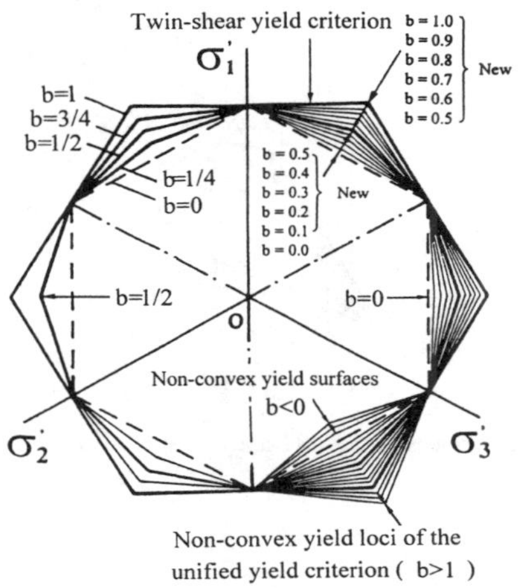

Fig. 4.8 A family of the yield loci of the unified yield criterion for $\alpha = \sigma_t/\sigma_c = 1$ materials in the π-plane (Yu et al. 1992)

4.9 Yield Surfaces of the Unified Strength Theory in Principal Stress Space

The yield surfaces in stress space of the unified strength theory are usually a semi-infinite hexagonal cone with unequal sides and a dodecahedron cone with unequal sides, as shown in Fig 4.9. The shape and size of the yield hexagonal cone depends on the parameter b and on the tension-compression strength ratio α.

In engineering practice, the compressive strength of materials σ_c is often much greater than the tensile strength σ_t for brittle materials, since the region in tension becomes smaller, while it becomes larger in compression. Assuming the compressive strength is positive, the yield surfaces of the UST (unified strength theory) with different values of b are shown in Figs. 4.9 and 4.10. Figs. 4.9 and 4.10 show the yield surfaces of the unified strength theory with b=0, b=1/4, and b=1/2, b=3/4 and b=1, respectively. The latter is the yield locus of the twin-shear strength theory.

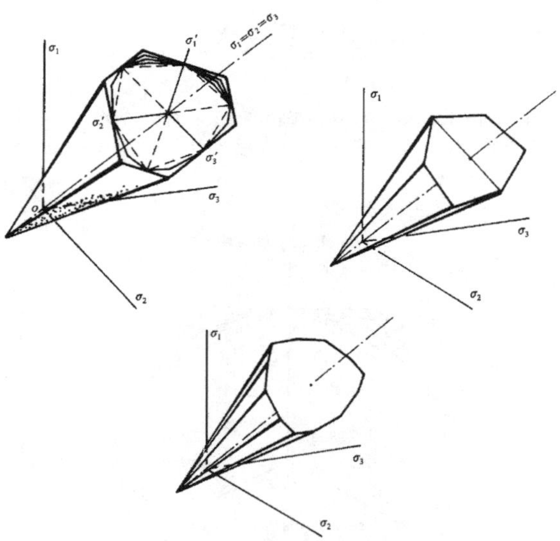

Fig. 4.9 Yield surfaces of the UST and two special cases (b=0 and b=1/4)

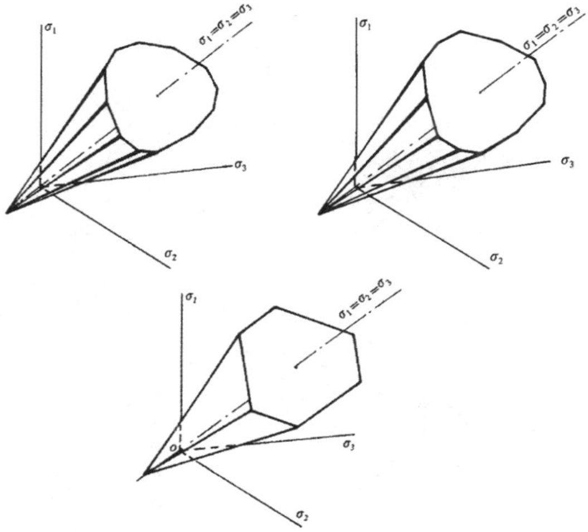

Fig. 4.10 Three yield surfaces of three special cases of UST with b=1/2, b=3/4 and b=1

The unified yield criteria can be deduced from the unified strength theory when $\alpha=1$, as follows.

$$f=\sigma_1-\frac{\alpha}{1+b}(b\sigma_2+\sigma_3)=\sigma_Y \text{ when } \sigma_2\geq\frac{1}{2}(\sigma_1+\sigma_3) \tag{4-47a}$$

$$f'=\frac{1}{1+b}(\sigma_1+b\sigma_2)-\sigma_3=\sigma_Y, \text{ when } \sigma_2\leq\frac{1}{2}(\sigma_1+\sigma_3) \tag{4-47b}$$

Their yield loci of the unified yield criterion for $\alpha=\sigma_t/\sigma_c=1$ materials in the π-plane have been shown in Fig. 4.8.

Six typical yield criteria for metallic materials can be obtained from Eqs. (4-47a) (4-47b) when $\alpha=1$ and b=1, b =3/4, b =1/2, b =1/4 and b =0. Their yield loci in the π-plane are shown in Figs 4.11 to 4.13. The five yield loci in deviatoric plane are shown in Fig. 4.13 (b). The middle yield locus is the linear approximation to the Huber-von Mises yield criterion.

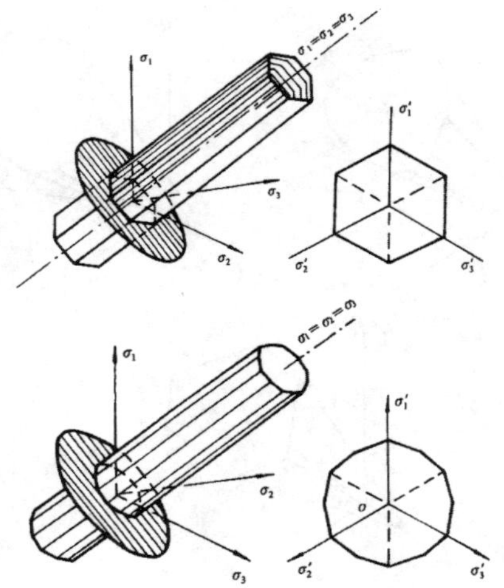

Fig. 4.11 Yield surface and yield loci of two cases of UYC with b=0 and b=1/4

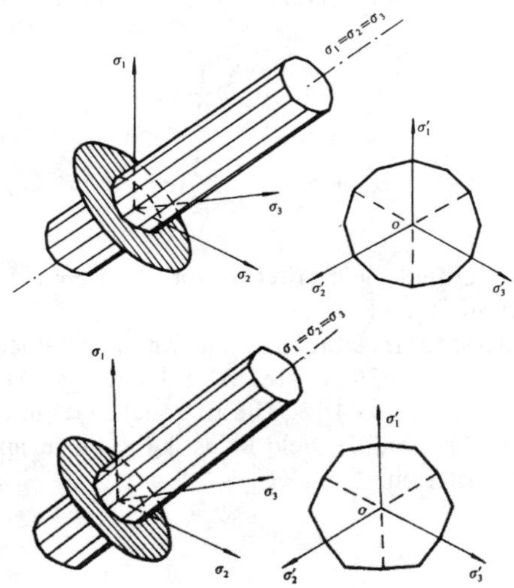

Fig. 4.12 Yield surface and yield loci of two cases of UYC with b=1/2 and b=3/4

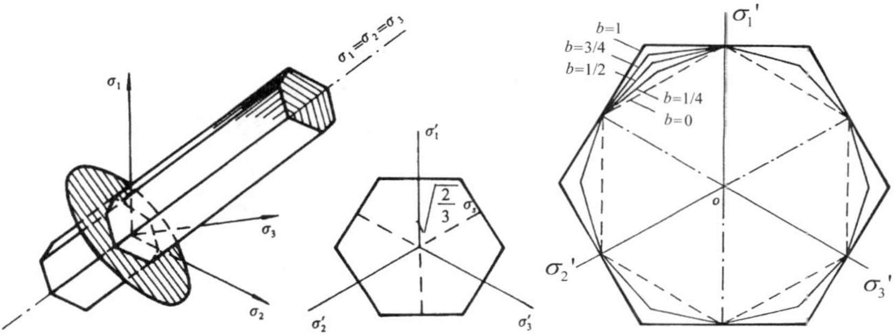

(a) Twin-shear yield loci (b=1) (b) Five typical yield loci

Fig. 4.13 Yield surface and yield loci of UYC with b=1 and five typical yield loci

4.10 Yield Loci of the Unified Strength Theory in the Plane Stress State

The yield loci of the unified strength theory in the plane stress state are the intersection line of the yield surface in principal stress space and the $\sigma_1-\sigma_2$ plane. Its shape and size depend on the values of b and α. It will be transformed into hexagon when b =0 or b=1, and into dodecagon when 0< b <1.

The equations of the 12 yield loci of the unified strength theory in the plane stress state can be given as follows. A series of new failure criteria and new yield loci in the plane stress state can be obtained from the unified strength theory

$$
\begin{aligned}
&\sigma_1-\frac{\alpha b}{1+b}\sigma_2=\sigma_t && \frac{\alpha}{1+b}(\sigma_1+b\sigma_2)=\sigma_t\\
&\sigma_2-\frac{\alpha b}{1+b}\sigma_1=\sigma_t && \frac{\alpha}{1+b}(\sigma_2+b\sigma_1)=\sigma_t\\
&\sigma_1-\frac{\alpha}{1+b}\sigma_2=\sigma_t && \frac{1}{1+b}\sigma_1-\alpha\sigma_2=\sigma_t\\
&\sigma_2-\frac{\alpha}{1+b}\sigma_1=\sigma_t && \frac{1}{1+b}\sigma_2-\alpha\sigma_1=\sigma_t\\
&\frac{\alpha}{1+b}(b\sigma_1+\sigma_2)=-\sigma_t && \frac{b}{1+b}\sigma_1-\alpha\sigma_2=\sigma_t\\
&\frac{\alpha}{1+b}(b\sigma_2+\sigma_1)=-\sigma_t && \frac{b}{1+b}\sigma_2-\alpha\sigma_1=\sigma_t
\end{aligned}
\qquad (4\text{-}48)
$$

The yield loci of the unified strength theory (UST) in the plane stress state with different values of b are shown in Fig. 4.14 (for α=1/2 material) and Fig.4.15 (for α=1 and α=1/4 materials).

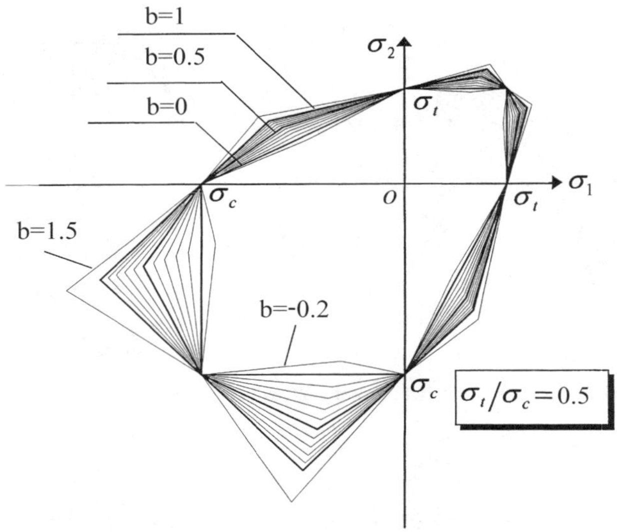

Fig. 4.14 Variation of yield loci of the UST in plane stress ($\alpha \neq 1$ materials)

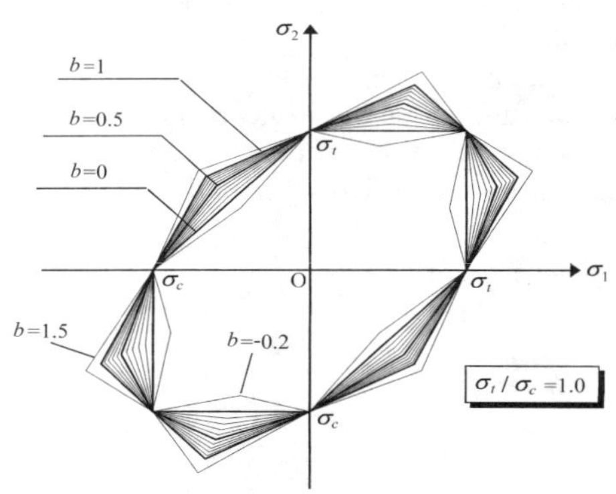

(a) α =1 material

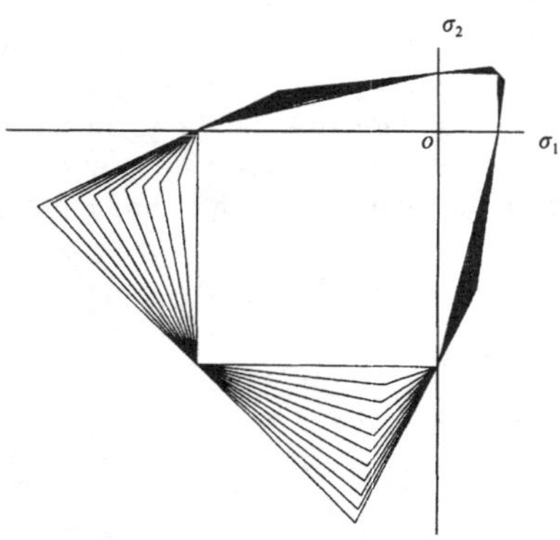

(b) α =1/4 material

Fig. 4.15 Yield loci of the UST in the plane stress state (α =1/4 material)

Various yield loci of the unified strength theory in the plane stress state are shown in Fig. 4.16. The unified yield criterion, the Mohr–Coulomb strength theory, the twin-shear strength theory and a series of new failure criteria as well as the non-convex failure loci can be obtained from the unified strength theory.

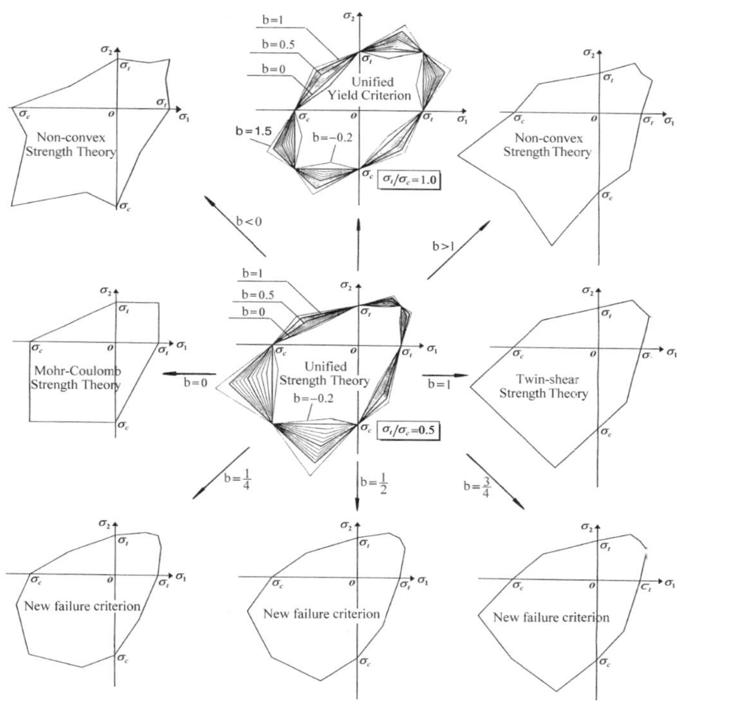

Fig. 4.16 Variation of the UST in the plane stress state

If the tensile strength is identical to the compressive strength, the unified strength theory will be transformed into the unified yield criterion. Its yield surfaces can be described in Fig.4.11, 4.12 and 4.13.

In the general case, The unified yield criterion for $\alpha = \sigma_t/\sigma_c = 1$ materials in the plane stress state (σ_1, σ_2) can be expressed by 12 equations as follows:

$$f_1 = \sigma_1 - \frac{b}{1+b}\sigma_2 = \pm\ \sigma y; \quad f_2 = \frac{b}{1+b}\sigma_1 - \sigma_2 = \pm\sigma_y$$

$$f_3 = \frac{1}{1+b}\sigma_1 + \frac{b}{1+b}\sigma_2 = \pm\sigma_y\ ; \ f_4 = \frac{b}{1+b}\sigma_1 + \frac{1}{1+b}\sigma_2 = \pm\sigma_y \quad (4\text{-}49)$$

$$f_5 = \sigma_1 - \frac{1}{1+b}\sigma_2 = \pm\sigma_y\ ; \ f_6 = \frac{1}{1+b}\sigma_1 - \sigma_2 = \pm\sigma_y$$

The yield loci of the unified yield criterion (α=1) in the plane stress state with different values of b are shown in Fig. 4.15(a).

A series of the yield loci of the unified yield criterion when $\alpha = \sigma_t/\sigma_c = 1$ in the plane stress state can be given. These yield loci cover all the regions of the convex yield criteria and also include the nonconvex yield criteria, which have never been formulated before. Varieties of the yield loci of the unified yield criterion in the plane stress states can be seen in Fig. 4.17.

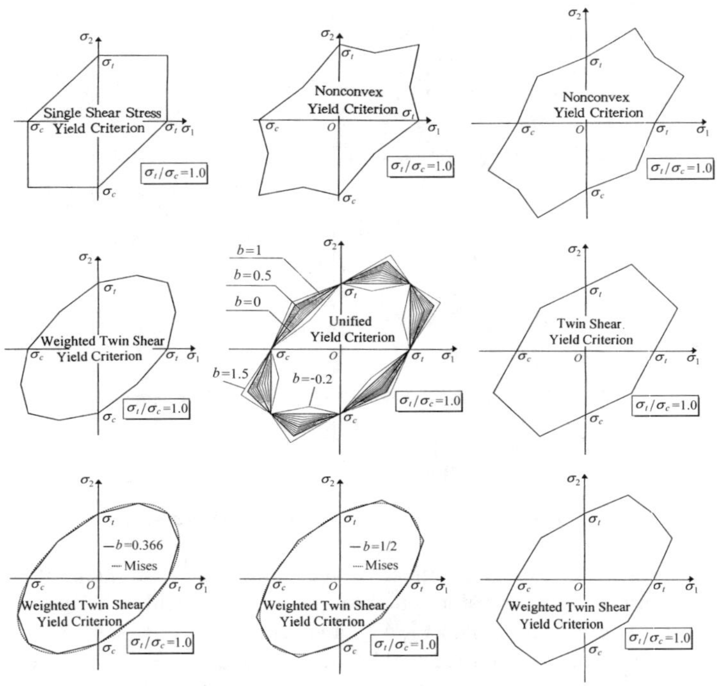

Fig. 4.17. Varieties of the UST for $\alpha=\sigma_t/\sigma_c=1$ materials in plane stress

4.11 Unified Strength Theory in Meridian Plane

The unified strength theory can also be expressed in other terms, such as by the octahedral normal stress σ_8 and octahedral shear stress τ_8 in plasticity, or by the generalized normal stress σ_g and the generalized shear stress τ_g (or q) in soil mechanics and geomechanics.

The relationships between the three principal stresses σ_1, σ_2, σ_3 and the cylindrical polar coordinates ξ, r, θ in the principal stress space are:

$$\begin{Bmatrix}\sigma_1\\ \sigma_2\\ \sigma_3\end{Bmatrix}=\frac{1}{\sqrt{3}}\xi+\sqrt{\frac{2}{3}}r\begin{Bmatrix}\cos\theta\\ \cos(\theta-2\pi/3)\\ \cos(\theta+2\pi/3)\end{Bmatrix} \tag{4-50}$$

in which ξ is the major coordinate axis in the stress space, and r is the length of the stress vector in the π-plane. They are given as follows:

$$\begin{aligned}\xi&=\frac{1}{\sqrt{3}}(\sigma_1+\sigma_2+\sigma_3)\\ r&=\frac{1}{\sqrt{3}}\sqrt{(\sigma_1-\sigma_2)^2+(\sigma_2-\sigma_3)^2+(\sigma_3-\sigma_1)^2}\end{aligned} \tag{4-51}$$

The relationship among the different variables is

$$\begin{aligned}\xi&=\frac{1}{\sqrt{3}}I_1=\sqrt{3}\sigma_8=\sqrt{3}p=\sqrt{3}\sigma_m\\ r&=\sqrt{2J_2}=\sqrt{3}\tau_8=\sqrt{\frac{2}{3}}q=2\tau_m\end{aligned} \tag{4-52}$$

The principal stress can be expressed as

$$\begin{Bmatrix}\sigma_1\\ \sigma_2\\ \sigma_3\end{Bmatrix}=\frac{1}{3}I_1+\frac{2}{\sqrt{3}}\sqrt{J_2}\begin{Bmatrix}\cos\theta\\ \cos(\theta-2\pi/3)\\ \cos(\theta+2\pi/3)\end{Bmatrix};$$

$$\begin{Bmatrix}\sigma_1\\ \sigma_2\\ \sigma_3\end{Bmatrix}=p+\frac{2}{3}q\begin{Bmatrix}\cos\theta\\ \cos(\theta-2\pi/3)\\ \cos(\theta+2\pi/3)\end{Bmatrix};$$

$$\begin{Bmatrix}\sigma_1\\ \sigma_2\\ \sigma_3\end{Bmatrix}=\sigma_8+\sqrt{2}\tau_8\begin{Bmatrix}\cos\theta\\ \cos(\theta-2\pi/3)\\ \cos(\theta+2\pi/3)\end{Bmatrix};$$

or

$$\begin{Bmatrix}\sigma_1\\ \sigma_2\\ \sigma_3\end{Bmatrix}=\sigma_m+\frac{2\sqrt{2}}{\sqrt{3}}\tau_m\begin{Bmatrix}\cos\theta\\ \cos(\theta-2\pi/3)\\ \cos(\theta+2\pi/3)\end{Bmatrix} \tag{4-53}$$

Substituting the above equations into the expression of the unified strength theory, the unified strength theory can then be expressed in other terms. Fig. 4.18 shows the yield loci of the unified strength theory in the meridian plane with θ =0° and θ =60°.

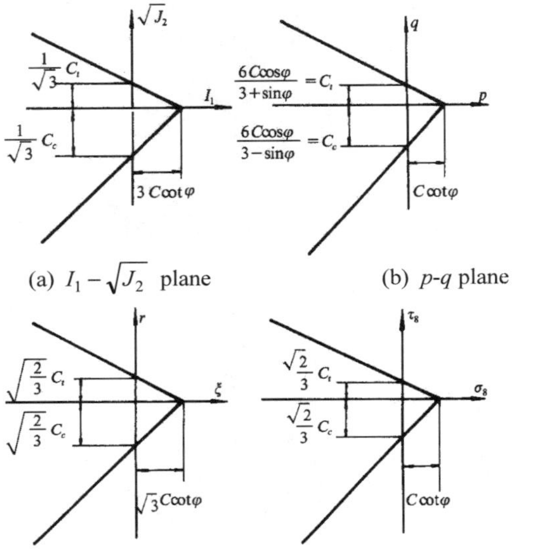

(a) $I_1-\sqrt{J_2}$ plane (b) p-q plane

(c) $\xi-r$ plane (d) $\sigma_8-\tau_8$ plane

Fig. 4.18 Yield loci of the UST in the meridian plane

The yield meridian loci with θ =0°and θ =60° are also called the tension yield meridian locus and the compression yield meridian locus, respectively. It is useful to understand the relationship for various kinds of equations, figures and tables.

In some books on soil mechanics and geomechanics, $(\sigma_1-\sigma_3)$ is often used as coordinate, then the figure $(\sigma_1-\sigma_3)\sim p$ can be drawn. In the case of triaxial confined pressure experiments, the stress state is axisymmetric, i.e., $\sigma_2=\sigma_3$. The generalized shear stress q is

$$q=\sqrt{\frac{1}{2}[(\sigma_1-\sigma_2)^2+(\sigma_2-\sigma_3)^2+(\sigma_3-\sigma_1)^2]}=\sigma_1-\sigma_3 \tag{4-54}$$

The $q\sim p$ coordinate and the $(\sigma_1-\sigma_3)\sim p$ coordinate are identical in the case of triaxial confined pressure. It is worth noting, however, that they are not identical in other cases.

4.12 Yield Surfaces of the Non-linear Unified Strength Theory

The unified strength theory can also be extended into various multiple–parameter criteria for more complex conditions. The expressions are

$$F = \tau_{13} + b\tau_{12} + \beta_1(\sigma_{13} + b\sigma_{12}) + A_1\sigma_m + B_1\sigma_m^2 = C\,, \qquad \text{(4-55a)}$$

$$F' = \tau_{13} + b\tau_{23} + \beta_2(\sigma_{13} + b\sigma_{23}) + A_2\sigma_m + B_2\sigma_m^2 = C\,, \qquad \text{(4-55a)}$$

or

$$F = (\tau_{13} + \beta\sigma_{13})^2 + b\,(\tau_{12} + \beta\sigma_{12})^2 + A_1\sigma_m^2 = \mathrm{C} \qquad \text{(4-56a)}$$

$$F' = (\tau_{13} + \beta\sigma_{13})^2 + b\,(\tau_{23} + \beta\sigma_{23})^2 + A_2\sigma_m^2 = \mathrm{C} \qquad \text{(4-56b)}$$

Equations. (4-55a) and (4-55b) can be simplified to the unified strength theory when $A_1=A_2=0$, $B_1=B_2=0$ and $\beta_1=\beta_2$. In this case, it becomes the single-shear strength theory (Mohr–Coulomb strength theory) when $b=0$, or the twin-shear strength theory when $b=1$.

When $A_1=A_2=0$, $B_1=B_2=0$ and $\beta_1=\beta_2=0$, Eqs. (4-55a) and (4-55b) are simplified to the unified yield criterion. In this case, the twin-shear yield criterion and the single–shear yield criterion (the Tresca yield criterion) are introduced when $b=1$ and $b=0$, respectively.

Equations (4-55a), (4-55b) and (4-56a), (4-56b) are nonlinear equations. It is not convenient for analytical solution in plasticity and engineering applications.

These formulations are the nonlinear unified strength theory. A yield surface of the nonlinear unified strength theory is shown in Fig. 4.7(b). The yield surface of the linear unified strength theory is the special case of the nonlinear unified strength theory, as shown in Fig. 4.7(a).

Example 4.2

Consider a metallic material with the same yield stress in tension and compression. If its shear yield stress is $\tau_y = 0.63\sigma_y$ and Poisson's ratio is $\nu = 0.3$, find an available yield criterion and draw its yield loci in the deviatoric plane, plane stress state and plane strain state.

Solution

1. Find an available yield criterion

The Tresca yield criterion (single-shear yield criterion) predicts the shear yield stress is $\tau_y = 0.5\sigma_y$, and the Huber-von Mises yield criterion (octahetral shear stress yield criterion) predicts the shear yield stress is $\tau_y = 0.677\sigma_y$. Obviously, these two yield criteria do not fit this kind of material with $\tau_y = 0.63\sigma_y$.

According to the unified yield criterion Eq. (4-21a) and (4-21b), the parameter b in the unified yield criterion can be determined as follows

$$b=\frac{2\tau_y-\sigma_y}{\sigma_y-\tau_y}=\frac{2\times 0.63-1}{1-0.63}=0.7 \tag{4-57}$$

Substituting b=0.7 into Eqs. (4-17a) and (4-17b), a new available yield criterion is obtained as follows

$$f=\sigma_1-\frac{1}{1.7}(0.7\sigma_2+\sigma_3)=\sigma_y, \quad \text{if } \sigma_2\le\frac{1}{2}(\sigma_1+\sigma_3) \tag{4-58a}$$

$$f'=\frac{1}{1.7}(\sigma_1+0.7\sigma_2)-\sigma_3=\sigma_y, \quad \text{if } \sigma_2\ge\frac{1}{2}(\sigma_1+\sigma_3) \tag{4-58b}$$

2. Draw the Yield Locus in the Deviatoric Plane

The π-plane is a special plane that makes equal angles with the coordinate σ_1, σ_2 and σ_3 axes. The projections of the axes upon this plane must make equal 120° angles with each other (Fig. 4.19). The equation of the π-plane is $\sigma_1+\sigma_2+\sigma_3=C$. The coordinates are threefold symmetric.

The relations between the threefold symmetric coordinates and the rectangular coordinate are

$$x=\frac{1}{\sqrt{2}}(\sigma_3-\sigma_2);\ y=\frac{1}{\sqrt{6}}(2\sigma_1-\sigma_2-\sigma_3);\ z=\frac{1}{\sqrt{3}}(\sigma_1+\sigma_2+\sigma_3) \tag{4-59}$$

The inverse relations are

$$\sigma_1=\frac{1}{3}(\sqrt{6}y+\sqrt{3}z);\ \sigma_2=\frac{1}{6}(2\sqrt{3}z-\sqrt{6}y-3\sqrt{2}x);$$

$$\sigma_3=\frac{1}{6}(3\sqrt{2}x-\sqrt{6}y+2\sqrt{3}z) \tag{4-60}$$

Substituting these relations (Eq. 4-60) in the equations of the unified yield criterion (Eq. 4-17), we obtain the expressions of the unified yield criterion in the rectangular coordinates as follows:

$$f=-\frac{\sqrt{2}(1-b)}{2(1+b)}x+\frac{\sqrt{6}}{2}y=\pm\sigma_y;\ f=\frac{\sqrt{2}(1-b)}{2(1+b)}x+\frac{\sqrt{6}}{2}y=\pm\sigma_y$$

$$f=-\frac{(2+b)\sqrt{2}}{2(1+b)}x-\frac{\sqrt{6}b}{2(1+b)}y=\pm\sigma_y\ ;\ f=-\frac{(1+2b)\sqrt{2}}{2(1+b)}-\frac{\sqrt{6}}{2(1+b)}y=\pm\sigma_y$$

$$f=\frac{\sqrt{2}(1+2b)}{2(1+b)}x-\frac{\sqrt{6}}{2(1+b)}y=\pm\sigma_y\ ;\ f=\frac{\sqrt{2}(2+b)}{2(1+b)}x-\frac{b\sqrt{6}}{2(1+b)}y=\pm\sigma_y \tag{4-61}$$

In the case of b=0.7, the expressions of the yield criterion are

$$\begin{aligned} &f=-0.125x+\frac{\sqrt{6}}{2}y=\pm\sigma_y\ ;\quad f=0.125x+\frac{\sqrt{6}}{2}y=\pm\sigma_y\\ &f=-1.123x-0.504y=\pm\sigma_y\ ;\quad f=-0.998x-0.7204y=\pm\sigma_y\\ &f=0.998x-0.7204y=\pm\sigma_y\ ;\quad f=1.123x-0.504y=\pm\sigma_y \end{aligned} \tag{4-62}$$

The yield locus of this yield criterion (b=0.7) in deviatoric plane is shown in Fig. 4.19.

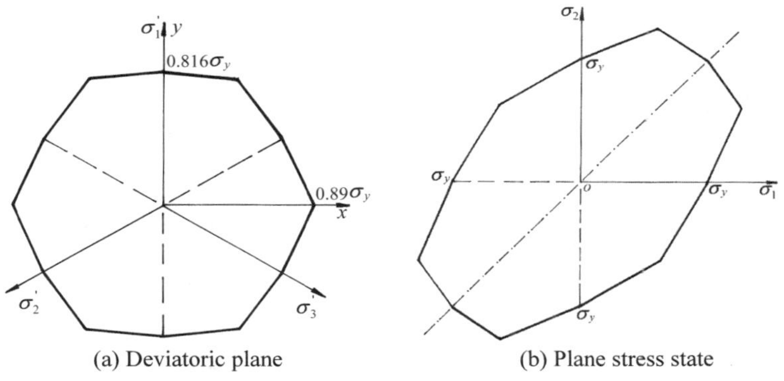

Fig. 4.19 Yield locus of a new yield criterion (b=0.7)

3. Draw the Yield Locus in the Plane Stress State

The unified yield criterion in plane stress state is given in Eq.(4-49). In the case of b=0.7, the expressions of this yield criterion are

$$\begin{aligned} &f=\sigma_1-0.412\sigma_2=\pm\sigma_y;\quad f=0.412\sigma_1-\sigma_2=\pm\sigma_y\\ &f=0.588\sigma_1+0.412\sigma_2=\pm\sigma_y\ ;\quad f=0.412\sigma_1+0.588\sigma_2=\pm\sigma_y\\ &f=\sigma_1-0.588\sigma_2=\pm\sigma_y\ ;\quad f=0.588\sigma_1-\sigma_2=\pm\sigma_y \end{aligned} \tag{4-63}$$

The yield locus of this yield criterion (b=0.7) in the plane stress state is shown in Fig. 4.19(b).

Example 4.3

Introduce an available yield criterion and draw its yield loci in deviatoric plane, plane stress state and plane strain state when the shear yield stress is $\tau_y = 0.59\sigma_y$ and the Poisson's ratio is ν =0.2.

Solution

The parameter b in the unified yield criterion can be determined by using the relationship

$$b=\frac{2\tau_y-\sigma_y}{\sigma_y-\tau_y}=\frac{2\times 0.59-1}{1-0.59}=0.44 \tag{4-64}$$

Substituting b=0.44 into Eq. (4-17a) and (4-17b), an available yield criterion is obtained

$$f=\sigma_1-\frac{1}{1.44}(0.44\sigma_2+\sigma_3)=\sigma_y, \quad \text{if } \sigma_2\le\frac{1}{2}(\sigma_1+\sigma_3) \tag{4-65a}$$

$$f'=\frac{1}{1.44}(\sigma_1+0.44\sigma_2)-\sigma_3=\sigma_y, \quad \text{if } \sigma_2\ge\frac{1}{2}(\sigma_1+\sigma_3) \tag{4-65b}$$

The yield loci of this yield criterion in deviatoric plane, plane stress state and plane strain state are illustrated in Figs. 4.20 and 4.21.

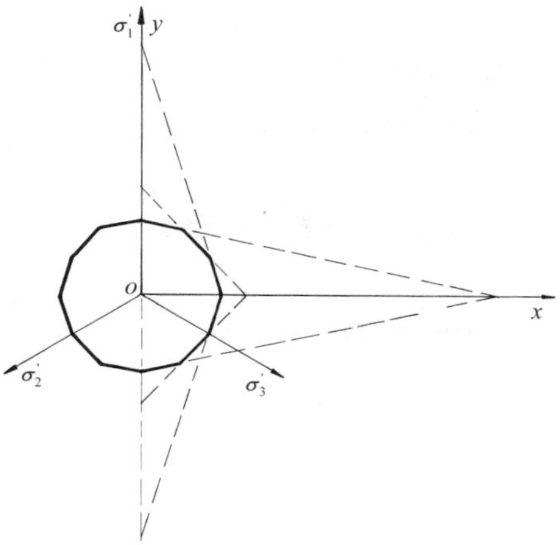

Fig. 4.20 Yield locus of a new yield criterion in the deviatoric plane (b=0.44)

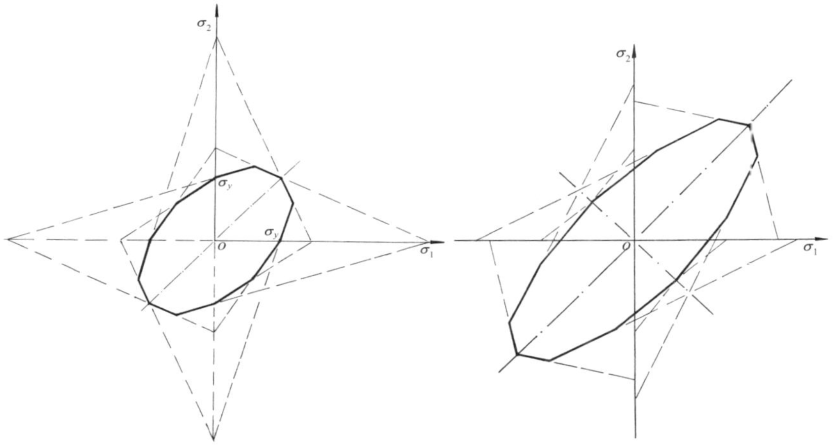

(a)plane stress state (b) plane strain state

Fig. 4.21 Yield loci of the unified yield criterion with b=0.44

Summary

Based on the concepts of the multiple slip mechanism, the twin-shear model and multi-shear model, a new unified strength theory was proposed by Yu in 1991 (Yu and He 1991; Yu 1992). This unified strength theory (unified strength theory) is not a single yield criterion suitable only for one kind of material, but a completely new system. It embraces many well-established criteria as its special or approximate cases, such as the Tresca yield criterion, the Huber-von Mises yield criterion, and the Mohr–Coulomb strength theory, as well as the twin-shear yield criterion (Yu 1961a), the generalized twin-shear strength theory (Yu et al. 1985), and the unified yield criterion. The unified strength theory forms an entire spectrum of convex and nonconvex criteria, which can be used to describe many kinds of engineering materials. The unified strength theory has a unified mechanical model and a simple and unified mathematical expression, which can be adapted to various experimental data. It is easy to use in both research and engineering.

The unified strength theory establishes very clear and simple relations among the various yield criteria, as show in Fig.4.22. It also provides a method to choose the appropriate yield criterion.

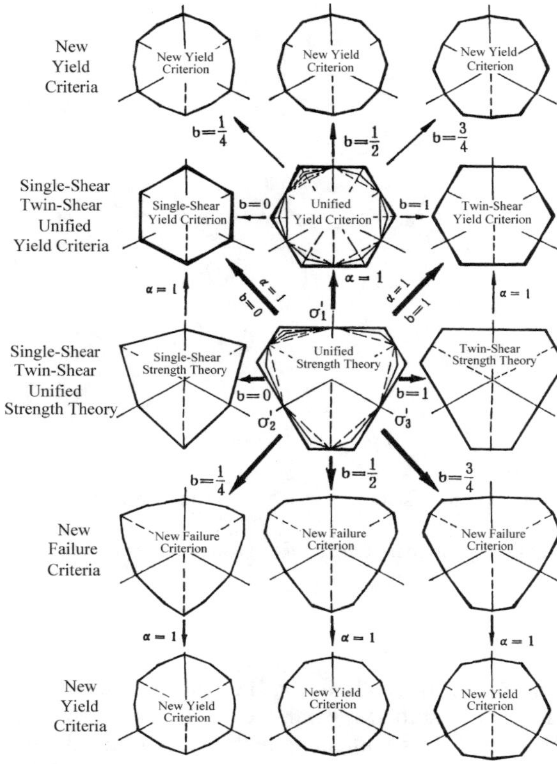

Fig. 4.22 Varation of the unified strength theory and the relationships among the criteria

The SD effect, hydrostatic stress effect, normal stress effect, effect of the Intermediate principal stress and the effect of intermediate principal shear stress are all taken into account in the unified strength theory.

The unified strength theory is a completely new theory system. The significance of the Yu unified strength theory is summarized as follows:

1. It is suitable for more kinds of isotropic materials.
2. It contains various strength theories and forms a new system of yield criteria and failure criteria. It gives a relation among the single-shear criterion, the twin-shear criterion, and a series of new criteria.
3. It gives good agreement with experimental results for various materials.
4. A series of new results can be obtained by using the unified strength theory.
5. The unified strength theory is easy to use for analytical solutions of plastic problems. The applications of the unified strength theory are described in Chapter 8 to Chapter 15. It is convenient for elastic limit design, elasto-plastic analysis and plastic limit analysis of structures.

The mathematical expression of the unified strength theory can be expressed into various forms. More than ten kinds of expressions are discussed in this chapter.

The parameters of unified strength theory are the same as the parameters used in the Mohr-Coulomb strength theory (1900), Drucker-Prager criterion (1952), the twin-shear strength theory (Yu et al. 1985). The two parameters, i.e. the tensile strength σ_t and the compressive strength σ_c (or σ_t , α) or friction angle φ and cohesion C_0 are the most widely used material parameters in engineering.

The yield function can be interpreted for an isotropic material in terms of a geometrical representation of the stress state obtained by taking the principal stresses as coordinates. The yield surface in a three-dimensional principal stress space was introduced by Haigh (1920) and Westergaard (1920). Sometimes, it is called the Haigh-Westergaard space. The advantage of such a space lies in its simplicity and visual presentation.

The yield surface of the unified strength theory in stress space and yield loci on plane stress, deviatoric plane, and meridian plane are illustrated in this chapter. Unified strength theory embraces many well-established yield surfaces and yield loci as its special or asymptotic cases, such as yield surfaces of the Tresca yield criterion, the Huber-von Mises yield criterion, and the Mohr–Coulomb strength theory, as well as the twin-shear yield criterion (Yu 1961a), the twin-shear strength theory, and the unified yield criterion. The unified strength theory forms an entire spectrum of convex and nonconvex criteria, which can be used to describe many kinds of engineering materials.

The yield surfaces and yield loci of the unified yield criterion, the twin-shear strength criterion, the twin-shear yield criterion, the single-shear strength criterion (Mohr–Coulomb theory), the single-shear yield criterion (Tresca yield criterion) and many empirical failure criteria are special cases or linear approximations of the yield surface of the unified strength theory. A series of new yield surfaces and yield loci are also can be drawn based on the unified strength theory.

Problems

Problem 4.1
Introduce the well-known Mohr–Coulomb strength theory from the unified strength theory when b =0.

Problem 4.2
Introduce a new failure criterion from the unified strength theory when b=1/4.

Problem 4.3
Introduce a new failure criterion from the unified strength theory when b=1/2.

Problem 4.4
Introduce a new failure criterion from the unified strength theory when b=3/4.

Problem 4.5
Introduce the twin-shear strength theory from the unified strength theory when b=1.

Problem 4.6
Compare the unified strength theory when b=1/2 with the Drucker–Prager criterion.

Problem 4.7
Introduce a new failure criterion from the unified strength theory taking any value of b.

Problem 4.8.
Introduce a new yield criterion from the unified yield criterion taking any value of parameter b and α=1.

Problem 4.9
Introduce the unified strength theory in terms of stress invariant $F(I_1, J_2, \theta)$ and materials parameters σ_t and α

Problem 4.10
Introduce the unified strength theory in terms of stress invariant $F(I_1, J_2, \theta)$ and material parameters c and φ

Problem 4.11
Introduce the unified strength theory by using the experimental condition of pure shear and uniaxial tension strength.

Problem 4.12
Introduce the unified strength theory by using the experimental condition of pure shear and uniaxial compression strength.

Problem 4.13
Introduce the unified strength theory by using the experimental condition of pure shear and biaxial compression strength.

Problem 4.14
Introduce the three-parameter unified strength theory.

Problem 4.15.
The five kinds of yield loci of the unified strength theory (UST) are shown in Fig.P4.1. Indicate the mathematical expressions of the parts AC and CB of the unified strength theory with b=1.

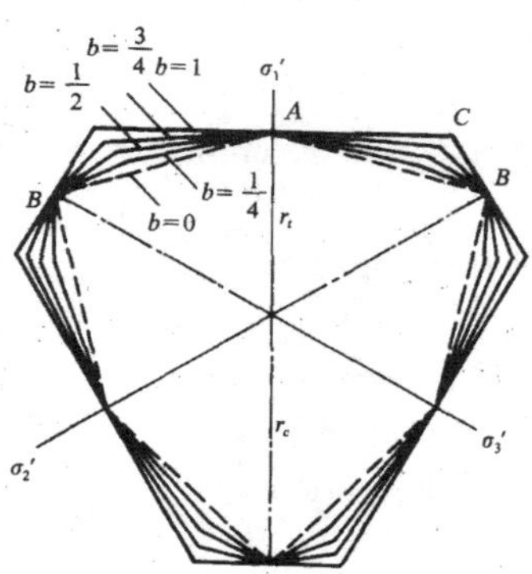

Fig.P4.1 Yield loci of UST in π-plane

Problem 4.16.

The five kinds of yield loci of the unified strength theory (UST) are shown in Fig.P4.1. Indicate the mathematical expressions of the parts AC and CB of the unified strength theory with b=3/4.

Problem 4.17.

The five kinds of yield loci of the unified strength theory (UST) are shown in Fig.P4.1. Indicate the mathematical expressions of the parts AC and CB of the unified strength theory with b=1/2.

Problem 4.18.

The five kinds of yield loci of the unified strength theory (UST) are shown in Fig.P4.1. Indicate the mathematical expressions of the parts AC and CB of the unified strength theory with b=1/4.

Problem 4.19.

The five kinds of yield loci of the unified strength theory (UST) are shown in Fig.P4.1. Indicate the mathematical expressions of the parts AC and CB of the unified strength theory with b=0.

Problem 4.20.

The yield equation in π-plane of the unified strength theory with b=0 (Mohr-Coulomb theory) is

$$F = F' = -\frac{\sqrt{2}}{2}\alpha x + \frac{\sqrt{6}}{6}(2+\alpha)y + \frac{\sqrt{3}}{3}(1-\alpha)z = \sigma_t$$

Draw the yield locus in π-plane.

Problem 4.21.

A new failure criterion equation in π-plane of the unified strength theory with b=1/4 is

$$F = -\frac{3\sqrt{2}}{10}ax + \frac{\sqrt{6}}{6}(2+\alpha)y + \frac{\sqrt{3}}{3}(1-\alpha)z = \sigma_t$$

$$F' = -(\frac{1}{5}+\alpha)\frac{\sqrt{2}}{2}x + (\frac{7}{5}+\alpha)\frac{\sqrt{6}}{6}y + \frac{\sqrt{3}}{3}(1-\alpha)z = \sigma_t$$

Draw the yield locus in π-plane.

Problem 4.22.

A new failure criterion equation in π-plane of the unified strength theory with b =1/2 is

$$F = -\frac{\sqrt{2}}{6}ax + \frac{\sqrt{6}}{6}(2+\alpha)y + \frac{\sqrt{3}}{3}(1-\alpha)z = \sigma_t$$

$$F' = -(\frac{1}{3}+\alpha)\frac{\sqrt{2}}{2}x + (1+\alpha)\frac{\sqrt{6}}{6}y + \frac{\sqrt{3}}{3}(1-\alpha)z = \sigma_t$$

Draw the yield locus in π-plane.

Problem 4.23.
A new failure criterion equation in π-plane of the unified strength theory with b =3/4 is

$$F=-\frac{\sqrt{2}}{14}ax+\frac{\sqrt{6}}{6}(2+\alpha)y+\frac{\sqrt{3}}{3}(1-\alpha)z=\sigma_t$$

$$F'=-(\frac{3}{7}+\alpha)\frac{\sqrt{2}}{2}x+(\frac{5}{7}+\alpha)\frac{\sqrt{6}}{6}y+\frac{\sqrt{3}}{3}(1-\alpha)z=\sigma_t$$

Draw the yield locus in π-plane.

Problem 4.24.
A new failure criterion equation in π-plane of the unified strength theory with b =1 (the twin-shear strength theory) is

$$F=\frac{\sqrt{6}}{6}(2+\alpha)y+\frac{\sqrt{3}}{3}(1-\alpha)z=\sigma_t$$

$$F'=-(\frac{1}{2}+\alpha)\frac{\sqrt{2}}{2}x+(\frac{1}{2}+\alpha)\frac{\sqrt{6}}{6}y+\frac{\sqrt{3}}{3}(1-\alpha)z=\sigma_t$$

Draw the yield locus in π-plane.

Problem 4.25.
Show the cross-sectional shapes of the unified strength theory when b=1 and α=1/3 (new strength criterion) on the meridian planes and on the $\sigma_1-\sigma_2$ plane with σ_3=0.

Problem 4.26.
Show the cross-sectional shapes of the unified strength theory when b=3/4 and α=1/3 (new strength criterion) on the meridian planes and on the $\sigma_1-\sigma_2$ plane with σ_3=0.

Problem 4.27.
Five kinds of yield loci of the unified yield criterion when $\alpha=\sigma_t/\sigma_c$=1 and in plane stress are shown in Fig.P4.2. These yield equations and yield loci of the unified yield criterion of $\alpha=\sigma_t/\sigma_c$=1 materials for any value of parameter b can be obtained. For example, the 12 yield equations of the unified yield criterion under the plane stress state when b=1/2 can be given as follows. The yield loci of this yield criterion are illustrated in Fig. P4.2.

$$f_{1,7}=\sigma_1-\frac{1}{3}\sigma_2=\pm\sigma_y;\quad f_{2,8}=2\sigma_1+\sigma_2=\pm\sigma_y$$

$$f_{3,9}=\frac{1}{3}(\sigma_1+2\sigma_2)=\pm\sigma_y;\ f_{4,10}=\frac{1}{3}\sigma_1-\sigma_2=\mp\sigma_y \qquad (4\text{-}59)$$

$$f_{5,11}=\frac{2}{3}\sigma_1-\sigma_2=\pm\sigma_y;\quad f_{6,12}=\sigma_1-\frac{2}{3}\sigma_2=\mp\sigma_y$$

Writing out the 12 yield equations of the unified yield criterion under the plane stress state when b=0, b=1/4, b=3/4, and b=1.

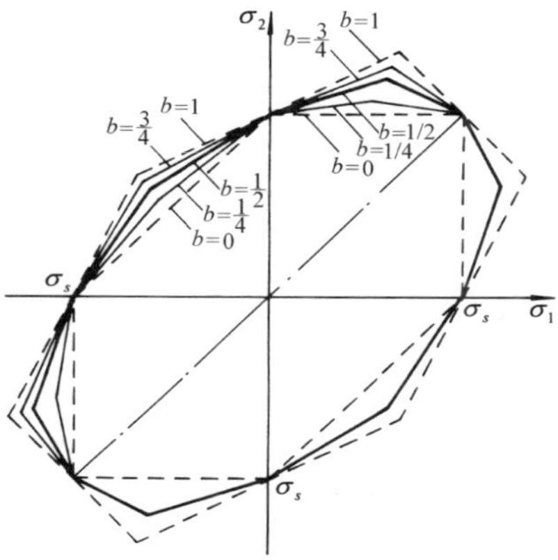

Fig.P4.2. Yield loci of UST in plane stress

Problem 4.28.
The unified yield criterion in plane stress state can be divided into three cases as follows:

1. $\sigma_1 \geq \sigma_2 > 0,\ \sigma_3 = 0$
The unified yield criterion with $\alpha = \sigma_t/\sigma_c = 1$ in the plane stress state is

$$f = \sigma_1 - \frac{b}{1+b}\sigma_2 = \sigma_y, \qquad \text{if} \quad \sigma_2 \leq \frac{1}{2}\sigma_1 \tag{4-55a}$$

$$f' = \frac{1}{1+b}\sigma_1 + \frac{b}{1+b}\sigma_2 = \sigma_y, \qquad \text{if} \quad \sigma_2 \geq \frac{1}{2}\sigma_1 \tag{4-55b}$$

2. $\sigma_1 \geq 0,\ \sigma_2 = 0,\ \sigma_3 < 0$

The unified yield criterion for $\alpha = \sigma_t/\sigma_c = 1$ materials in plane stress state is

$$f = \sigma_1 - \frac{1}{1+b}\sigma_3 = \sigma_y, \qquad \text{if} \quad \frac{1}{2}(\sigma_1 + \sigma_3) \geq 0 \tag{4-56a}$$

$$f' = \frac{1}{1+b}\sigma_1 - \sigma_3 = \sigma_{y}, \qquad \text{if} \quad \frac{1}{2}(\sigma_1 + \sigma_3) \geq 0 \tag{4-56b}$$

3. $\sigma_1 = 0$, $\sigma_2 \geq \sigma_3 < 0$

The unified yield criterion for $\alpha = \sigma_t/\sigma_c = 1$ materials in plane stress state is

$$f = -\frac{1}{1+b}(b\sigma_2 + \sigma_3) = \sigma_{y}, \qquad \text{if} \quad \sigma_2 \leq \frac{1}{2}\sigma_3 \tag{4-57a}$$

$$f' = \frac{b}{1+b}\sigma_2 - \sigma_3 = \sigma_{y}, \qquad \text{if} \quad \sigma_2 \geq \frac{1}{2}\sigma_3 \tag{4-57b}$$

Draw a yield locus in plane stress state for b=1 and $\alpha = \sigma_t/\sigma_c = 1$ material.

Problem 4.29.
Draw a yield locus in plane stress state for b=3/4 and $\alpha = \sigma_t/\sigma_c = 1$ material.

Problem 4.30.
Draw a yield locus in plane stress state for b=0.6 and $\alpha = \sigma_t/\sigma_c = 1$ material.

Problem 4.31.
Draw a yield locus in plane stress state for b=1/2 and $\alpha = \sigma_t/\sigma_c = 1$ material.

Problem 4.32.
Draw a yield locus in plane stress state for b=1/4 and $\alpha = \sigma_t/\sigma_c = 1$ material.

Problem 4.33.
Draw a yield locus in plane stress state for b=0 and $\alpha = \sigma_t/\sigma_c = 1$ material.

Problem 4.34.
Show the cross-sectional shapes of the unified strength theory when $\alpha = 1$ and b=1/2 (new yield criterion) on the deviatoric planes and on the meridian planes.

Problem 4.35.
Show the cross-sectional shapes of the unified strength theory when $\alpha = 1$ and b=3/4 (new yield criterion) on the deviatoric planes and on the meridian planes.

Problem 4.36.
Show the cross-sectional shapes of the unified strength theory when b=1 and α=1/3 (twin-shear strength theory) on the deviatoric planes, on the meridian planes and on the $\sigma_1 - \sigma_2$ plane with σ_3 =0.

Problem 4.37.
Compare the unified strength theory when b=1/2 and α=1/3 with the Drucker–Prager criterion (and α=1/3).

Problem 4.38.
Draw a yield locus of a new failure criterion from the unified strength theory taking any value of b.

References and Bibliography

Besseling JF and van der Giessen E (1994) *Mathematical Modelling of Inelastic Deformation*, Chapman & Hall, London..

Chen WF and Saleeb AF (1994) *Constitutive Equations for Engineering Materials. Vol.1: Elasticity and Modeling, Revised edn.* Elsevier, Amsterdam. 259–304, 462–489.

Chen WF et al. (1994) *Constitutive Equations for Engineering Materials. Vol. 2 : Plasticity and modeling,* Elsevier, Amsterdam.

Drucker DC and Prager W (1952). Soil mechanics and plastic analysis for limit design. *Quart. Appl. Math.*, **10**, 157–165.

Encyclopedia of China (1985): Mechanics. Encyclopedia Press, Beijing.

Haigh BT (1920) The strain energy function and the elastic limit. *Engineering*, **109**, 158-160

Huber MT (1904) Przyczynek do podstaw wytorymalosci. *Czasopismo Technizne* 1904, **22**, 81 (Lwow, 1904); Pisma, 2, PWN, Warsaw, 1956.

Meyer WJ (1985) *Concepts of Mathematical Modeling*, McGraw-Hill Book Company,

von Mises R (1913). Mechanik der festen Körper im plastisch deformablen Zustand. *Nachrichten von der Königlichen Gesellschaft der wissenschaften zu Göettinger, Mathematisch-Physikalische Klasse,* 582–592.

Mohr O (1900) Welche Umstande bedingen die Elastizitatsgrenze und den Bruch eines Materials? *Zeitschrift des Vereins deutscher Ingenieure*, **44**, 1524–1530; 1572–1577.

Mohr O (1905) *Abhandlungen aus den Gebiete der Technischen Mechanik.* Verlag von Wilhelm Ernst and Sohn, 1905, 1913Third edn. 1928.

Tayler AB (1986) *Mathematical Models in Applied echanics.* Clarendon Press, Oxford

Timoshenko SP (1953) History of Strength of Materials: With a Brief Account of the History of Theory of Elasticity and Theory of Structures. McGraw-Hill, New York

Tresca H (1864) Sur I'e coulement des corps solids soumis a de fortes pression. *Comptes Rendus hebdomadaires des Seances de l' Academie des Sciences*, Rend 1864, **59**, 754–758.

Westergaard HM (1920) On the resistance of ductile materials to combined stresses. *J. Franklin Inst.*,**189**, 627-640.

Wu KKS, Lahav O and Rees MJ (1999) The large-scale smoothness of the Universe, *Nature,* **397,** 225-230.

Yu Mao-Hong (1961) General behaviour of isotropic yield function (in Chinese), *Res. Report of Xi'an Jiaotong University.*Xi'an.

Yu Mao-Hong (1961) Plastic potential and flow rules associated singular yield criterion (in Chinese), *Res. Report of Xi'an Jiaotong University.*Xi'an.

Yu Mao-Hong (1983) Twin shear stress yield criterion. *Int. J. Mech.* Sci. 25 (1) 71–74.

Yu Mao-Hong, He Li-nan and Song Ling-yu (1985) Twin shear stress theory and its generalization, *Scientia Sinica* (*Sciences in China*), English Edition, Series A, **28**(11), 1174–1183.

Yu MH and Liu FY (1988) Twin shear three-parameter criterion and its smooth ridge model. *China Civil Engng. J.* **21**(3), 90–95 (in Chinese, English abstract).

Yu MH and Liu FY (1990) Smooth ridge model of generalized twin shear stress criterion. *Acta Mechanica Sinica* **22**(2), 213–216 (in Chinese, English abstract)..

Yu MH and He LN (1991) A new model and theory on yield and failure of materials under the complex stress state, in: *Mechanical Behaviour of Materials—6,* (ICM-6). Jono, M. and Inoue, T. Ed, Pergamon Press, Oxford, Vol. 3, 841-846.

Yu Mao-Hong (1992) *New System of Strength Theories.* Xi'an Jiaotong University Press, Xi'an (in Chinese).

Yu Mao-Hong (1994) Unified strength theory for geomaterials and its application. *Chinese J. of Geotech. Eng.,* **16**(2):1-10 (in Chinese, English Abstract).

Yu Mao-hong. (1998) *Twin-shear Theory and Its Applications.* Science Press, Beijing (in Chinese).

Yu MH, Zan YW, Zhao J and Yoshimine M (2002) A unified strength criterion for rock. *Int. J. of Rock Mechanics and Mining Science* **39**, 975–989.

Yu Mao-Hong (2002) Advance in strength theory of material and complex stress state in the 20th Century. *Applied Mechanics Reviews*, 55(3):169–218.

Yu Mao-Hong (2004) *Unified Strength Theory and Its Applications.* Springer, Berlin.

5 Reasonable Choice of a Yield Function

5.1 Introduction

A yield criterion is a function of stress state and material parameters. The suitability of any proposed yield criterion must be examined by experiment results.

A great many experiments and investigations on yield criteria of metallic materials were done by Guest (1900,1940), Scoble (1906, 1910), Hancock (1908), Smith (1909a, 1909b), Lode (1926), Taylor and Quinney (1931), Ivey (1961), Mair and Pugh (1964), Paul (1968), Bell (1973), Michno and Findley (1976), Pisarenko and Lebedev (1976), Winstone (1984), Wu and Yeh (1991), Ellyin (1993), and others.

The experimental results of rock, concrete, soil and other geomaterials were given by many researchers. The discrepancies between different experiments and different materials are great.

The reasonable choice of a yield criterion is very important for calculation and applications. An example is given in the following:

Four sets of the initial yield surfaces of the cast nickel superalloy Mar-M002 at 750°C were given by Winstone (1984) in UK, as shown in Fig. 5.1. They lie within a tight scatter band. The ratio of shear yield stress τ_s to the tensile yield stress σ_s is $\tau_s/\sigma_s = 0.7$.

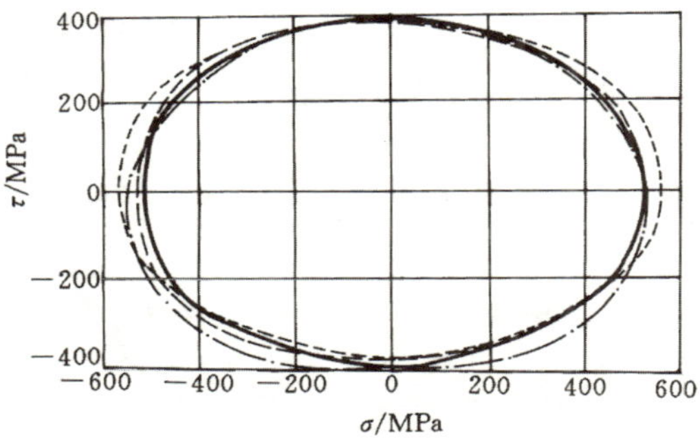

Fig. 5.1 Initial yield surface of Mar-M002 alloy (Winstone 1984)

Winstone pointed out that this value was surprisingly high when compared with the values of τ_s/σ_s =0.58 and τ_s/σ_s =0.5 expected from the Huber-von Mises yield criterion and the Tresca yield criterion, respectively. Clearly neither of these criteria can accurately model the yield behaviour of this material. These results, however, are close to the value of the twin-shear stress yield criterion. The ratio of shear yield stress τ_y to the tensile yield stress σ_y for the twin-shear yield criterion is $\tau_s / \sigma_s = 0.667$.

The initial yield surfaces of the cast nickel superalloy Mar-M002 indicated a ratio of shear yield stress to tensile yield stress of 0.7. The comparisons of this experimental result with the three yield criteria are as follows. The deviation are:

Tresca yield criterion: $\frac{0.7-0.5}{0.5} = 40\%$

von Mises yield criterion: $\frac{0.7-0.577}{0.577} = 21\%$

Twin-shear yield criterion: $\frac{0.7-0.667}{0.667} = 4.9\%$

Many experimental results for various materials under complex stress states have been published, and they are very valuable. These experimental results and the unified strength theory give us a possibility to choose a reasonable yield criterion for research and applications.

5.2 Some Experimental Data of Metallic Materials

A great many experiments and investigations on yield criteria of metallic materials were done in the 20th Century.

The initial yield locus of aluminum alloy 19S found by Ivey (1961) is shown in Fig. 5.2. The experiment data for mild steel found by Taylor and Quinney (1931) are also shown in Fig. 5.2.

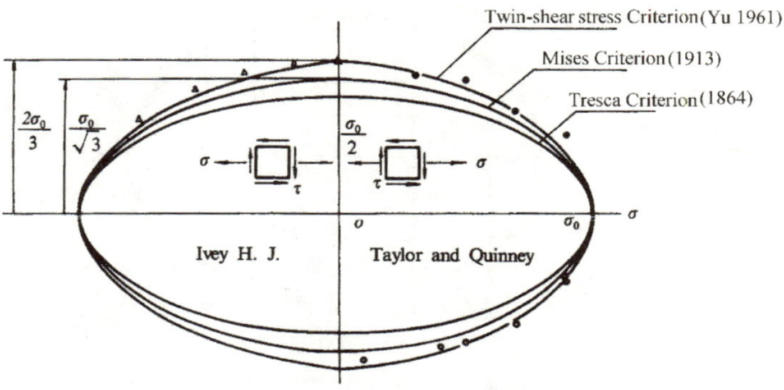

Fig. 5.2 Experimental results of Ivey (1961) and Taylor and Quinney (1931)

The ratio of the shear yield stress to the tensile yield stress equals approximately two-thirds. This agrees well with the twin-shear yield criterion, i.e., a special case of the unified strength theory when $\alpha=b=1$.

Winstone (1984) presented some new research results using the acoustic emission technique, which provides an accurate and sensitive method for determining yield surfaces. Combined tension and torsion tests were carried out on a servohydraulic testing machine capable of applying a maximum tensile load of 50 kN and a maximum torque of 200 Nm. The testing was undertaken using constant rates of tension and torsion. An acoustic emission transducer was used. All the yield surface tests used tubular test pieces.

The sequence of a typical test to determine the initial yield surface of the cast nickel superalloy Mar-M002 at 750°C was given by Winstone in UK, four sets of test results are shown in Fig. 5.3. This material is usually used for gas turbine blades. The yield surface was obtained by probing the plastic region under various combinations of tensile and torsional loads. Experimental points were obtained in the first and third quadrants of the yield surface, but the second and forth quadrants have been completed by symmetry (Winstone 1984).

The yield loci of the Tresca yield criterion and the von Mises yield criterion are also shown. The deviations of the experimental result from the Tresca yield criterion and the Huber-von Mises yield criterion are significant.

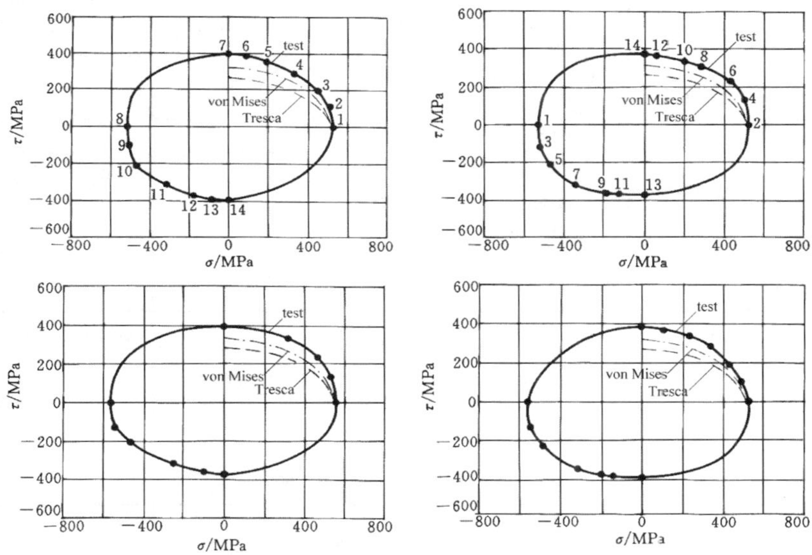

Fig. 5.3 Initial yield surface of Mar-M002 alloy at 750°C (Winstone 1984)

Numerous experiments of metallic materials under complex stresses have been carried out. The experimental data are summarized in Table 5.1. Some data before 1975 are taken from the historical survey article of Michno and Findley (1976). The ratio τ_s/σ_s of the shear yield strength τ_s with tensile yield strength σ_s are also given.

Table 5.1. Summary and comparison of the yield criteria with experimental results

Researchers	Materials	Specimen	τ_s / σ_s	Suitable criterion
Guest 1900	Steel, brass, etc.	Tubes	0.474, 0.727	Tresca, no one agreed
Hancock 1906 1908	Mild steel, unannealed	Solid rods, tube	0.50–0.82	No one agreed
Scoble 1906	Mild steel	Solid rods	0.45–0.57	Tresca
Smith 1909	Mild steel	Solid rods	0.55–0.56	> Tresca
Turner 1909, 1911	Annealed steel	Tubes	0.460–0.572	
Turner 1909, 1911	Steels	Review work	0.55–0.65	von Mises to twin-shear
Mason 1909	Mild steel	Tubes	0.64	–
Scoble 1910	Steel	–	0.38–0.45	No one agreed
Becker 1916	Mild steel	Tubes	–	No one agreed
Seeley & Putnam 1919	Steels	Bars & tubes	0.6	> von Mises
Seigle & Cretin 1925	Mild steel	Solid bars	0.45–0.49	Tresca
Lode 1926	Iron, mild steel, nickle, copper	Tubes	–	von Mises
Ros & Eichinger 1926	Mild steel	Tubes	–	von Mises
Taylor & Quinney 1931	Aluminum, copper, Mild steel	Tubes	–	von Mises von Mises near twin shear
Marin 1936	Mild steel	Review work	–	No one agreed
Morrison 1940, 1948	Mild steel	Tubes	–	Tresca, von Mises
Davis 1945 Davis and Parker 1948	Copper, medium carbon steel	Tubes	–	von Mises
Osgood 1947	Aluminum alloy	Tubes	–	von Mises
Cunningham *et al.* 1947	Magnesium alloy	Tubes	–	von Mises
Bishop and Hill 1951	Polycrystals	Tubes	0.54	von Mises
Fikri and Johnson 1955	Mild steel	Tubes	–	> von Mises
Marin and Hu 1956	Mild steel	Tubes	–	von Mises
Naghdi et al. 1958	Aluminum alloy	Tubes	–	> von Mises
Hu and Bratt 1958	Aluminum alloy	Tubes	–	von Mises

Table 5.1. (Continue)

Researchers	Materials	Specimen	τ_s/σ_s	Suitable Criterion
Ivey 1961	Aluminum alloy	Tubes	o.66	Twin shear
Bertsch and Findley 1962	Aluminum alloy	Tubes	–	von Mises
Mair and Pugh 1964	Copper	Tubes	–	von Mises Twin shear
Miastkowski 1965	Brass	–	–	von Mises
Rogan 1969	Steel	tubes	0.5	Tresca
Pisarenko et al. 1969	Copper, Cr-steel	Low temperature	–	von Mises
Dawson 1970	Polycrystals	–	0.64	near Twin shear
Phillips et al. 1970, 1972	Aluminum	Elevated temperature	0.53	between Tresca and von Mises
Deneshi et al. 1976	Aluminum, Copper	Low temperature	0.6	> von Mises
Winstone 1984	Nickel alloy	Elevated temperature	0.7	Twin shear
Ellyin 1989	Titanium	Tubes	0.66	Twin shear
Wu and Yeh 1991	Aluminum stainless steel	Tubes	0.58 0.66–0.7	von Mises Twin shear
Ellyin 1993	Titanium	Tubes	0.62–0.7	–
Ishikawa 1997	Stainless steel	Tubes	0.6–0.63	> von Mises

The unified strength theory under $\sigma-\tau$ combined stress can be expressed as

$$f=\frac{2+b}{2+2b}\sqrt{\sigma^2+4\tau^2}+\frac{b}{2+2b}\sigma=\sigma_s, \quad \text{when } \sigma\geq 0 \tag{5-1a}$$

$$f=\frac{2+b}{2+2b}\sqrt{\sigma^2+4\tau^2}-\frac{b}{2+2b}\sigma=\sigma_s, \quad \text{when } \sigma\geq 0 \tag{5-1b}$$

The corresponding yield loci are shown in Fig. 5.4.

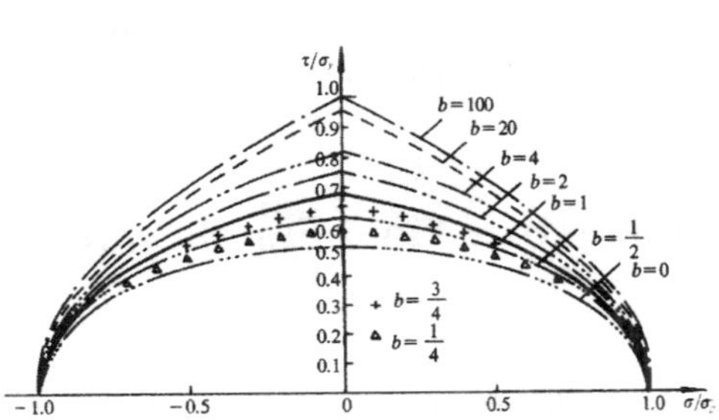

Fig. 5.4 Yield loci of the unified yield criterion under $\sigma-\tau$ combined stress state

It is seen that:

1. The unified yield criterion almost encompasses various yield criteria of materials under $(\sigma-\tau)$ combined stresses.
2. The yield loci are convex when the parameter b varies from 0 to 1. It should be noted that the yield loci in the three–dimensional stress state and the plane stress state will be nonconvex when $b>1$ or $b<0$. The nonconvex yield surfaces have been discussed in Chap. 3.
3. The differences between various yield functions are obvious under simple shear stress. The ratios of pure shear yield stress τ_s to tensile yield stress σ_s of some typical yield criteria are shown as follows: single-shear yield criterion (Tresca yield criterion): $\tau_s = 0.5\sigma_s$; Huber-von Mises yield criterion: $\tau_s=0.577\sigma_s$; twin-shear yield criterion: $\tau_s = 0.667\sigma_s$; maximum tensile strain theory: $\tau_s=0.769\sigma_s$ ($\nu=0.3$); maximum normal stress theory: $\tau_y=1.0\sigma_y$.

After the comparison of the shear yield strength and tensile yield strength among the 30 materials, Kishkin and Ratner (see Onksov 1963) divided the metals into four kinds according to the ratio of the shear yield strength to tensile yield strength τ_s/σ_s as follows:

1. $\tau_s/\sigma_s \cong 0.50$ (0.48 - 0.53, five materials), which agrees with the single-shear yield criterion (Tresca yield criterion) or with the unified yield criterion when b=0.
2. $\tau_s/\sigma_s \cong 0.58$ (0.54 - 0.62, nine materials), which agrees with the Huber-von Mises yield criterion, or with the unified yield criterion when $b=1+1/(1+\sqrt{3})$ or b=1/2.
3. $\tau_s/\sigma_s \cong 0.68$ (0.67 - 0.71, eight materials), which agrees with the twin-shear yield criterion, or with the unified yield criterion when $b=1$.
4. $\tau_s/\sigma_s < 0.40$ (0.31 - 0.41, eight materials), which gives a nonconvex result that does not agree with existing criteria. This kind of result is not indicated by any theoretical criterion, but matches the unified yield criterion when b= - 1/3.

With regard to the ratio of shear strength to tensile strength of metals, the values in some manuals are 0.52~0.63 (carbon steel) and 0.65~0.78 (alloy steel).

Generally, the ratio of shear yield stress τ_y to the tensile yield stress σ_s is different for different materials. According to the convexity of yield surface, shear yield strengths are in the range of $\tau_s=(0.5\sim0.667)\sigma_s$ for those metallic materials that have same strength both in tension and compression. Yield stresses of metallic materials are higher than $\tau_s/\sigma_s=0.5$.

Many yield stresses of high-strength steel, high-strength alloys, and non-metallic materials are higher than τ_s=0.667. These kinds of materials may be the SD (strength difference in tension and in compression) materials.

5.3 Reasonable Choice of a Yield Function for Non-SD Materials

Because of the significant differences in shear strengths of materials, the yield function can be selected on the basis of the ratio of the shear yield stress τ_s to the tensile yield stress σ_s.

According to the unified yield criterion

$$f=\sigma_1-\frac{1}{1+b}(b\sigma_2+\sigma_3)=\sigma_s,\quad \text{if}\quad \sigma_2\le\frac{1}{2}(\sigma_1+\sigma_3) \tag{5-2a}$$

$$f'=\frac{1}{1+b}(\sigma_1+\sigma b_2)-\sigma_3=\sigma_s,\quad \text{if}\quad \sigma_2\ge\frac{1}{2}(\sigma_1+\sigma_3) \tag{5-2b}$$

where b is a yield criterion parameter that represents the effect of the intermediate principal shear stress on the yield of materials and lies in the range of $0\le b\le 1$ when the yield surfaces are convex.

In the case of critical state in pure shear $\sigma_1=\tau$, $\sigma_2=0$, $\sigma_3=-\tau$ and $\tau=\tau_s$, we have

$$\sigma_1=\tau_s,\ \sigma_2=0,\sigma_3=-\tau_s,\ \text{and}\ \sigma_2=(\sigma_1+\sigma_3)/2=0 \tag{5-3}$$

hence, we can use the anyone of Eqs. (5-2a) or (5-2b).

Substituting Eq. (5-3) into the equation of the unified yield criterion Eq.(5-2a) or Eq. (5-2b), the relation among the tensile yield stress σ_s, shear yield stress τ_s and the parameter b in the unified yield criterion is obtained as

$$(2+b)\tau_s=(1+b)\sigma_s \tag{5-4}$$

The parameter b can be determined from the ratio of shear yield stress to tensile yield stress as follows:

$$b=\frac{2\tau_s-\sigma_s}{\sigma_s-\tau_s}=\frac{2\alpha_\tau-1}{1-\alpha_\tau}, \tag{5-5}$$

Inversely, the ratio of shear yield stress to tensile yield stress can be given as

$$\alpha_\tau=\frac{\tau_s}{\sigma_s}=\frac{b+1}{b+2} \tag{5-6}$$

The unified yield criterion establishes clear and simple relations among the various yield criteria. It also provides a method to choose the appropriate yield criterion. It can be obtained from the unified yield criterion as follows:

(1) First, we need know the tensile yield stress σ_s and shear yield stress τ_s;
(2) Second, the ratio of the shear yield stress to the tensile yield stress $\alpha_\tau=\tau_s/\sigma_s$ can be determined ;
(3) then, substituting α_τ into Eq. (5-5), the parameter b is determined;
(4) at last, substituting b into Eq. (5-1a) and (5-1b), an appropriate yield criterion is obtained.

For example, if the ratio of the shear yield stress to the tensile yield stress is $\alpha_\tau=\tau_s/\sigma_s=0.6$, the parameter b can be determined from Eq. (5-5) is b=1/2. Substituting

b=1/2 into Eq.(5-2a) and (5-2b), then an appropriate yield criterion is given as follows:

$$f=\sigma_1-\frac{1}{3}(\sigma_2+2\sigma_3)=\sigma_y, \qquad \text{when } \sigma_2\le\frac{1}{2}(\sigma_1+\sigma_3) \tag{5-7a}$$

$$f'=\frac{1}{3}(2\sigma_1+\sigma_2)-\sigma_3=\sigma_y, \qquad \text{when } \sigma_2\ge\frac{1}{2}(\sigma_1+\sigma_3) \tag{5-7b}$$

This means that a new yield criterion is introduced. This new yield criterion can be approximated to the Huber-von Mises yield criterion.

If the ratio of the shear yield stress to the tensile yield stress is $\alpha_\tau=\tau_s/\sigma_s=0.667$, the parameter b can be determined from Eq. (5-5) is b=1. Substituting b=1 into Eq.(5-2a) and (5-2b), the yield criterion is given as follows:

$$f=\sigma_1-\frac{1}{2}(\sigma_2+\sigma_3)=\sigma_s, \quad \text{if} \quad \sigma_2\le\frac{1}{2}(\sigma_1+\sigma_3) \tag{5-8a}$$

$$f'=\frac{1}{2}(\sigma_1+\sigma_2)-\sigma_3=\sigma_s, \quad \text{if} \quad \sigma_2\ge\frac{1}{2}(\sigma_1+\sigma_3) \tag{5-8b}$$

It is the twin-shear yield criterion proposed by Yu in 1961.

Various new yield criteria may be introduced when the ratio of the shear yield stress to the tensile yield stress is different.

5.4 Experiments for Iron under σ–τ Stress State

A combined σ–τ stress state can be produced in thin tube under axial force and torsion. For brittle materials, such as iron, the limit locus of the materials is asymmetrical in the τ–coordinate because its tensile strength is different from its compressive strength. Since the ratio of the tensile-compressive strength is different, two material constants are required, such as tensile strength σ_t and compressive strength σ_c, tensile strength σ_t and the tensile-compressive ratio $\alpha=\sigma_t/\sigma_c$, or compressive strength σ_c and the compression-tension strength ratio $m=\sigma_c/\sigma_t$. For example, the limit loci corresponding to different tension-compression strength ratios of twin-shear yield function are shown in Fig. 5.5. The comparison of the experimental results of iron with the twin-shear yield function are given.

A series of experiments of iron under σ–τ combined stress states were conducted by Grassir and Cornet (1949), Mair (1968) and Coffin (1950). Their experimental results are shown in Fig. 5.6.

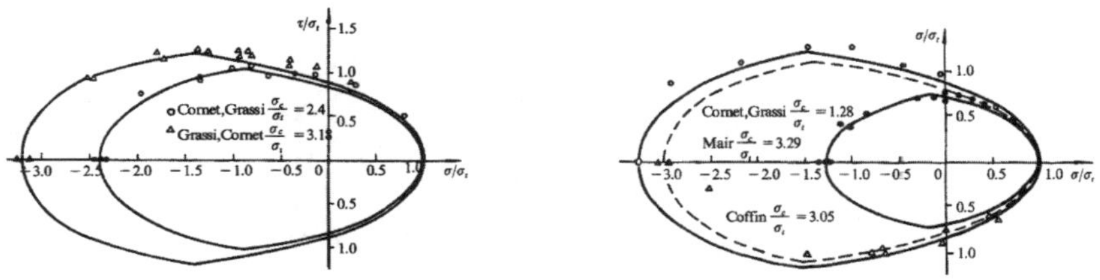

Fig. 5.5 Comparison between experimental results for iron and twin-shear strength theory (Grossir and Cornet 1949; Coffin 1950)

5.5 Experiments for Concrete under Complex Stress

The experimental results of concrete with the twin-shear yield function are shown in Fig.5.6.

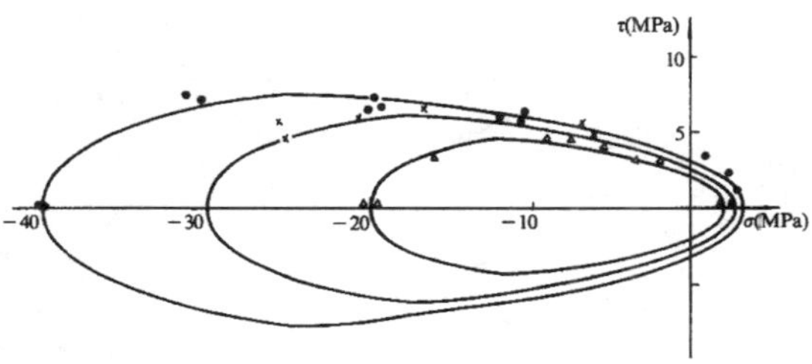

Fig. 5. 6 Comparison between experimental results for concrete and twin-shear stress theory

The primary experiments on limit loci in the π-plane on concrete were conducted by Gachon and Launay of the French National Institute of Technology. A series of limit loci in the π-plane for concrete are given under different hydrostatic pressures. Figs. 5.7 and 5.8 show the limit loci in the π-plane for concrete (Gachon 1972, Launay and Gachon 1973).

It can be seen from Figs. 5.7 and 5.8 that although smooth curves can approximate the results, the polygonal line in Fig. 5.8 is closer to the experimental results. Fig. 5.8 shows three groups of experimental limit loci and compares them with yield function. The limit locus of yield function shown in Fig. 5.8 is the unified strength theory with b=1/2. In order to enable comparison, the limit locus of the unified strength theory with b=0 and b=1 under $3p$=4 and 8 are also given. Although there are only two data points when $3p$=1, which cannot be compared with yield function, other experimental results all agree with the estimation of the unified strength theory with b=1/2. As to the experimental results in Fig. 5.8, they also agree with the unified strength theory with b=1/2.

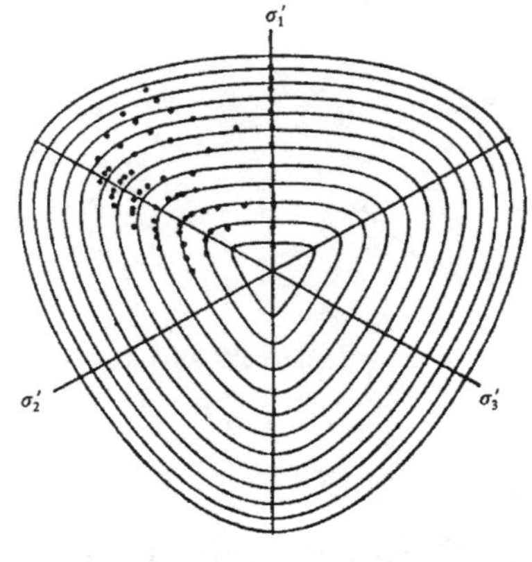

Fig. 5.7 Limit loci in the π-plane for concrete (Gachon 1972)

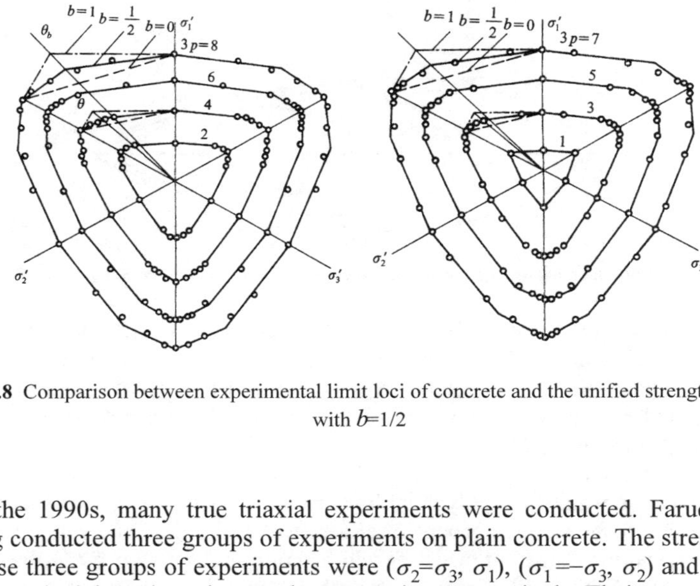

Fig. 5.8 Comparison between experimental limit loci of concrete and the unified strength theory with b=1/2

In the 1990s, many true triaxial experiments were conducted. Faruque and Chang conducted three groups of experiments on plain concrete. The stress states of these three groups of experiments were ($\sigma_2=\sigma_3$, σ_1), ($\sigma_1=-\sigma_3$, σ_2) and ($\sigma_1=\sigma_2$, σ_3), i.e., triaxial tension, shear and compression, respectively. Their experimental results can be connected as shown in Fig. 5.9. It can be seen that the experimental results agree with the unified strength theory with b=1. The experimental results for steel fiber concrete obtained by Dong et al. (1993) also agree with twin-shear stress theory.

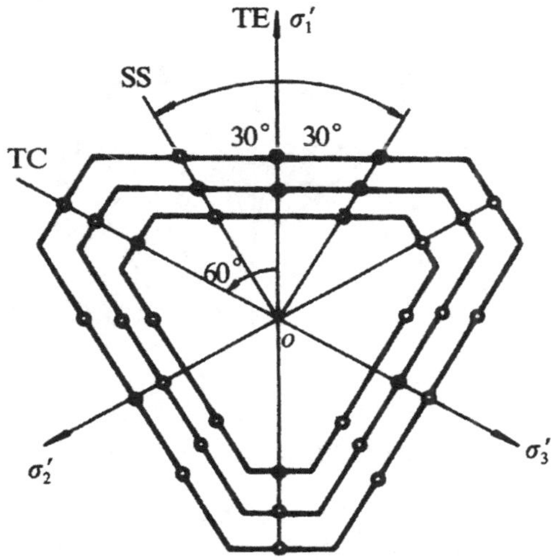

Fig. 5.9 Concrete (Faruque and Chang 1990)

From the above experimental results, it can be deduced that the limit loci of concrete in the π-plane lies between the unified strength theory when b=1/2 and b=1. There are lots of experimental data on meridian lines for concrete (see Chen WF 1982). It can be expressed approximately by a two-parameter criterion, giving a straight limit meridian line. Generally, it must be expressed by three-parameter, four-parameter or five-parameter criteria, giving curved limit meridian lines.

5.6 Experiments for Rock under Complex Stress

Michelis (1985, 1987) conducted many fine true triaxial experiments on rock and concrete. The results show that the effect of intermediate principal stress σ_2 is an important characteristic for geomaterials. According to the three series of experimental data (Michelis 1985, 1987) three limit loci may be drawn as shown in Fig. 5.10.

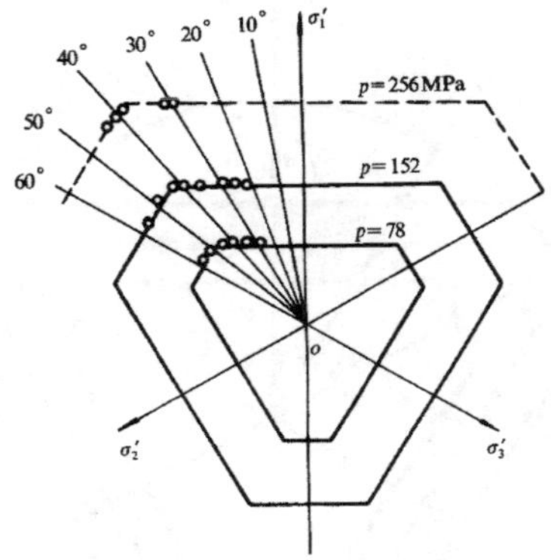

Fig. 5.10 Marble (Michelis 1987)

The biaxial experiments of rock have also been conducted at the same time. Fig. 5.11 shows the experimental result published by Amadei and Kuberan in the 24th Conference of Rock Mechanics of American in 1984. The dotted line in the figure is the limit line of Mohr-Coulomb strength theory in plane stress state. The solid line is the limit line of the unified strength theory with b=3/4. Compared with the experimental results, there is a large difference between the Mohr-Coulomb theory and the experimental results. Mohr-Coulomb strength theory is too conservative. The limit surface of the unified theory with b=3/4 is close to the experimental results, but in the region of biaxial compressive stress state, it is also too conservative.

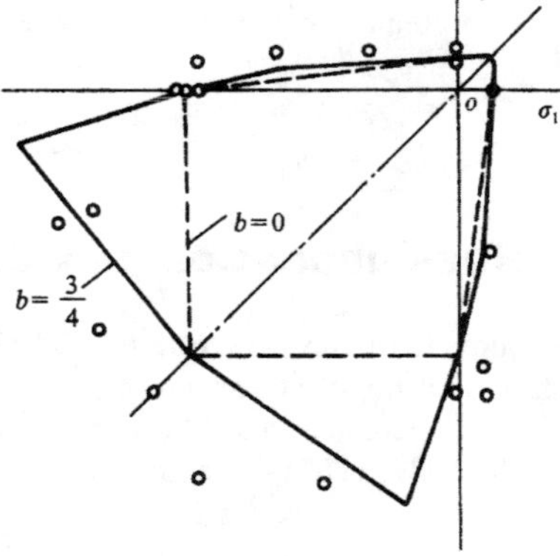

Fig. 5.11 Experimental results under biaxial condition for limestone (Amadei et al, 1984)

A granite with properties such as high hardness, high density, high strength and high elastic modulus can be found in the upstream of the Yellow River (China). Its uniaxial compressive strength is σ_c=157 MPa; it has elastic modulus G=50 Gpa, tensile strength σ_t=7.8 MPa, shear strength parameter C_0=16 MPa, $tg\varphi$=0.96, the unit weight r=2680 kg/m^3 and Possion's ratio $\nu = 0.2$.

A series of experiments on rock corresponding to different high pressures under the same stress angle were carried out by the Northwest Design and Research Institute (China) and the Wuhan Institute of Rock and Soil Mechanics of the Chinese Academy of Science. The failure strength of rock corresponding to different hydraulic pressures with the same stress angle can be obtained, and then the limit meridian lines of rock under this stress angle can then be obtained. Five stress angles θ =0°, 13.9°, 30°, 46.1°, 60° are adopted, and the corresponding five limit meridian lines are shown in Fig. 5.12. Six sections are adopted in Fig. 5.12 within the range of p=80–200 MPa, and then six limit loci in the π-plane corresponding to different values of p are obtained as shown in Fig. 5.12.

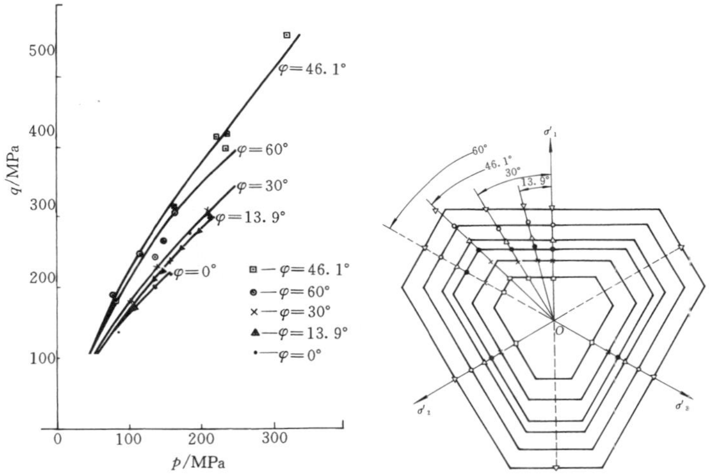

Fig. 5.12 Limit loci in meridian lines and the π-plane

It can be seen that:

1. The strength of granite increases with increase of the hydraulic pressure p.
2. The length of vector q differs corresponding to different stress angles θ in the π-plane when the hydraulic pressure p is constant. Granite shows an obvious stress angle effect. There is an obvious distinction between the circular limit loci of the Drucker–Prager criterion and the experimental results.
3. All the experimental points are located outside the limit loci of the Mohr–Coulomb yield function, and they are closer to that of the twin-shear yield function.
4. The limit meridian line changes with the stress angle.
5. In the process of varying stress angle from θ=0° to θ=60°, the value of q increases and reaches q=262.2 MPa, and then decreases to q =255 MPa. This result agrees with the twin-shear stress theory.

5.7 Experiments on Clay and Loess under Complex Stress

Soil is a widely used material in structural engineering and geotechnical engineering. The yield function is the basis of soil mechanics and foundation engineering. At present, the axisymmetric triaxial experiments on soil are the elementary tests in soil mechanics and have gradually developed into true triaxial experiments.

The early research on failure criteria for soils under true triaxial stress states or plane strain states was done by Shibata and Karube (1965) at Kyoto University, Roscoe (1968), Roscoe and Burland (1968), Wood and Roth (1972) at Cambridge University, Ko and Scott (1967) at Colorado State University, Brown and Casbarian (1965), Sutherland and Mesdary (1969) at the University of Glasgow, Bishop (1971) and Green (1972a) at Imperial College. Fig. 5.13 indicates the experimental results for normally consolidated soil carried out by Shibata and Karube (1965). The experimental results on normally consolidated soil lie almost between the unified strength theory with b=1/4 and b=1/2. They do not agree with the Mohr-Coulomb yield function.

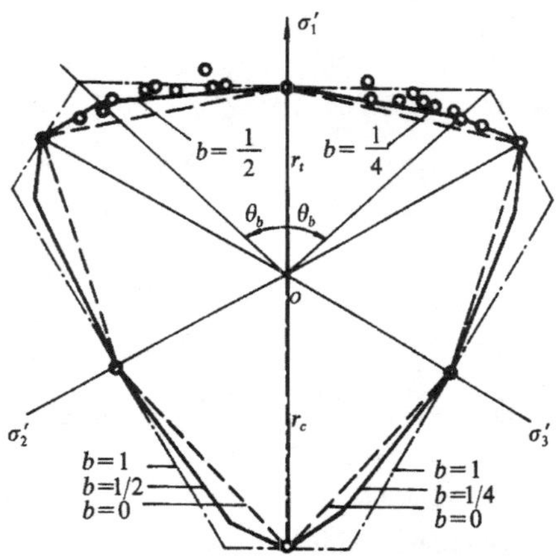

Fig. 5.13 Soil (Shibata and Karube 1965)

Fig. 5.14 shows the experimental results for compactive loess given by Fang in 1986. Fig. 5.15 shows the experimental results under true triaxial conditions for undisturbed loess and remolded loess given by Xing et al. at Xi'an Science and Technological University. The experimental results all agree with the unified strength theory with b=1/2.

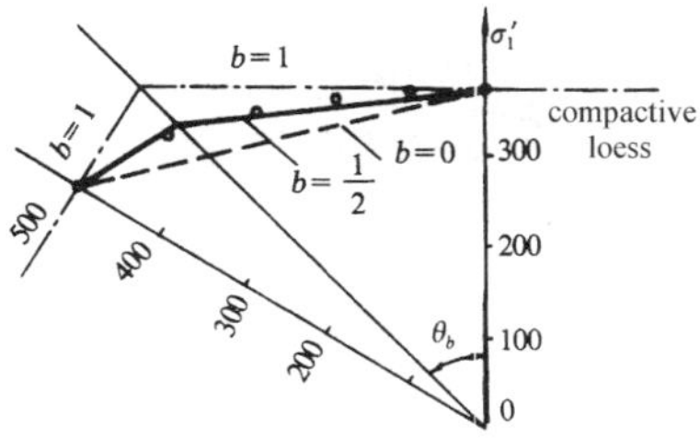

Fig. 5.14 Compacted loess (Fang 1986)

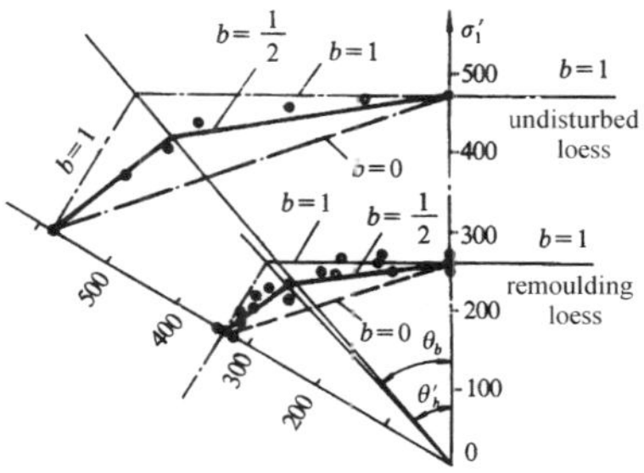

Fig. 5.15 Loess (Xing et al. 1992)

Recently, the limit loci in π-plane of undisturbed loess soils is obtained by Yoshimine et al. at Tokyo Metropolitan University, Tokyo, Japan. It is shown in Fig.5.16. It is in agreement with the twin-shear strength theory or the unified strength theory with b=1.

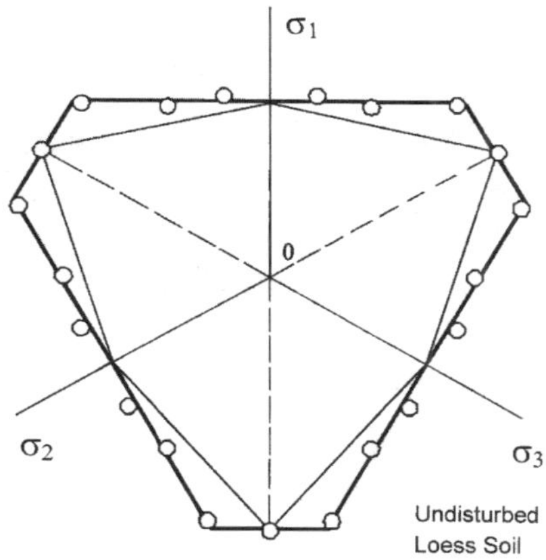

Fig. 5.16 Loci in π-plane of undisturbed loess soils (Yoshimine et al. 2004)

5.8 Experiments on Sand under Complex Stress

A well-known experiment on sand was carried out by Green and Bishop (1969). Fig. 5.17 shows the experimental results. Fig. 5.18 shows the dynamic strength for sand obtained by Zhang and Shao (1988). The results are close to the unified strength theory with b=3/4 and b=2/3, respectively.

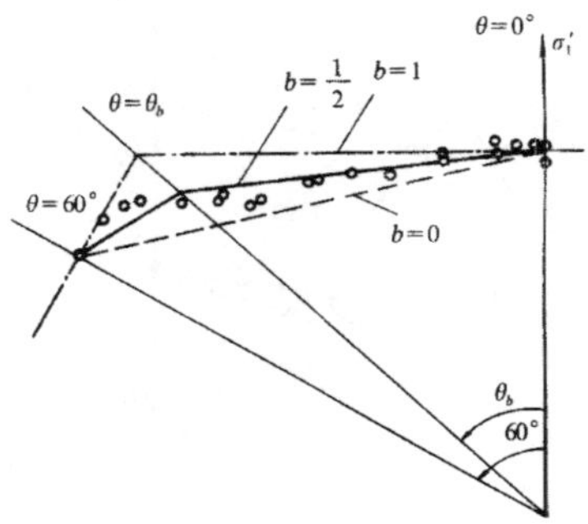

Fig. 5.17 Sand (Green and Bishop 1969)

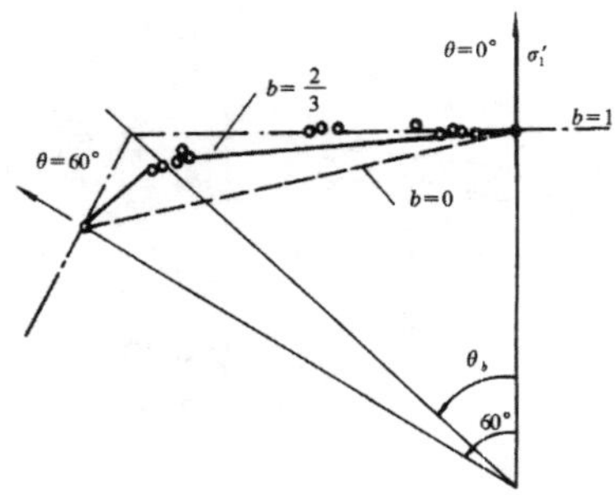

Fig. 5.18 Dynamic strength of sand (Zhang and Shao)

Fig. 5.19 shows the comparison between the unified strength theory (b=3/4) with experimental limit locus in the π-plane for sand obtained by Nakai and Matsuoda (1983) in Japan.

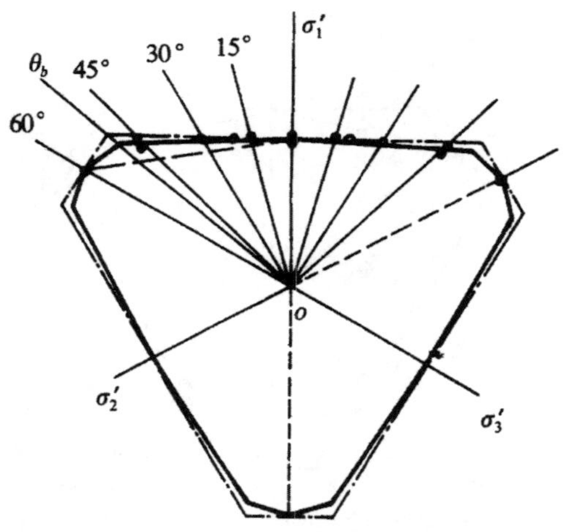

Fig. 5.19 Limit locus in the π-plane for Toyoura Sand (Nakai and Matsuoka)

Fig. 5.20 a, b shows the experimental results on loose sand and dense sand, given by Dakoulas and Sun (1992). The results agree with the unified strength theory with b=1/2 and b=3/4, respectively.

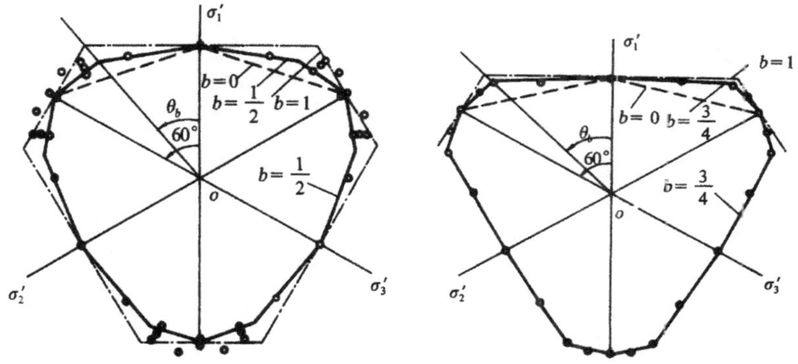

(a) loose sand (b) dense sand

Fig. 5.20 Limit loci for Ottawa fine sand: (a) loose sand; (b) dense sand (Dakoulas and Sun 1992)

The limit loci in the π-plane for sand obtained by Ko and Scott (1968) are shown in Figs. 5.21. The results agree with the unified strength theory with b=1/2.

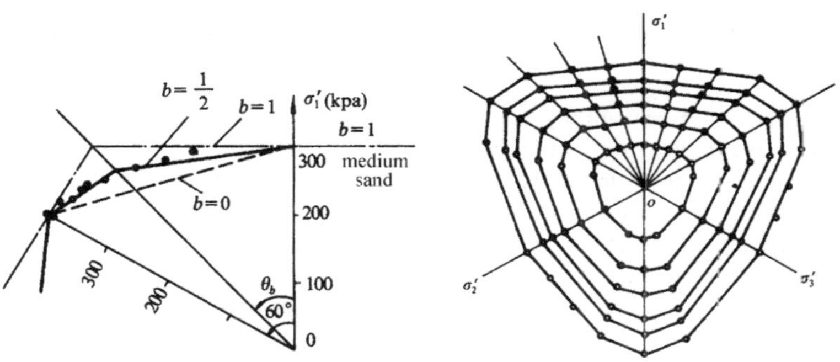

Fig. 5.21 Limit locus in the π–plane for medium sand (Ko and Scott 1968)

Fig. 5.22 shows the static and dynamic strength for saturated sand under complex stress states obtained by Zhang and Shao (1988).

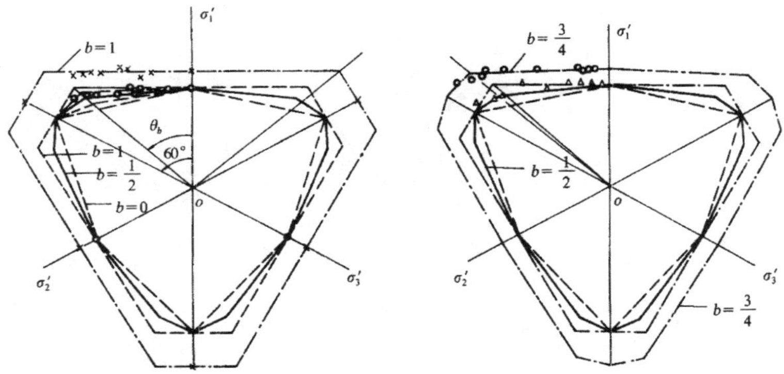

Fig. 5.22 Static and dynamic limit loci for saturated sand (Zhang and Shao 1988)

These experimental data are situated between the limit loci of the unified strength theory with b=1/2 and b=1.

5.9 Reasonable Choice of a Yield Function for SD- Materials

It is difficult to choose an appropriate failure criterion for SD (Strength Difference in tension and in compression) materials.

The unified strength theory (unified strength theory) establishes a clear and simple relation among the various yield criteria. It also provides a method to choose the appropriate yield criterion. The reasonable choice method can be obtained from the unified strength theory as follows.

First, we need know the tensile strength σ_t and compressive strength σ_c ;

Second, the ratio $\alpha = \sigma_t / \sigma_c$, i.e. the ratio of tensile strength σ_t to compressive strength σ_c, can be determined

Third, if the ratio $\alpha = \sigma_t / \sigma_c = 1$, it means that the tensile strength equals the compressive strength. We can use the unified yield criterion and choose a resonable parameter b, as described in section 5.3 this chapter.

At last, if the ratio $\alpha = \sigma_t / \sigma_c \neq 1$, or $\sigma_t \neq \sigma_c$, it is the material with SD effect (effect of Strength Difference in tension and in compression). We have to use the two-parameters yield function. The unified strength theory can be used as a two-parameters yield function. The mathematical equation is expressed as follows:

$$F = \sigma_1 - \frac{\alpha}{1+b}(b\sigma_2 + \sigma_3) = \sigma_t, \quad \text{when } \sigma_2 \le \frac{\sigma_1 + \alpha\sigma_3}{1+\alpha} \tag{5-9a}$$

$$F = \frac{1}{1+b}(\sigma_1 + b\sigma_2) - \alpha\sigma_3 = \sigma_t, \quad \text{when } \sigma_2 \ge \frac{\sigma_1 + \alpha\sigma_3}{1+\alpha} \tag{5-9b}$$

It is linear and simple. The difficulty is how to choose the failure criterion parameter b. A method is using the true triaxial test for every material, however, it is expensive and difficult. One simple method is choosing the intermediate value of b, which is a mediated criterion between the lower bound (the Mohr-Coulomb strength theory) and upper bound (the twin-shear strength theory). The mathematical equation is also linear and simple, it is shown as follows:

$$F = \sigma_1 - \frac{\alpha}{3}(\sigma_2 + 2\sigma_3) = \sigma_t\,, \ \sigma_2 \le \frac{\sigma_1 + \alpha\sigma_3}{1+\alpha} \tag{5-10a}$$

$$F' = \frac{1}{3}(2\sigma_1 + \sigma_2) - \alpha\sigma_3 = \sigma_t\,, \ \sigma_2 \ge \frac{\sigma_1 + \alpha\sigma_3}{1+\alpha} \tag{5-10b}$$

This is a new failure criterion deduced from the unified strength theory with b=1/2. Since the Drucker–Prager criterion cannot match with the practice, this criterion is more reasonable and can be substituted for the Drucker–Prager criterion.

According to the experimental data described in sections 5.4 to 5.8. The strength theory parameter b can be recommended as follows:

b=1/2 to b=1 for concrete, rock and high strength steel and alloy; i.e. $1/2 \le b \le 1$

b=1/4 to b=3/4 for soils and soft rock, i.e. $1/4 \le b \le 3/4$.

5.10 The Beauty of the Unified Strength Theory

What is the beauty of a thing? Beauty is the qualities that give pleasure to the senses or lift up the mind or spirit (Longman Dictionary of Contemporary English 1978). Beauty is the combination of qualities that give pleasure to the senses or to the moral sense or the intellect (The Oxford Advanced learner's Dictionary of Current English 1963). A thing of beauty is a joy for ever (Keats, John 1795–1821).

What is the beauty of science? Six elements of the beauty of mathematics and physics were described by Tzanakis (1997), Kosso (1999), Shen (2004) and others. Six relevant points about the beauty of mathematics and physics are:

(a) Clarity: the conceptual clarity in the foundations and the development of a theory.
(b) Simplicity, Elegant and economic reasoning. It is also said that: "Science advocates simplicity" (Shen ZJ 2004).
(c) Unification of a priori unrelated concepts, methods, theories or phenomena.
(d) The "naturality" of a theory.
(e) Symmetry.
(f) Analogy

The beauty of the unified strength theory was first proposed and studied by Fan, et al. (2001) of Nanyang Technological University in Singapore. Fan, et al. (2001) a lecture at an International Congress on Computational Mechanics held at Sydney in 2001, Australia. The power and unification of the unified strength theory are discussed.

A closing lecture was delivered by Yu et al. at the International Symposium on the Developments in Plasticity and Fracture, Centenary of M.T.Huber Criterion, in Cracow, Poland (Yu et al. 2004). The Beauty of Strength Theories was discussed.

The unified strength theory was derived based on the concept of multiple slip mechanism and the multi-shear element model or twin-shear element. Multi-shear element and twin-shear element are spatial equipartition available for continuum mechanics, which are shown in Fig. 5.23.

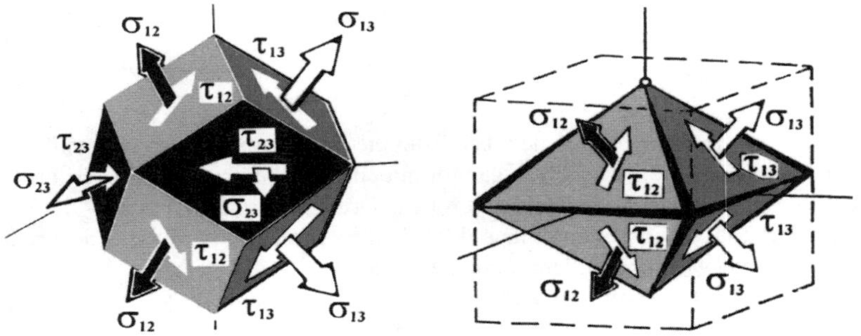

Fig. 5.23 Multi-shear element and twin-shear element

It is obvious that this element model is different from the principal stress element used in common continuum mechanics. There are three sets of principal shear stress and normal stress acted on the sections, on which the principal shear stress is acted respectively. However, there are only two independent components in three principal shear stresses, because the maximum shear stress τ_{13} equals the sum of the other two, *i.e.* $\tau_{13} = \tau_{12} + \tau_{23}$.

Considering the two larger principal shear stresses and the corresponding normal stress and their different effects on failure of materials, the mathematical modeling of the unified strength theory can be formulated as (Yu-He 1991; Yu 1992, 1994)

$$F = \tau_{13} + b\tau_{12} + \beta(\sigma_{13} + b\sigma_{12}) = C \text{, when } \tau_{12} + \beta\sigma_{12} \geq \tau_{23} + \beta\sigma_{23} \quad (5\text{-}11a)$$

$$F' = \tau_{13} + b\tau_{23} + \beta(\sigma_{13} + b\sigma_{23}) = C \text{, when } \tau_{12} + \beta\sigma_{12} \leq \tau_{23} + \beta\sigma_{23} \quad (5\text{-}11b)$$

where b is a coefficient of the effect of the other principal shear stresses on the strength of materials. Introducing a tension-compression strength ratio $\alpha = \sigma_t / \sigma_c$. The unified strength theory is expressed in terms of three principal stresses as follows

$$F = \sigma_1 - \frac{\alpha}{1+b}(b\sigma_2 + \sigma_3) = \sigma_t, \quad \text{when } \sigma_2 \leq \frac{\sigma_1 + \alpha\sigma_3}{1+\alpha} \quad (5\text{-}12a)$$

$$F' = \frac{1}{1+b}(\sigma_1 + b\sigma_2) - \alpha\sigma_3 = \sigma_t, \quad \text{when } \sigma_2 \geq \frac{\sigma_1 + \alpha\sigma_3}{1+\alpha} \quad (5\text{-}12b)$$

The mathematical expression of the unified strength theory is simple and linear, but it has rich and varied contents, and can be easily changed to suit many new conditions. It possesses fundamentally all the expected characteristics for a unified strength theory.

A series of limit loci of the unified strength theory on the deviatoric section are shown in Fig. 2 and Fig. 3 in Preface of this book. Several characteristics of the unified strength theory are discussed as follows:

(1) Linearity and Simplicity

It is physically meaningful and can be expressed by mathematically simple equation. The formulation of the unified strength theory is linear and convenient to use in applications to analytical solution of plasticity and plastic analysis of structure and computational implementation for numerical solution.

(2) Unification

According to Tzanakis (1997) one of six relevant points about the beauty of mathematics and physics is unification of a priori unrelated concepts, methods, theories or phenomena.

All the yield criteria and failure criteria are single criterion having various model and expression. The limiting loci of the unified strength theory cover all regions from the Mohr-Coulomb theory to the twin-shear strength theory. The unified strength theory is not a single criterion. It is a series of failure criteria, a system of strength theory. This theory gives a series of new failure criteria, establishes a relationship among various failure criteria, and encompasses previous yield criteria, failure models and other smooth criteria or empirical criteria as special cases or linear approximations. The famous Tresca yield criterion, the Mohr-Coulomb strength theory, the twin-shear yield criterion, the twin-shear strength theory and some others are special criteria of the unified strength theory.

The unification can be seen from Figs. 2 and 3 in the Preface and Fig. 4.22 in Chapter 4. The relation among the single-shear theory, the twin-shear theory and a series of new failure criteria can be seen in Fig. 5.24.

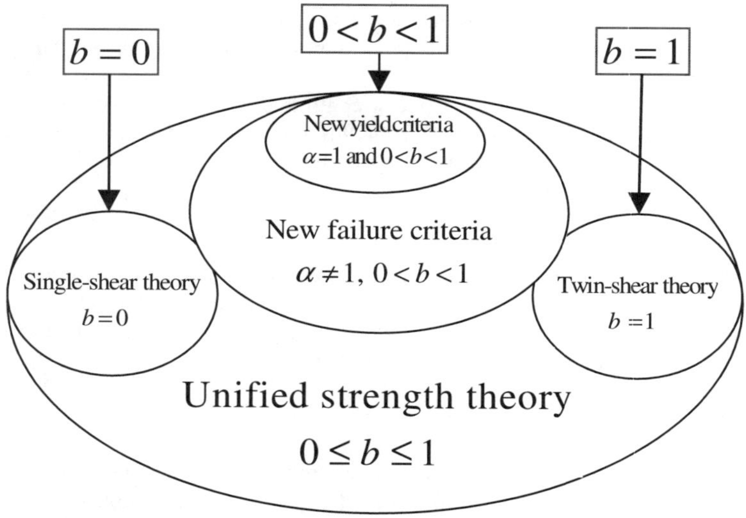

Fig. 5.24 Relation among various yield criteria

(3) Clarity and Naturality

The unified strength theory is a result of continual studies for 30 years from 1961 to 1991. It is the natural developments of the twin-shear yield criterion (1961), generalized twin-shear strength theory (1985) and multi-parameter twin-shear criteria (1988-1990) to the unified strength theory (1991). The SD effect (strength difference at tension and compression), the effect of hydrostatic stress, the effect of intermediate principal stress and the effect of intermediate principal shear-stress on failure of materials are taken into account in the unified strength theory.

The limit locus of the twin-shear strength theory at π plane is the upper (external) bound of all the convex limit loci.

The non-convex failure surfaces can also be given from the unified strength theory when b<0 or b>1. This kind of failure criterion has not been studied before.

(4) Symmetry

The mechanical model of the unified strength theory can be illustrated by a multi-shear element model or twin-shear element, they are symmetric as shown in Fig.5.22. The yield surface and yield loci of the unified strength theory are also symmetric.

(5) Analogy

There is an analogy between the unified strength theory and previous failure criteria. It is the same when we make a comparison between the Mohr-Coulomb theory and the unified strength theory with b=0. It is also the same when we make a comparison between the twin-shear strength theory and the unified strength theory with b=1. A series of yield criteria and failure criteria can be introduced from the unified strength theory. They are ranged between the single-shear theory and the twin-shear theory.

This unified strength theory agrees with experimental results over a wide range of stress state for many materials including metal (when α=1), rock, soil, concrete and others.

The unified strength theory forms an entire spectrum of convex and non-convex criteria, which can be used to describe many kinds of engineering materials.

The unified strength theory is beauty in its clarity, its simplicity and linear, its analogy, its symmetry and its unification.

Summary

The experimental verification and reasonable choice of failure criteria for various materials are discussed in this chapter.

The suitability of any proposed yield criterion must be examined by experiment results. A comprehensive description of the experimental investigations on yield surfaces may be found in the literatures. Experimental results from different researchers from 1900 to 1997 are briefly summarized in Table 4.1. The differences between various materials are great. No single yield criterion was found to be fully adequate. The single-shear yield criterion (Tresca yield criterion) can be adopted only for non-SD materials and the ratio of shear yield stress τ_y to the tensile yield stress σ_y is τ_y/σ_y=0.5. The Huber-von Mises yield criterion can be adopted only for non-SD materials with the ratio τ_y/σ_y =0.577. The twin-shear yield criterion (Yu 1961) can be adopted only for non-SD materials with the ratio τ_y/σ_y =0.667. The values of the ratio of shear yield stress to tensile yield stress (τ_y/σ_y=0.4~0.7) differ from the values of 0.50, 0.577 and 0.667 expected from the the Tresca yield criteria, the Huber-von Mises yield criterion and the twin-shear yield criterion, respectively. Clearly, no single yield criteria can accurately model the yield behaviour of various materials. All the yield criteria, including the Tresca yield criterion, the Huber-von Mises yield criterion, the twin-shear yield criterion and the unified yield criterion, are one-parameter criteria.

The yield surfaces of the unified yield criterion (Yu et al. 1991, 1992) cover all the convex regions and are extended to the non-convex region. Therefore, it can match most experimental data. More experimental data regarding the yield of materials under complex stress are expected.

High-strength steel, high-strength alloys, and most nonmetallic materials, such as polymers, ceramics, rock, concrete and soil are dependent on hydrostatic pressure, and their yield stresses in tension and compression are not identical. Therefore a generalized yield function or two-parameters failure criteria are needed for these materials. Reasonable choice of a yield function for geomaterial is more complex than the metallic materials with the identical strength in tension and in compression. The unified strength theory is compared with various experimental data. The choice of failure criteria parameter b and the beauty of the unified strength theory are discussed.

The study of the strength of materials under complex stress states is complicated both in theory and in tests. The experimental verification of strength theories is of paramount importance. If one proposes a failure criterion, a material model or a yield function, it is better that it is verified by others. The independent proofs of strength theories are of great importance.

Problems

Problem 5.1.
What is the lower bound of the convex failure criteria adopted for hydrostatic stress-sensitive material?

Problem 5.2.
What is the upper bound of the convex failure adopted for hydrostatic stress-sensitive material?

Problem 5.3.
What is the lower bound of the convex yield criteria for those materials in which the yield stress in tension and compression are equal?

Problem 5.4.
Compare the limit loci for various strength theories on the deviatoric plane.

Problem 5.5.
What is the upper bound of the convex yield criteria for those materials in which the yield stress in tension and compression are equal?

Problem 5.6.
Why can we not use failure criteria arbitrarily?

Problem 5.7

Explain why we have to determine the stress state condition $\sigma_2 \le \frac{\sigma_1 + \alpha\sigma_3}{1+\alpha}$ or $\sigma_2 \ge \frac{\sigma_1 + \alpha\sigma_3}{1+\alpha}$ when we using the unified strength theory?

Problem 5.8.
How do you choose between the two equations in the unified strength theory?

Problem 5.9.
What is the result if you use the second equation of the unified strength theory for the stress state of $\sigma_2 \le \frac{\sigma_1 + \alpha\sigma_3}{1+\alpha}$.

Problem 5.10.

What is the result if you use the first equation of the unified strength theory for the stress state of $\sigma_2 \geq \frac{\sigma_1 + \alpha\sigma_3}{1+\alpha}$.

References and Bibliography

Annin BD and Zygalkin BM (1999) Behaviour of Materials under Complex Loading. Science Academic of Russia, Novosibirsk (in Russia)

Bell JF (1973) Mechanics of Solids, Vol.1: The experimental foundations of solid mechanics. In:*Encyclopedia of Physics,*Vol.6a/1,Springer, Berlin, 483-512, 666-690.

Butterfield R and Harkness RM (1971) The kinematics of Mohr-Coulomb materials. In :*Stress-Strain Behaviour of Soils,* Parry RHG ed., Foulis Co.Ltd, 220-233.

Buyukozturk O, Nilson AH and Slate FO (1971) Stress-strain response and fracture of a concrete model in biaxial loading, *ACI Journal,***68**(8), 590-599.

Coffin LF (1950) The flow and fracture of a brittle material. *J.Appl. Mech. ASME,***72**, 233-248.

Cook G (1932) The elastic limit of metals exposed to tri-axial stress. *Proc. Roy. Soc. London* **137**, 559.

Davis EA (1945) Yielding and fracture of medium-carbon steel under combined stress. *J. Appl. Mech.,***12**(1), 13-24.

Davis HE and Parker ER (1948) Behavior of steel under biaxial stress as determined by tests on tubes. *J. Appl. Mech.* **15**, A201.

Dimaggio FL and Sandler IS (1971) Material model for granular soils, *J. Engrg. Mechanics,* ASCE, **97**(3), 935-950.

Dorn JE (1948) Effect of stress state on the fracture strength of metals. In: *Fracturing of Metals,* ASM, 1948, 32-50.

Ellyin F (1993) On the concept of initial and subsequent yield loci. In: Boehler JP ed. (1993) *Failure Criteria of Structured Media.* A.A. Balkema, Rotterdam, 293–304.

Fan SC and Qiang HF (2001) Normal high-velocity impact concrete slabs-a simulation using the meshless SPH procedures. *Computational Mechanics–New Frontiers for New Millennium.* Valliappan S and Khalili N eds. Elsevier, Amsterdam, pp 1457–1462.

Faruque MO and Chang CJ.(1990) A constitutive model for pressure sensitive materials with particular reference to plain concrete. *Int.J. Plasticity*, 6(1),29-43.

Gensamer M (1940) Strength of metals under combined stresses.*Trans.Am.Soc. Metals*, **28**, 38-60.

Grassir RC, and Cornet I (1949) Fracture of gray cast-iron tubes under biaxial stresses, *J. Appl. Mech.* ASME, **71**, 178–182.

Guest JJ (1900) On the strength of ductile materials under combined stress. *Phil. Mag. and J. Sci.* 1900, 69–133.

Guest JJ (1940) Yield surface in combined stress. *Phil.Mag.*Ser.7, **30**, 349–369.

Green GE (1971) Strength and deformation of sand measured in an independent stress control cell. In : *Stress-Strain Behaviour of Soils,*Parry RHG ed., Foulis Co.Ltd, 285–323.

Harkness RM (1971) An essay on 'Mohr-Coulomb'. In : *Stress-Strain Behaviour of Soils,* Parry R.H.G. ed., Foulis Co.Ltd, 212–219.

Hjelm HE (1994) Yield surface for grey cast iron under biaxial stress, *J. Engrg. Materials and Technology,* ASME , 116, 148–154

Hancock EL (1908) Results of tests on materials subjected to combined stresses. *Phil. Mag.* **11**, 276 (1906); 12, 418 (1906); 15, 214 (1908); 16, 720 (1908)

Ivey HJ (1961) Plastic stress-strain relations and yield surfaces for aluminium alloys, *J. of Mech. Eng. Sci.,* **3**(1), 15–31.

Ko HY and Scott RF (1967) A new soil testing apparatus. *Geotechnique* **17**(1).

Kosso P (1999) Symmetry Arguments in Physics. *Studies in History and Philosophy of Sciences,* 30(3), 479-492.

Launay P and Gachon H (1972) Strain and ultimate strength of concrete under triaxial stress. Am. Concrete Inst.Spec. Publ. 34, paper 13 ; in: *Chinese translation Collect of Strength and Failure of Concrete* , Hydraulic Press, Beijing, 1982, 247–265.

Li XC and Xu DJ (1990) Experimental verification of the twin shear strength theory--true triaxial teest research of strength of the granite in a large power station at Yellow River (in Chinese), Institute of Rock and Soil Mechanics, Chinese Academy of Sciences, *Research Report (Rock and Soil)*,1990–52

Lode W (1926) Versuche ueber den Einfluss der mittleren Hauptspannung auf das fliessen der metals eisen kupfer und nickel. *Z. Physik* **36**, 913–939.

Mair WM (1968) Fracture criterion for cast iron under biaxial stresses. *J. of Strain Analysis,* **3**, 254–263.

Marin J (1935) Failure theories of materials subjected to combined stresses, *Proc. Am. Soc. Civ. Engrs.*, **61**, 851-867.

Matsuoka H, Hoshikawa T and Ueno K (1990) A general failure criterion and stress-strain relation for granular materials to metalls, *Soils and Foundations,* JSSMFE, 30(2),119-127.

Michelis P (1985) Polyaxial yielding of granular rock, *J. of Eng. Mech.,* ASCE, **111**(8), 1049-1066.

Michelis P (1985) A true triaxial cell for low and high-pressure experiments.. *Int.J. Rock Mechanics and Geomech, Abstract,* **22**, 183–188.

Michelis P (1987) True triaxial cycle behavior of concrete and rock in compression, *Int. J. of Plasticity,* **3**, 249–270.

Michino MJ and Findley WN (1976) An historical perspective of yield surface investigation for metals, *Int.J.Non-linear Mechanics,* **11**(1), 59-82.

Morrison JLM (1948) The criterion on yield of gun steels, *Proc. Institution of Civil Engineers,* **159**,81-94.

Nakai T and Matsuoka H (1983) Shear behaviors of sand and clay under three-dimensional stress condition. *Soils and Foundations* (*Japan*), **23**(2), 26-42.

Paul B (1968) Macroscopic criteria for plastic flow and brittle fracture. In : *Fracture, An Advanced Treatise,* ed by Liebowitz,H.,Vol.2 Academic, New York, 313-496.

Pisarenko GS and Lebedev AA (1976) *Deformation and strength of material under complex stressed state* (in Russian), Naukova Dumka, Kiev.

Pugh H.D.LI, Mair WM and Rapier AC (1964) An apparatus for combined-stress testing in the plastic range, *Exptl Mech.* 4, 281.

Roscoe KH and Burland JB (1968) On the generalized stress-strain behaviour of wet clay. In: *Engineering Plasticity*, Cambridge University Press, Cambridge, 535-609.

Roscoe KH ed (1968) *Engineering Plasticity*, Cambridge University Press, Cambridge.

Scoble WA (1906) The strength and behavior of ductile materilas under combined stress. *Phil. Mag. and J.Sci.* 1906,533-547.

Scoble WA (1910) Ductile materilas under combined stress.*Phil. Mag. and J.Sci.* 1910, 116-128.

Shen ZJ (2004) Science advocates simplicity. *Chinese Journal of Geotechnical Engineering,* 26(2):299–300 (in Chinese).

Smith CAM (1909) Compound stress experiments, *Proc. Instn. Mech. Engrs* **4,** 1237.

Smith CA (1909) Some experiments on solid steel bars under combined stress. *Engineering,* **20,**238-243.

Sutherland HB and Mesdary MS (1969) The influence of the intermediate principal stress on the strength of sand. In : *Proceedings of 7th Int.Conf. on Soil Mechanics and Foundation Engineering* (Mexico City),Vol.1 , 391-399.

Taylor GI and Quinney H (1931) The plastic distortion of metals. *Phil. Trans. Roy. Soc.* A **230**, 323–362.

Tzanakis C (1997) The quest of beauty in research and teaching of mathematics and physics: A historical approach. *Non-linear Analysis, Theory, Methods & Applications*, 30(4): 2097-2105.

Winstone MR (1984) Influence of prestress on the yield surface of the cast nickel superalloy Mar-M002 at elevated temperature. *Mechanical Behavour of Materials-4* (ICM-4), Carlsson J and Ohlson NG eds. Pergamon Press, Vol. 1, pp 199–205.

Wu HC and Yeh WC (1991) On the experimental determination of yield surface and some results of annealed 304 stainless steel, *Int. J. of Plasticity*,7, 803 .

Xing RC, Liu ZD and Zheng YR (1992) A failure criterion of loess (English Abstract), *J. of Hydraulic Engineering* (1),12-19.(in Chinese) .

Yu Mao-Hong (2002) Advance in strength theory of material and complex stress state in the 20th Century. *Applied Mechanics Reviews* 55(3):169–218.

Yu Mao-Hong (2004) *Unified Strength Theory and Its Applications*. Springer, Berlin, New York,

Yu Mao-hong, Yoshimine M, Oda Y et al. (2004) The Beauty of Strength Theories— Closing Lecture at the *International Symposium on the Developments in Plasticity and Fracture*, Centenary of M.T.Huber, Cracow, Poland .

Zhang JM and Zhao SJ (1988) Dynamic strength criterion on sands under the three-dimmentional condition (in Chinese, English Abstract). *J.of Hydraulic Eng.,* 1988, (3):54-59.

6 Elasto-plastic Constitutive Relations

6.1 Introduction

The previous chapters described the yield functions and the corresponding yield surfaces that characterize material initial yielding. The post-yielding, loading-and-unloading behavior of the materials is related to the stress-strain relation for plastically deformed solids, namely the constitutive relations for plastic deformation of engineering materials. Classical plasticity theory discusses plastic flow rules such as Levy-Mises and Prandtl-Reuss equations, Drucker's stability postulate, isotropic, kinematic and combined hardening rules, and derives general stress-strain relations for plastic deformation of different materials.

In this chapter, the solving of singularity of piece-wise linear yield functions, the associated flow rule and the incremental constitutive relations for the unified strength theory will be mainly developed.

6.2 Plastic Deformation in Uniaxial Stress State

For a stress-strain curve shown in Fig. 6.1 of an elasto-plastic material, the behavior can be characterized an elastic region with an elastic modulus E until yielding commences at the axial yield stress σ_Y, and a plastic region with a continually varying local tangent E_T to the curve. E_T is termed as the elasto-plastic tangent modulus. The hardening law $k = k(\kappa)$ can be readily derived in terms of the plastic work done for the material for strain hardening hypothesis. In the elastic region, the stress-strain relation has a linear form,

$$\sigma = E\varepsilon \tag{6-1}$$

In the plastic region, the total strain increment strain in the uniaxial stress state is the sum of the elastic strain increment and the plastic strain increment, i.e.

$$d\varepsilon = d\varepsilon^e + d\varepsilon^p \tag{6-2}$$

Assuming that the plastic deformation is rate insensitive, the stress increment is linearly related to the elastic strain increment in the plastic region and can be expressed by

$$d\sigma = Ed\varepsilon^e = E(d\varepsilon - d\varepsilon^p) \tag{6-3}$$

The initial yield point σ_Y differentiates the elastic and plastic regions. The stress in the plastic region can be determined by a hardening rule,

$$\sigma = \sigma_Y(\kappa) \tag{6-4}$$

where κ is a hardening parameter. In the uniaxial stress state, the plastic strain ε^p is usually used for the hardening parameter, i.e. $\kappa=\varepsilon^p$. The plastic strain ε^p is history or path dependent, it can be calculated by

$$\varepsilon^p = \int d\varepsilon^p \tag{6-5}$$

Considering the strain decomposition Eq. (6-2), it can be derived that

$$d\varepsilon^p = d\varepsilon - d\varepsilon^e = \frac{d\sigma}{E_T} - \frac{d\sigma}{E} = \left(\frac{1}{E_T} - \frac{1}{E}\right)d\sigma \tag{6-6}$$

The tangent modulus E_T is considered to be a function of stress and plastic strain ε^p,

$$E_T = E_T(\sigma, \varepsilon^p) \tag{6-7}$$

which should be determined experimentally from a simple uniaxial yield test. Based on Eqs. (6-6) and (6-7), an incremental constitutive relation of the material can thus be derived.

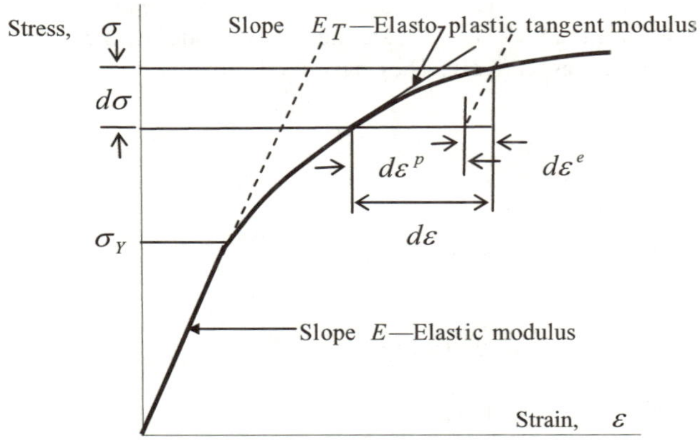

Fig. 6.1 Uniaxial elasto-plastic stress strain curve

The constitutive relation differs for plastic loading and elastic loading or unloading. It is necessary to identify the process as belonging to plastic loading or elastic unloading. Loading and unloading represent a deformation process starting from a plastic state and continuing to deform plastically and then returning to the elastic region. A previous plastic state is always implied. The loading and unloading criterion in the uniaxial stress state can be represented as

$$\sigma d\sigma \geq 0 \quad \text{for loading} \tag{6-8}$$

$$\sigma d\sigma < 0 \quad \text{for unloading} \tag{6-9}$$

It is seen that the fundamental elements of the plastic deformation includes initial yielding of the material, strain hardening and subsequent yielding, incremental constitutive equation and loading and unloading criterion, etc. In three-dimensional case, the constitutive equations can be represented in tensor notation, which will be discussed in the following sections.

6.3 Three-dimensional Elastic Stress-strain Relation

The plastic theory in the uniaxial stress can be extended to the three-dimensional case. The elastic stress-strain relationship in the three-dimensional case can be given by the generalized Hooke's law in the Cartesian coordinate system for isotropic materials. In tensor notations, it has the form of

$$\varepsilon_{ij} = \frac{\sigma_{ij}}{2G} - \frac{3v}{E} p\delta_{ij} \tag{6-10}$$

where σ_{ij} is the Kronecker delta and $p=\sigma_{kk}/3$ is the mean stress or hydrostatic pressure in the material. E, G and v are the elastic modulus, the shear modulus and the Poisson's ratio, respectively. They have the following relation,

$$G = \frac{E}{2(1+v)} \tag{6-11}$$

Eq. (6-10) can be rewritten as

$$\begin{aligned} \varepsilon_{ij} &= \frac{1}{E}\left[(1+v)\sigma_{ij} - 3vp\delta_{ij}\right] \\ &= \frac{1}{E}\left[(1+v)\delta_{ik}\delta_{jl} - v\delta_{ik}\delta_{kl}\right]\sigma_{kl} \\ &= M_{ijkl}\sigma_{kl} \end{aligned} \tag{6-12}$$

where $M_{ijkl} = \frac{1}{E}\left[(1+v)\delta_{ik}\delta_{jl} - v\delta_{ik}\delta_{kl}\right]$. The elastic stress tensor can then be deduced from Eq. (6-12) as

$$\begin{aligned}\sigma_{ij} &= \frac{E}{1+\nu}\left[\frac{3\nu}{1-2\nu}\varepsilon_m\delta_{ij}+\varepsilon_{ij}\right]\\ &= \frac{E}{1+\nu}\left[\delta_{ik}\delta_{il}+\frac{\nu}{1-2\nu}\delta_{ij}\delta_{kl}\right]\varepsilon_{kl}\\ &= C_{ijkl}\varepsilon_{kl}\end{aligned} \tag{6-13}$$

in which C_{ijkl} is the stiffness tensor of the fourth order, and, $\varepsilon_m=\varepsilon_{kk}/3$is the mean strain, and

$$C_{ijkl} = \frac{E}{1+\nu}\left[\delta_{ik}\delta_{il}+\frac{\nu}{1-2\nu}\delta_{ij}\delta_{kl}\right] \tag{6-14}$$

The stiffness tensor C_{ijkl} has 21 independent components for anisotropic materials. The components depend on two independent constants only for isotropic materials.

The stress tensor can also be written using Lame constant,

$$\sigma_{ij} = 2G\varepsilon_{ij}+\lambda\varepsilon_{kk}\delta_{ij} \tag{6-15}$$

where λ is the Lame constant and can be expressed in terms of E and ν as,

$$\lambda = \frac{E\nu}{(1+\nu)(1-2\nu)} \tag{6-16}$$

The elastic stress strain relation can also be expressed by an incremental form, i.e.,

$$d\varepsilon_{ij} = M_{ijkl}d\sigma_{ij} \tag{6-17}$$

or

$$d\sigma_{ij} = C_{ijkl}d\varepsilon_{ij} \tag{6-18}$$

6.4 Plastic Work Hardening and Strain Hardening

The total strain increments in three-dimensional case can be generalized as

$$d\varepsilon_{ij} = d\varepsilon_{ij}^e + d\varepsilon_{ij}^p \tag{6-19}$$

where $d\varepsilon_{ij}^e$ is the elastic strain increments and $d\varepsilon_{ij}^p$ is the plastic strain increments. The decomposition is correct for cases of infinitesimal strain only in the case of finite strain, there will be geometrical elasto-plastic coupling between the elastic and plastic strain measures so that the Eq. (6-19) will lose its conventional physical meaning.

After initial yielding, the stress level at which further plastic deformation occurs may be dependent on the current degree of plastic straining. Such a phenomenon is termed work hardening or strain hardening. Thus the yield surface will vary at each stage of the plastic deformation, with the subsequent yield surfaces being dependent on the plastic strains in some way. Due to its dissipation feature the plastic deformation process is history or path dependent. In other words, there will not be a one-to-one correspondence between stress-strain during plastic deformation. Some alternative models which describe strain hardening in a material are illustrated in Fig. 6.2. A perfect plastic material means that the yield stress level does not depend in any way on the degree of plastic deformation. If the subsequent yield surfaces are a uniform expansion of the original yield curve, without translation, the strain hardening model is said to be isotropic. On the other hand if the subsequent yield surfaces preserve their shape and orientation but translate in the stress space as a rigid body, kinematic hardening is said to take place. Such a hardening model gives rise to the experimentally observed Bauschinger effect on cyclic loading.

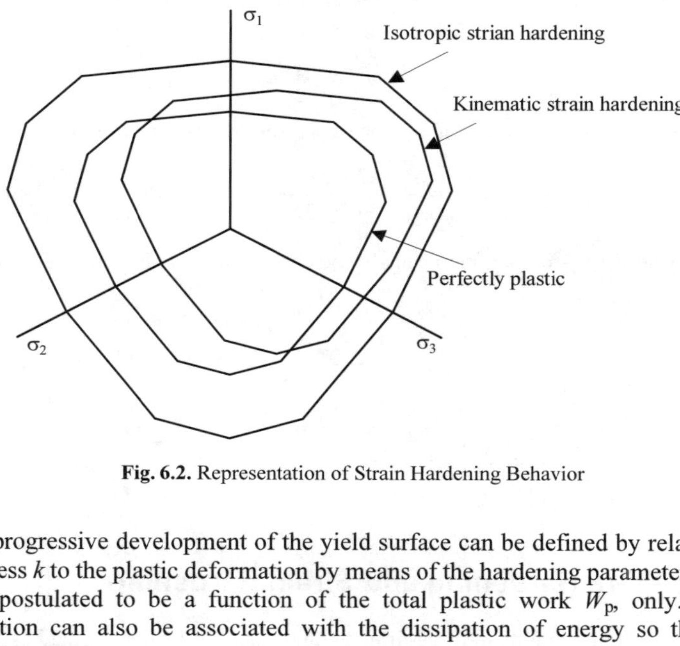

Fig. 6.2. Representation of Strain Hardening Behavior

The progressive development of the yield surface can be defined by relating the yield stress k to the plastic deformation by means of the hardening parameter κ. This can be postulated to be a function of the total plastic work W_p, only. Plastic deformation can also be associated with the dissipation of energy so that it is irreversible. Then

$$\kappa = W_p \tag{6-20}$$

where

$$W_p = \int \sigma_{ij} \, (d\varepsilon_{ij}^p) \tag{6-21}$$

in which $d\varepsilon_{ij}^{p}$ are the plastic components of strain occurring during a strain increment. Alternatively κ can be related to a measure of the total plastic deformation termed the effective, generalized or equivalent plastic strain which is defined incrementally as

$$d\bar{\varepsilon}_p = \sqrt{\frac{2}{3}}\left[(d\varepsilon_{ij}^{p})(d\varepsilon_{ij}^{p})\right]^{\frac{1}{2}} \tag{6-22}$$

A physical insight of this definition is proved where uniaxial yielding is considered. For situations where the assumption that yielding is independent of any hydrostatic stress is valid, $d\varepsilon_{ii}^{p} = 0$ and hence $d\varepsilon_{ij}'^{p} = d\varepsilon_{ij}^{p}$, where $d\varepsilon_{i}'^{p}$ is the deviatoric plastic strain increments. Consequently the above equation can be rewritten as

$$d\bar{\varepsilon}_p = \sqrt{\frac{2}{3}}\left[(d\varepsilon_{ij}'^{p})(d\varepsilon_{ij}'^{p})\right]^{\frac{1}{2}} \tag{6-23}$$

Then the hardening parameter κ, is assumed to be defined as

$$\kappa = \bar{\varepsilon}_p \tag{6-24}$$

where $\bar{\varepsilon}_p$ is the result of integrating $d\bar{\varepsilon}_p$ over the strain path. This behavior is termed strain hardening.

Strain states for which $f = k$ represent plastic states, while elastic behavior is characterized by $f < k$. At a plastic state, $f = k$, the incremental change in the yield function due to an incremental stress change is

$$df = \frac{\partial f}{\partial \sigma_{ij}} d\sigma_{ij} \tag{6-25}$$

Then if

$df < 0$ elastic unloading occurs and the stress point returns inside the yield surface.

$df = 0$ neutral loading and the stress point remains on the yield surface.

$df > 0$ plastic loading for a strain hardening material.

6.5 Plastic Flow Rule

The general mathematical treatment of the constitutive equation for plastic deformation or flow was proposed by Huber-von Mises in 1928. In elastic theory the strain tensor was related to the stress tensor through an elastic potential function, the complementary strain energy U such that

$$\varepsilon_{ij} = \frac{\partial U}{\partial \sigma_{ij}} \tag{6-26}$$

By extending this idea to plasticity theory, Mises proposed that there existed a plastic potential function Q (σ_{ij}), and the plastic strain increments $d\varepsilon_{ij}{}^{P}$ can be derived similar to Eq. (6-26),

$$d\varepsilon_{ij}^{P} = d\lambda \frac{\partial Q}{\partial \sigma_{ij}} \tag{6-27}$$

where $d\lambda$ is a proportional positive scalar factor. To determine $d\lambda$, the yield function should be used. The plastic flow rule shown in Eq. (6-27) is called plastic potential theory. The plastic potential Q (σ_{ij}) $=C$, or a constant, represents a surface in the six-dimensional stress space, and the plastic strain $d\varepsilon_{ij}{}^{P}$ can be represented to a vector which is perpendicular to the surface Q (σ_{ij}) $=C$.

A common approach in plasticity theory is to assume that the plastic potential function Q (σ_{ij}) is the same as the yield function F (σ_{ij}),

$$Q(\sigma_{ij}) = F(\sigma_{ij}) \tag{6-28}$$

Eq. (6-27) can then be rewritten as

$$d\varepsilon_{ij}^{P} = d\lambda \frac{\partial F}{\partial \sigma_{ij}} \tag{6-29}$$

and the plastic flow vector is normal to the yield surface. This is called the associated flow rule. On the other hand, if $Q \neq F$, the flow rule is called nonassociated.

The association of Q with F is based on an assumption whose validity can be verified empirically. Experimental observations show that the plastic deformation of metals can be characterized quite well by the associated flow rule, but for some porous materials such as rocks, concrete, and soils, the nonassociated flow rule may provide a better representation of their plastic deformation. Mathematically it can be proved by using Drucker's stability postulate that if the material is stable in Drucker's sense, the flow vector must be associated.

The Prandtl-Reuss equation is a special case of the associated flow rule. Indeed applying the Huber-von Mises yield criterion yields

$$\frac{\partial J_2}{\partial \sigma_{ij}} = S_{ij} \tag{6-30}$$

where S_{ij} is the deviatoric stress tensor. Eq. (6-29) then gives

$$d\varepsilon_{ij}^{P} = d\lambda \frac{\partial F}{\partial \sigma_{ij}} = d\lambda S_{ij} \tag{6-31}$$

which is the Prandtl-Reuss equation, or the Levy-Mises equation if the elastic strain rate is ignored. Thus, within the general frame of the plastic potential theory, the Prandtl-Reuss or the Levy-Mises equation implies the Huber-von Mises yield function and the associated flow rule.

The complete incremental relationship between stress and strain for elasto-plastic deformation is found to be

$$d\varepsilon_{ij} == \frac{1}{E}\left[(1+\nu)\delta_{ik}\delta_{jl} - \nu\delta_{ik}\delta_{kl}\right]d\sigma_{kl} + d\lambda\frac{\partial Q}{\partial\sigma_{ij}} \tag{6-32}$$

6.6 Drucker's Postulate – Convexity of the Loading Surface

Drucker (1951) proposed a unified approach based on his stability postulate to establish the general plastic stress-strain relations. One major consequence of Drucker's posulate is that the flow rule for stable materials is associated (i.e., $Q = F$). For a stable material, it can be proved that the yield surface must be convex (Drucker 1952, Mendelson, 1968).

Considering a material element in equilibrium with a given state of stress σ_{ij}^{0} inside the loading surface as shown in Fig. 6.3, first it is necessary to define a stress cycle or a closed loading-unloading path in stress space. Let some external agency add stresses along some arbitrary path inside the surface. Only elastic changes have taken place so far. Now suppose the external agency to add a very small outward pointing stress increment $d\sigma_{ij}$ which produces small plastic strain increments $d\varepsilon_{ij}^{P}$, as well as elastic increments. The external agency then releases the $d\sigma_{ij}^{P}$ and the state of stress is returned to σ_{ij}^{0} along an elastic path.

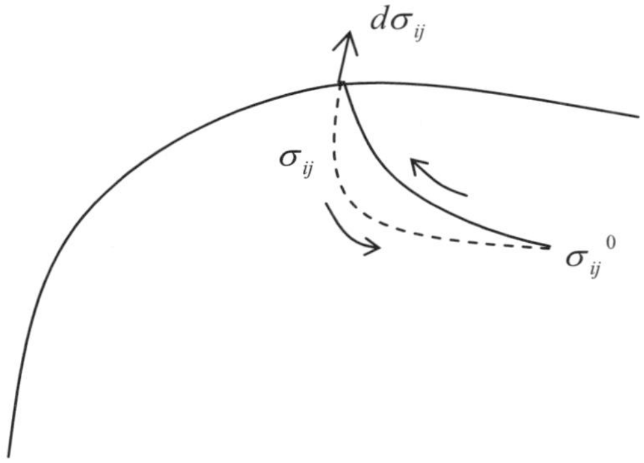

Fig. 6.3 A closed loading-unloading path

The work done by the external agency over the cycle is

$$\delta W = (\sigma_{ij} - \sigma_{ij}^{0}) d\varepsilon_{ij}^{p} + d\sigma_{ij} d\varepsilon_{ij}^{p} \tag{6-33}$$

If the plastic strain coordinates are superimposed on the stress coordinates as in Fig. 6.4, δW may be interpreted as the scalar product of the vector σ_{ij}-σ_{ij}^{0} and the vector $d\varepsilon_{ij}^{p}$ plus the scalar product of $d\sigma_{ij}$ and $d\varepsilon_{ij}^{p}$. Now, from the strain-hardening definition,

$$d\sigma_{ij}\, d\varepsilon_{ij}^{p} \geq 0 \tag{6-34}$$

or

$$\left|d\sigma_{ij}\right|\left|d\varepsilon_{ij}^{p}\right| \cos\theta \geq 0 \tag{6-35}$$

and $-\pi/2 \leq \theta \leq \pi/2$.

That is, the vector $d\sigma_{ij}$ and $d\varepsilon_{ij}^{p}$ make an acute angle with each other. In a similar fashion, since the magnitude of σ_{ij}-σ_{ij}^{0} can always be made larger than the magnitude of $d\sigma_{ij}$, it follows that

$$(\sigma_{ij} - \sigma_{ij}^{0}) d\varepsilon_{ij}^{p} \geq 0 \tag{6-36}$$

or

$$\left|\sigma_{ij} - \sigma_{ij}^{0}\right|\left|d\varepsilon_{ij}^{p}\right| \cos\psi \geq 0 \tag{6-37}$$

hence $-\pi/2 \leq \psi \leq \pi/2$.

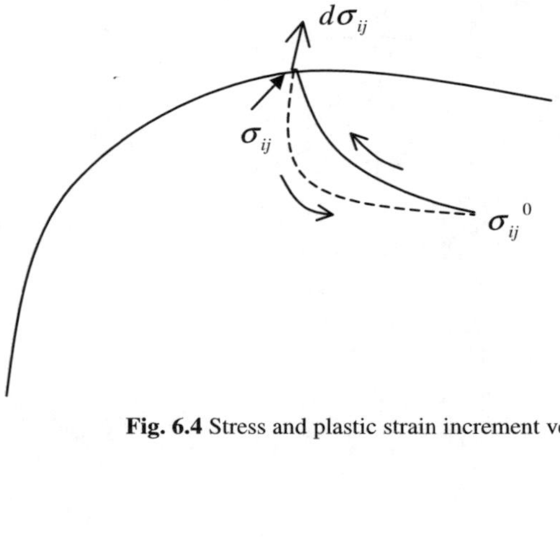

Fig. 6.4 Stress and plastic strain increment vectors

Thus the vector $\sigma_{ij}-\sigma_{ij}^0$ makes an acute angle with the vector $d\varepsilon_{ij}^P$ for all choices of σ_{ij}^0. Therefore, all points σ_{ij}^0 must lie on one side of a plane perpendicular to $d\varepsilon_{ij}^P$, and, since $d\varepsilon_{ij}^P$ is normal to the yield surface, this plane will be tangent to the yield surface. This must be true for all points σ_{ij} on the yield surface, so that no vector $\sigma_{ij}-\sigma_{ij}^0$ can pass outside the surface intersecting the surface twice, as shown in Fig. 6.5. The surface must therefore be convex. On the other hand, if the surface is not convex, there exist some points σ_{ij} and σ_{ij}^0 such that the vector $\sigma_{ij}-\sigma_{ij}^0$ forms an obtuse angle with vector $d\sigma_{ij}$, as shown in Fig. 6.6. This completes the convexity proof.

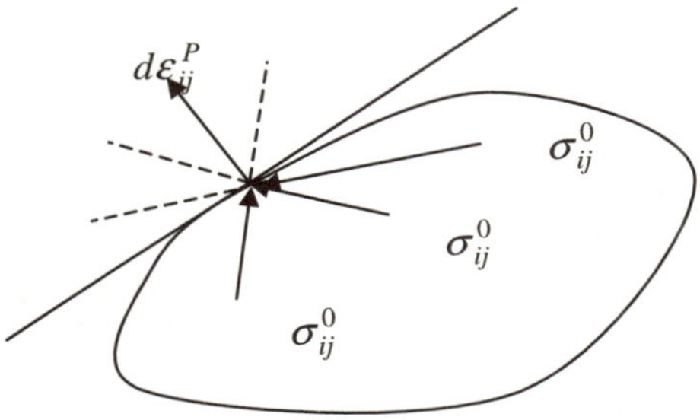

Fig. 6.5 Convex yield surface

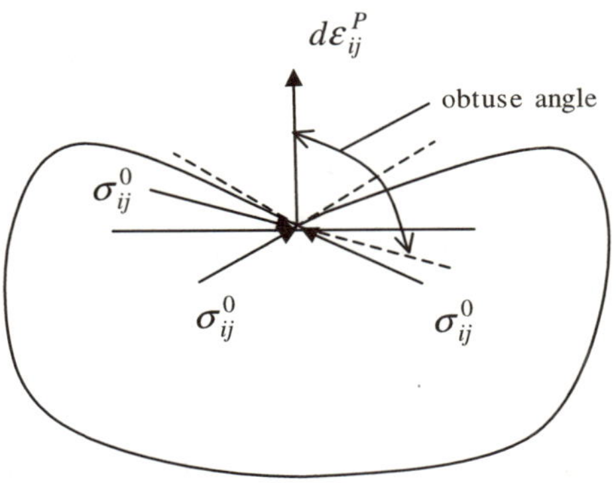

Fig. 6.6 Non-convex yield surface

The associate plastic flow rule implies that the yield surface has a unique gradient. It may happen, however, that the yield surface has vertices or corners where the gradient is not defined. For example, the Tresca hexagon has no unique normal at the corners, where two of the stresses are equal. Such points are called singular points or singular yield conditions. Process of the singularity of the yield functions will be discussed in Sections 6.9 - 6.12.

6.7 Incremental Constitutive Equations in Matrix Formulation

To various engineering materials, a yield criterion indicating the stress level at which plastic flow commences must be postulated. A relationship between stress and strain must be developed for post-yield behavior, i.e. when the deformation is made up of both elastic and plastic components. The yield surface separates the plastic region from the elastic region. The change of the stress state from the yield surface toward its interior will cause elastic unloading. Plastic loading will occur only if the increment of the stress is directed toward the outside of the yield surface.

The yield function can be represented by

$$F(\sigma_{ij}, k) = 0 \quad \text{or} \quad f(\sigma_{ij}) = k \tag{6-38}$$

where k is a material parameter which may be determined experimentally. It can be a function of a few material strength coefficients, or a constant for elastic-perfect-plastic material. The term k can be defined a function of a hardening parameter κ; thus the yield function can be extended to describe post-yield of the material, or $k = k(\kappa)$.

For simplicity, the yield function in Eq. (6-38) can be rewritten in terms of the three principal stresses,

$$F(\sigma_1, \sigma_2, \sigma_3, k) = 0 \tag{6-39}$$

For isotropic material, the yield function is independent of the orientation of the coordinate system employed, therefore, it is usually presented by a function of the three invariants, i.e.,

$$F(I_1, J_2, J_3, k) = 0 \tag{6-40}$$

where I_1 is the first invariant of the stress tensor, J_2 and J_3 are respectively the second and the third invariants of the deviatoric stress tensor. Alternatively, the above yield function is represented as

$$F(I_1, J_2, \theta, k) = 0 \tag{6-41}$$

in which is a Haigh-Westergaard coordinate or Lode angle termed in the geotechnical engineering.

For the elasto-perfect-plastic material, the parameter k in the yield functions is a constant. It means that the yield surface is independent of the plastic strain, thus the geometry and the size of the yield surface will not change with the successive deformation of the material. The post-yielding surface is exactly the initial one. The stress point retains at the yield surface means under loading condition, and the stress point moves into the inside of the yield surface implies unloading. For plastic hardening material, k can be defined by a work-hardening or strain-hardening parameter . The post-yielding surface is thus different from the initial yielding surface.

The equation is termed the normality condition since $\partial F/\partial \sigma_{ij}$ is a vector directed normal to the yield surface at the stress point under consideration.

Differentiating the a hardening yield function, it has

$$dF = \frac{\partial F}{\partial \sigma_{ij}} d\sigma_{ij} + \frac{\partial F}{\partial k} dk = \frac{\partial F}{\partial \sigma_1} d\sigma_1 + \frac{\partial F}{\partial \sigma_2} d\sigma_2 + \frac{\partial F}{\partial \sigma_3} d\sigma_3 + \frac{\partial F}{\partial k} d\kappa = 0 \quad (6\text{-}42)$$

Introducing a parameter A, where $A = -\frac{1}{d\lambda}\frac{\partial F}{\partial k} dk$, the above equation can be rewritten to a matrix form,

$$\frac{\partial F}{\partial \sigma_{ij}} d\sigma_{ij} - A d\lambda = 0 \quad (6\text{-}43)$$

Converting Eq. (6-43) to a vector form by denoting

$\{\sigma\}^T = \{\sigma_x, \sigma_y, \sigma_z, \tau_{xy}, \tau_{yz}, \tau_{zx}\}$, it has

$$\{a\}^T = \frac{\partial F}{\partial\{\sigma\}} = \left\{ \frac{\partial F}{\partial \sigma_x}, \frac{\partial F}{\partial \sigma_y}, \frac{\partial F}{\partial \sigma_z}, \frac{\partial F}{\partial \tau_{xy}}, \frac{\partial F}{\partial \tau_{yz}}, \frac{\partial F}{\partial \tau_{zx}} \right\} \quad (6\text{-}44)$$

where $\{a\}$ is termed the flow vector. Thus,

$$\{a\}^T d\{\sigma\} - A d\lambda = 0 \quad (6\text{-}45)$$

The strain increments can then be derived as follows,

$$d\{\varepsilon\} = [D]^{-1} d\{\sigma\} + \{a\} d\lambda \quad (6\text{-}46)$$

where $[D]$is the usual matrix of elastic constants. Premultiplying both sides of Eq. (6-46) by $\{a\}^T[D]$.

$$\begin{aligned} \{a\}^T [D] d\{\varepsilon\} &= \{a\}^T d\{\sigma\} + d\lambda \{a\}^T [D]\{a\} \\ &= A d\lambda + d\lambda \{a\}^T [D]\{a\} \end{aligned} \quad (6\text{-}47)$$

The plastic multiplier is then obtained as

$$d\lambda = \frac{\{a\}^T [D] d\{\varepsilon\}}{A + \{a\}^T [D]\{a\}} \tag{6-48}$$

Substituting $d\lambda$ into Eq. (6-46), the complete elasto-plastic incremental stress-strain relation can be derived to be

$$d\{\sigma\} = [D]_{ep} d\{\varepsilon\} \tag{6-49}$$

in which the elasto-plastic stiffness matrix $[D]_{ep}$ is

$$[D]_{ep} = [D] - \frac{[D]\{a\}\{a\}^T [D]}{A + \{a\}^T [D]\{a\}} \tag{6-50}$$

It now remains to determine the explicit form of the scalar term A. A is a function of the hardening parameter κ. Employing the work hardening hypothesis and the normality condition,

$$d\kappa = \{\sigma\}^T d\{\varepsilon^p\} = \{\sigma\}^T d\lambda\{a\} = d\lambda\{a\}^T \{\sigma\} \tag{6-51}$$

For uniaxial case, $\sigma = \bar{\sigma} = \sigma_Y$, $d\varepsilon_p = d\bar{\varepsilon}_p$, $\bar{\sigma}$ and $\bar{\varepsilon}_p$ are respectively the effective stress and plastic strain.

$$d\kappa = \sigma_Y \, d\bar{\varepsilon}_p = d\lambda\{a\}^T \{\sigma\} \tag{6-52}$$

The effective stress $\bar{\sigma}$ is a function of $\bar{\varepsilon}_p$, i.e. $\bar{\sigma} = H(\bar{\varepsilon}_p)$, differentiating it, we obtain

$$\frac{d\bar{\sigma}}{d\bar{\varepsilon}_p} = \frac{d\sigma_Y}{d\bar{\varepsilon}_p} = H' \tag{6-53}$$

Using Euler's theorem applicable to all homogeneous functions of order one, it has

$$\{a\}^T \{\sigma\} = \sigma_Y \tag{6-54}$$

Substituting Eqs. (6-53) and (6-54) into Eq. (6-52), we obtain

$$d\lambda = d\bar{\varepsilon}_p \quad \text{and} \quad A = H' \tag{6-55}$$

The parameter A is determined by the local slope of the uniaxial stress strain curve as

$$H' = \frac{E_T}{1 - E_T / E} \tag{6-56}$$

6.8 Determination of Flow Vector for Different Yield Functions

For convenience of numerical simulation implementation, the yield functions in Eqs. (6-39)-(6-41) are often used (Nayak and Zienkiewicz 1972; Owen and Hinton, 1980). The principal stresses can be calculated by

$$\begin{Bmatrix} \sigma_1 \\ \sigma_2 \\ \sigma_3 \end{Bmatrix} = \frac{2\sqrt{J_2}}{\sqrt{3}} \begin{Bmatrix} \cos\theta \\ \cos\left(\theta - \frac{2\pi}{3}\right) \\ \cos\left(\theta + \frac{2\pi}{3}\right) \end{Bmatrix} + \frac{I_1}{3} \begin{Bmatrix} 1 \\ 1 \\ 1 \end{Bmatrix} \tag{6-57}$$

with $\sigma_1 \geq \sigma_2 \geq \sigma_3$ and $0 \leq \theta \leq \pi/3$.
The flow vector can thus be expressed as,

$$\begin{aligned}\{a\}^T &= \frac{\partial F}{\partial\{\sigma\}} = \frac{\partial F}{\partial I_1}\frac{\partial I_1}{\partial\{\sigma\}} + \frac{\partial F}{\partial\sqrt{J_2}}\frac{\partial\sqrt{J_2}}{\partial\{\sigma\}} + \frac{\partial F}{\partial\theta}\frac{\partial\theta}{\partial\{\sigma\}} \\ &= C_1\{a_1\} + C_2\{a_2\} + C_3\{a_3\}\end{aligned} \tag{6-58}$$

where

$$C_1 = \frac{\partial F}{\partial I_1},$$

$$C_2 = \left(\frac{\partial F}{2\sqrt{J_2}} + \frac{\cot 3\theta}{\sqrt{J_2}}\frac{\partial F}{\partial\theta}\right),$$

$$C_3 = -\frac{\sqrt{3}}{2\sin 3\theta\sqrt{J_2^3}}\frac{\partial F}{\partial\theta}$$

$$\{a_1\}^T = \frac{\partial I_1}{\partial\{\sigma\}} = \{1,1,1,0,0,0\}$$

$$\{a_2\}^T = \frac{\partial\sqrt{J_2}}{\partial\{\sigma\}} = \frac{1}{2\sqrt{J_2}}\{\sigma'_x, \sigma'_y, \sigma'_z, 2\tau_{yz}, 2\tau_{zx}, 2\tau_{xy}\}$$

$$\begin{aligned}\{a_3\}^T = \frac{\partial J_3}{\partial\{\sigma\}} = \Bigg\{&\left(\sigma'_y\sigma'_z - \tau_{yz}^2 + \frac{J_2}{3}\right), \left(\sigma'_x\sigma'_z - \tau_{xz}^2 + \frac{J_2}{3}\right), \left(\sigma'_x\sigma'_y - \tau_{xy}^2 + \frac{J_2}{3}\right), \\ &2\left(\tau_{xz}\tau_{xy} - \sigma'_x\tau_{yz}\right), \left(\tau_{xy}\tau_{yz} - \sigma'_y\tau_{xz}\right), \left(\tau_{yz}\tau_{xz} - \sigma'_z\tau_{yz}\right)\Bigg\}\end{aligned}$$

and

$$\frac{\partial\theta}{\partial\{\sigma\}} = \frac{\sqrt{3}}{2\sin 3\theta}\left[\frac{1}{\sqrt{J_3^{\,2}}}\frac{\partial J_3}{\partial\{\sigma\}} - \frac{3J_3}{J_2^{\,2}}\frac{\partial\sqrt{J_2}}{\partial\{\sigma\}}\right]$$

For different yield function, the vectors $\{a_1\}$, $\{a_2\}$, and $\{a_3\}$ are consistent and only the constants C_1, C_2 and C_3 need to be determined. In Table 6.1, the constants for six different yield functions are given.

Table 6.1 Parameters C_1, C_2 and C_3 for 6 different yield functions

Yield function		C_1	C_2	C_3
Tresca		0	$2\left[\sin\left(\theta+\frac{\pi}{3}\right)+\cos\left(\theta+\frac{\pi}{3}\right)\cot 3\theta\right]$	$-\frac{\sqrt{3}\cos\left(\theta+\frac{\pi}{3}\right)}{J_2\sin 3\theta}$
Mises		0	$\sqrt{3}$	0
Yu $\alpha=1$ $b=1$	$\theta\le\theta_b$	0	$\sqrt{3}\sin\theta(\cot\theta-\cot 3\theta)$	$\frac{3\sin\theta}{2J_2\sin 3\theta}$
	$\theta\ge\theta_b$	0	$\sqrt{3}\left[\cos\left(\theta-\frac{\pi}{3}\right)-\sin\left(\theta-\frac{\pi}{3}\right)\cot 3\theta\right]$	$\frac{3\sin\left(\theta-\frac{\pi}{3}\right)}{2J_2\sin 3\theta}$
Mohr-Coulomb		$\frac{\sin\varphi}{3}$	$\left(\frac{\sin\varphi}{\sqrt{3}}+\cot 3\theta\right)\cos\left(\theta+\frac{\pi}{3}\right)+\left(1-\frac{1}{\sqrt{3}}\cot 3\theta\sin\varphi\right)\sin\left(\theta+\frac{\pi}{3}\right)$	$\frac{\sin\left(\theta+\frac{\pi}{3}\right)\sin\varphi-\sqrt{3}\cos\left(\theta+\frac{\pi}{3}\right)}{2J_2\sin 3\theta}$
Drucker-Prager		α	1	0
Yu $\alpha\ne1$ $b=1$	$\theta\le\theta_b$	$\frac{2\sin\varphi}{3}$	$\frac{1}{\sqrt{3}}(3+\sin\varphi)\sin\theta(\cot\theta-\cot 3\theta)$	$\frac{(3+\sin\varphi)\sin\theta}{2J_2\sin 3\theta}$
	$\theta\ge\theta_b$	$\frac{2\sin\varphi}{3}$	$\frac{1}{\sqrt{3}}(3-\sin\varphi)\sin\left(\theta-\frac{\pi}{3}\right)\cdot\left[\cot\left(\theta-\frac{\pi}{3}\right)-\cot 3\theta\right]$	$\frac{(3-\sin\varphi)\sin\left(\theta-\frac{\pi}{3}\right)}{2J_2\sin 3\theta}$

6.9 Singularity of Piecewise-linear Yield Functions

The plastic flow vector exists singular points for the piecewise-linear yield functions, such as the Tresca yield function, the Mohr-Coulomb yield function and the unified strength theory. The flow vector at the corners is not unique when the normality condition is applied. Some smooth corner models have been proposed to eliminate the singularity of plastic flow for piecewise-linear yield functions. These smooth corner models can be divided into two categories (Koiter and Hinton, 1953; Nayak and Zienkiewicz, 1972; Zienkiewicz and Pande, 1977; and Owen, 1980).

One category is that the projection on the deviatoric plane of the yield surface is simply approximated by a circle (the Huber-von Mises and Drucker-Prager criteria fall in this category). It assumes same tensile and compressive meridians, thus, can not agree the experimental results for geomaterials which always give different vector lengths for different meridians with different angle on the deviatoric plane, as shown in Fig. 6.7.

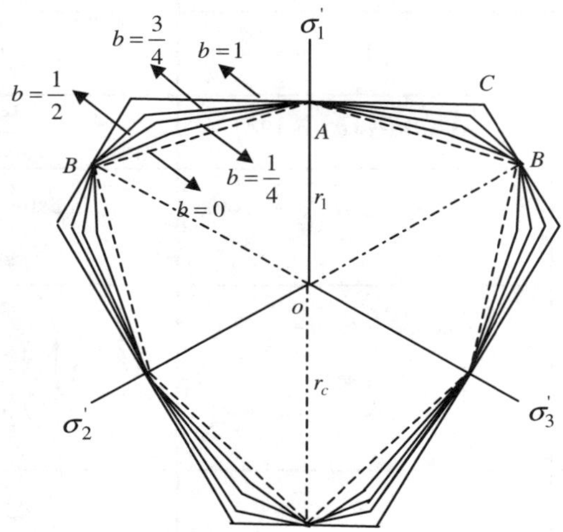

Fig. 6.7 Different vector lengths for different angle on deviatoric plane

The other category is a kind of smoothening model, which smoothen the corners using very complex mathematical models. It lacks physical concept, and is not convenient to be used in analytical and numerical derivations. The following will introduce some convenient ways to solve the corner singularity problem for the piecewise-linear yield functions, which can be readily implemented into elasto-plastic finite element analysis.

Considering the associated flow rule, the flow vector is normal to the yield surface. At the corners of the piecewise-linear yield functions, the flow vectors from different sides are thus not consistent. For the unified strength theory, there are three corners on the deviatoric plane as shown in Fig. 6.8, i.e. = 0 (point A), = $_b$ (point C), and = 60° (point B). At the corners, the derivative of the yield function does not exist, the value and the direction of plastic strain increment vector of corner can not be determined directly from the plastic flow rule.

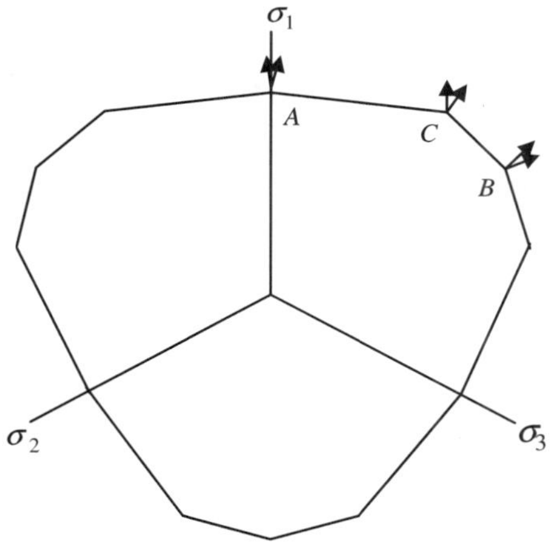

Fig. 6.8 Singular points on the deviatoric plane

The process of the singularity function includes

1. Vector summation method (Koiter and Hinton 1953). Its mathematical expression is

$$d\varepsilon_{ij}{}^{p} = d\lambda_1 \frac{\partial F_1}{\partial \sigma_{ij}} + d\lambda_2 \frac{\partial F_2}{\partial \sigma_{ij}} \tag{6-59}$$

2. Partially smoothening method (Nayak and Zienkiewicz, 1972). When dealing with the singularity of the Tresca and the Mohr-Coulomb functions, Nayak and Zienkiewicz (1972) expressed the yield function as $F = F(I_1, J_2, \theta, k)$, where I_1 is first stress invariant, J_2 is second partial stress invariant, θ is the angle of deviatoric plane. The plastic strain increment of the corner ($\theta=\theta_0$) can be expressed as

$$d\varepsilon_{ij}{}^{p} = d\lambda\left(\frac{\partial F_{\theta=\theta_0}}{\partial I_1}\frac{\partial I_1}{\partial \sigma_{ij}} + \frac{\partial F_{\theta=\theta_0}}{\partial \sqrt{J_2}}\frac{\partial \sqrt{J_2}}{\partial \sigma_{ij}} + \frac{\partial F_{\theta=\theta_0}}{\partial \theta}\frac{\partial \theta}{\partial \sigma_{ij}}\right) \tag{6-60}$$

The third item of Eq. (6-60) is simply set to 0.

3. Linear combination method, i.e.

$$d\varepsilon_{ij}{}^{p} = \mu d(\varepsilon_{ij}{}^{p})_1 + (1-\mu) d(\varepsilon_{ij}{}^{p})_2 = \mu d\lambda_1 \frac{\partial f_1}{\partial \sigma_{ij}} + (1-\mu) d\lambda_2 \frac{\partial f_2}{\partial \sigma_{ij}} \tag{6-61}$$

where $0 \le \mu \le 1$, the direction of $d\varepsilon_{ij}^{p}$ is between $d(\varepsilon_{ij}^{p})_1$ and $d(\varepsilon_{ij}^{p})_2$.

These three methods can eliminate the singularity of yield surface in some cases; however, each of them has some drawbacks and sometimes may introduce additional errors. For method (1), when the piecewise-linear function is the special case $F_1 = F_2$, the plastic strain increment is two times of the real value, which is not unreasonable. An average of the vector summation is suggested to solve the problem. Method (2) eliminates the singularity of $\partial\theta/\partial\sigma_{ij}$, but it made the assumption that $\left.\frac{\partial F}{\partial \theta}\right|_{\theta=\theta_0} = 0$. Thus its application is limited. Method (3) introduced an uncertain parameter μ, and from a practical analysis, this method can not eliminate the singularity in some cases. Besides, Zienkiewicz (1972) and Zienkiewicz and Pande (1977) proposed some smooth corner model to replace the piecewise-linear yield function.

Invariant expressions of the unified yield theory has the following form,

$$F = \left(1+\frac{\alpha}{2}\right)\frac{2}{\sqrt{3}} J_2^{1/2} \cos\theta + \frac{\alpha(1-b)}{1+b} J_2^{1/2} + \frac{I_1}{3}(1-\alpha) = \sigma_t \quad (6\text{-}62a)$$

$$F' = \left(\frac{2-b}{1+b}+\alpha\right)\frac{1}{\sqrt{3}} J_2^{1/2} \cos\theta + \left(\alpha + \frac{b}{1+b}\right) J_2^{1/2} \sin\theta + \frac{I_1}{3}(1-\alpha) = \sigma_t \quad (6\text{-}62b)$$

The projection curve on the deviatoric plane of Eqs. (6-62a and b) is shown in Fig. 6.7. From $F = F'$, it has $\theta_b = \arctan\frac{\sqrt{3}}{2\alpha+1}$. According to the symmetrical condition, only the singularity of the three points A, B, and C should be discussed. For the Tresca criterion and the Mohr-Coulomb criterion, there are two singular points A, B only.

Using the plastic flow vector defined in Eq. (6-58), when $F \geq F'$, or $0 \leq \theta \leq \theta_0$, it has

$$C_1 = \frac{1}{3}(1-\alpha) \quad (6\text{-}63a)$$

$$C_2 = \left(1+\frac{\alpha}{2}\right)\frac{2}{\sqrt{3}}\cos\theta + \frac{\alpha(1-b)}{1+b}\sin\theta + \cot 3\theta\left[-\left(1+\frac{\alpha}{2}\right)\frac{2}{\sqrt{3}}\sin\theta + \frac{\alpha(1-b)}{1+b}\cos\theta\right] \quad (6\text{-}63b)$$

$$C_3 = -\frac{\sqrt{3}}{2J_2\sin 3\theta}\left[-\left(1+\frac{\alpha}{2}\right)\frac{2}{\sqrt{3}}\sin\theta + \frac{\alpha(1-b)}{1+b}\cos\theta\right] \quad (6\text{-}63c)$$

When $F < F'$, or $\theta_0 < \theta \leq \pi/3$, it has

$$C_1' = \frac{1}{3}(1-\alpha) \quad (6\text{-}64a)$$

$$C_2' = \left(\frac{2-b}{1+b}\right)\frac{1}{\sqrt{3}}\cos\theta + \left(\alpha + \frac{b}{1+b}\right)\sin\theta + \cot 3\theta\left[-\left(\frac{2-b}{1+b}\right)\frac{1}{\sqrt{3}}\sin\theta + \left(\alpha + \frac{b}{1+b}\right)\cos\theta\right] \tag{6-64b}$$

$$C_3' = -\frac{\sqrt{3}}{2J_2\sin 3\theta}\left[-\left(\frac{2-b}{1+b}+\alpha\right)\frac{1}{\sqrt{3}}\sin\theta + \left(\alpha + \frac{b}{1+b}\right)\cos\theta\right] \tag{6-64c}$$

On the two lines AC and BC except the corners, C_1, C_2, C_3 or C_1', C_2', C_3' have unique value, and the plastic strain increment is unique. The singularity at the three corner points A, B, C is discussed separately below.

Point A, the corresponding stress state is $\sigma_1 \geq \sigma_2 = \sigma_3$, and $\theta = 0$
when $b \neq 1$

$$\sin\theta = 0, \cos\theta = 1, \cot 3\theta \to \infty \tag{6-65}$$

So, $C_2 \to \infty$, $C_3 \to \infty$, the plastic vector is a singular function.
when $b = 1$

$$C_1 = \frac{1}{3}(1-\alpha), \lim_{\theta\to(\pi/3)} C_2 = \frac{2}{3\sqrt{3}}(2+\alpha), \lim_{\theta\to(\pi/3)} C_3 = \frac{1}{6J_2}(2+\alpha) \tag{6-66}$$

Then

$$C_2\big|_{\theta=0} = \frac{2}{3\sqrt{3}}(2+\alpha), C_3\big|_{\theta=0} = \frac{1}{6J_2}(2+\alpha) \tag{6-67}$$

So, when b=1, there is no singularity for point A.

Point B, the stress state is $\sigma_1 \geq \sigma_2 = \sigma_3$, and $\theta = \pi/3$.
When $b \neq 1$,

$$\sin\left(\theta - \frac{\pi}{3}\right) = 0, \cos\left(\theta - \frac{\pi}{3}\right) = 1, \cot 3\theta \to \infty \tag{6-68}$$

So, $C'_2 \to \infty$, $C'_3 \to \infty$, there is singularity for the plastic flow.
when b =1,

$$C_1' = \frac{1}{3}(1-\alpha), \lim_{\theta\to(\pi/3)} C_2' = \frac{2}{3\sqrt{3}}(2\alpha+1), \lim_{\theta\to(\pi/3)} C_3' = -\frac{1}{6J_2}(2\alpha+1) \tag{6-69}$$

So that,

$$C_2'\Big|_{\theta=\pi/3} = \frac{2}{3\sqrt{3}}(2\alpha+1), C_3'\Big|_{\theta=\pi/3} = -\frac{1}{6J_2}(2\alpha+1) \tag{6-70}$$

So, when $b = 1$, there is no singularity for point B.

Point C, $F = F', \theta = \theta_0 = \arctan\dfrac{\sqrt{3}}{2\alpha+1}$

when $b \neq 0$,

$$C_1 = C_1', C_2 \neq C_2'. C_3 \neq C_3' \tag{6-71}$$

There is singularity for the plastic flow.

when $b = 0$, $F = F'$, it is the Mohr-Coulomb criterion,

$$C_1 = C_1', C_2 = C_2'. C_3 = C_3' \tag{6-72}$$

There is no singularity for the plastic flow.

So, for the three singular points A, B, C, when b=1, there is no singularity for points A, B; when $b = 0$, there is no singularity for point C.

6.10 Process of the Plastic Flow Singularity

The three different methods discussed in Section 6.9 are first adopted to solve the singularity at the three point A, B, C.

Point A

If method (1) is used,

$$C_1^{\mathrm{I}} = \frac{2}{3}(1-\alpha) \tag{6-73a}$$

$$C_2^{\mathrm{I}} = \lim_{\theta\to 0^+} C_2 + \lim_{\theta\to 0^-} C_2 = \frac{4}{3\sqrt{3}}(2+\alpha) \tag{6-73b}$$

$$C_3^{\mathrm{I}} = \lim_{\theta\to 0^+} C_3 + \lim_{\theta\to 0^-} C_3 = \frac{1}{3J_2}(2+\alpha) \tag{6-73c}$$

When b=1, there is no singularity for point A. When $b\to 1$, $C_1{}^{\mathrm{I}}$, $C_2{}^{\mathrm{I}}$, $C_3{}^{\mathrm{I}}$ should be equal to $C_1\big|_{\substack{b=0\\\theta=0}}, C_2\big|_{\substack{b=0\\\theta=0}}, C_3\big|_{\substack{b=0\\\theta=0}}$. However, $C_1{}^{\mathrm{I}} = 2C_1\big|_{\substack{b=0\\\theta=0}}$, $C_2{}^{\mathrm{I}} = 2C_2\big|_{\substack{b=0\\\theta=0}}$, $C_3{}^{\mathrm{I}} = 2C_3\big|_{\substack{b=0\\\theta=0}}$, so, method (1) is actually unreasonable.

If method (2) is used, it has

$$C_1{}^{\mathrm{II}} = \frac{1}{3}(1-\alpha) \tag{6-74a}$$

$$C_2{}^{\mathrm{II}} = \frac{4}{3}(2+\alpha) \tag{6-74b}$$

$$C_3{}^{\mathrm{II}} = 0 \tag{6-74c}$$

i.e., $C_1{}^{\mathrm{II}} = C_1\big|_{\substack{b=1\\\theta=0}}$, $C_2{}^{\mathrm{II}} \neq C_2\big|_{\substack{b=1\\\theta=0}}$, $C_3{}^{\mathrm{II}} \neq C_3\big|_{\substack{b=1\\\theta=0}}$, so, method (2) is also unreasonable.

If method (3) is used, it has

$$C_1{}^{\mathrm{III}} = \frac{1}{3}(1-\alpha) \tag{6-75a}$$

$$C_2{}^{\mathrm{III}} = \mu \lim_{\theta\to 0^+} C_2 + (1-\mu) \lim_{\theta\to 0^-} C_2 \tag{6-75b}$$

$$C_3{}^{\mathrm{III}} = \mu \lim_{\theta\to 0^+} C_3 + (1-\mu) \lim_{\theta\to 0^-} C_3 \tag{6-75c}$$

When μ=1/2, $C_1{}^{\mathrm{III}} = C_1\big|_{\substack{b=1\\\theta=0}}, C_2{}^{\mathrm{III}} = C_2\big|_{\substack{b=1\\\theta=0}}, C_3{}^{\mathrm{III}} = C_3\big|_{\substack{b=1\\\theta=0}}$. So, the singularity of point A is eliminated. When $\mu\neq 1/2$, $C_2{}^{\mathrm{III}} \to \infty$, $C_3{}^{\mathrm{III}} \to \infty$, the singularity of point A can not be eliminated. So, method (3) is not suitable for some instances.

Point B, after some derivation, it can be obtained

$$C_1{}^{'\mathrm{I}} = \frac{2}{3}(1-\alpha),\ C_2{}^{'\mathrm{I}} = \frac{4}{3\sqrt{3}}(2\alpha+1),\ C_3{}^{'\mathrm{I}} = -\frac{1}{3J_2}(2\alpha+1) \tag{6-76a}$$

$$C_1{}^{'\mathrm{II}} = \frac{1}{3}(1-\alpha),\ C_2{}^{'\mathrm{II}} = \frac{1}{\sqrt{3}}(2\alpha+1),\ C_3{}^{'\mathrm{II}} = 0 \tag{6-76b}$$

$$C_1^{'\mathrm{III}} = \frac{1}{3}(1-\alpha)\,,\quad C_2^{'\mathrm{III}} = \mu \lim_{\theta\to(\pi/3)^+} C_2^{'} + (1-\mu) \lim_{\theta\to(\pi/3)^-} C_2^{'}$$

$$C_3^{'\mathrm{III}} = \mu \lim_{\theta\to(\pi/3)^+} C_3^{'} + (1-\mu) \lim_{\theta\to(\pi/3)^-} C_3^{''} \tag{6-76c}$$

It can come out similar conclusions as for point A.

Point C
If method (1) is used, it has

$$C_1^{\mathrm{I}} = \frac{2}{3}(1-\alpha) \tag{6-77a}$$

$$C_2^{\mathrm{I}} = C_2\big|_{\theta=\theta_0} + C_2^{'}\Big|_{\theta=\theta_0} \tag{6-77b}$$

$$C_3^{\mathrm{I}} = C_3\big|_{\theta=\theta_0} + C_3^{'}\Big|_{\theta=\theta_0} \tag{6-77c}$$

It can be seen $\lim\limits_{b\to 0} C_1^{\mathrm{I}} = 2C_1\big|_{b=0}$, $\lim\limits_{b\to 0} C_2^{\mathrm{I}} = 2C_2\big|_{b=0}$, $\lim\limits_{b\to 0} C_3^{\mathrm{I}} = 2C_3\big|_{b=0}$. However, when b=0, there is no singularity for point C, so, method (1) is unreasonable.
If method (2) is used,

$$C_1^{\mathrm{II}} = C_1^{'\mathrm{II}} = \frac{1}{3}(1-\alpha) \tag{6-78a}$$

$$C_2^{\mathrm{II}} = \left(1+\frac{\alpha}{2}\right)\frac{2}{\sqrt{3}}\cos\theta_0 + \left(\frac{\alpha(1-b)}{1+b}\right)\sin\theta_0 \tag{6-78b}$$

$$C_2^{'\mathrm{II}} = \left(\frac{2-b}{1+b}+\alpha\right)\frac{1}{\sqrt{3}}\cos\theta_0 + \left(\alpha+\frac{b}{1+b}\right)\sin\theta_0 \tag{6-78c}$$

$$C_3^{\mathrm{II}} = C_3^{'\mathrm{II}} = 0 \tag{6-78d}$$

For $C_2^{\mathrm{II}} \neq C_2^{'\mathrm{II}}$ (b $\neq$ 0), so the method (2) can not eliminate the singularity at point C.
If method (3) is used, it has

$$C_1^{\mathrm{III}} = \frac{1}{3}(1-\alpha) \tag{6-79a}$$

$$C_2^{\mathrm{III}} = \mu C_2\big|_{\theta=\theta_0} + (1-\mu)C_2^{'}\Big|_{\theta=\theta_0} \tag{6-79b}$$

$$C_3^{\mathrm{III}} = \mu C_3\big|_{\theta=\theta_0} + (1-\mu)C_3^{'}\Big|_{\theta=\theta_0} \tag{6-79c}$$

This method can eliminate the singularity of point C and $\lim_{b\to 0} C_2^{\text{III}} = C_2\big|_{b=0}$, $\lim_{b\to 0} C_3^{\text{III}} = C_3\big|_{b=0}$.

6.11 Suggested Methods

From the above discussion, method (1) and (2) are unreasonable, method (3) can not eliminate the singularity in some cases. Two simple methods are suggested to eliminate the singularity for the piecewise-linear yield functions.

Method (4), using the average of the flow vectors, i.e. assuming $\mu = 1/2$ in method 3. At point A, it has

$$
\begin{aligned}
C_1^{(A)} &= \frac{1}{3}(1-\alpha) \\
C_2^{(A)} &= \frac{1}{6J_2}(2+\alpha) \\
C_3^{(A)} &= \frac{1}{6}(1-\alpha)
\end{aligned}
\tag{6-80}
$$

At point B, the constants are

$$
\begin{aligned}
C_1^{'(A)} &= \frac{1}{3}(1-\alpha) \\
C_2^{'(A)} &= \frac{1}{3\sqrt{3}}(2\alpha+1) \\
C_3^{'(A)} &= \frac{1}{6J_2}(2\alpha+1)
\end{aligned}
\tag{6-81}
$$

They are equal to the corresponding parameters when $b = 1$.
At point C, the constants become

$$
\begin{aligned}
C_1^{(A)} &= \frac{1}{2}\left(C_1\big|_{\theta=\theta_0} + C_1^{'}\big|_{\theta=\theta_0} \right) = \frac{1}{3}(1-\alpha) \\
C_2^{(A)} &= \frac{1}{2}\left(C_2\big|_{\theta=\theta_0} + C_2^{'}\big|_{\theta=\theta_0} \right) \\
C_3^{(A)} &= \frac{1}{2}\left(C_3\big|_{\theta=\theta_0} + C_3^{'}\big|_{\theta=\theta_0} \right)
\end{aligned}
\tag{6-82}
$$

And it satisfies that $C_1^{(A)} = C_1\big|_{b=0}$, $\lim_{b\to 0} C_2^{(A)} = C_2\big|_{b=0}$, $\lim_{b\to 0} C_3^{(A)} = C_3\big|_{b=0}$.
The constants based on method (4) are reasonable because they can be degraded to that for the special cases when $b = 1$ at corners A and B, and when $b = 0$ at corner C.

Method (5): simply use the constants of the case of $b=1$ for points A and B, and the constants of $b=0$ for point C.
For point A, it has

$$
\begin{aligned}
C_1^{(B)} &= \frac{1}{3}(1-\alpha) \\
C_2^{(B)} &= \frac{1}{3\sqrt{3}}(2+\alpha) \\
C_3^{(B)} &= \frac{1}{6J_2}(2+\alpha)
\end{aligned}
\tag{6-83}
$$

For point B, the constants are

$$
\begin{aligned}
C_1'^{(B)} &= \frac{1}{3}(1-\alpha) \\
C_2'^{(B)} &= \frac{1}{3\sqrt{3}}(2\alpha+1) \\
C_3'^{(B)} &= \frac{1}{6J_2}(2\alpha+1)
\end{aligned}
\tag{6-84}
$$

For point C, they are

$$
\begin{aligned}
&C_1^{(B)} = \frac{1}{3}(1-\alpha) \qquad C_2^{(B)} = \left(1+\frac{\alpha}{2}\right)\frac{2}{\sqrt{3}}\cos\theta_0 + \alpha\sin\theta_0 \\
&C_3^{(B)} = -\frac{\sqrt{3}}{2J_2\sin 3\theta_0}\left[-\left(1+\frac{\alpha}{2}\right)\frac{2}{\sqrt{3}}\sin\theta_0 + \alpha\cos\theta_0\right]
\end{aligned}
\tag{6-85}
$$

These two methods are shown in Fig. 6.9, average of the piecewise-linear yield function flow vector is used for the method (4); the corner of the yield function is 'cut' by using the method (5).

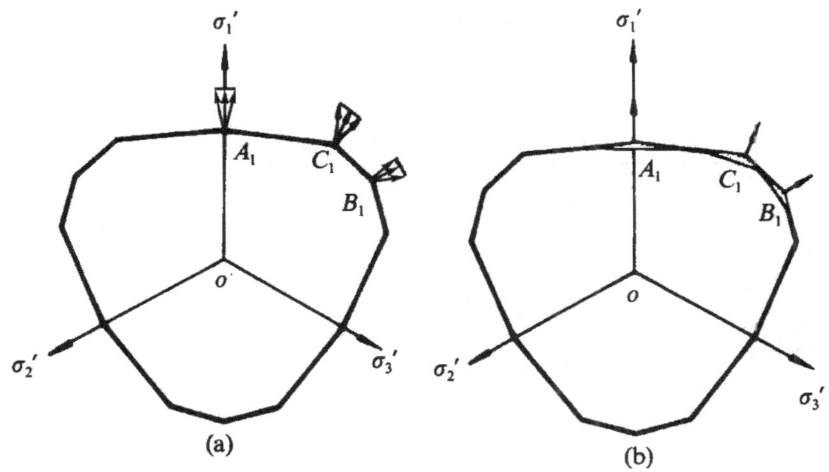

Fig. 6.9 Process of singular points

Table 6.2 shows the values of the parameters C_1, C_2, C_3 for different methods. From the table, it can be found that the parameters C_1, C_2, C_3 of points *A*, *B* for method (4) and method (5). These two methods are more reasonable than method (1) and method (2) and more applicable than method (3). It is very simple and physics concept is clear.

Table 6.2 Values of the parameters

Method	Point A			Point B		
	C_1	C_2	C_3	C_1	C_2	C_3
(1)	$\frac{2}{3}(1-\alpha)$	$\frac{4}{3\sqrt{3}}(2+\alpha)$	$\frac{1}{3J_2}(2+\alpha)$	$\frac{2}{3}(1-\alpha)$	$\frac{4}{3\sqrt{3}}(2\alpha+1)$	$-\frac{1}{3J_2}(2\alpha+1)$
(2)	$\frac{1}{3}(1-\alpha)$	$\frac{1}{\sqrt{3}}(2+\alpha)$	0	$\frac{1}{3}(1-\alpha)$	$\frac{1}{\sqrt{3}}(2\alpha+1)$	0
(3)	$\frac{1}{3}(1-\alpha)$	$\mu\neq\frac{1}{2},\infty$ $\mu=\frac{1}{2}$ $\frac{2}{3\sqrt{3}}(2+\alpha)$	$\mu\neq\frac{1}{2},\infty$ $\mu=\frac{1}{2}$ $\frac{1}{6J_2}(2+\alpha)$	$\frac{1}{3}(1-\alpha)$	$\mu\neq\frac{1}{2},\infty$ $\mu=\frac{1}{2}$ $\frac{2}{3\sqrt{3}}(2\alpha+1)$	$\mu\neq\frac{1}{2},\infty$ $\mu=\frac{1}{2}$ $-\frac{1}{6J_2}(2\alpha+1)$
(4)	$\frac{1}{3}(1-\alpha)$	$\frac{2}{3\sqrt{3}}(2+\alpha)$	$\frac{1}{6J_2}(2+\alpha)$	$\frac{1}{3}(1-\alpha)$	$\frac{2}{3\sqrt{3}}(2\alpha+1)$	$-\frac{1}{6J_2}(2\alpha+1)$
(5)	$\frac{1}{3}(1-\alpha)$	$\frac{2}{3\sqrt{3}}(2+\alpha)$	$\frac{1}{6J_2}(2+\alpha)$	$\frac{1}{3}(1-\alpha)$	$\frac{2}{3\sqrt{3}}(2\alpha+1)$	$-\frac{1}{6J_2}(2\alpha+1)$

Table 6.3 shows the parameters of method (4) and (5) when α, b are given different values. From this table, it can be found that the method (5) is independent of the parameter b, and is close to method (4) when b is near 0.

Table 6.3 Comparison of the parameters

α	Method (5)		Method (4)									
	$0\le b\le 1$		$b=0$		$b=0.25$		$b=0.5$		$b=0.75$		$b=1$	
	c_2	$c_3\cdot J_2$	c_2	$c_3\cdot J_2$	c_2	$c_3\cdot J_2$	c_2	$c_3\cdot J_2$	c_2	$c_3\cdot J_2$	c_2	$c_3\cdot J_2$
$\alpha=1$	2	0	2	0	1.8	0	1.67	0	1.6	0	1.5	0
$\alpha=0.75$	1.81	0.26	1.8	0.26	1.62	0.21	1.5	0.17	1.4	0.15	1.3	0.13
$\alpha=0.5$	1.78	0.58	1.8	0.58	1.58	0.47	1.44	0.39	1.4	0.33	1.3	0.29
$\alpha=0.25$	2.32	1.31	2.3	1.31	1.98	1.05	1.76	0.88	1.6	0.75	1.5	0.66

6.12 Unified Process of the Corner Singularity

From the above analysis, it can be found

1. The expression of the piece-linear yield function is simple and easy to use. It can be adopted for various close-form analyses for various classical elasto-plastic problems. However, the singularity of plastic flow for the piecewise-linear yield function may cause some troubles to the elasto-plastic flow vector calculation. Process of singular points for the piece-wise functions is necessary.
2. Traditional methods used to eliminate singularity are unreasonable or can not be used in all cases. There exist obvious errors and drawbacks.
3. The two suggested methods are easy to handle the singularity problem and can be used conveniently to eliminate the singularity of the present piecewise-linear yield function plastic flow.
4. Using the unified strength theory as the yield potential function and adopting the suggested unified process of the flow vector singularity can solve the singularity of all kinds of corner which improves the calculation efficiency. The processing method of the singularity problem can be implemented into a computer program conveniently and in a unified form.

   Different results of single yield function can be obtained as follows:

Tresca Yield Criterion

When $\theta=0°$, $\theta=60°$, it has

$$f=\sqrt{3}\sqrt{J_2}-\sigma_s=0 \tag{6-86}$$

Derivation as Eqs. (6-80) - (6-82), it has

$$C_1 = 0, C_2 = \sqrt{3}, C_3 = 0 \tag{6-87}$$

Comparing with Table 6.1, it can be found that in the corner of Tresca, this result is the same with the one of Huber-von Mises.

Mohr-Coulomb Yield Criterion

$$\begin{aligned}\theta = 0^\circ, F &= \frac{1}{3}I_1 \sin\varphi + \frac{1}{2}\sqrt{\frac{J_2}{3}}(3+\sin\varphi) - C_0\cos\varphi = 0 \\ \theta = 60^\circ, F &= \frac{1}{3}I_1 \sin\varphi + \frac{1}{2}\sqrt{\frac{J_2}{3}}(3-\sin\varphi) - C_0\cos\varphi = 0\end{aligned} \tag{6-88}$$

Derivation as Eqs. (6-80) - (6-82), it has

$$\begin{aligned}\theta = 0^\circ, C_1 &= \frac{1}{3}\sin\varphi, C_2 = \frac{1}{2\sqrt{3}}(3+\sin\varphi), C_3 = 0 \\ \theta = 60^\circ, C_1 &= \frac{1}{3}\sin\varphi, C_2 = \frac{1}{2\sqrt{3}}(3-\sin\varphi), C_3 = 0\end{aligned} \tag{6-89}$$

Twin-shear Yield Criterion

At the corner

$$\theta_b = \arctan\frac{\sqrt{3}}{3} = \frac{\pi}{6} \tag{6-90}$$

$$f = f' = \frac{3}{2}\sqrt{J_2} - \sigma_3 = 0 \tag{6-91}$$

Then

$$C_1 = 0, C_2 = \sqrt{3}, C_3 = 0 \tag{6-92}$$

Generalized Twin-shear Stress Yield Criterion

The corner is not at $\theta = \pi/6$ for the generalized twin-shear stress yield criterion. From $F = F'$, the angle can be obtained as Fig. (6-9).

$$\theta_b = \arctan\left(\frac{\sqrt{3}(1+\sin\varphi)}{3-\sin\varphi}\right) \tag{6-93}$$

At the corner

$$F = F' = \frac{2}{3} I_1 \sin\varphi + \sqrt{\frac{J_2}{3}} (3 + \sin\varphi) \cos\theta_b - 2C_0 \cos\varphi \tag{6-94}$$

From Eqs. (6-80) - (6-82), it has

$$C_1 = \frac{2}{3} \sin\varphi, C_2 = \frac{1}{\sqrt{3}} (3 + \sin\varphi) \cos\theta_b, C_3 = 0 \tag{6-95}$$

After the handling of the generalized twin-shear yield criterion, the direction and relative value of plastic flow vector on the deviatoric plane are shown in Fig. 6.10.

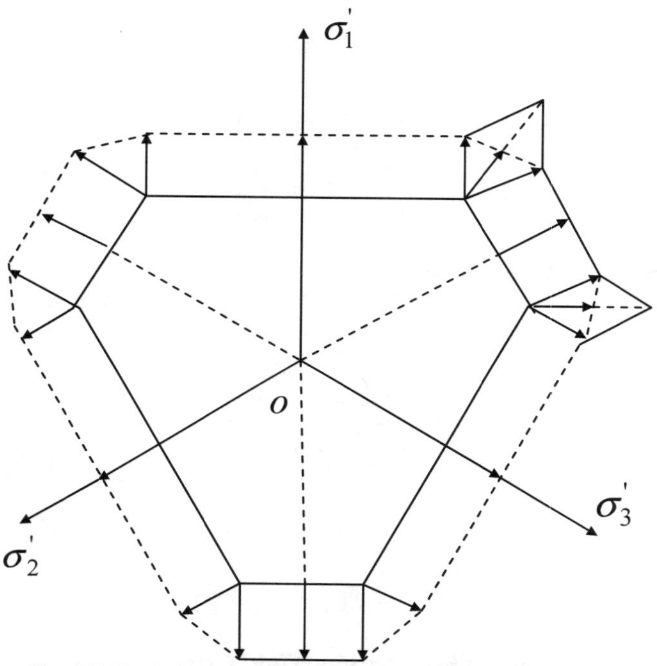

Fig. 6.10 Plastic flow of the unified strength theory (b=1)

From the above singularity process, it can get a certain value of the flow vector for any single criterion.

For the unified strength theory, the constants can be derived with unified solution. The constant C_i of the unified strength theory is shown in Table 6.4.

The direction and the relative value of plastic flow vector for the unified strength theory are shown in the solid lines of Fig. 6.11. It can be seen, the singularity processing method can lead to a reasonable, uniformly, continually variable flow vector on the yield surface.

Table 6.4 Parameters for the unified strength theory

θ	b	C_1	C_2	C_3
$\theta=0^\circ$	$b=1$	$\frac{1}{3}(1-\alpha)$	$\frac{2}{3\sqrt{3}}(2+\alpha)$	$\frac{1}{6J_2}(2+\alpha)$
	$0\le b<1$	$\frac{1}{3}(1-\alpha)$	$\frac{1}{\sqrt{3}}(2+\alpha)$	0
$\theta=60^\circ$	$b=1$	$\frac{1}{3}(1-\alpha)$	$\frac{2}{3\sqrt{3}}(2\alpha+1)$	$-\frac{1}{6J_2}(2+\alpha)$
	$0\le b<1$	$\frac{1}{3}(1-\alpha)$	$\frac{1}{\sqrt{3}}(1+2\alpha)$	0
$\theta=\theta_b$	$0<b\le 1$	$\frac{1}{3}(1-\alpha)$	$\frac{(2+\alpha)}{\sqrt{3}}\cos\theta - \frac{(2+\alpha)\cos 3\theta}{\sqrt{3}(3\cos^2\theta-\sin^2\theta)}$	$\frac{(2+\alpha)}{2J_2(3\cos^2\theta-\sin^2\theta)}$

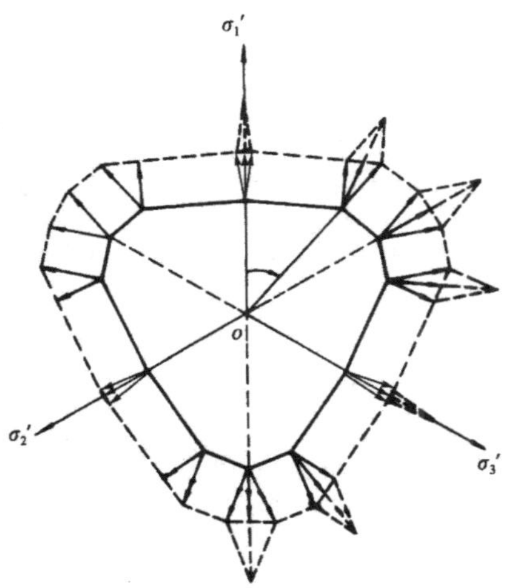

Fig. 6.11 Plastic strain increments of the unified strength theory (0<b<1, Yu 1998)

Summary

The plastic stress-strain relation is an important part of plasticity. The associated flow rule and the incremental constitutive relations for the unified strength theory are described. The mathematical expression of the unified strength theory is simple and linear, however, the yield surface of the unified strength theory is piece-wise linear. The plastic flow vector exists singular points for the piecewise-linear yield functions. The flow vector at the corners is not unique when the normality condition is applied. A simple and unified method is suggested for solving the singularity. The plastic strain increments of the unified strength theory in whole region are shown in Fig.6.11.

The plastic deformation in uniaxial stress state, three-dimensional elastic stress strain relation, plastic work hardening and strain hardening, plastic flow rule, Drucker postulate and convexity of the loading surface, incremental constitutive equations in matrix formulation, determination of flow vector for different yield functions, singularity of piecewise-linear yield functions, process of the plastic flow singularity, and the suggested unified process of the corner singularity are described in this chapter.

Problems

Problem 6.1.
Express the Tresca criterion (b=0, =1) and the Huber-von Mises criterion in the form of $f(I_1, J_2, \ ) = k$.

Problem 6.2.
Express the Mohr-Coulomb criterion (b=0, 1) and the Drucker-Prager criterion in the form of $f(I_1, J_2, \ ) = k$.

Problem 6.3.
Express the twin-shear criterion (b=1, =1) in the form of $f(I_1, J_2, \ ) = k$.

Problem 6.4.
Express the generalized twin-shear criterion (b=1, 1) in the form of $f(I_1, J_2, \ ) = k$.

Problem 6.5.
Express the unified strength theory in the form of $f(I_1, J_2, \ ) = k$.

Problem 6.6.
Derive the incremental elasto-plastic stiffness matrix for the Tresca and the Huber-von Mises criteria.

Problem 6.7.
Derive the incremental elasto-plastic stiffness matrix for the Mohr-Coulomb criteria.

Problem 6.8.
Derive the incremental elasto-plastic stiffness matrix for the twin-shear yield criterion.

Problem 6.9.
Derive the incremental elasto-plastic stiffness matrix for the generalized twin-shear yield criterion.

Problem 6.10.
Derive the incremental elasto-plastic stiffness matrix for the unified strength theory.

References and Bibliography

Chakrabarty J (1987) *Theory of Plasticity*. McGraw-Hill, New York.

Drucker DC (1952) A more fundamental approach to plastic stress-strain relations, 1st U.S. Congress of Applied Mechanics, ASME, New York, pp.116–126.

Hill R (1950) *The Mathematical Theory of Plasticity*. Clarendon, Oxford.Johnson W and Mellor PB (1962) *Plasticity for Mechanical Engineers*. Van Nostrand, Lonton and New York.

Khan AS (1995) *Continuum Theory of Plasticity*. John Wiley & Sons, New York.

Koiter WT (1953) Stress-strain relations, uniqueness and variational theorems for elastic-plastic materials with singular yield surface, Quart. Appl. Math., 11, 350–354.

Martin J.B (1975) *Plasticity: Fundamentals and General Results*. The MIT Press.

Mendelson A (1968) *Plasticity: Theory and Application*. MaCmillan, New York.

Nayak GC and Zienkiewicz OC (1972) Convenient form of stress invariants for plasticity, Journal of the Structure Division, Proc. Of ASCE, 949–953.

Owen DRJ and Hinton E (1980) *Finite Elements in Plasticity: Theory and Practice*. Pineridge Press Limited, Swansea, UK.

Yu MH (1983) Twin shear stress yield criterion, Int. J. of Mechanical Science, 25(1), 71–74.

Yu MH, He LN, and Song LY (1985) Twin shear stress strength theory and its generalization, Scientia Sinica (Sciences in China), series A, 28 (11), 1174–1183.

Yu MH, and He LN (1991) A new model and theory on yield and failure of materials under the complex stress state, Mechanical Behavior of Materials-6, Pergamon Press, Oxford, vol.3, pp 841–846.

Yu Mao-Hong, Zeng Wen-Bing, (1994) New theory of engineering structural analysis and its application. *J. of Engineering Mechanics,* **11**(1),9–20. (in Chinese, English abstract)

Yu Mao-hong (1998) *Twin-shear Theory and Its Applications*. Science Press, Beijing (in Chinese)

Yu MH (1999) *Engineering Strength Theory*, Higher Education Press, Beijing (in Chinese).

Yu MH, Yang SY and Fan SC (1999) Unified elasto-plastic associated and non-ass0ciated constitutive model and its application. *Computers and Structures,* **71**(6), 627–636.

Yu Mao-hong (2004) *Unified Strength theory and Its Applications*. Springer, Berlin.

Zienkiewicz OC and Pande GN (1977) Some useful forms of isotropic yield surfaces for soil and rock mechanics, Finite Elements in Geomechanics, ed. Gudehus, John Wiley & Sons Ltd. pp 179–190.

7 Concrete Plasticity

7.1 Introduction

The unified strength theory and its associated flow rule have been introduced in details in the preceding chapters. The unified strength theory has the advantage to be applied to different materials. The yield function for a certain material, however, must satisfy experimental result obtained in some special stress states. For geomaterials, there is a large amount of test data reported by different researchers. To fit the test data, various multi-parameter yield functions have been proposed. In this chapter, a multi-parameter unified yield criterion based on the unified strength theory (Wang 1998; Fan and Wang 2002) will be introduced. Its associated flow rule as well as strain softening simulation will be presented.

Strength criterion determines the stress level at which the material behavior changes drastically. Due to infinity of combinations of stress components, it is impossible to determine all strengths of a certain material by experiments in every stress combination. So assumptions based on material deformation and failure characteristics must be promoted. Generally, limit strength surface of an isotropic material is convex, its tensile and compressive meridians are different for brittle materials such as concrete and rock (Chen 1998; Yu 1998). The yield surface is three-fold symmetric about hydrostatic axis in principal stress space because the three principal stresses have reciprocal relationship. To meet these requirements, trajectories of a strength theory must pass through three symmetric radii r_t and r_c respectively and go between the two hexagons as shown in Fig 7.1. It is seen that any trajectories outside this range will break the rule of convexity. The upper limit hexagon in Fig. 7.1 is the twin shear strength criterion (Yu et al. 1985), while the lower limit hexagon is the Mohr-Coulomb criterion. Many strength theories that satisfy the aforementioned requirements have been suggested. With introduction of a strength criterion coefficient '*b*', this strength theory embraces or approximates all convex trajectories between the two hexagons in Fig. 7.1 with different values of *b*. It is suitable for any isotropic materials. So it can be said that the unified twin shear strength theory unified the exiting isotropic material strength criteria.

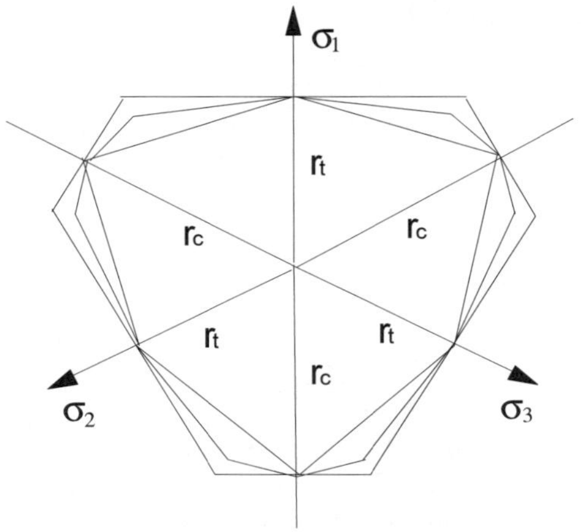

Fig. 7.1 General upper and lower limit strength surfaces of an isotropic material

This strength theory has a clear mechanical background because its definition is that when a function of all the independent stress components on orthogonal octahedral twin shear element reaches a certain value, material will yield or fail. Its shape function has two pieces of linear segments in the deviatoric plane makes the strength theory comparatively easier to use. However, there comes out the problem of the determination for meridians. The unified twin shear strength criterion may take on different form of formula for its meridians to cater for different material failure behaviour. Some materials like concrete may require a curvilinear formulation. Till now, the application of the general form of the unified strength criterion is limited to linear formulation for meridians only. The application of multi-parameter unified twin shear strength criterion to take in non-linear term of the hydrostatic stress is restrained to certain form of unified strength criterion with specific value of b=1, i.e. twin shear strength criterion instead of unified strength theory. Herein, unified form of multi-parameter unified strength criterion with curvilinear term of hydrostatic stress is derived for concrete material. The determination of coefficients in the criterion is very complicated, it is suggested that the concept of unified twin shear strength theory is applied for deviatoric shape functions only with meridians take complete separate formulae. The meridians take different form for different materials, it may be determined by experimentally-fitted curves. In the following sections, discussion on the multi-parameter unified yield criterion is mainly with respect to concrete material. The multi-parameter unified yield criterion based on the unified strength theory can be conveniently extended to other geomaterials if the required test data is available.

Considering the traditional yield criteria for geomaterials, the Mohr-Coulomb strength criterion is an irregular hexagonal pyramid. Its deviatoric sections are irregular hexagons as show in Fig. 7.2. The Drucker-Prager strength criterion is a simple adaptation of the Huber-von Mises criterion for materials that has different tensile, compressive properties by introduction of an additional term to reflect the influence of the hydrostatic stress component on failure (see Fig. 7.3). The disadvantage of this criterion is that its projection on the deviatoric plane is a circle. Its tensile and compressive meridians are therefore the same straight line, implying that the Lode angle θ has no effect on the material strength. Thus its application is very limited.

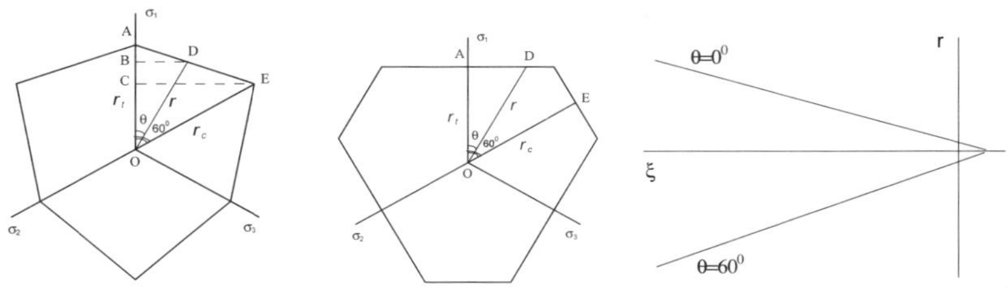

Fig. 7.2 Meridians and deviatoric loci for the Mohr-Coulomb criterion and twin-shear criterion

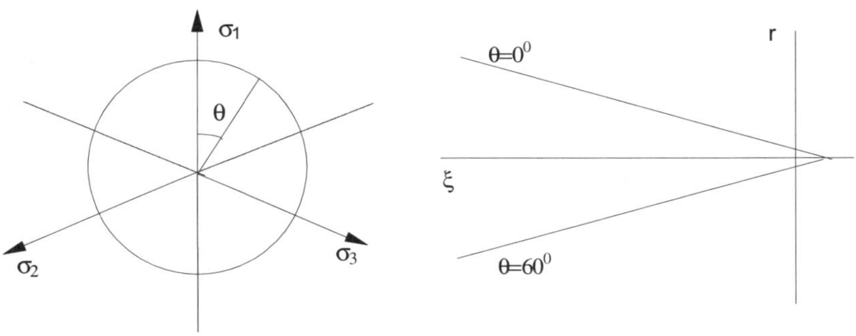

Fig. 7.3 Meridians and deviatoric section for Drucker-Prager criterion

7.2 Multi-Parameter Yield Criteria

7.2.1 Ottosen Four-Parameter Criterion

To meet the complicated geometric requirements of the failure surface for concrete materials, Ottosen (1977) suggested the following criterion accounting for the effect of all the three stress invariants,

$$\frac{aJ_2}{f_c^2}+\frac{\lambda\sqrt{J_2}}{f_c}+\frac{bI_1}{f_c}=1 \tag{7-1}$$

where λ is a function of cos3θ,

$$\lambda=\begin{cases} n_1\cos\left[\frac{1}{3}\cos^{-1}(n_2\cos3\theta)\right] & for \quad \cos3\theta\geq 0 \\ n_1\cos\left[(\frac{\pi}{3}-\frac{1}{3}\cos^{-1}(-n_2\cos3\theta)\right] & for \quad \cos3\theta\geq 0 \end{cases} \tag{7-2}$$

here a, b, n_1, n_2 are material constants. The failure surface of this criterion has curved meridians and noncircular cross sections on the deviatoric plane as shown in Fig. 7.4. The cross sections have the geometric characteristics of three-fold symmetry, convexity and changing shapes from nearly triangular to nearly circular along the hydrostatic stress axis, which meet the concrete failure characteristics. It has several strength criteria as its special cases, e.g., when a=b=0, λ=constants, it becomes the Huber-von Mises criterion, when a=0, b≠0 and λ=constants, the Durcker-Prager criterion is obtained. In general, this four-parameter criterion is suitable for concrete in a wide range of stress combinations. However, the expression for the λ-function makes the criterion very inconvenient. And it gives over-estimated predictions at high hydrostatic stress.

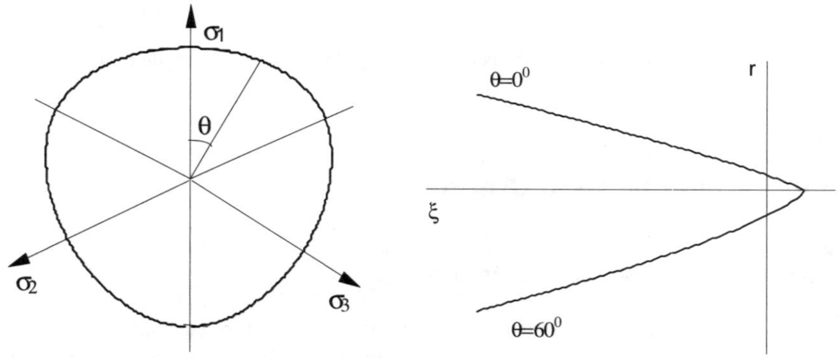

Fig. 7.4 Meridians and deviatoric section for Ottosen criterion

7.2.2 William-Wranke Five-Parameter Criterion

William-Wranke five-parameter criterion has separate expression for meridians and shape functions. It has parabolic tensile and compressive meridians expressed by

$$r_t = a_0 + a_1\sigma_m + a_2\sigma_m^2 \tag{7-3}$$

$$r_c = b_0 + b_1\sigma_m + b_2\sigma_m^2 \tag{7-4}$$

in which $a_0, a_1, a_2, b_0, b_1, b_2$ are material constants. Since tensile and compressive meridians should intersect the hydrostatic axis at the same point, only five stress points are required to determine the six parameters. William-Wranke's suggestion for the strength constants is that α=0.15, $\overline{\alpha} = 1.8$ and two strength points on tensile and compressive meridians respectively, i.e., (ξ_1, r_1, θ_1)=(−3.67, 1.5, 0°), (ξ_2, r_2, θ_2)=(-3.67, 1.94, 60°). Here α is the ratio of uniaxial tensile strength f_t to uniaxial compressive strength f_c, $\overline{\alpha}$ is the ratio of equal bi-axial compressive strength f_{cc} to uniaxial compressive strength f_c. On the above assumption, the material constants are derived as a_0 =0.081143, a_1 =-0.52553, a_2 =-0.03785, b_0=0.11845, b_1=-0.76444, b_2=-0.07305. William-Wranke five-parameter criterion has a critical hydrostatic stress at about $\xi = -12.02$. Its trajectories on deviatoric planes have elliptical expression of

$$r(\theta) = \frac{2r_c(r_c^2 - r_t^2)\cos\theta + r_c\,(2r_t - r_c)\left[\sqrt{4(r_c^2 - r_t^2)\cos^2\theta + 5r_t^2 - 4r_c r_t}\,\right]}{4(r_c^2 - r_t^2)\cos^2\theta + (r_c - 2r_t)^2} \tag{7-5}$$

The failure surfaces of this criterion are convex and smooth everywhere (see Fig. 7.5). The limitation of this criterion is that the ellipse degenerates into a circle when r_t/r_c=1 (similar to the deviatoric trance of the von Mises and Drucker-Prager criteria), and when the ratio r_t/r_c approaches the value of 0.5, the deviatoric trace becomes nearly triangular (similar to that of maximum tensile-stress criterion). Besides its expression for the shape function on the deviatoric plane is very complicated and not convenient to be used in analytical and numerical analyses.

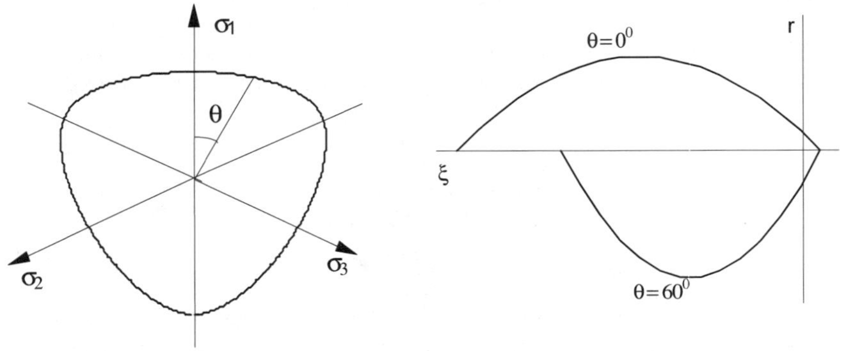

Fig. 7.5. Meridians and deviatoric section for William-Wranke criterion

7.2.3 Podgorski Concrete Criterion

Studying on the existing failure criteria, Podgorski found out that the shape of deviatoric cross section of the failure surface of existing criteria just adopt one parameter, which is not enough to capture the complex features of concrete strength surface. He suggested a general form of yield or failure criterion for most of the materials such as metal, rock, concrete, clay, etc. Specific form of failure criterion for concrete is also given. In his concrete criterion, two parameter, i. e. λ, φ are used in the shape function, which are the ratio of radius at $\theta = 0^0$ and $\theta = 30^0$ to that at $\theta = 60^0$ respectively.

The expression of the Podgoski concrete criterion is

$$\sigma_0 - C_0 + C_1 P \tau_0 + C_2 \tau_0^2 = 0 \tag{7-6}$$

in which

$$P = \cos((\arccos \alpha' J)/3 - \beta') \tag{7-7}$$

And C_0, C_1, C_2, α, β are constants, $J = \cos 3\theta$, σ_0 is octahedral normal stress and

$$\sigma_0 = \frac{\xi f_c}{\sqrt{3}} \tag{7-8}$$

τ_0 is octahedral stress and

$$\tau_0 = \frac{rf_c}{\sqrt{3}} \tag{7-9}$$

The demerit of this criterion is that its derivation of parameters is very complicated. Fig. 7.6 shows characteristics for the shape functions and meridians of the Podgorski concrete criterion.

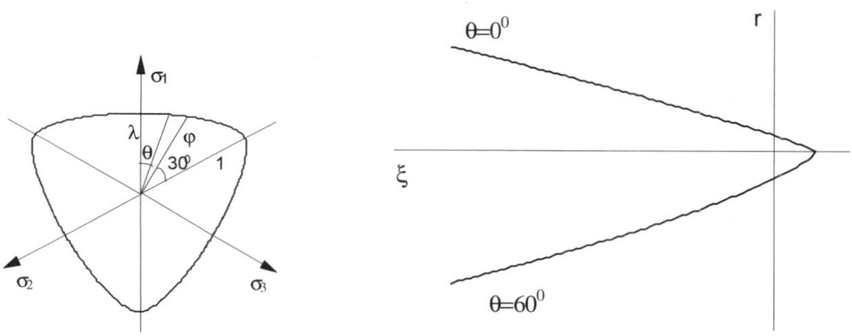

Fig. 7.6 Meridians and deviatoric section for Podgorski criterion

7.2.4 Kotsovos Concrete Criterion

Kotsovos and Pavlovic (1995) derived a concrete strength criterion by fitting curves to experimental results. The compressive and tensile meridians and the deviatoric section of this criterion take from of potential function as shown in Fig. 7.7. Their expressions in terms of octahedral stresses are

$$\frac{\tau_{0c}}{f_c} = 0.944\left(\frac{\sigma_0}{f_c} + 0.05\right)^{0.724} \tag{7-10}$$

$$\frac{\tau_{0t}}{f_c} = 0.633\left(\frac{\sigma_0}{f_c} + 0.05\right)^{0.857} \tag{7-11}$$

Eqs. (7-10) and (7-11) represent two open-ended convex envelops, the slopes of which tend to become equal to that of the space diagonal in principal stress space as σ_0 tends to infinity. Once the values of τ_{0t}, τ_{0c} (the octahedral shear stress at θ=0°, θ=60°

respectively) are determined for various levels of the hydrostatic stress, the value of τ_{0u} with respect to a angle θ in between 0° and 60° is obtained as follows,

$$\tau_{ou} = \frac{2\tau_{oc}(\tau_{oc}^2 - \tau_{ot}^2)\cos\theta + \tau_{oc}(2\tau_{ot} - \tau_{oc})\sqrt{4\tau_{oc}^2 - \tau_{ot}^2)\cos^2\theta + 5\tau_{ot}^2 - 4\tau_{oc}\tau_{ot}}}{4(\tau_{oc}^2 - \tau_{ot}^2)\cos^2\theta + (\tau_{oc}^2 - 2\tau_{ot}^2)^2} \tag{7-12}$$

This expression is elliptical, which describes a smooth convex curve with tangents perpendicular to the directions of τ_{0t}, τ_{0c} at 0° and 60° respectively. Eqs. (7-10), (7-11) and (7-12) define an ultimate strength surface which conforms with generally accepted shape requirements such as three-fold symmetry, convexity with respect to the space diagonal, open-end shape which tends to be cylindrical as σ_0 tends to be infinite (Franklin 1970) (see Fig. 7.7). The accuracy of the whole strength model was assessed by comparing the predictions with wide body of experimental data (Kotsovos and Pavlovic 1995). The comparison shows that Kotsovos's model can give fair average to scattered test data of two-dimensional compression-compression and tension-compression stress states. However, its expression on the deviatoric plane is too complicated.

Through analysis of the aforementioned several representative multi-parameter concrete criteria, we can see that for all their merits, they still have their own demerits. The parameters in the Podgorski concrete criterion are very complicated to calculate. Although Kotsovos criterion may be the most preferable concrete criterion, it lacks a theoretical background and its shape function is too complicated.

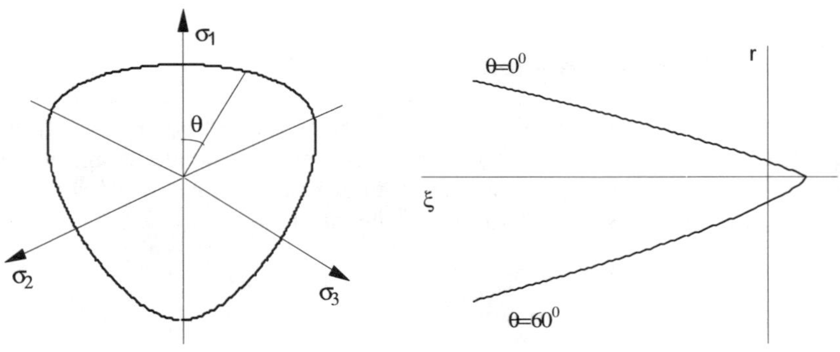

Fig. 7.7 Meridians and deviatoric section for Kotsovos criterion

7.3 Multi-Parameter Unified Yield Criterion

Although the derived multi-parameter unified twin shear concrete strength criterion has piece-wise linear trajectory and it adopts two parameter in the shape function, the derivation of its tensile and compressive meridians are very tedious and the meridians are not so convenient for use because the introduction of the revising tangential lines. Reminiscent of the essence of unified strength theory in the previous chapters, its fundamental is that it unites the existing criteria by introduction of a weighted coefficient b in the mathematical expressions. It can be united into any form of the tensile and compressive meridians as it can define its meridians by adopting any form of expression of hydrostatic stress in its formulae. It is suggested to generalise the unified twin shear strength theory by adopting a general function of the hydrostatic stress only in it (Wang 1998; Fan and Wang 2002). Now the formulae of this theory becomes

$$F=\tau_{13}+b\tau_{12}+\beta(\sigma'_{13}+b\sigma'_{12})+A_0g_0(\sigma_m)+bA_1g_1(\sigma_m)=C \quad \text{when } F\geq F' \tag{7-13a}$$

$$F'=\tau_{13}+b\tau_{23}+\beta(\sigma'_{13}+b\sigma'_{23})+A_0g_0(\sigma_m)+bA_2g_0(\sigma_m)=C \quad \text{when } F\leq F' \tag{7-13b}$$

The functions of the hydrostatic stress in Eqs. (7-13a), (7-13b) may be expressed in any form of formula and change with different materials. However, it is very difficult to derive the coefficients in the generalised unified twin shear strength theory expressed in Eqs. (7-13a), (7-13b) except for that $g_0(\sigma_m)$, $g_1(\sigma_m)$, $g_2(\sigma_m)$ are linear functions. So it is further suggested that for any material, the meridians and the shape functions are expressed separately. The expression of tensile and compressive meridians is functions of the hydrostatic stress only and may be determined by catering for experimental results. For metals, it can adopt linear formulation and for geomaterials, it may take up the form of polynomials. For shape function, the piecewise-linear formulation in line with the concept of the unified twin shear strength theory can give very successful approximation. The weighted coefficient b is still used in the formulation of the suggested shape function. The formula is determined as that when b=1, it also give the upper limit of the convex shape function and when b=0, it gives the lower limit convex shape function.

In Fig. 7.8, the triangle *ABC* is similar to that of *ACF*, so we obtain

$$\frac{\overline{BE}}{\overline{CF}}=\frac{\overline{AB}}{\overline{AC}} \tag{7-14}$$

and

$$\overline{BE}=r\sin\theta \tag{7-15}$$

$$\overline{CF}=r_c\sin 60^0 \tag{7-16}$$

$$\overline{AB}=r_t-r\cos\theta \tag{7-17}$$

$$\overline{AC} = r_t - r_c \cos 60^0 \tag{7-18}$$

Substitute Eqs. (7-15) - (7-18) into (7-14) and rewrite Eq. (7-14), it is derived that

$$r = \frac{r_t r_c \sin 60^0}{r_t \sin\theta + r_c \sin(60 - \theta)} \quad \text{for } b=0 \tag{7-19}$$

And it is easy to derive the following formulae for b=1 as shown in Fig. 7.8.

$$\begin{cases} r = \dfrac{r_t}{\cos\theta} & when \quad 0^o \leq \theta \leq \theta_b \\ r = \dfrac{r_c}{\cos(60^o - \theta)} & when \quad \theta_b \leq \theta \leq 60^0 \end{cases} \quad \text{for } b=1 \tag{7-20}$$

And when $0 < b < 1$, the formula is determined by a linear interpolation of the those of b=1 and b=0 as follows,

$$\begin{cases} r = \dfrac{r_t r_c \sin 60^o}{r_t \sin\theta + r_c \sin(60^o - \theta)}(1-b) + b\dfrac{r_t}{\cos\theta} & when \quad 0^o \leq \theta \leq \theta_b \\ r = \dfrac{r_t r_c \sin 60^o}{r_t \sin\theta + r_c \sin(60^o - \theta)}(1-b) + b\dfrac{r_c}{\cos(60^o - \theta)} & when \quad \theta_b \leq \theta \leq 60^0 \end{cases}$$

(7-21)

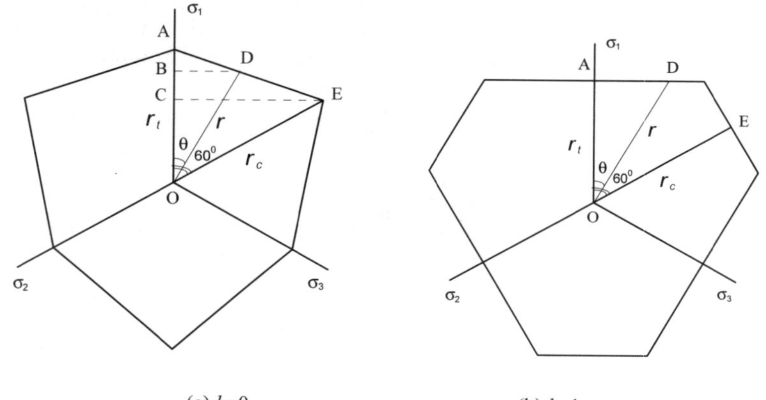

(a) b=0 (b) b=1

Fig. 7.8 Derivation of generalised unified strength model

The value θ_b is derived from Eq. (7-20) by equating the two vector length as follows,

$$\theta_b = \arctan\left[\frac{1}{\sqrt{3}}\left(\frac{2r_c}{r_t} - 1\right)\right] \tag{7-22}$$

The value of b is different for different materials and it may vary over the hydrostatic level for a specific material such that the deviatoric sections along the hydrostatic axis are different. As for meridians, we can take up a linear formulation of the Mohr-Coulomb criterion, the William-Wranke's second-order polynomials, the Podgorski's parabolas or the Kotsovos's potential functions.

The meridians of the Mohr-Coulomb criterion is

$$r_t = a_0 + a_1\xi \tag{7-23a}$$

$$r_c = b_0 + b_1\xi \tag{7-23b}$$

where

$$\begin{cases} a_0 = \dfrac{\sqrt{3}\alpha}{\sqrt{2}\left(1+\dfrac{\sqrt{3}}{2}\alpha\right)}, & a_1 = \dfrac{\alpha \quad -1}{\sqrt{2}\left(1+\dfrac{\sqrt{3}}{2}\alpha\right)} \\ b_0 = \dfrac{\sqrt{3}\alpha}{\sqrt{\dfrac{3}{2}+\alpha}}, & b_1 = \dfrac{\alpha \quad -1}{\sqrt{\dfrac{3}{2}+\alpha}} \end{cases} \tag{7-24}$$

The meridians of the William-Wranke's five-parameter criterion takes the form of

$$r_t = a_0 + a_1\xi + a_2\xi^2 \tag{7-25a}$$

$$r_c = b_0 + b_1\xi + b_2\xi^2 \tag{7-25b}$$

According to William-Wranke's suggestion, the constants in (7-25), (7-25b) are derived as

$$\begin{cases} a_0 = 0.081143, \quad a_1 = -0.3034, \quad a_2 = 0.01262 \\ b_0 = 0.11845, \quad b_1 = -0.44135, \quad b_2 = -0.02435 \end{cases} \tag{7-26}$$

The expressions of the meridians for the Podgorski's concrete criterion is complicated. It is written in a form that the hydrostatic stress is a function of the radii on the deviatoric planes as follows,

$$\xi = \frac{\sqrt{3}C_0}{f_c} - C_1 P_0 r_t + \frac{C_2 f_c}{\sqrt{3}} r_t^2 \tag{7-27a}$$

$$\xi = \frac{\sqrt{3}C_0}{f_c} - C_1 P_1 r_c + \frac{C_2 f_c}{\sqrt{3}} r_c^2 \tag{7-27b}$$

Here C_0, C_1, C_2 are material constants, and P_0, P_1 are expressed as

$$P_0 = \cos\left(\frac{1}{3}\arccos(\alpha'\cos 0^0) - \beta'\right) \tag{7-28a}$$

$$P_1 = \cos\left(\frac{1}{3}\arccos(\alpha'\cos 60^0) - \beta'\right) \tag{7-28b}$$

The value of α', β' are given with respect to different ratios of the tensile and compressive strength.

The formulae of the meridians for the Kotsovos' criterion, however, are very simple. The expressions are

$$r_t = 0.633\sqrt{3}\left(\frac{-\xi}{\sqrt{3}} + 0.05\right)^{0.857} \tag{7-29a}$$

$$r_c = 0.944\sqrt{3}\left(\frac{-\xi}{\sqrt{3}} + 0.05\right)^{0.724} \tag{7-29b}$$

where are curve-fitted from experimental test data.

Figs. 7.9 and 7.10 give the comparison of the generalised unified concrete criterion with Podgorski concrete criterion and Kotsovos criterion respectively. 'Guc' stands for the generalised unified concrete criterion in these two figures. It is seen that the shape functions of the suggested piece-wise formulae with that of the Podgorski criterion agrees very well in comparatively low hydrostatic stress while it gives very close approximation to Kotsovos criterion till very high hydrostatic stress is obtained. The value of b (b=0.6) herein adopted in these figures are calculated with the value of α=0.06 and α =1.20, which are average values for common concrete. The meridians of the Kotsovos criterion are shown in Eqs. (7-29a) and (7-29b) are adopted for the generalized unified concrete criterion. On the deviatoric plane, a generalized form of the multi-parameter unified yield function is used. In this way, the new criterion incorporates the unified strength theory on the deviatoric sections and accurate experimental results of the meridians. The expressions are relatively simple, however, the physical and geometrical meanings of the parameters are very clear. This criterion is supposed to provide a most preferable concrete criterion, which unifies the existing concrete criteria. Also the generalised piece-wise formulation is adaptable for all other materials when different meridians are adopted.

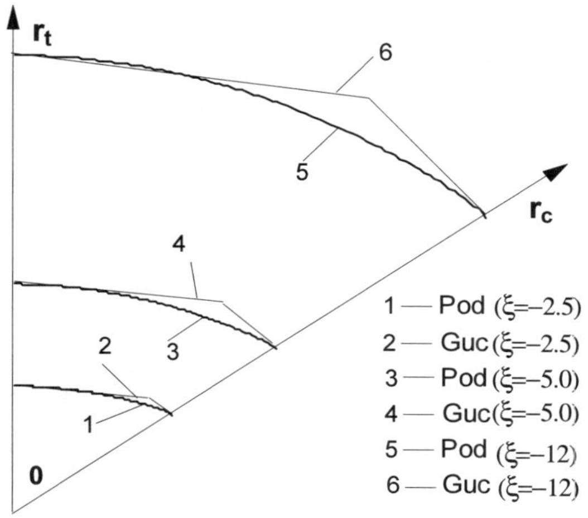

Fig. 7.9 Comparison of the generalised unified concrete with Podgorski concrete criterion

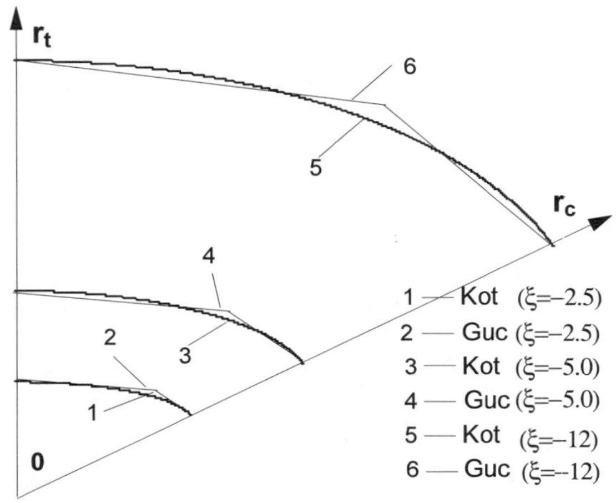

Fig. 7.10 Comparison of the generalised unified concrete with Kotsovos criterion

7.4 Yield and Loading Functions

In 1993, Labbane et al. analyzed hardening curves, using the Huber-von Mises criterion, the Drucker-Prager criteiron, the William-Wranker five-parameter criterion, the Bresler-Pister criterion and the Tsieh-Ting-Chen criterion as the failure criterion respectively. They found that a hardening curve based on effective plasticity strain only, together with the assumption of isotropic hardening under different loading conditions, cannot merge into a single curve regardless of the fracture criterion. As a result, the curve cannot satisfy the uniqueness condition. They also observed that, in the presence of high confining pressure, it is difficult to model the hardening behaviour using a single hardening parameter because large volumetric strains dominate the behaviour of concrete. Hardening, which is based on total effective plastic strain only, may overestimate the hardening capacity of concrete. Therefore, they suggested considering the hydrostatic and deviatoric effective strain components separately. Following Labbane's suggestion, Wang (1998) employed the plastic octahedral normal and shear strains ε_{0p}, γ_{0p} as hardening parameters, which are equivalencies of the hydrostatic and deviatoric effective plastic strain components. Kotsovos and Pavlovic's three-dimensional stress-strain relationship up to ultimate strength is used to characterize stress-strain relationships. It is worth noting that the effect of the Lode angle θ on the yield function is not accounted for in order to simplify the formulation, while its effect on plastic strain flow is taken into account by flow vectors. The effect of the deviatoric stress on the hydrostatic straining is also considered.

The multi-parameter unified yield function is adopted to govern the initial yield condition and hardening functions. The tensile and compressive meridians herein just adopt a general form as a function of the hydrostatic stress to generalize the use of unified strength theory for any kind of material besides concrete. To obtain the initial yielding and subsequent loading functions for concrete stress-strain relationship with the isotropic hardening assumption, a loading function $\Gamma(\varepsilon_{0p}, \gamma_{0p})$ will be characterized by Kotsovos's experimentally-fitted three-dimensional stress-strain relationship for concrete. In line with the isotropic hardening concept, the following yielding function is suggested,

$$\frac{r}{\Gamma(\varepsilon_{0p},\gamma_{0p})}=\frac{r_t r_c \sin 60^o}{r_t \sin\theta + r_c \sin(60^o-\theta)}(1-b)+b\frac{r_t}{\cos\theta} \quad \text{when } 0^o \le \theta \le \theta_b \quad (7\text{-}30a)$$

$$\frac{r}{\Gamma(\varepsilon_{0p},\gamma_{0p})}=\frac{r_t r_c \sin 60^o}{r_t \sin\theta + r_c \sin(60^o-\theta)}(1-b)+b\frac{r_c}{\cos(60^o-\theta)}$$

$$\text{when } \theta_b \le \theta \le 60^o \quad (7\text{-}30b)$$

in which ε_{0p}, γ_{0p} are the plastic octahedral normal and shear strains respectively.

Denoting

$$\vartheta=\frac{r_t r_c \sin 60^o}{r_t \sin\theta + r_c \sin(60^o-\theta)}(1-b)+b\frac{r_t}{\cos\theta} \quad \text{when } 0^o \le \theta \le \theta_b \quad (7\text{-}31a)$$

$$\vartheta=\frac{r_t r_c \sin 60^o}{r_t \sin\theta + r_c \sin(60^o-\theta)}(1-b)+b\frac{r_c}{\cos(60^o-\theta)} \quad \text{when } \theta_b \le \theta \le 60^o \quad (7\text{-}31b)$$

and substituting Eqs. (7-31a) and (7-31b) into (7-30a) and (7-30b) leads to

$$r=\Gamma(\varepsilon_{0p},\gamma_{0p})\vartheta \quad (7\text{-}32)$$

in which r_t and r_c are functions of the hydrostatic stress only. Hence, ϑ is a function of the hydrostatic stress and the load angle θ.

The loading functions and the stress function can be rewritten as

$$f(\sigma_{ij})=\frac{r}{\vartheta} \quad (7\text{-}33)$$

$$\Psi(\sigma_{ij},\varepsilon_0,\gamma_0)=f(\sigma_{ij})-\Gamma(\varepsilon_{0p},\gamma_{0p})=0 \quad (7\text{-}34)$$

The flow vector can then be derived based on the loading function

$$\frac{\partial I_1}{\partial \sigma_{ij}}=\{1 \quad 1 \quad 1 \quad 0 \quad 0 \quad 0\} \quad (7\text{-}35)$$

$$\frac{\partial (J_2)^{1/2}}{\partial \sigma_{ij}} = \frac{1}{2(J_2)^{1/2}} \{\sigma_{11} - \sigma_m \quad \sigma_{22} - \sigma_m \quad \sigma_{33} - \sigma_m \quad 2\sigma_{12} \quad 2\sigma_{13} \quad 2\sigma_{23}\}$$

(7-36)

$$\frac{\partial J_3}{\partial \sigma_{ij}} = \Big\{(\sigma_{22} - \sigma_m)(\sigma_{33} - \sigma_m) - \sigma_{23}^2 + \frac{J_3}{3} \quad (\sigma_{11} - \sigma_m)(\sigma_{33} - \sigma_m) - \sigma_{13}^2 + \frac{J_3}{3}$$

$$(\sigma_{11} - \sigma_m)(\sigma_{22} - \sigma_m) - \sigma_{12}^2 + \frac{J_3}{3} \quad 2(\sigma_{13}\sigma_{12} - (\sigma_{11} - \sigma_m)\sigma_{23})$$

$$2(\sigma_{12}\sigma_{23} - (\sigma_{22} - \sigma_m)\sigma_{13}) \quad 2(\sigma_{23}\sigma_{13} - (\sigma_{33} - \sigma_m)\sigma_{12}\Big\}$$

(7-37)

Assume

$$\{a\} = \{a_{11} \quad a_{22} \quad a_{33} \quad a_{12} \quad a_{13} \quad a_{23}\} = C_1\{a_1\} + C_2\{a_2\} + C_3\{a_3\} \tag{7-38}$$

where

$$\{a_1\}^T = \frac{\partial \xi}{\partial \sigma_{ij}} = \frac{\partial \xi}{\partial I_1}\frac{\partial I_1}{\partial \sigma_{ij}} = \frac{1}{\sqrt{3} f_c}\frac{\partial I_1}{\partial \sigma_{ij}} \tag{7-39}$$

$$\{a_2\}^T = \frac{\partial r}{\partial \sigma_{ij}} = \frac{\partial r}{\partial (J_2)^{1/2}}\frac{\partial (J_2)^{1/2}}{\partial \sigma_{ij}} = \frac{\sqrt{2}}{f_c}\frac{\partial (J_2)^{1/2}}{\partial \sigma_{ij}}\} \tag{7-40}$$

$$\{a_3\}^T = \frac{\partial \theta}{\partial \sigma_{ij}} = \frac{\partial \theta}{\partial J_3}\frac{\partial J_3}{\partial \sigma_{ij}} + \frac{\partial \theta}{\partial J_2^{1/2}}\frac{\partial J_2^{1/2}}{\partial \sigma_{ij}} = -\frac{\sqrt{3}}{2\sqrt{J_2^3}\sin 3\theta}\frac{\partial J_3}{\partial \sigma_{ij}} + \frac{3\sqrt{3}J_3}{2J_2^2 \sin 3\theta}\frac{\partial J_2^{1/2}}{\partial \sigma_{ij}}$$

(7-41)

The expression for C_1 is

$$C_1 = \frac{\partial \Psi}{\partial \xi} = \frac{\partial \Psi}{\partial r_t}\frac{\partial r_t}{\partial \xi} + \frac{\partial \Psi}{\partial r_c}\frac{\partial r_c}{\partial \xi} \tag{7-42}$$

Using Eqs. (7-31a), (7-31b), and (7-32), we get

$$\frac{\partial \Psi}{\partial r_t} = \frac{\partial \Psi}{\partial \vartheta}\frac{\partial \vartheta}{\partial r_t} = -\frac{r}{\vartheta^2}\frac{\partial \vartheta}{\partial r_t} \tag{7-43}$$

$$\frac{\partial \vartheta}{\partial r_t} = \frac{r_c^2 \sin 60^0 \sin(60^0 - \theta)}{\left[r_t \sin\theta + r_c \sin(60^0 - \theta)\right]^2}(1-b) + \frac{b}{\cos\theta} \quad \text{when} \quad 0^o \le \theta \le \theta_b \tag{7-44}$$

$$\frac{\partial \vartheta}{\partial r_t} = \frac{r_c^2 \sin 60^0 \sin(60^0 - \theta)}{\left[r_t \sin\theta + r_c \sin(60^0 - \theta)\right]^2}(1-b) \quad \text{when} \quad \theta_b < \theta \le 0^o \tag{7-45}$$

$$\frac{\partial \Psi}{\partial r_c} = \frac{\partial \Psi}{\partial \vartheta}\frac{\partial \vartheta}{\partial r_c} = -\frac{r}{\vartheta^2}\frac{\partial \vartheta}{\partial r_c} \tag{7-46}$$

$$\frac{\partial \vartheta}{\partial r_c} = \frac{r_t^2 \sin 60^0 \sin\theta}{\left[r_t \sin\theta + r_c \sin(60^0 - \theta)\right]^2}(1-b) \quad \text{when} \quad 0^o \le \theta \le \theta_b \tag{7-47a}$$

$$\frac{\partial \vartheta}{\partial r_c} = \frac{r_t^2 \sin 60^0 \sin\theta}{\left[r_t \sin\theta + r_c \sin(60^0 - \theta)\right]^2}(1-b) + \frac{b}{\cos\left(60^0 - \theta\right)} \quad \text{when}\, \theta_b < \theta \le 0^o \tag{7-47b}$$

$$C_2 = \frac{\partial \Psi}{\partial r} = \frac{1}{\vartheta} \tag{7-48}$$

$$C_3 = \frac{\partial \Psi}{\partial \theta} = \frac{\partial \Psi}{\partial \vartheta}\frac{\partial \vartheta}{\partial \theta} = -\frac{r}{\vartheta^2}\frac{\partial \vartheta}{\partial \theta} \tag{7-49}$$

$$\frac{\partial \vartheta}{\partial \theta} = -\frac{r_t r_c \sin 60^0 \left[r_t \cos\theta + r_c \cos(60^0 - \theta)\right]}{\left[r_t \sin\theta + r_c \sin(60^0 - \theta)\right]^2}(1-b) + \frac{b r_t \sin\theta}{\cos^2\theta}$$

$$\text{when} \quad 0^o \le \theta \le \theta_b \tag{7-50a}$$

$$\frac{\partial \vartheta}{\partial \theta} = -\frac{r_t r_c \sin 60^0 \sin\theta \left[r_t \cos\theta + r_c \cos(60^0 - \theta)\right]}{\left[r_t \sin\theta + r_c \sin(60^0 - \theta)\right]^2}(1-b) - \frac{b r_c \sin(60^0 - \theta)}{\cos^2(60^0 - \theta)}$$

$$\text{when} \quad \theta_b < \theta \le 60^o \tag{7-50b}$$

Differentiating Eq. (7.34) yields

$$d\Psi(\sigma_{ij}, \varepsilon_0, \gamma_{0p}) = \frac{\partial f\left(\sigma_{ij}\right)}{\partial \sigma_{ij}} d\sigma_{ij} - \frac{\partial \Gamma(\varepsilon_{0p}, \gamma_{0p})}{\partial \varepsilon_{0p}} d\varepsilon_{0p} - \frac{\partial \Gamma(\varepsilon_{0p}, \gamma_{0p})}{\partial \gamma_{0p}} d\gamma_{0p} = 0 \tag{7-51}$$

Adopting the associated flow theory of plasticity, the plastic strain components are obtained as

$$(d\varepsilon_{ij})_p = d\lambda \frac{\partial Q}{\partial \sigma_{ij}} = d\lambda \frac{\partial f}{\partial \sigma_{ij}} \tag{7-52}$$

The scalar A in the elasto-plastic stiffness matrix $[D]_{ep}$ can be written as

$$A = -\frac{1}{d\lambda}\frac{\partial \Psi}{\partial \Gamma}\left(\frac{\partial \Gamma}{\partial \varepsilon_{0p}} d\varepsilon_{0p} + \frac{\partial \Gamma}{\partial \gamma_{0p}} d\gamma_{0p}\right) \tag{7-53}$$

To deduce the elasto-plastic matrix, A must be derived first.

From Eq. (7-34), we have

$$A = \frac{1}{d\lambda}\left(\frac{\partial f}{\partial \sigma_{ij}}\frac{\partial \sigma_{ij}}{\partial \varepsilon_{0p}} d\varepsilon_{0p} + \frac{\partial f}{\partial \sigma_{ij}}\frac{\partial \sigma_{ij}}{\partial \gamma_{0p}} d\gamma_{0p}\right) \tag{7-54}$$

in which $f(\sigma_{ij})$ is an implicit function of ε_{0p}, γ_{0p} through which the experimentally-fitted relationship given by Kotsovos and Pavlovic (1995) will be introduced into the hardening function.

From Eq. (7-34), we derive

$$\begin{aligned} A &= \frac{1}{\vartheta d\lambda}\left(\frac{\partial r}{\partial \varepsilon_{0p}} d\varepsilon_{0p} - \frac{r}{\vartheta}\frac{\partial \vartheta}{\partial \varepsilon_{0p}} d\varepsilon_{0p} + \frac{\partial r}{\partial \gamma_{0p}} d\gamma_{0p}\right) \\ &= \frac{1}{\vartheta d\lambda}\left(\frac{\partial r}{\partial \varepsilon_{0p}} d\varepsilon_{0p} - \frac{r}{\vartheta}\frac{\partial \vartheta}{\partial \xi}\frac{\partial \xi}{\partial \varepsilon_{0p}} d\varepsilon_{0p} + \frac{\partial r}{\partial \gamma_{0p}} d\gamma_{0p}\right) \end{aligned} \tag{7-55}$$

Unfold $d\varepsilon_{0p}$, we get

$$d\varepsilon_{0p} = \frac{d(\varepsilon_{11})_p + d(\varepsilon_{22})_p + d(\varepsilon_{33})_p}{3} \tag{7-56}$$

By considering the associate flow rule, it is obtained

$$d\varepsilon_{0p} = \frac{d\lambda}{3}\left(\frac{df}{d\sigma_{11}} + \frac{df}{d\sigma_{22}} + \frac{df}{d\sigma_{33}}\right) \tag{7-57}$$

Substituting the components of the flow vector $\{a\}$ into Eq. (7-57) leads to

$$d\varepsilon_{0p} = \frac{d\lambda}{3}\left(a_{11} + a_{22} + a_{33}\right) \tag{7-58}$$

The expression of γ_{op} can be written as

$$d\gamma_{0p}=\sqrt{\frac{2}{3}\left[(d\varepsilon_{11})_p^2+(d\varepsilon_{22})_p^2+(d\varepsilon_{33})_p^2+(d\varepsilon_{12})_p^2+(d\varepsilon_{13})_p^2+(d\varepsilon_{23})_p^2\right]} \qquad (7\text{-}59)$$

It can be further derived by considering the associated flow rule as

$$d\gamma_{0p}=\frac{\sqrt{2}d\lambda}{\sqrt{3}}\sqrt{\left(\frac{df}{d\sigma_{11}}\right)^2+\left(\frac{df}{d\sigma_{22}}\right)^2+\left(\frac{df}{d\sigma_{33}}\right)^2+\left(\frac{df}{d\sigma_{12}}\right)^2+\left(\frac{df}{d\sigma_{13}}\right)^2+\left(\frac{df}{d\sigma_{23}}\right)^2} \qquad (7\text{-}60)$$

Rewrite Eq. (7-60) in terms of the components of flow vector $\{a\}$, then

$$d\gamma_{0p}=\frac{\sqrt{2}d\lambda}{\sqrt{3}}\sqrt{a_{11}{}^2+a_{22}{}^2+a_{33}{}^2+a_{12}{}^2+a_{13}{}^2+a_{23}{}^2} \qquad (7\text{-}61)$$

The octahedral stresses σ_0, τ_0 adopted in the Kotsovos's stress-strain relationship have the following relation with the Haigh-Westergaard variables (ξ, r, θ),

$$\sigma_0=-\frac{f_c}{\sqrt{3}}\xi\,,\quad \tau_0=\frac{f_c}{\sqrt{3}}r \qquad (7\text{-}62)$$

In Kotsovos and Pavlovic's material model, the bulk and shear modulus K_t, G_t are given as

$$K_t=\frac{1}{3}\frac{d\sigma_0}{d\varepsilon_0}\,,\quad G_t=\frac{1}{2}\frac{d\tau_0}{d\gamma_0} \qquad (7\text{-}63)$$

Substituting Eq. (7-62) into Eq. (7-63) yields

$$K_t=-\frac{f_c}{3\sqrt{3}}\frac{d\xi}{d\varepsilon_0}\,,\quad G_t=\frac{f_c}{2\sqrt{3}}\frac{dr}{d\gamma_0} \qquad (7\text{-}64)$$

By decomposing the octahedral strains into elastic and plastic portions, we obtain

$$d\varepsilon_0=d\varepsilon_{0p}+d\varepsilon_{0e} \qquad (7\text{-}65)$$

$$d\gamma_0=d\gamma_{0p}+d\gamma_{0e} \qquad (7\text{-}66)$$

The elastic portions have the following relationship with initial bulk K_e and shear modulus G_e

$$K_e = \frac{1}{3}\frac{d\sigma_0}{d\varepsilon_{0e}} = -\frac{f_c}{3\sqrt{3}}\frac{d\xi}{d\varepsilon_{0e}}, \quad G_e = \frac{d\tau_0}{2d\gamma_{0e}} = \frac{f_c}{2\sqrt{3}}\frac{dr}{\gamma d_{0e}} \tag{7-67}$$

Hence,

$$\begin{aligned} \frac{d\xi}{d\varepsilon_{0p}} &= -\frac{\sqrt{3}}{f_c}\frac{d\sigma_0}{d\varepsilon_{0p}} = -\frac{3\sqrt{3}K_t K_e}{f_c(K_t + K_e)}, \\ \frac{dr}{d\gamma_{0p}} &= \frac{f_c}{\sqrt{3}}\frac{d\tau_0}{d\gamma_{0p}} = \frac{2\sqrt{3}G_t G_e}{f_c(G_t + G_e)} \end{aligned} \tag{7-68}$$

Denoting

$$K_{tp} = \frac{3K_t K_e}{K_t + K_e}, \qquad G_{tp} = \frac{2G_t G_e}{(G_t + G_e)} \tag{7-69}$$

Eq. (7-68) can be rewritten as

$$\frac{d\xi}{d\varepsilon_{0p}} = -\frac{\sqrt{3}}{f_c}K_{tp}, \qquad \frac{dr}{d\gamma_{0p}} = \frac{\sqrt{3}}{f_c}G_{tp} \tag{7-70}$$

The coupling effect of the octahedral shear stress on normal stress-strain is given by equivalent normal stress σ_{id} caused by the deviatoric component. It has the expression of (Kotsovos and Pavlovic 1995)

$$\frac{\sigma_{id}}{f_c} = M\left(\frac{\tau_0}{f_c}\right)^n \tag{7-71}$$

Differentiating Eq. (7.71) with respect to ε_0 leads to

$$\frac{d\tau_0}{d\varepsilon_0} = \frac{1}{Mn\left(\frac{\tau_0}{f_c}\right)^{n-1}}\frac{d\sigma_{id}}{d\varepsilon_0} \tag{7-72}$$

Denoting

$$K_t' = \frac{1}{3}\frac{d\sigma_{id}}{d\varepsilon_0} \tag{7-73}$$

it can be derived,

$$K'_{tp} = \frac{3K'_t K_e}{K'_t + K_e} \tag{7-74}$$

where K'_t is the tangential bulk modulus at the hydrostatic stress value of σ_{id}.

Referring to Eqs. (7-62), (7-72) and (7-74), we obtain the following expression

$$\frac{dr}{d\varepsilon_{0p}} = \frac{3\sqrt{3}}{f_c Mn\left(\frac{\tau_0}{f_c}\right)^{n-1}} K'_{tp} \tag{7-75}$$

Substituting Eqs. (7-58), (7-61), (7-70) and (7-75) into Eq. (7-55), we get

$$A = \frac{1}{\vartheta}\left(\frac{\sqrt{3}K'_{tp}}{f_c Mn\left(\frac{\tau_0}{f_c}\right)^{n-1}} + \frac{K_{tp}}{\sqrt{3}f_c}\frac{r}{\vartheta}\frac{\partial\vartheta}{\partial\xi}\frac{\partial\xi}{\partial\varepsilon_{0p}}\right)(a_{11} + a_{22} + a_{33}) \tag{7-76}$$
$$+ \frac{\sqrt{2}G_{tp}}{\vartheta f_c}\sqrt{a_{11}^2 + a_{22}^2 + a_{33}^2 + a_{12}^2 + a_{13}^2 + a_{23}^2}$$

The expressions for $\frac{\partial\vartheta}{\partial r_t}$ and $\frac{\partial\vartheta}{\partial r_c}$ have been derived previously. The differentiations of the tensile and compressive meridians with respect to ξ can be derived accordingly. The following gives the two partial derivatives for different meridian formulae.

linear formulae,

$$r_t = a_0 + a_1\xi \tag{7-77a}$$

$$r_c = b_0 + b_1\xi \tag{7-77b}$$

hence

$$\frac{\partial r_t}{\partial\xi} = a_1 \tag{7-78a}$$

$$\frac{\partial r_c}{\partial \xi} = b_1 \tag{7-78b}$$

parabolic formulae

$$r_t = a_0 + a_1\xi + a_2\xi^2 \tag{7-80a}$$

$$r_c = b_0 + b_1 + b_2\xi^2 \tag{7-80b}$$

then

$$\frac{\partial r_t}{\partial \xi} = a_1 + 2a_2\xi \tag{7-80c}$$

$$\frac{\partial r_c}{\partial \xi} = b_1 + 2b_2\xi \tag{7-80d}$$

Kotsovos approximation

$$r_c = 0.944\sqrt{3}\left(\frac{-\xi}{\sqrt{3}} + 0.05\right)^{0.724} \tag{7-81a}$$

$$r_t = 0.633\sqrt{3}\left(\frac{-\xi}{\sqrt{3}} + 0.05\right)^{0.857} \tag{7-81b}$$

then

$$\frac{\partial r_t}{\partial \xi} = -0.542\left(\frac{-\xi}{\sqrt{3}} + 0.05\right)^{-0.143} \tag{7-82a}$$

$$\frac{\partial r_c}{\partial \xi} = -0.638\left(\frac{-\xi}{\sqrt{3}} + 0.05\right)^{-0.276} \tag{7-82b}$$

7.5 Processing of Corner Singularity

For the derived yield surface, the flow vector $\{a\}$ is not uniquely defined at three angular points. One of them is the stress point at $\theta=\theta_b$ and the other two are the stress points when two principal stresses equal, i.e. $\theta=0°$ and $\theta=60°$. At these stress combinations, the directions of plastic straining are indeterminate. Numerical difficulty will be encountered as stress state approaches these points. Koiter (1953) suggested that when two yield functions are active, the plastic strain increment can be written as

$$d\varepsilon_{ij}^{p} = d\lambda_1 \frac{\partial f_1}{\partial \sigma_{ij}} + d\lambda_2 \frac{\partial f_2}{\partial \sigma_{ij}} \tag{7-83}$$

when associated flow rule is used. Here f_1 and f_2 are the loading functions on different sides of singular points. Now we have two proportionality multipliers $d\lambda_1$ and $d\lambda_2$ instead of only one. We suppose that during an infinitesimally small increment the stress point remains in the corner of the yield surface. When the stress point remains in a singular point of the yield surface the consistency condition for the first as well as for the second yield function must be satisfied.

$$\dot{f}_1 = 0 \quad \text{and} \quad \dot{f}_2 = 0 \tag{7-84}$$

Via Eq. (7-83), Eq. (7-84) can be expanded as

$$\left(\frac{\partial f_1}{\partial \sigma_{ij}}\right)^T d\sigma_{ij} + \frac{\partial f_1}{\partial \varepsilon_{ij}^{p}} \left(\frac{\partial \varepsilon_{ij}^{p}}{\partial \varepsilon_{0p}} + \frac{\partial \varepsilon_{ij}^{p}}{\partial \gamma_{0p}}\right)^T \left(d\lambda_1 \frac{\partial f_1}{\partial \sigma_{ij}} + d\lambda_2 \frac{\partial f_2}{\partial \sigma_{ij}}\right) = 0 \tag{7-85a}$$

$$\left(\frac{\partial f_2}{\partial \sigma_{ij}}\right)^T d\sigma_{ij} + \frac{\partial f_2}{\partial \varepsilon_{ij}^{p}} \left(\frac{\partial \varepsilon_{ij}^{p}}{\partial \varepsilon_{0p}} + \frac{\partial \varepsilon_{ij}^{p}}{\partial \gamma_{0p}}\right)^T \left(d\lambda_1 \frac{\partial f_1}{\partial \sigma_{ij}} + d\lambda_2 \frac{\partial f_2}{\partial \sigma_{ij}}\right) = 0 \tag{7-85b}$$

The incremental stress strain relationship at these stress points then can be written as

$$d\sigma_{ij} = [D]_e \left(d\varepsilon_{ij} - d\lambda_1 \frac{\partial f_1}{\partial \sigma_{ij}} - \lambda_2 \frac{\partial f_2}{\partial \sigma_{ij}}\right) \tag{7-86}$$

Premultiplying Eq. (7-86) with the gradients to f_1 and f_2, Eqs. (7-85a) and (7-85b) become

$$\mu_1 d\lambda_1 + \mu_2 d\lambda_2 = \left(\frac{\partial f_1}{\partial \sigma_{ij}}\right)^T [D]_e \, d\varepsilon_{ij} \tag{7-87a}$$

$$\mu_3 d\lambda_1 + \mu_4 d\lambda_2 = \left(\frac{\partial f_2}{\partial \sigma_{ij}}\right)^T [D]_e \, d\varepsilon_{ij} \tag{7-87b}$$

where μ_1, μ_2, μ_3 and μ_4 are defined as

$$\mu_1 = \left(-\frac{\partial f_1}{\partial \varepsilon_{ij}^p} \left(\frac{\partial \varepsilon_{ij}^p}{\partial \varepsilon_{op}} + \frac{\partial \varepsilon_{ij}^p}{\partial \gamma_{op}} \right) + [\mathrm{D}]_e \frac{\partial f_1}{\partial \sigma_{ij}} \right)^T \frac{\partial f_1}{\partial \sigma_{ij}} \tag{7-88}$$

$$\mu_2 = \left(-\frac{\partial f_1}{\partial \varepsilon_{ij}^p} \left(\frac{\partial \varepsilon_{ij}^p}{\partial \varepsilon_{op}} + \frac{\partial \varepsilon_{ij}^p}{\partial \gamma_{op}} \right) + [\mathrm{D}]_e \frac{\partial f_1}{\partial \sigma_{ij}} \right)^T \frac{\partial f_2}{\partial \sigma_{ij}} \tag{7-89}$$

$$\mu_3 = \left(-\frac{\partial f_2}{\partial \varepsilon_{ij}^p} \left(\frac{\partial \varepsilon_{ij}^p}{\partial \varepsilon_{op}} + \frac{\partial \varepsilon_{ij}^p}{\partial \gamma_{op}} \right) + [\mathrm{D}]_e \frac{\partial f_2}{\partial \sigma_{ij}} \right)^T \frac{\partial f_1}{\partial \sigma_{ij}} \tag{7-90}$$

$$\mu_4 = \left(-\frac{\partial f_2}{\partial \varepsilon_{ij}^p} \left(\frac{\partial \varepsilon_{ij}^p}{\partial \varepsilon_{op}} + \frac{\partial \varepsilon_{ij}^p}{\partial \gamma_{op}} \right) + [\mathrm{D}]_e \frac{\partial f_2}{\partial \sigma_{ij}} \right)^T \frac{\partial f_2}{\partial \sigma_{ij}} \tag{7-91}$$

Solving the simultaneous equations (7-87a) and (7-87b) for λ_1 and λ_2 yields

$$\lambda_1 = \frac{\mu_4 (\partial f_1 / \partial \sigma_{ij})^T [\mathrm{D}]_e \, \mathrm{d}\varepsilon_{ij} - \mu_2 (\partial f_2 / \partial \sigma_{ij})^T [\mathrm{D}]_e \, \mathrm{d}\varepsilon_{ij}}{\mu_1 \mu_4 - \mu_2 \mu_3} \tag{7-92}$$

$$\lambda_2 = \frac{\mu_1 (\partial f_2 / \partial \sigma_{ij})^T [\mathrm{D}]_e \, \mathrm{d}\varepsilon_{ij} - \mu_2 (\partial f_1 / \partial \sigma_{ij})^T [\mathrm{D}]_e \, \mathrm{d}\varepsilon_{ij}}{\mu_1 \mu_4 - \mu_2 \mu_3} \tag{7-93}$$

Hence, the incremental stress strain relation for the singular points is

$$d\sigma_{ij} = \left\{ [D] - \frac{[D]_e \left[\mu_1 \frac{\partial f_2}{\partial \sigma_{ij}} \left(\frac{\partial f_2}{\partial \sigma_{ij}} \right)^T + \mu_4 \frac{\partial f_1}{\partial \sigma_{ij}} \left(\frac{\partial f_1}{\partial \sigma_{ij}} \right)^T \right] [D]_e}{\mu_1 \mu_4 - \mu_2 \mu_3} - \frac{[D]_e \left[\mu_2 \frac{\partial f_1}{\partial \sigma_{ij}} \left(\frac{\partial f_2}{\partial \sigma_{ij}} \right)^T + \mu_3 \frac{\partial f_2}{\partial \sigma_{ij}} \left(\frac{\partial f_1}{\partial \sigma_{ij}} \right)^T \right] [D]_e}{\mu_1 \mu_4 - \mu_2 \mu_3} \right\} d\varepsilon_{ij} \tag{7-94}$$

7.6 Strain Softening Phenomena and Material Damage

The material behavior of concrete is assumed to be isotropic and homogeneous before macroscopic cracks occur. However, it shows apparent anisotropy in the post-crack stress state. It is not a straightforward process and very costly to simulate the strain-softening induced by an individual crack or defect in numerical computation. Therefore, a reliable equivalent continuum model for concrete will be extremely useful.

Continuum damage mechanics (Lemaitre and Chaboche 1978; Krajcinovic 1996) provides systematic approaches to interrelate districuted defects with the observed macroscopic behaviour. The fundamental notion of damage mechanics is to represent the damage state of the material be come internal variables. The variables directly characterize the distribution of cracks formed during the loading process. Compared with the fracture mechanics used in the context of discrete cracks (Saouma et al. 1980), the continuum damage models with fixed mesh have the advantage of avoiding remeshing when finite element methods are adopted.

Different damage model establishes different mechanical equations to describe the evolution of the internal variables and the mechanical behaviour of damaged material. The isotropic damage mechanics model uses a single scalar parameter and is based on Lemaitre's hypothesis of strain equivalence, which has been widely used in creep analysis of ductile materials (Lemaitre 1986). The isotropic damage model is very simple. However, it is argued that only one damage scalar is not enough to model the damage properties of brittle material like concrete (Mazars and Pijaudier-Cabot, 1989). Some damage-based constitutive relationship has been suggested for concrete (Mazars 1986; Yazdani and Schreyer 1988, 1990; Lubarda, et al 1994). In Mazars' model, damage is described by coupling the compression and tension effects to define a single damage variable ω. The calculation of this variable is based on a certain measure of the strain field. Yazdani and Schreyer (1988) coupled damage and plasticity and assumed that once the limit damage surface is reached, the softening regime is solely controlled by elastic and inelastic damage processes and that no plastic flow occurs. On the other hand, in the hardening regime of deformation, both damage and plasticity surfaces are used and plastic flow in their model is controlled by Huber-von-Mises criterion. Those models defines damage scalars all based on uniaxial tensile and uniaxial compressive experiments, the threshold strain is determined by the strain in uniaxial ulitimate state. Plastic flow with pressure-sensitive strength criterion is not included.

When the stresses evolve beyond the ultimate stress envelope, observations show that concrete material exhibits some residual stress. In other words, the damage is not hundred percent. In the present model, the partial damage is described by a damage scalar ω, which is defined by a two-parameter Weibull distribution function as follows (Mazars 1986),

$$\omega = 1 - \exp\left[-\alpha\left(\frac{\varepsilon - \varepsilon_u}{\varepsilon_u}\right)^m\right] \tag{7-95}$$

Where ε_u is the threshold strain. The value of the damage scalar equals zero when the strain ε is less than the threshold strain ε_u. α and m are the two Weibull parameters. In the uniaxial sense, the stress-strain relationship changes from the form of $\sigma = E\varepsilon$ for pre-damage concrete to $\tilde{\sigma} - \varepsilon$ for the damaged concrete as follows,

$$\tilde{\sigma} = E\varepsilon\,(1-\omega), \quad \varepsilon \geq \varepsilon_u \tag{7-96a}$$

Writing Ω=(1-ω), we get

$$\tilde{\sigma} = (1-\omega)\sigma = \Omega\sigma, \quad \varepsilon \geq \varepsilon_u \tag{7-96b}$$

Fig. 7.11 shows the effect of the two Parameters α and m on the stress-strain curve in uniaxial stress state.

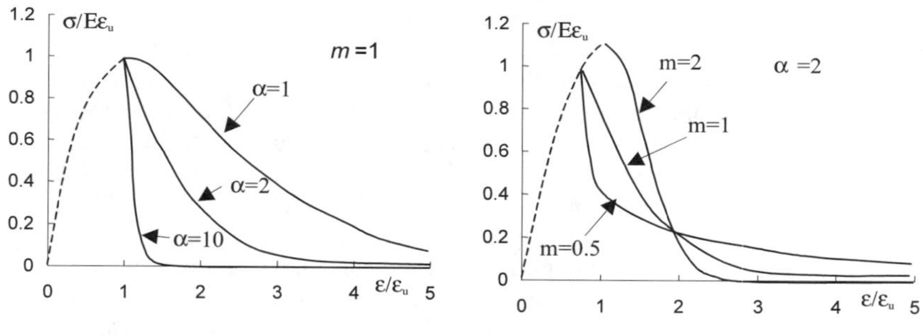

Fig. 7.11 Effect of the two parameters α and m on stress-strain curves

In the general three-dimensional stress states, once the ultimate strength envelope is reached, material exhibits anisotropy. The anisotropic stress strain evolution is regarded as a combination of damage and plastic-hardening. In the present three-dimensional elasto-plastic damage model, different damage scalars (ω_1, ω_2, ω_3) are adopted in the respective principal stress directions. The value of ω_i (i=1,2,3) is a function of the threshold strain ε_{ui} (i=1,2,3), which is defined as the specific principal strain when the ultimate strength envelope is reached. Assumption is made that the directions of the principal stress and strain coincide. The propagation of damage is described as below.

When the ultimate strength envelope is reached the first time and at least one of the principal stresses is tensile, crack is assumed to occur. A damage scalar ω_1 is then introduced in this principal stress direction and the associated shears as well. The degenerated stress $[\tilde{\sigma}]$ is

$$[\tilde{\sigma}]=[\Omega]\ [\sigma] \tag{7-97}$$

where

$$\{\sigma\}^T=[\sigma_{11}\ \sigma_{22}\ \sigma_{33}\ \sigma_{12}\ \sigma_{13}\ \sigma_{23}] \tag{7-98}$$

$$[\Omega]=[\Omega_1]=\begin{bmatrix} 1-\omega_1 & 0 & 0 & 0 & 0 & 0 \\ 0 & 1 & 0 & 0 & 0 & 0 \\ 0 & 0 & 1 & 0 & 0 & 0 \\ 0 & 0 & 0 & 1-\omega_1 & 0 & 0 \\ 0 & 0 & 0 & 0 & 1-\omega_1 & 0 \\ 0 & 0 & 0 & 0 & 0 & 1 \end{bmatrix} \tag{7-99}$$

where $[\Omega]$ is the general damage matrix and $[\Omega_1]$ is the damage matrix when the first damage scalar is introduced. Hence, the stress state returns to somewhere within the ultimate strength envelope and the stress-strain relationship is governed by the elasto-plastic constitutive law again.

In incremental form,

$$d\{\tilde{\sigma}\}=[D]_{ep}\,d\{\varepsilon\} \tag{7-100}$$

in which, $[D]_{ep}$ is the elasto-plastic stiffness matrix.

Upon further loading, the stress state may reach the ultimate strength envelope again, a second damage scalar is then introduced. And the general damage takes the form of

$$[\Omega]=[\Omega_2]=\begin{bmatrix} 1-\omega_1 & 0 & 0 & 0 & 0 & 0 \\ 0 & 1-\omega_2 & 0 & 0 & 0 & 0 \\ 0 & 0 & 1 & 0 & 0 & 0 \\ 0 & 0 & 0 & (1-\omega_1)(1-\omega_2) & 0 & 0 \\ 0 & 0 & 0 & 0 & (1-\omega_1) & 0 \\ 0 & 0 & 0 & 0 & 0 & (1-\omega_2) \end{bmatrix} \tag{7-101}$$

Where ω_2 is obtained by the two-parameter Weibull function in terms of the second principal tensile strain.

Subsequently, the stress state returns to somewhere within the ultimate strength envelope again. Upon further loading, when the stress state reaches the ultimate strength envelope once again, a third damage scalar ω_3 is introduced. And we

$$[\Omega] = [\Omega_3] = \begin{bmatrix} 1-\omega_1 & 0 & 0 & 0 & 0 & 0 \\ 0 & 1-\omega_2 & 0 & 0 & 0 & 0 \\ 0 & 0 & 1-\omega_3 & 0 & 0 & 0 \\ 0 & 0 & 0 & (1-\omega_1)(1-\omega_2) & 0 & 0 \\ 0 & 0 & 0 & 0 & (1-\omega_1)(1-\omega_3) & 0 \\ 0 & 0 & 0 & 0 & 0 & (1-\omega_2)(1-\omega_3) \end{bmatrix} \tag{7-102}$$

7.7 Applications

The multi-parameter yield criterion has been applied to analyze two groups of reinforced concrete slabs and a parabolic cylindrical shell. The nonlinear finite-element-analysis code for plates and shells written by Huang (1988) and his predecessors (Owen and Hilton 1980) is modified to incorporate the present material model for concrete (Wang 1998).

The two parameters in the Weibull's function are chosen, i. e., $\alpha = 2$ and $m = 1$ in the following examples. Three-dimensional moderately thick shell elements and layered model are employed. Different thickness for different layers can be defined such that any combinations of layers of reinforcements and concrete across the depth are allowed.

Example I - Duddeck's Reinforced Concrete Slabs

Duddeck et al. (1978) carried out experimental tests on 3 square reinforced concrete slabs, which are simply supported along the edges and loaded at the centre of the slabs. The three slabs constitute of the same concrete material and steel reinforcements, having the same dimensions and boundary conditions. The total amount of reinforcement is constant in each slab. The configurations of reinforcement layers are different so that it resulted in different degrees of orthogonality along the edge of the slabs. The two orthogonal direction are denoted as x for the direction of relatively more reinforcement, whilst y for the other direction. The dimensions and reinforcement arrangement of the slabs are shown in Fig. 7.12. The following material parameters taken by Figueiras and Owen (1984) and later adopted by Gervera and Hinton (1986) are applied in the present study,

E_c= 16400. MPa, E_s= 201000. MPa, f_c= 43.0 MPa, f_t= 3.0 MPa, f_y= 670.0 MPa, ν = 0.2

where E_c, E_s are the initial Young's modulus for concrete and steel reinforcement respectively; f_c and f_t are the uniaxial compressive and tensile strengths of concrete; f_y is the yield strength of steel reinforcements, and ν is the Poisson's ratio for concrete.

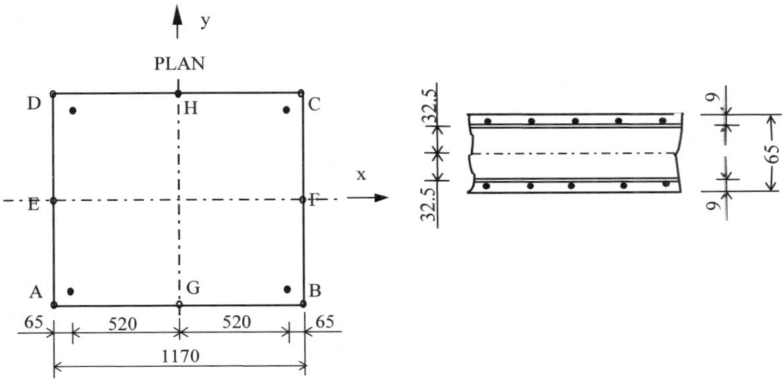

Fig. 7.12 Dimensions (in mm) and Reinforcement Arrangement in Duddeck's Slabs

The equivalent thickness of the reinforcement layers near the top and bottom surfaces of the slabs are tabulated in Table 7.1.

Table 7.1 Equivalent Thickness of Reinforcement Layers in Duddeck Slabs

Slabs	Top Layer (mm)		Bottom Layer (mm)	
	x-dir	*y*-dir	*x*-dir	*y*-dir
Slab S1	0.193	0.193	0.397	0.397
Slab S2	0.252	0.133	0.520	0.273
Slab S3	0.283	0.103	0.582	0.212

In Table 7.1 slabs S1, S2 and S3 represent, respectively, the first, second and third slab specimen tested by Duddeck et al. (1978). As seen in Table 7.1, S1 has the same amount of reinforcement in x and y directions, while S3 has the most orthogonality of reinforcement. In the present numerical simulation, only one quarter of the slabs is analysed by use of symmetry of the slabs and the loading, boundary conditions. In the finite-element discretization, the quarter slab is divided into 3×3 of 8 -node Serendipity elements. The element-size is smaller near the centre of the slab where higher stresses are expected. Integration across the depth goes through twelve layers.

Each load-increment is set to be 5 percent of the ultimate load. The maximum number of iterations for convergence within each incremental step is assigned to be 20 and the equilibrium tolerance of 0.5 percent is applied.

The calculated ultimate load-carrying capacities for the three slabs are given in Table 7.2. The load-deflection at different load levels for the central point are plotted and compared with the experimental results for slabs S1, S2 and S3 in Fig.5. It can be seen from Table 7.2 and Fig. 7.13 that the present model yields rather accurate estimations for the ultimate load-carrying capacities. The percentage errors of the calculated ultimate load-carrying capacities are all within 3% of the experimental results. The predicted deflections at different load levels are also in good agreement with the test data.

Table 7.2 Ultimate Load-carrying Capacity of Duddeck Slabs (KN)

Slabs	Present Estimations (kN)	Experimental Results (kN) [Duddeck]	Percentage Error
Slab S1	62.65	61.06	+2.06%
Slab S2	44.49	43.46	+2.37%
Slab S3	34.60	34.25	+1.02%

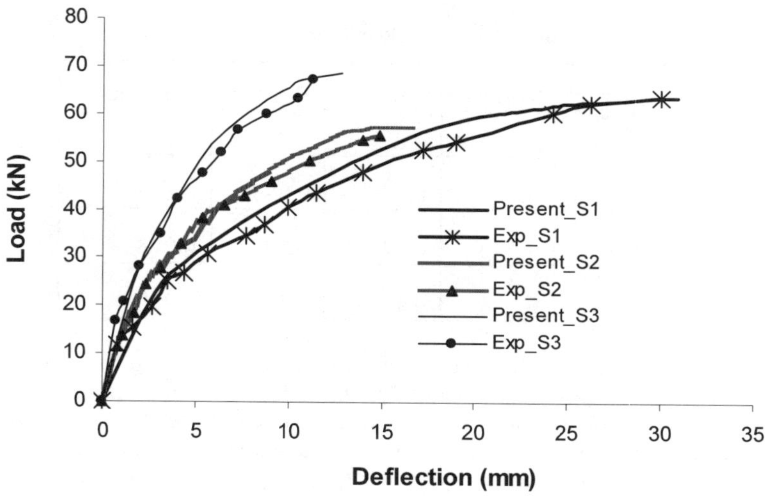

Fig.7.13 Load-deflection Curves at Centre of Slabs S1, S2 and S3

Figs. 7.14-7.16 show the damage distributions of the bottom layer at failure for slabs S_1, S_2, S_3, respectively. It can be seen from these figures that the damage level in the direction of less reinforcement (y-direction), as expected, is higher than that in the other orthogonal direction (x-direction) for slabs S_2 and S_3. The more different in the reinforcement amount, the more pronounced the difference of orthogonal damage distribution. The damage herein refers to the damage scalar ω_1 in the direction that the first crack occurs.

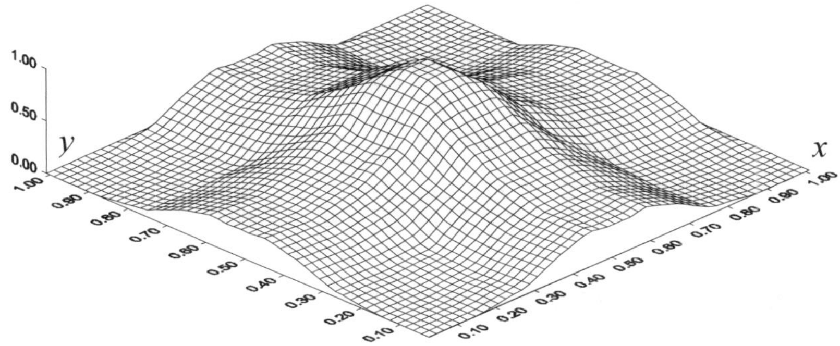

Fig. 7.14 Damage distribution of bottom layer at failure for slab S_1

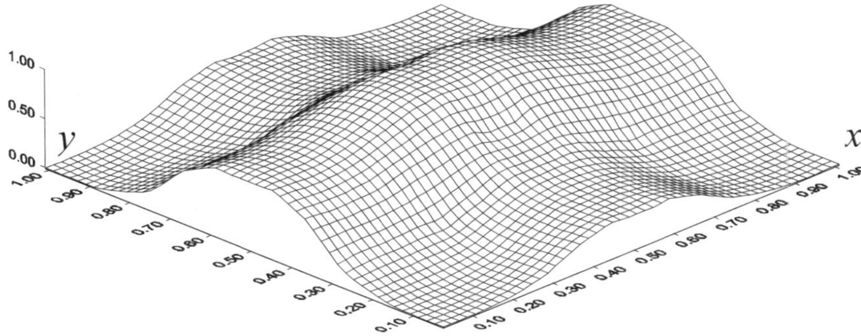

Fig. 7.15 Damage distribution of bottom layer at failure for slab S_2

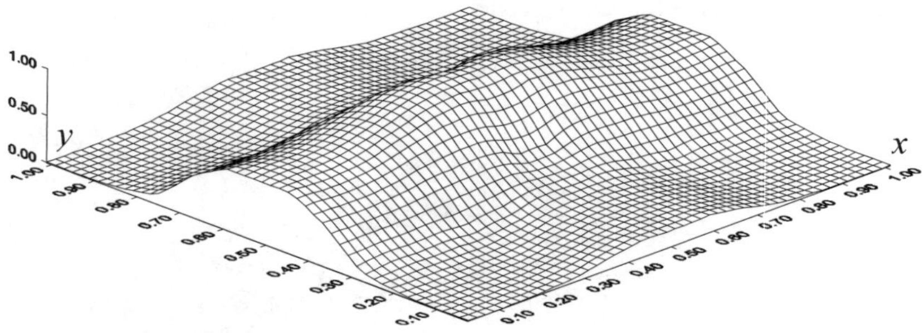

Fig. 7.16 Damage distribution of bottom layer at failure for slab S_3

Example 2 High-Strength Concrete Slabs

Mazouk and Hussein (1991) carried out a series of tests on common and high-strength reinforced concrete slabs with various steel ratios. The slabs are simply supported along all edges and loaded centrally through a column stub. The dimensions and reinforcement arrangement of a typical specimen is shown in Fig.7.17. Three slabs among the seventeen tested high-strength concrete slabs are analysed in the present study. Material parameters for concrete and details for the steel reinforcement for these three slabs are shown in Table 7.3.

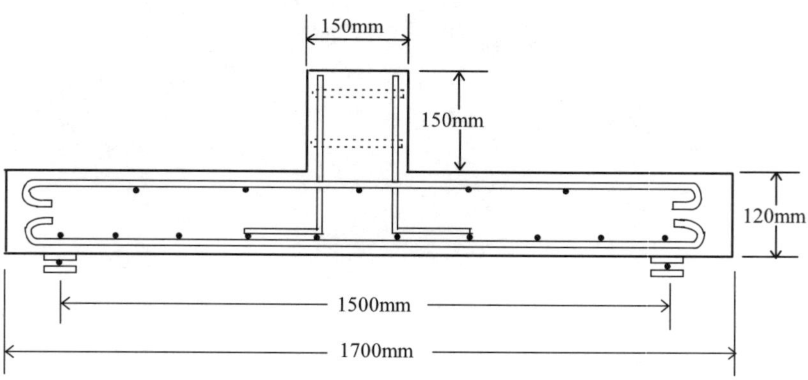

Fig. 7.17 Dimensions and Reinforcement Details in the High-Strength Concrete Slabs

Table 7.3 Details of 3 High-Strength Reinforced Concrete Slabs

Slabs	Uniaxial Compressive Strength (MPa)	Steel Bar Size (mm)	Steel Bar Spacing (mm)	Slab Thickness (mm)	Slab Depth (mm)	Steel Ratio (%)
HS1	67	M10	214.3	120.0	95.0	0.491
HS3	69	M10	71.4	120.0	95.0	1.473
HS7	74	M10	88.2	120.0	95.0	1.193

The reinforced bars are Grade 400 steel conforming to CSA standards with an actual yield strength of 490MPa and ultimate strength of 690MPa. Making use of symmetry, only one quarter of the slab is analysed. A mesh comprising 6×6 of 8-node Serendipity elements is used. The rigidity effect of the column stub is ignored and the load is applied as a uniformly distributed pressure over the small area equivalent to the cross-section of the stub. Across the depth, each slab is sub-divided into eight layers. Load-increment is set to be 5 percent of the ultimate load. The maximum number of iterations for convergence within each incremental step is 20 and the equilibrium tolerance of 0.5% is applied.

The calculated ultimate load-carrying capacity and the percentage errors with regard to the experimental results are listed in Table 7.4. The load-deflection curves for the central point are plotted and compared with the experimental results for slabs HS1, HS3 and HS7 in Fig. 7.18.

Table 7.4 Ultimateload-carrying Capaciry of High-strength Slabs

Slab	Present Estimations (kN)	Experimental Results [Mazouk] (kN)	Percentage Error
HS1	173.39	178	–2.37%
HS3	357	356	+0.28%
HS7	340	356	–4.49%

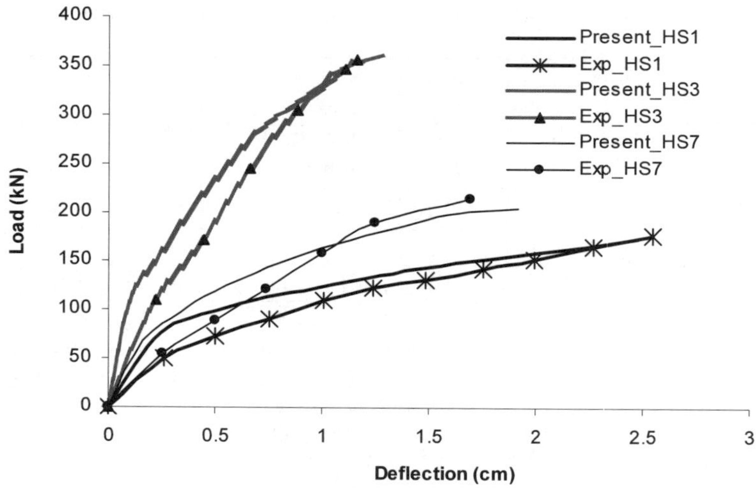

Fig. 7.18 Load-deflection Curve at Central Point of Slab HS7

It can be seen from Table 7.4 and Fig. 7.17 that the present estimations of ultimate loads agree well with the experimental results. The percentage error of predicted ultimate loads to the respective experimental data fall in a range of –5% and 5% of the experimental data. On the other hand, the predicted deflections at different levels of load are smaller than the corresponding experimental results for all the three high-strength slabs. It implies that the slabs behave stiffer in the simulation. The reason may be that the Young's modulus adopted for the concrete or the steel reinforcement material is higher than the practical value.

The damage distributions at failure are shown in Figs. 7.19, 7.21, and 7.23. The stress distributions in the bottom reinforcement at failure are shown in Figs. 7.20, 7.22 and 7.24, respectively. It can be seen from Fig. 7.18 that the damage or the crack for slab HS1 expands to almost the whole area of the slab. And the yield region in the bottom layer of reinforcement in slab HS1 (above the level of 490 MPa in Fig. 7.20) is very wide and almost all the bottom reinforcement yields. The damage distribution and the bottom steel stress indicate a flexure failure mode. The failure type of flexure for slab HS1 is confirmed by the large deflection of 2.7 cm at failure (see Fig. 7.18) compared with the relatively much lower deflections of 1.45 cm and 1.9 cm of Slabs HS3 and HS7 at failure. Compared with slab HS1, the deflection, the bottom reinforcement yield area and damage distribution of slab HS3 at failure are rather small (see Figs. 7.18, 7.21 and 7.22). The damage area is concentrated at the central area of the slab implying that the slab fails in pure shear. On the other hand, the deflection, damage area and the bottom reinforcement yield area for slab HS7 at failure go between the counterparts of slab HS1 and HS3, which indicates that the failure mode is most probably ductile shear failure. The failure patterns for the three slabs predicted in the present study conform to those reported by Marzouk and Hussein (1991).

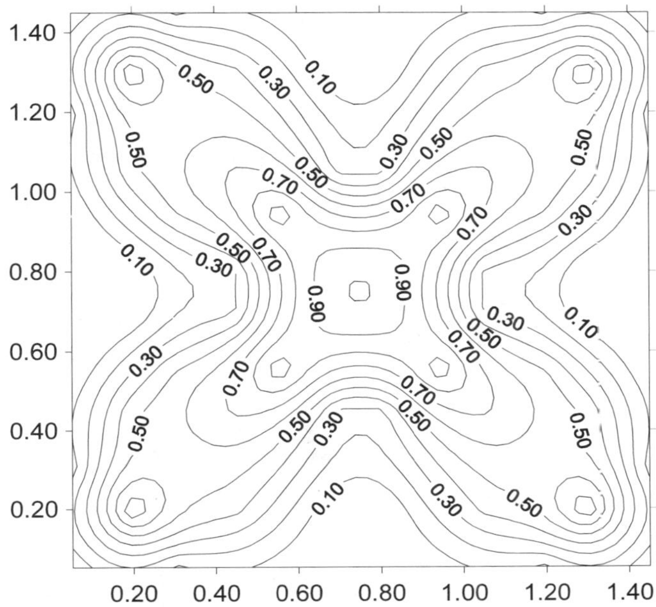

Fig. 7.19 Damage Distribution of Bottom Layer at Failure for Slab HS1

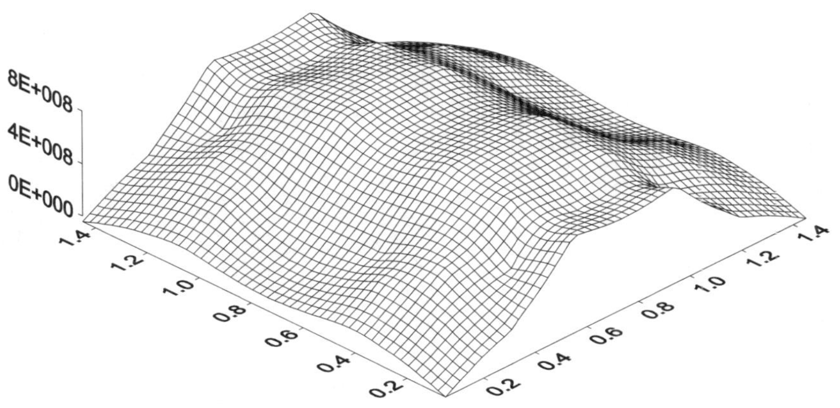

Fig. 7.20 Stress Distribution of Bottom Reinforcement at Failure for Slab HS1

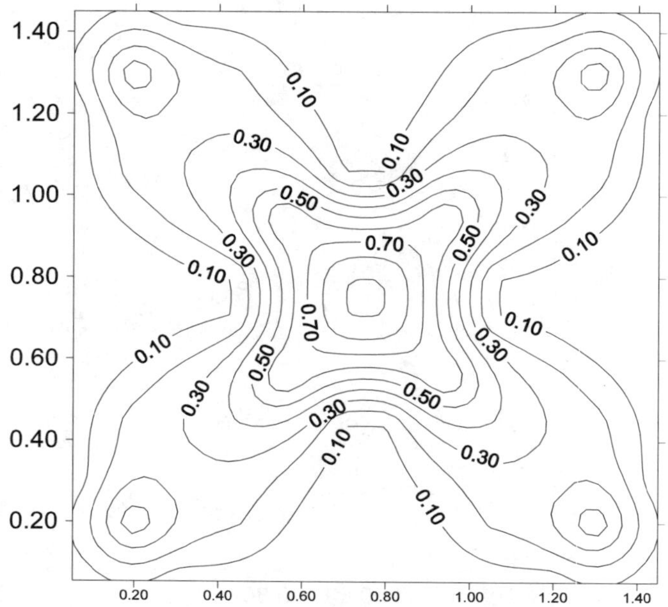

Fig. 7.21 Damage Distribution of Bottom Layer at Failure for Slab HS3

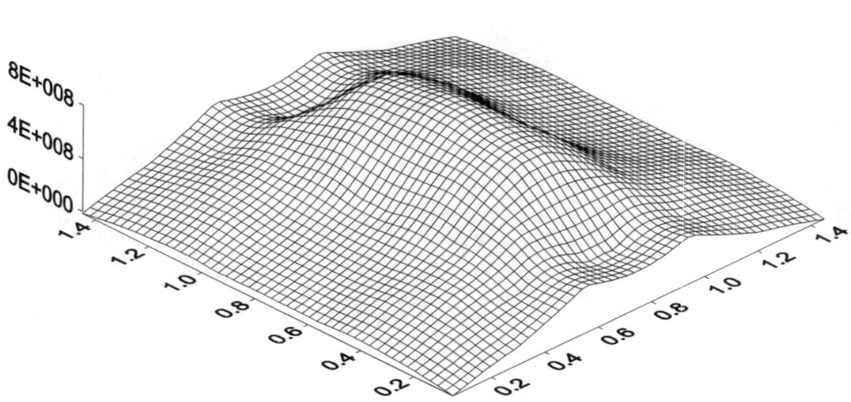

Fig. 7.22 Stress Distribution of Bottom Reinforcement at Failure for Slab HS3

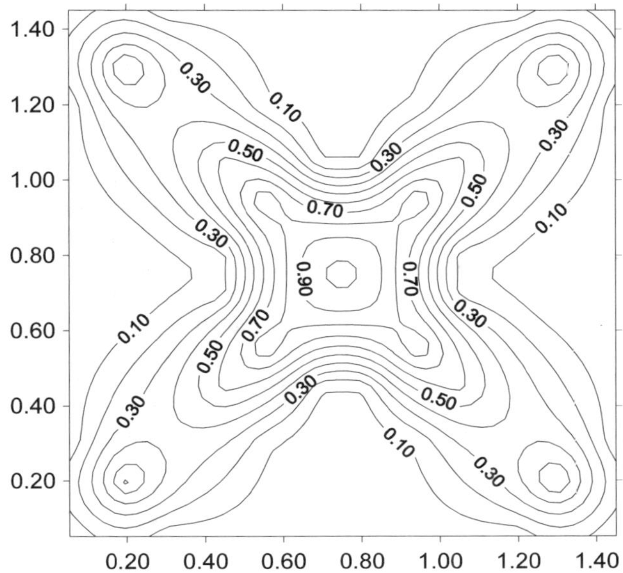

Fig. 7.23 Damage Distribution of Bottom Layer at Failure for Slab HS7

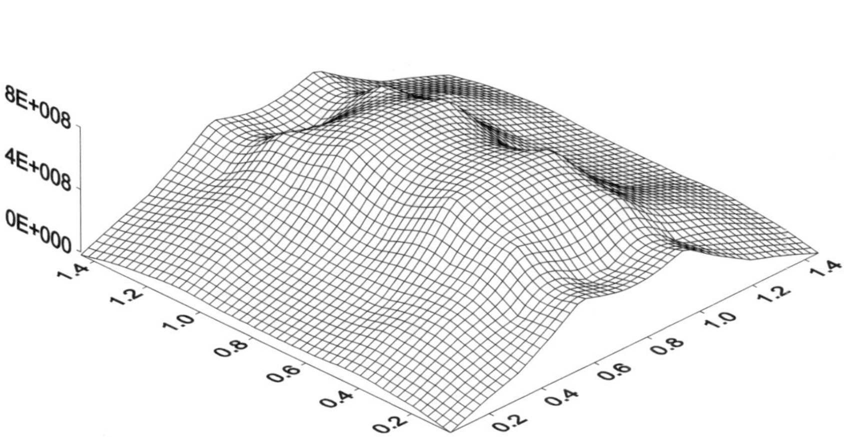

Fig. 7.24 Stress Distribution of Bottom Reinforcement at Failure for Slab HS7

Summary

Through comparison of FE simulation results and the experimental data, conclusions can be drawn that a new three-dimensional elasto-plastic-damage constitutive model for concrete is successfully established. It is a generalization of the unified strength theory, and referred to the Generalized Unified Theory (GUT). The derived load-carrying capacities for all the slabs and the shell are in good agreement with the experimental data. Generally, the calculated deflections at different levels of load for all the slabs and the shell also reflect the real deformation procedure. The only exception is that the predicted deflections for the high-strength slabs are smaller than the experimental counterparts, which implies that the high-strength slabs in the simulation are stiffer than the actual slabs.

The unified strength theory is also generalized to rock material. A non-linear unified strength criterion for rock material was proposed by Yu, Zan, Zhao and Yoshimine (2002). The yield surface of the nonlinear unified strength criterion for rock materials is shown in Fig.7.25. The mathematical modeling of the non-linear unified strength criterion is discussed in section 4.14 of chapter 4.

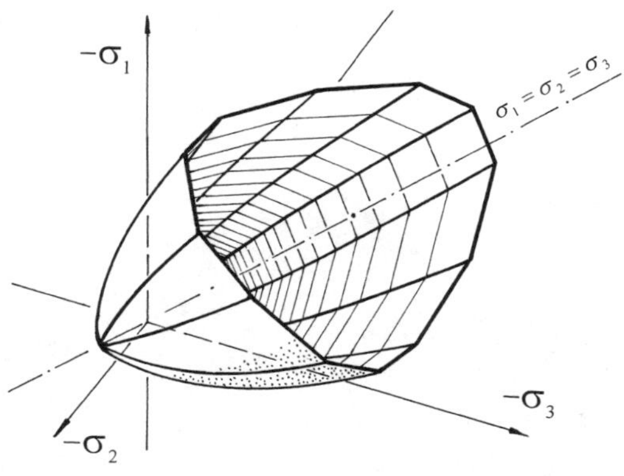

Fig. 7.25 Yield surface of the nonlinear unified strength criterion for rock material

Damage distributions and the reinforcement stress distributions predict well the reinforcement anisotropy of the common concrete slabs and also the failure patterns for the high-strength concrete slabs. The plastic FE method and relevant Program will be described in detail in another book: “Computational Plasticity”.

Problems

Problem 7.1

Compare the meridians of the Ottosen four parameter criterion, the William-Wranke five-parameter criterion, the Podgorski concrete criterion, the Kotsovos concrete criterion, and the generalized unified yield criterion. Discuss the merits and demerits of each criterion.

Problem 7.2

Compare the deviatoric sections of the Ottosen four parameter criterion, the William-Wranke five-parameter criterion, the Podgorski concrete criterion, the Kotsovos concrete criterion, and the generalized unified yield criterion. Discuss the merits and demerits of each criterion.

Problem 7.3

Discuss the applicability of the generalized unified yield criterion for other geomaterials.

References and Bibliography

Chen WF (1982) *Plasticity in Reinforced Concrete.* McGraw–Hill, New York

Chen WF (1998) Concrete plasticity: past, present and future. In: *Strength Theory: Applications, Developments and Prospects for the 21st Century,* Yu MH and Fan SC eds, Science Press, Beijing, New York, 7-48

Duddeck H, Griebenow G, and Schaper G (1978) Material and Time-Dependent Nonlinear Behavior of Cracked Slabs. In: Nonlinear Behaviour of Reinforced Concrete Spatial Structures. Vol. 1, Preliminary Report, IASS Smp., Darmstadt. Darmstadt (Eds: Mehlhorn G, Ruhle H. and Zerna W). Werner-Verlag Dusseldorf, 101–113.

Fan SC, Wang F (2002) A new strength criterion for concrete. ACI Structural Journal, 99, May-June, 317–326.

Figueigas JA, Owen DRJ (1984) Analysis of Elasto-Plastic and Geometrically Nonlinear Anisotropic Plates and Shells. In: Finite Element Software for Plates and Shells (eds. Hinton E., Owen D.R.J.). Prineridge Press, Swansea, U. K, 235–326.

Franklin JA (1970) Classification of rock according to its mechanical properties, Rock Mechanics Research Report, No. T. 1, Imperical College, London.

Gervera M, and Hinton E (1986) Nonlinear Analysis of Reinforced Concrete Plates and Shells Using a Three Dimensional Model. In: Computational Modelling of Reinforced Concrete Structures (eds. Hinton E. and Owen D.R.J.). Prineridge Press, Swansea, U. K.

Huang HC (1988) Static and Dynamic Analysis of Plates and Shells: Theory Software and Application. Pineridge Press Limited, Swansea, U. K.

Koiter WT (1953) Stress-strain relations, uniqueness and variational theorems for elastic-plastic materials with singular yield surface, Quart. Appl. Math., 11, 350-354.

Kotsovos MD, Pavlovic MN (1995) Structural Concrete: Finite element analysis for limit-state design. Thomas Tedford Publications, Thomas Telford Services Ltd., 1 Heron Quay, London E14 4JD.

Krajcinovic D (1996) Damage mechanics, Elsevier.

Labbane M, Saha NK, and Ting EC (1993) Yield Criterion and Loading Function for Concrete Plasticity. International Journals of Solids and Structures, 30, 1269–1288.

Lemaitre J (1985) Coupled elasto-plasticity and damage constitutive equations, Comp. Methods in Appl. Mech. and Engng, 51, 31-49.

Lemaitre J and Chaboche JL (1978) Aspect phenomenologique De la rupture par endommagement, J. Mech, Applique, 2, 317-365.

Lubarda VA, Krajcinovic D, and Mastilovic S (1994) Damage model for brittle elastic solids with unequal tensile and compressive strengths, *Eng. Fracture Mech.* 49, 681-697.

Mazars JA (1986) Description of Micro- and Macro-scale Damage of Concrete Structures. Engineering Fracture Mechanics, 1986, 25, No. 5/6, 729–737.

Mazars J and Pijaudier-Cabot G (1989) Continuum damage theory—application to concrete, *J. of Engineering Mechanics*, 115(2), 345-365.

Marzouk HM, and Hussein A (1991) Experimental Investigation on the Behavior of High-Strength Concrete Slabs. ACI Structural Journal, 88, 701–713.

Nielsen M.P (1984, 1999) *Limit Analysis and Concrete Plasticity*. CRC Press, London.

Ottosen NS (1977) A failure criterion for concrete, ASCE, EM. 4, vol.103.

Owen DJR, Hinton H (1980) *Finite Elements in Plasticity: Theory and Practice*. Pineridge Press Limited, Swansea, U. K.

Saouma VE, Ingraffea AR, and Catalano DM (1980) Fracture toughness of concrete-K_{IC} revisited, Report 80-9, Department of Structural Engineering, Cornell University.

Wang F (1998) *Nonlinear Finite Element Analysis of RC Plate and Shell Using Unified Strength Theory*. Ph.D. Thesis, Nanyang Technological University, Singapore.

Yazdani S and Schreyer HL (1988) An anisotropic damage model with dilatation for concrete, *Mech. Meter*. 7: 231-244.

Yu MH, He LN and Song LY (1985) Twin shear stress theory and its generalization, *Scientia Sinica* (Sciences in China), English Edition, Series A, 28(11), 1174-1183.

Yu MH and He LN (1991) A new model and theory on yield and failure of materials under the complex stress state. In: *Mechanical Behavior of Materials-6*, (ICM-6). Jono M. and Inoue T. ed., Pergamon Press, Oxford, vol. 3, 841-846.

Yu MH (2002) *Concrete Strength Theory and Its Applications* (in Chinese). Higher Education Press, Beijing..

Yu MH, Zan YW, Zhao J. Yoshimine M (2002) A unified strength criterion for rock material. *Int.J. of Rock Mechanics and Mining Science*, 39(6): 975-989.

Yu MH (2004) *Unified Strength Theory and Its Applications*. Springer, Berlin.

8 Twin-Shear Slip-Line Field for Plane Strain Problem

8.1 Introduction

Plane stress, plane strain and axisymmetric problems are three important problems in plasticity and engineering. Figure 8.1 shows an example of these three kinds of structure. Figure 8.1(a) is a plane stress structure with an uniform thickness thin lamina deformed under the action of force which lie in its median plane. Figure 8.1(b) is a plane strain problem with zero strain at z deriction (length direction) in a very large thickness structure. Figure 8.1(c) is a axisymmetrical problems which are symmetrical in terms of geometry, boundary conditions and external loading about an axis.

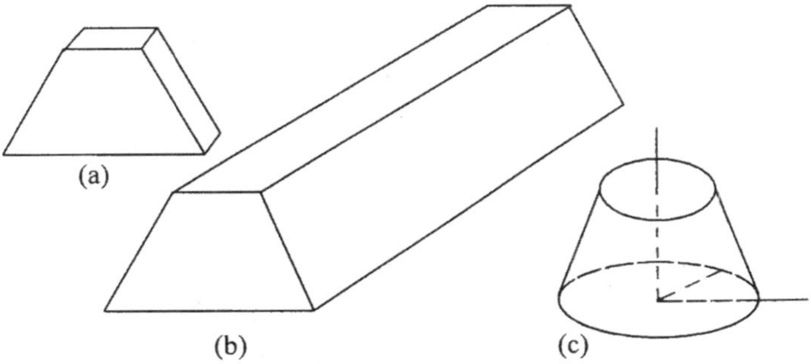

Fig. 8.1 Three kinds of structures

These three kinds of structures shown in Fig.8.1 have an identical section, a trapezoid, but different stress states. The stresses normal to the solution domain σ_z in plane stress state is zero principal stress and nonzero principal stress in plane strain; hoop stress σ_θ in axisymmetrical problems is also a principal stress. The stress state of the three kinds of structures will be discussed in Chapters 8, 10 and 12.

In this chapter, a general theory, known as the slip-line field theory of plane strain problem is described to analyse the bearing capacity of plane strain structures. The characteristics line theory for plane stress and axisymmetrical problems will be discussed in Chapters 10, 11 and Chapter 12 respectively.

The slip line field theory and characteristics theory deal with non-strain hardening or softening. Clearly, these idealization are not realzed in engineering materials, but they can be made to give very good first approximations to bearing capacity of structures, and provide indications of the manner in which material deforms. There is no allowance for time or strain rate effects. Also, all inertia forces are neglected and the problems treated as quasi-static.

Slip line field theory for plane strain problems can be obtained by combining the characteristic line theory with the plastic plane strain equations. The theoretical frame of characteristic line field theory was proposed by Kotter (1903), and developed by Prandtl (1923), Hencky (1923), Gecteinger (1930), Hill (1950), Prager(1949), Березанлев (1953), Соколовский (1960) and Johnson and Mellor (1982), et al. Plane-strain slip line fields for metal deformation processes were summarized by Johnson, Sowerby and Venter in 1982. Plane-strain slip line fields for geomaterials and civil engineering were summarized by Sokolovsky in 1960. The theory is now becoming an important content for plasticity, metal processing mechanics, and geomechanics. It is widely applied in the fields of metal plastic forming, the limit analysis of structure, civil engineering and mechanical engineering.

The Prandtl slip field and the Hill slip field of a non-SD material under footing are shown in Fig. 8.2(a) and (b). The Prandtl slip field for SD material under footing is shown in Fig. 8.3.

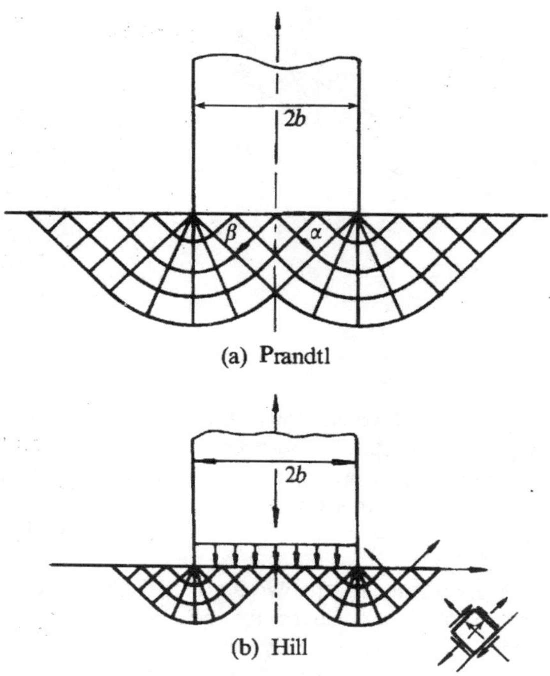

Fig. 8.2 Prandtl slip field and the Hill slip field

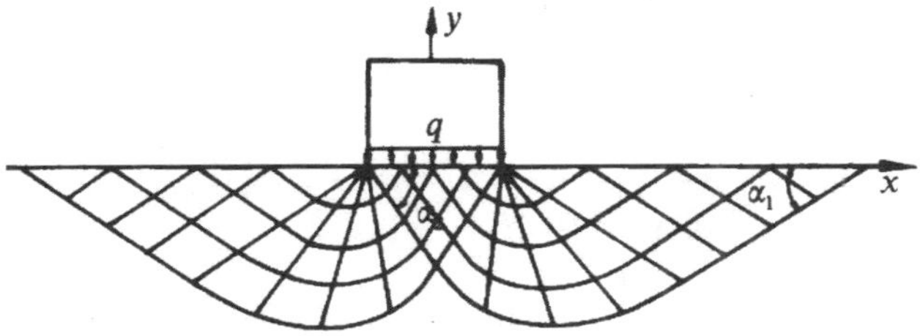

Fig. 8.3 Slip field for SD materials

The slip field theory is supported by other observations. Figure 8.4 is a experimental observation on a strip footing. It is obtained by using a series of aluminium bar under the pressure of a strip. The displacement of aluminium bar shows a slip zone (from Matsuoka).

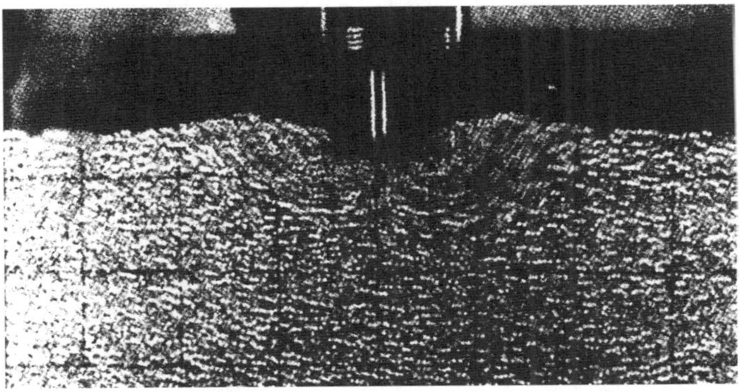

Fig. 8.4 The experimental observation of flow of strip footing (Matsuoka 2000)

The conventional slip line field theories for plane strain problems are based on the single-shear yield function such as the Tresca (Maximum shear stress criterion) criterion and the Mohr-Coulomb strength theory. The effect of intermediate principal stress σ_2 was not taken into account in this kind of slip line field theories. Zenkiewicz et al. (1992) indicated that the effect of intermediate principal stress to the material's strength is so prominent that it should not be neglected especially for some particular materials such as high strength steel, alloy, iron, ceramics, rock and soil.

Slip line field theories based on single-shear strength theories can not be adapted for those materials in which intermediate principal stress plays an important role. A new slip field theory based on the twin-shear yield criterion and generalized twin-shear criterion was proposed by Yu et al. in 1994. A new orthogonal and non-orthogonal slip line field theory for plane strain problems was proposed. The effects of intermediate principal stress on the materials' yield or failure is taken into account in the twin-shear slip field, for plane strain problems.

8.2 Stress State in Plane Strain

In plane strain condition the displacements all occur in parallel planes in the body. It means that planes parallel to the xy plane, and all stresses and strains are independent of z, i.e.,

$$\varepsilon_z = \varepsilon_{xz} = \varepsilon_{yz} = \tau_{xz} = \tau_{yz} = v_z = 0, \tag{8-1}$$

where z is a principal direction, and σ_z is a principal stress. In plane strain state we have only four steses components σ_x, σ_y, τ_{xy} and σ_z.

$$\sigma_x = \sigma_x(x,y), \quad \sigma_y = \sigma_y(x,y), \quad \sigma_z = \sigma_z(x,y), \quad \tau_{xy} = \tau_{xy}(x,y) \tag{8-2}$$

$$\varepsilon_x = \varepsilon(x,y), \quad \varepsilon_y = \varepsilon_y(x,y), \quad \gamma_{xy} = \gamma_{xy}(x,y) \tag{8-3}$$

$$v_x = v_x(x,y), \quad v_y = v_y(x,y), \tag{8-4}$$

The strain rate tensor is

$$\dot{\varepsilon}_x = \begin{bmatrix} \dfrac{\partial v_x}{\partial x} & \dfrac{1}{2}\left(\dfrac{\partial v_x}{\partial y} + \dfrac{\partial v_y}{\partial x}\right) & 0 \\ \dfrac{1}{2}\left(\dfrac{\partial v_x}{\partial y} + \dfrac{\partial v_y}{\partial v_x}\right) & \dfrac{\partial v_y}{\partial y} & 0 \\ 0 & 0 & 0 \end{bmatrix} \tag{8-5}$$

The maximum principal stress σ_1 and minimum principal stress σ_3 ($\sigma_1 \geq \sigma_2 \geq \sigma_3$) of plane strain problem are

$$\begin{matrix} \sigma_1 = \sigma_{max} \\ \sigma_3 = \sigma_{min} \end{matrix} = \frac{\sigma_x + \sigma_y}{2} \pm \sqrt{\left(\frac{\sigma_x - \sigma_y}{2}\right)^2 + \tau_{xy}} \tag{8-6}$$

Introducing a parameter m, the intermediate principal stress σ_2 can be expressed as follows

$$\sigma_2 = \frac{m}{2}(\sigma_1 + \sigma_3), \tag{8-7}$$

where $0 < m \leqslant 1$ for plane strain problem, and m can be referred as the intermediate principal stress state parameter. It can be determined by both theoretical and experimental analysis. Generally, $m<1$ for elastic region (this can be explained by Hooke law, where m/2 is equal to Poisson ratio) i.e. $\sigma_z = m(\sigma_1 + \sigma_3)/2 < (\sigma_1 + \sigma_3)/2$, and $m \rightarrow 1$ for plastic region.

The stress state in plane strain (σ_x, σ_y, τ_{xy} and σ_z) can be expressed by three principal stresses (σ_1, σ_2 and σ_3); the stress state can also be characterized by superposition of the hydrostatic stress (mean stress) p on the pure shear stress R as follows:

$$p = \frac{1}{2}(\sigma_1 + \sigma_3) = \frac{1}{2}(\sigma_x + \sigma_y) = \frac{1}{m}\sigma_2 = \frac{1}{m}\sigma_z, \tag{8-8}$$

$$R = \frac{1}{2}(\sigma_1 - \sigma_3) = \sqrt{\left(\frac{\sigma_x - \sigma_y}{2}\right)^2 + \tau_{xy}^2}, \tag{8-9}$$

The stress state is illustrated as in Fig. 8.5

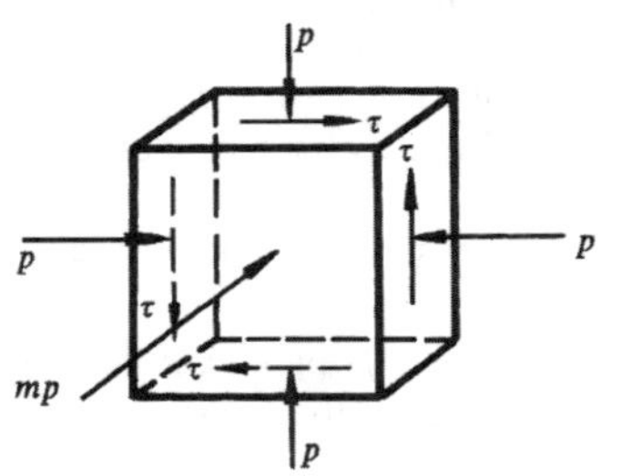

Fig. 8.5 Stresses in plane strain state

The relations between these two kinds of stresses (σ_x, σ_y, τ_{xy}, σ_z; and p, R, σ_2) are

$$\begin{aligned} \sigma_x &= p + R\cos 2\theta \\ \sigma_y &= p - R\cos 2\theta \\ \tau_{xy} &= R\sin 2\theta \\ &= (p + C_0 \cdot \cot\varphi)\sin\varphi \cdot \sin 2\varphi \end{aligned} \tag{8-10}$$

where θ is introduced when we define by the angle from x coordinate to the direction of σ_1.

The three states, i.e. principal stress state σ_1, σ_2, σ_3, pure shear stress state or maximum shear stress state p, R, σ_2 and general stress state σ_x, σ_y, τ_{xy}, σ_z in plane strain can be illustrated in Fig. 8.6.

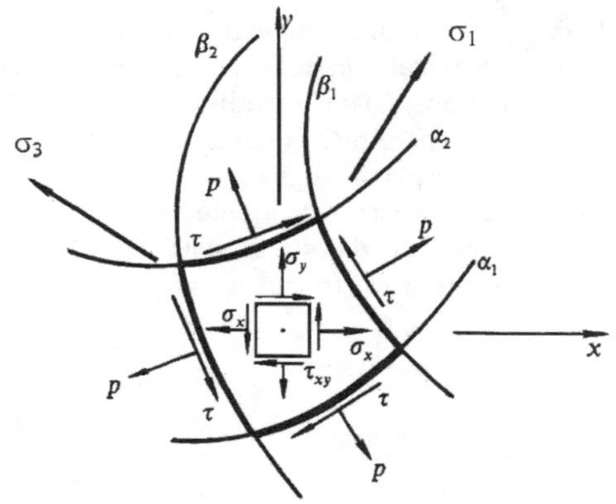

Fig. 8.6 Relation between stresses in plane strain state

There are two families of curve α and β in Fig. 8.6. The directions of the surfaces on which the maximum tangential stresses act make angles $\pm\pi/4$ with the principal directions. They are the maximum shear stress line which is tangent at every point to the surface of maximum tangential stress. The α-line and β-line are inclined to the first principal direction at 45° (Fig. 8.6). It is obvious that there are two orthogonal families of slip lines, characterized by the equations

$$x=x(\alpha,\beta), \qquad y=y(\alpha,\beta)$$

The lines of the first family (α-lines) correspond to fixed values of the parameter β (β=const.); along β-lines the parameter α is constant. The α-lines and β-lines are referred to as the slip lines. Slip line is a important concept in plasticity.

8.3 Twin-Shear Strength Theory for Plane Strain

The twin-shear strength theory (Yu 1985) has been described in Chapter 4. The mathematical formulae of the twin-shear strength theory can be expressed in two equations as follows:

$$F=\sigma_1-\frac{\alpha}{2}(\sigma_2+\sigma_3)=\sigma_t, \quad \text{when } \sigma_2\le\frac{\sigma_1+\alpha\sigma_3}{1+\alpha} \tag{8-11a}$$

$$F'=\frac{1}{2}(\sigma_1+\sigma_2)-\alpha\sigma_3=\sigma_t, \quad \text{when } \sigma_2\ge\frac{\sigma_1+\alpha\sigma_3}{1+\alpha} \tag{8-11b}$$

Where σ_t and σ_c are uniaxial tensile and compressive strength of material respectively. Parameter $\alpha=\sigma_t/\sigma_c$ is tensile-compressive strength ratio, which reflects material's Strength Difference (SD) effect.

The twin-shear strength theory can also be expressed in terms of shear strength parameter C_0 and friction angle φ_0 as follows:

$$F=\sigma_1-\frac{\sigma_2+\sigma_3}{2}+(\sigma_1+\frac{\sigma_2+\sigma_3}{2})\sin\varphi_0=2C_0\cos\varphi_0$$

$$\text{when} \quad \sigma_2\leq\frac{\sigma_1+\sigma_3}{2}+\frac{\sigma_1-\sigma_3}{2}\sin\varphi_0,$$

or

$$\text{when} \quad \sigma_2\leq p+R\sin\varphi_0 \tag{8-12a}$$

$$F'=\frac{\sigma_1+\sigma_2}{2}-\sigma_3+(\frac{\sigma_1+\sigma_2}{2}+\sigma_3)\ \sin\varphi_0=2C_0\cos\varphi_0$$

$$\text{when} \quad \sigma_2\geq\frac{\sigma_1+\sigma_3}{2}+\frac{\sigma_1-\sigma_3}{2}\sin\varphi_0.$$

or

$$\text{when} \quad \sigma_2\leq p+R\sin\varphi_0 \tag{8-12b}$$

Because the intermediate principal stress equals

$$\sigma_2=mp\leq p+R\sin\varphi_0=\frac{1}{2}(\sigma_1+\sigma_3)+\frac{1}{2}(\sigma_1-\sigma_3)\sin\varphi_0, \tag{8-13}$$

It accords with the first condition of the twin-shear strength theory equation (8-12). Hence, we adopt the first equation of the twin-shear strength theory (equations 8-12a), combining equations (8-7), the twin-shear strength theory for plane strain is obtained as follows:

$$\frac{\sigma_1-\sigma_3}{4}(3+\sin\varphi_0)+\frac{\sigma_1+\sigma_3}{4}[(1-m)+\sin\varphi_0(3+m)]=2C_0\cdot\cos\varphi_0 \tag{8-14}$$

or

$$R=\frac{4C_0\cdot\cos\varphi_0}{2+(1+\sin\varphi_0)}+\frac{(1-m)+(3+bm)\sin\varphi_0}{2+(1+\sin\varphi_0)}p \tag{8-15}$$

When m=1 the twin-shear strength theory for plane strain (Eq.8-15) can be simplified to

$$R=\frac{4C_0\cdot\cos\varphi_0}{2+(1+\sin\varphi_0)}+\frac{4\sin\varphi_0}{2+(1+\sin\varphi_0)}p \tag{8-16}$$

We introduce two new parameters φ_t and C_t for twin-shear strength theory in plane strain state. They are referred to as the twin-shear friction angle φ_t and the twin-shear cohesion C_t respectively (Yu et al. 1994).

$$\sin\varphi_t = \frac{(1-m)+(3+bm)\sin\varphi_0}{2+(1+\sin\varphi_0)} \quad \text{when } m \neq 1, \tag{8-17a}$$

$$\sin\varphi_t = \frac{4\sin\varphi_0}{2+(1+\sin\varphi_0)} \quad \text{when } m=1, \tag{8-17b}$$

$$C_t = \frac{4C_0 \cdot \cos\varphi_0}{2+(1+\sin\varphi_0)} \frac{1}{\cos\varphi_t}, \tag{8-18}$$

Substituting equations (8-17) and (8-18) into equation (8-15), we obtain

$$R = p\sin\varphi_t + C_t\cos\varphi_t, \tag{8-19}$$

where the compressive stress is usually defined positive.

8.4 Twin-Shear Slip Line Field Theory for Plane Strain Problem (Statically Admissible Field)

The equilibrium differential equations for plane strain can be written as (see: Hill 1950 or Johnson and Mellor 1962)

$$\begin{aligned} \frac{\partial\sigma_x}{\partial x} + \frac{\partial\tau_{yx}}{\partial y} &= \gamma\sin\theta_0 \\ \frac{\partial\tau_{xy}}{\partial x} + \frac{\partial\sigma_y}{\partial y} &= -\gamma\cos\theta_0 \end{aligned} \tag{8-20}$$

where γ is specific weight and θ_0 is the angle between the direction of gravity and the minus direction of y coordinate, as shown in Fig. 8.2.

Solving equations (8-20), (8-10), and (8-13), the stress governing equations are obtained as follows:

$$\begin{aligned} &\frac{\partial p}{\partial x}\left(1+\frac{(1-m)+(3+m)\sin\varphi_0}{2+(1+\sin\varphi_0)}\cos 2\theta\right)+\frac{\partial p}{\partial y}\frac{(1-m)+(3+m)\sin\varphi_0}{2+(1+\sin\varphi_0)}\sin 2\theta + \\ &\qquad +2R\left(\frac{\partial\theta}{\partial y}\cos 2\theta - \frac{\partial\theta}{\partial x}\sin 2\theta\right) = \gamma\sin\theta_0 \\ &\frac{\partial p}{\partial x}\frac{(1-m)+(3+m)\sin\varphi_0}{2+(1+\sin\varphi_0)}\sin 2\theta + \frac{\partial p}{\partial y}\left(1-\frac{(1-m)+(3+m)\sin\varphi_0}{2+(1+\sin\varphi_0)}\cos 2\theta\right) + \\ &\qquad +2R\left(\frac{\partial\theta}{\partial x}\cos 2\theta + \frac{\partial\theta}{\partial y}\sin 2\theta\right) = -\gamma\cos\theta_0 \end{aligned} \tag{8-21}$$

Equations (8-21) satisfy the twin-shear strength theory, equations (8-19), and plane strain static equilibrium equations (8-20).

Equations (8-21) may be elliptic-type (two different real roots), parabolic-type (two same real roots) or hyperbolic-type (no real roots) pseudo-linear differential equations which can be solved by the method of characteristics

$$\alpha \text{ family:} \quad \frac{dy}{dx} = tg(\theta - \mu), \tag{8-22a}$$

$$\beta \text{ family:} \quad \frac{dy}{dx} = tg(\theta + \mu), \tag{8-22b}$$

in which $\mu = \pi/4 - \varphi_t/2$.

We can see that the equations are elliptic-type pseudo-lineal differential equations. Equations (8-22) are not only two different real roots of the pseudo-linear equations (8-21), but also two characteristic lines at the angle of 2μ with each other.

By using the Eqs. (8-17) and (8-18), the governing equations (8-21) can be rewritten as

$$\begin{aligned} &\frac{\partial p}{\partial x}(1+\sin\varphi_t\cdot\cos 2\theta)+\frac{\partial p}{\partial y}\sin\varphi_t\cdot\sin 2\theta+2R\left(\frac{\partial\theta}{\partial y}\cos 2\theta-\frac{\partial\theta}{\partial x}\sin 2\theta\right)=\gamma\sin\theta_0 \\ &\frac{\partial p}{\partial x}\sin\varphi_t\cdot\sin 2\theta+\frac{\partial p}{\partial y}(1-\sin\varphi_t\cdot\cos 2\theta)+2R\left(\frac{\partial\theta}{\partial x}\cos 2\theta+\frac{\partial\theta}{\partial y}\sin 2\theta\right)=-\gamma\cos\theta_0 \end{aligned} \tag{8-23}$$

We can see from equations (8-21) and (8-23) that it is the substitution of C_t and φ_t for C_0 and φ_0, which embodies the effect of intermediate principal stress in the twin-shear slip line field theory.

Choose the curvilinear coordinates system S_α and S_β coinciding with the characteristic (slip) lines α and β. According to the rules of directional derivative, stress governing equations can be expressed in terms of curvilinear coordinates S_α and S_β

α family:

$$-\sin 2\mu\frac{\partial p}{\partial S_\alpha}+2R\frac{\partial\theta}{\partial S_\alpha}+\gamma\left[\sin(\theta_0+2\mu)\frac{\partial x}{\partial S_\alpha}+\cos(\theta_0+2\mu)\frac{\partial y}{\partial S_\alpha}\right]=0 \tag{8-24a}$$

β family:

$$\sin 2\mu\frac{\partial p}{\partial S_\beta}+2R\frac{\partial\theta}{\partial S_\beta}+\gamma\left[\sin(\theta_0-2\mu)\frac{\partial x}{\partial S_\beta}+\cos(\theta_0-2\mu)\frac{\partial y}{\partial S_\beta}\right]=0\cdot \tag{8-24b}$$

Equations (8-21) to (8-24) are the stress governing equations (statically admissible) of the twin-shear slip line field theory for plane strain problem. The equations differ from ordinary stress governing equations by substituting φ_t and C_t for φ_0 and C_0.

8.5 Twin-Shear Slip Line Field Theory for Plane Strain Problem (Kinematically Admissible Field)

Adopting the associated flow rule,

$$d\varepsilon_{ij}{}^{p}=d\lambda\frac{\partial g}{\partial\sigma_{ij}}=d\lambda\frac{\partial f}{\partial\sigma_{ij}}, \tag{8-25}$$

The twin-shear yield function equations is

$$f=\frac{1}{2}(\sigma_x+\sigma_y)\sin\varphi_t-\sqrt{\frac{1}{4}(\sigma_y-\sigma_x)^2+\tau_{xy}^2}+C_t\cdot\cos\varphi_t=0, \tag{8-26}$$

and plastic strain rate under small deformation and rigid plastic condition

$$\dot{\varepsilon}_x=\frac{\partial\dot{u}_x}{\partial x}=\frac{\partial v_x}{\partial x}$$

$$\dot{\varepsilon}_y=\frac{\partial\dot{u}_y}{\partial y}=\frac{\partial v_y}{\partial y},$$

$$\dot{\gamma}_{xy}=\frac{\partial\dot{u}_x}{\partial y}+\frac{\partial\dot{u}_y}{\partial x}=\frac{\partial v_x}{\partial y}+\frac{\partial v_y}{\partial x}, \tag{8-27}$$

The velocity governing equations of twin-shear slip line field theory for plane strain problem are

$$\alpha\text{ family:}\quad dv_\alpha+[v_\alpha\cdot ctg(\frac{\pi}{2}-\varphi_t)-v_\beta\cdot\csc(\frac{\pi}{2}-\varphi_t)]d\psi=0, \tag{8-28a}$$

$$\beta\text{ family:}\quad dv_\beta+[v_\alpha\cdot\csc(\frac{\pi}{2}-\varphi_t)-v_\beta\cdot ctg(\frac{\pi}{2}-\varphi_t)]d\psi=0. \tag{8-28b}$$

In equations (8-28), u_x and u_y are displacement components, $v_x=du_x/dt$ and $v_y=du_y/dt$ are velocity components along x and y directions, whereas v_α and v_β are velocity components along α and β slip lines respectively. Superimposed dot means derivative with respect to time.

Substituting the twin-shear yield function for plane strain into associated flue rule equation (8-25), and taking the rigid plastic postulate into consideration, we have

$$\begin{aligned} d\varepsilon_x &= -d\lambda\frac{\partial f}{\partial\sigma_x} = \frac{-d\lambda}{2}\left\{\sin\varphi_t + \frac{1/2(\sigma_y-\sigma_x)}{\sqrt{\frac{1}{4}(\sigma_y-\sigma_x)^2+\tau_{xy}^2}}\right\} \\ d\varepsilon_y &= -d\lambda\frac{\partial f}{\partial\sigma_y} = -\frac{d\lambda}{2}\left\{\sin\varphi_t + \frac{1/2(\sigma_y-\sigma_x)}{\sqrt{\frac{1}{4}(\sigma_y-\sigma_x)^2+\tau_{xy}^2}}\right\} \\ d\gamma_{xy} &= 2d\varepsilon_{xy} = -2d\lambda\frac{\partial f}{\partial\tau_{xy}} = d\lambda\frac{xy}{\sqrt{\frac{1}{4}(\sigma_y-\sigma_x)^2+\tau_{xy}^2}} \end{aligned} \tag{8-29}$$

where the relation between shear strain ε_{xy} and engineering shear strain γ_{xy}, $\gamma_{xy}=2\varepsilon_{xy}$, and the definition of positive compressive normal stress are used.

Figure 8.7 gives the plane strain Mohr circle, were Ψ is the angle between α slip line and x direction, i.e. $\Psi=\theta-\mu=\theta-(\pi/4-\varphi_t/2)$. Following this definition, equation (8-22) can be simplified to

$$\begin{aligned} d\varepsilon_x &= -\frac{d\lambda}{2}[\sin\varphi_t+\sin(2\psi-\varphi_t)] \\ d\varepsilon_y &= -\frac{d\lambda}{2}[\sin\varphi_t+\sin(2\psi-\varphi_t)] \\ d\gamma_{xy} &= d\lambda\cos(2\psi-\varphi_t) \end{aligned} \tag{8-30}$$

When the direction of slip line is along x direction, *i.e.* $\Psi = -(\pi/2-\varphi_t)$ (β line) or $\Psi = 0$ (α line), equation (8-22) can be expressed

$$d\varepsilon_x\big|_{\psi=0} = d\varepsilon_x\big|_{\psi=-(\pi/2-\varphi_t)} = 0. \tag{8-31}$$

This equation indicates that the rigid strain rate vanishes along slip lines.

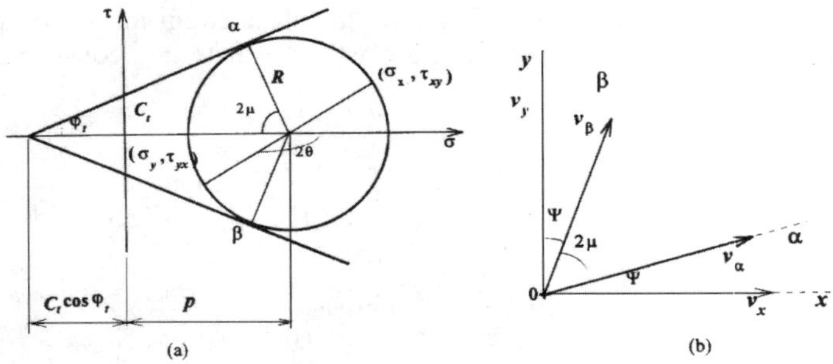

Fig. 8.7 Stresses and velocity in plane strain state

The relation between velocity components along slip lines $v_\alpha = \dot{u}_\alpha$ ($v_\beta = \dot{u}_\beta$) and velocity components along x (y) axis $v_x(v_y)$, can be expressed as

$$\begin{aligned} v_x = \dot{u}_x = \frac{v_\alpha \sin(\psi+2\mu) - v_\beta \sin\psi}{\sin 2\mu} \\ v_y = \dot{u}_y = \frac{v_\alpha \cos(\psi+2\mu) - v_\beta \cos\psi}{-\sin 2\mu} \end{aligned} . \tag{8-32}$$

Combination of equations (8-30), (8-31) and (8-32) yields,

$$d\varepsilon_x \big|_{\psi=0} = \dot{\varepsilon}_x \big|_{\psi=0} dt = \frac{\partial v_x}{\partial x}\Big|_{\psi=0} dt$$

$$= \{dv_\alpha + [v_\alpha ctg(\pi/2 - \varphi_t) - v_\beta \csc(\pi/2 - \varphi_t)]\, d\psi\}\, dt = 0 \tag{8-33}$$

In the above equation, we have used the relation $\mu = \pi/4 - \varphi_t/2$ and definitions

$$dv_\alpha \equiv (\partial v_\alpha / \partial x)\big|_{\psi=0} \text{ and } d\psi \equiv (\partial \psi / \partial x)\big|_{\psi=0}.$$

When slip line field generates plastic flow, by integrating equations (8-21), we can work out the velocity field that satisfies the boundary conditions of the problem. The velocity field solution complies with the unified yield function, associated flow rule and rigid plastic small deformation condition. When $\varphi_t = \varphi_0$, the case of orthogonal slip line field, equations (8-21) can be simplified to the famous Geiringer velocity equations (Johnson and Mellor 1962).

$$\alpha \text{ family:} \quad dv_\alpha - v_\beta \, d\theta = 0, \tag{8-34a}$$

$$\beta \text{ family:} \quad dv_\beta + v_\alpha \, d\theta = 0. \tag{8-34b}$$

By now, we have extended the twin-shear yield function to slip line field theory for plane strain by means of the introduction of twin-shear effective parameters φ_t and C_t and get the governing equations of the twin-shear slip line field theory for plane strain problem (equations 8-14, 8-18 and 8-21).

Twin-shear slip line field theory can be used to materials with obvious intermediate principal stress effect and give the linkage of various slip line field theories without theoretical difficulty. Moreover, the twin-shear slip field theory contains infinite number of orthogonal and non-orthogonal slip line field theories. With the different choice of parameter (α, β, γ) or φ_t and C_t, we can get a series special cases of the twin-shear slip field theory.

When twin-shear slip line field theory for plane strain problem is applied to materials where gravity is left out of consideration, the velocity governing equations are the same as equations (8-21), whereas the stress governing equations (8-18) can be simplified to

$$\alpha \text{ family:} \quad -\sin 2\mu \frac{\partial p}{\partial S_\alpha} + 2R \frac{\partial \theta}{\partial S_\alpha} = 0, \tag{8-35a}$$

$$\beta \text{ family:} \quad \sin 2\mu \frac{\partial p}{\partial S_\beta} + 2R \frac{\partial \theta}{\partial S_\beta} = 0 \tag{8-35b}$$

Integrating the above equations, after simplification, we can get the solutions

$$\alpha \text{ family:} \quad p = C_\alpha \exp(2\theta \cdot ctg 2\mu) - C_t \cdot ctg\varphi_t, \tag{8-36a}$$

$$\beta \text{ family:} \quad p = C_\beta \exp(-2\theta \cdot ctg 2\mu) - C_t \cdot ctg\varphi_t, \tag{8-36b}$$

where C_α and C_β are integral constants, which will be determined by different boundary conditions. This is a new type of non-orthogonal slip line field.

When φ_t=0 the twin-shear slip field simplized to a orthogonal twin-shear slip line field theory because of the perpendicularly of two (α and β) slip lines, $2\mu = \pi/2 - \varphi_t = \pi/2$. In this case, from equations (8-16) and (8-17) that

$$\varphi_t = \varphi_0 = 0, \quad R = C_t = 4C_0/3, \quad \alpha = 1. \tag{8-37}$$

Substituting $2\mu = \pi/2$ into equations (8-18) and (8-21), we get the stress and velocity governing equations of this case

α family
$$-\frac{\partial p}{\partial S_\alpha}+2R\frac{\partial \theta}{\partial S_\alpha}+\gamma[\cos\theta_0\frac{\partial x}{\partial S_\alpha}-\sin\theta_0\frac{\partial y}{\partial S_\alpha}]=0,$$
$$dv_\alpha - v_\beta d\theta = 0 \quad (8\text{-}38a)$$

β family
$$\frac{\partial p}{\partial S_\beta}+2R\frac{\partial \theta}{\partial S_\beta}+\gamma[-\cos\theta_0\frac{\partial x}{\partial S_\beta}+\sin\theta_0\frac{\partial y}{\partial S_\beta}]=0.$$
$$dv_\beta + v_\alpha d\theta = 0 \quad (8\text{-}38b)$$

Integrating the stress governing equations (8-31), we obtain

α family: $$p-\gamma\cos\theta_0 x+\gamma\sin\theta_0 y-2R\theta=C_\alpha, \quad (8\text{-}39a)$$

β family: $$p-\gamma\cos\theta_0 x+\gamma\sin\theta_0 y+2R\theta=C_\beta, \quad (8\text{-}39b)$$

When φ_t=0 and γ =0, the gravity is negleted. The orthogonal twin-shear slip line field theory is given. From Equations (8-16) and (8-17), we can get the restrictions of this case

$$\varphi_t=\varphi_0=0,\ R=C_t=\frac{4}{3}C_0,\ \gamma=0,\ \alpha=1. \quad (8\text{-}40)$$

Substituting above equations into equations (31), we will have the stress and velocity governing equations of this case

$$\frac{\partial}{\partial S_\alpha}(p-2R\theta)=0,$$
$$dv_\alpha - v_\beta d\theta = 0 \quad \text{α family} \quad (8\text{-}41a)$$

$$\frac{\partial}{\partial S_\beta}(p+2R\theta)=0.$$
$$dv_\beta - v_\alpha d\theta = 0 \quad \text{β family} \quad (8\text{-}41b)$$

where the condition that R is constant with respect to S_α and S_β is used when material and strength theory are fixed

Integrating Equations (8-34), we can get the solutions of this case

α family: $p-2R\theta=C_\alpha$, (8-42a)

β family: $p+2R\theta=C_\beta$. (8-42b)

where C_α and C_β are integral constants. Substitute γ=0 into equations (8-28), we can also get above equations.

The form of equations (8-34) and (8-35) is similar to that of traditional solutions of orthogonal slip line field theories. But the definition of R is different. In this discussion, R is twin-shear parameter.

8.6 Applications of the Twin-Shear Slip Line Field Theory for Plane Strain Problems

8.6.1 Example 1: Strip Footing

In geotechnical engineering, the ultimate bearing capacity of soil under strip footing is an important problem (Fig. 8.8). The least pressure that will cause complete shear failure of the soil in the vicinity of the foundation is defined as the ultimate bearing capacity (q_{limit}). If the pressure on a foundation is steadily increased to the value qf the soil in the vicinity of the foundation changes from the state of elastic equilibrium to the state of plastic equilibrium. The change starts at the edges of the foundation, gradually spreading downwards then outwards on each side of the foundation. Eventually all the soil between the failure surfaces and ground level reaches the state of plastic equilibrium and complete shear failure takes place with the foundation breaking into the soil.

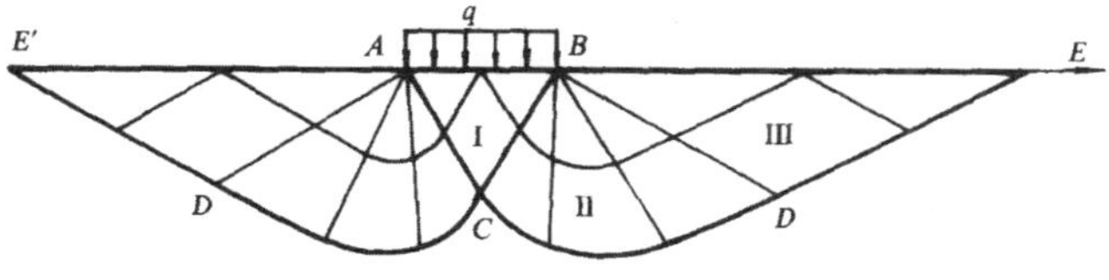

Fig. 8.8 Slip lines field of foundation under strip footing

Prandtl used the the Mohr-Coulomb theory and considered conditions of equilibrium just before the flow starts and found that the ultimate bearing capacity may be expressed in terms of the internal cohesion C_0 of the given material and its friction angle φ, as follows

$$q_0 = C_0 \cdot ctg\varphi_0 [\frac{1+\sin\varphi_0}{1-\sin\varphi_0} \exp(\pi \cdot tg\varphi_0) - 1] \tag{8-43}$$

In the case of α=1 materials, or φ=0 materials, equation 8-8 furnishes q_0=5.14 C_0, or q_0=2.57σ_s, in which the Tresca criterion was used.

In the study of foundation problems, it is generally assumed that there are three regions. They are an active prism ABC and a passive prism BDE connected with a curve CD, lines AC and DE being tangent to this curve. Curve CD, bounding zone

BCD and ACD from below and passing through the tip of all radii vectors, is a logarithmic spiral.

the surfaces AB are smooth. So that there is no friction. It is also assumed that there is a constant pressure on the top. The rest of the boundary is stress-free.The slip line field is given in Fig. 8.8 and Fig. 8.9.

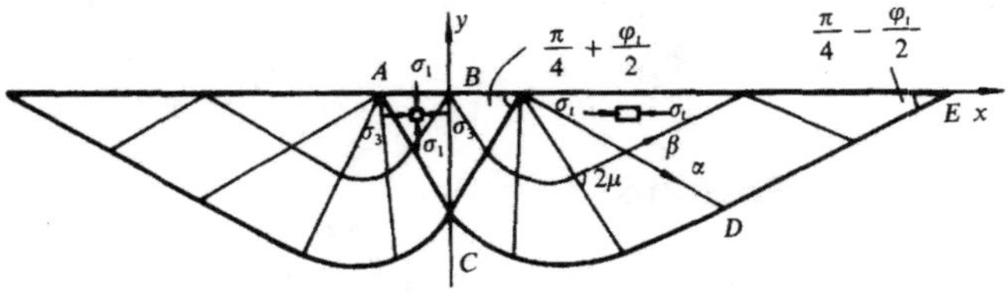

Fig. 8.9 Slip lines field of foundation under strip footing

The slip line field can be divided into three regions:I, II, and III. In which I and III are uniform stress regions, II is a centered fan region.

(1) The triangular region III (BDE)

It is formed by free boundary BE and the slip lines BD and DE. This is a constant stress region. The slip lines are straight line with $\alpha_1=\pi/4-\varphi/2$. The mean stress is a constant and must satisfy condition throughout this region: $\sigma_n=\tau_n=0$, $\sigma_1=\sigma_t$, $\sigma_3=\sigma_n=0$. along the β lines, we have

$$p=\frac{-C_t\cdot\cos\varphi_t}{1-\sin\varphi_t}=C_\beta\exp(\pi\cdot tg\varphi_t)+C_t\cdot ctg\varphi_t \tag{8-44}$$

The integrate constant is determined by

$$C_\beta=\frac{-C_t\cdot\cos\varphi_t}{1-\sin\varphi_t}\exp(-\pi\cdot tg\varphi_t) \tag{8-45}$$

(2) The active region I (ABC)

It is formed by loading boundary AB and the slip lines AC and BC. This is a constant stress region. The slip lines are straight line with angle $(\pi/4+\varphi_t/2)$ for α lines and $(\pi/4+\varphi_t/2)$ for β lines.

The integrate constant is determined by

$$C_\beta=\frac{-q-C_t\cdot ctg\varphi_t}{1+\sin\varphi_t} \tag{8-46}$$

(3) Centered fan region II (BCD and ACD)

This is a slip region. The α lines are straight line with different angle acrossed at point A (and B). β lines are logarithmic spirals. The expression of the logarithmic spirals is

$$r_1 = r_0 \exp(\pi \cdot tg\varphi_t)$$

The length AC (and BC) of a side of center fan is

$$r_{BC} = \frac{\overline{AB}}{2\cos(\frac{\pi}{4} + \frac{\varphi_t}{2})}$$

The length BD (and AD) of another side of center fan is

$$r_{BD} = \frac{\overline{AB} \cdot \exp(\pi \cdot tg\varphi_t)}{2\cos(\frac{\pi}{4} + \frac{\varphi_t}{2})}$$

The limit loading of strip footing is obtained as follows

$$q_t = C_t \cdot ctg\varphi_t \left[\frac{1+\sin\varphi_t}{1-\sin\varphi_t} \exp(\pi \cdot tg\varphi_t) - 1 \right] \quad (8\text{-}47)$$

With different values of the friction angle, some results of limit loading are obtained as shown in Table 8.1.

Table 8.1 Comparison of twin-shear and single-shear slip field

φ_0	5°	10°	15°	20°	25°	30°
φ_t	6.48°	12. 64°	18.52°	24. 16°	29. 60°	34.85°
C_0	C_0	C_0	C_0	C_0	C_0	C_0
C_t	1.30 C_0	1.27 C_0	1.25 C_0	1.23 C_0	1.22 C_0	1.21 C_0
q_0	6.49 C_0	8.34 C_0	12.98 C_0	14.8 C_0	20.7 C_0	30.14 C_0
q_t	9.06 C_0	12.2 C_0	15.9 C_0	24.0C_0	35.6 C_0	55.05 C_0

It is seen from the Table 8.1 that the limit loading of twin-shear slip field is greater than that of single-shear slip field (Mohr-Coulomb material). It is due to the effect of intermediate principal stress, which is taken into account in the twin-shear slip field.

If the friction angle equals zero, i.e. φ_0=0, it is the case of non-SD materials, and the slip lines field changes to two families of orthogonal curves, as shown in Fig.8.10. The results of limit loading by using different slip field are shown in Table 8.2.

Table 8.2 Comparison of three slip fields for non-SD materials

	Single-shear slip	Three-shear slip	Twin-shear slip
	Tresca	Huber-Mises	Yu
φ_0	0°	0°	0°
φ_t	0°	0°	0°
C_0	C_0	C_0	C_0
Q	q_0=5.14 C_0	5.94C_0	6.86C_0
q_t/q_0	1	1.155	1.335

The limit loading obtained from the twin-shear slip field equals q_t=1.335q_0. It is greater than the conventional result.

We can seen that the twin-shear solution is reduced to the Tresca solution (φ_t=φ_0=0) and the Mohr-Coulomb solution (φ_t=φ_0, $\alpha \neq 1$) when φ_t=φ_0 and C_t=C_0 (equations (8-16) and (8-17)).

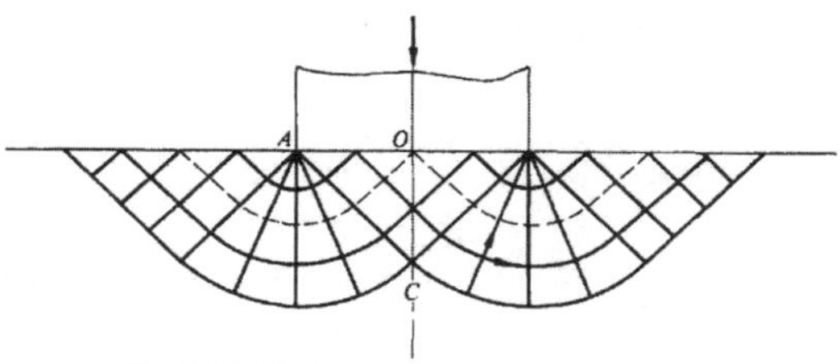

Fig. 8.10 Slip lines field of orthgonal curves

8.6.2. Example 2: Trapezoid Structure

The trapezoid structure is an important structure in engineering. Figure 8.11 shows a typical structure in railway and high-road engineering. It can be simplified as a plane strain problem. The uiform distributed load is applied on the top.

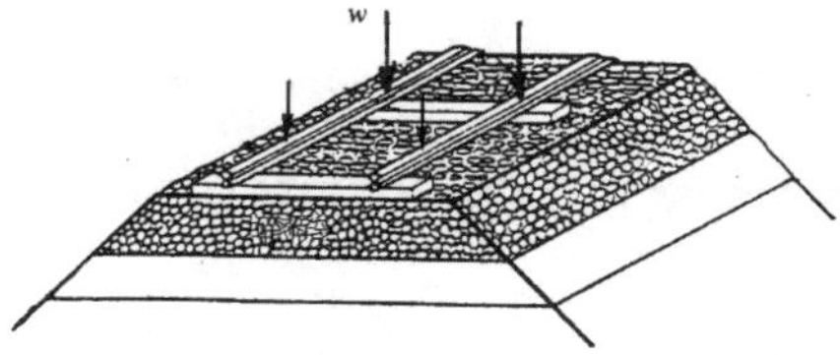

Fig. 8.11 Base of railroad

The slip lines field of the trapezoid structure with a top angle 2θ_1 is shown in Fig. 8.12.

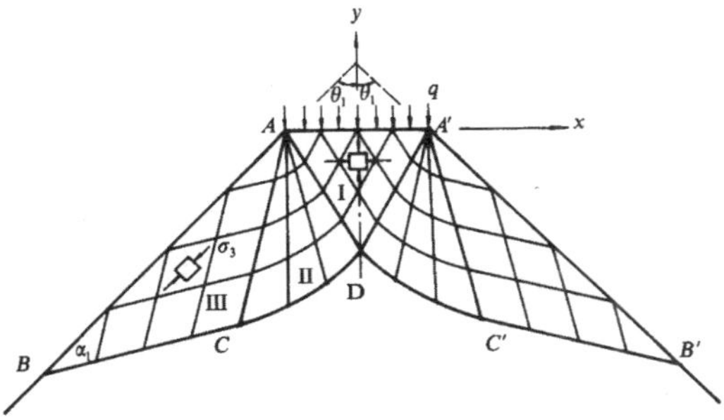

Fig. 8.12 Slip lines field of a trapezoid structure

The limit loading of trapezoid structure can be obtained by using the twin-shear slip lines field method as follow

$$q_t = C_0 \cdot ctg\varphi_0 \left[\frac{3+5\sin\varphi_0}{3-3\sin\varphi_0} \times \exp(2\theta_1 \frac{4\sin\varphi_0}{\sqrt{9+6\sin\varphi_0 - 15\sin^2\varphi_0}}) - 1 \right] \tag{8-48}$$

The limit loading of trapezoid for Mohr-Coulomb material is

$$q_0 = C_0 \cdot ctg\varphi_0 [\frac{1+\sin\varphi_0}{1-\sin\varphi_0} \exp(2\theta_1 \cdot tg\varphi_0) - 1] \tag{8-49}$$

The calculation results on the basis of the twin-shear slip lines field method and the conventional slip lines method (solution for Mohr-Coulomb material) are given for a trapezoid structure with a top angle $\theta_1 = \pi/4$ as shown in Table 8.3.

Table 8.3. Comparison of twin-shear and single-shear slip field

φ_0	5°	10°	15°	20°	25°	30°
φ_t	6.48°	12.64°	18.52°	24.16°	29.6°	34.85°
q_0	4.19 C_0	4.95 C_0	5.92 C_0	7.18 C_0	8.85 C_0	11.14 C_0
q_t	4.40 C_0	5.44 C_0	6.77 C_0	8.53 C_0	10.92 C_0	14.29 C_0

It is seen that the higher of friction angle of material, the greater of the difference of these two solutions.

The two formulae of the twin-shear solution and the single-shear solution can be unified to a unified expression as follows:

$$q = C_t \cdot ctg\varphi_t [\frac{1+\sin\varphi_t}{1-\sin\varphi_t} \exp(2\theta_1 \cdot tg\varphi_t) - 1] \tag{8-50}$$

where twin-shear cohesive C_t and friction angle φ_t can be determined from Eqs. 8.17 and 8.18. The formulae are:

$$\sin\varphi_t = \frac{4\sin\varphi_0}{2+(1+\sin\varphi_0)} \tag{8-17}$$

$$C_t = \frac{4C_0\cdot\cos\varphi_0}{2+(1+\sin\varphi_0)}\frac{1}{\cos\varphi_t}, \tag{8-18}$$

8.6.3. Example 3: Obtuse wedge

An obtuse wedge with angle $\angle BAE = \gamma > \pi/2$ is in a plane strain state. Boundary surface AE subjected to uniform pressure q, as shown in Fig.8.13. Determine the limit load p_{uni} on the surface AB.

The slip lines field is illustrated in Fig.8.13. The twin-shear limit load can be expressed as follows. It was given by Zhang,Li and Yin in 1998.

$$p_t = (q + C_t\cdot ctg\varphi_t)\cdot tg^2(\frac{\pi}{4}+\frac{\varphi_t}{2})\cdot \exp[(2\gamma-\pi)tg\varphi_t] - C_t\cdot ctg\varphi_t \tag{8-51}$$

where

$$\sin\varphi_t = \frac{4\sin\varphi_0}{2+(1+\sin\varphi_0)} \tag{8-17}$$

$$C_t = \frac{4C_0\cdot\cos\varphi_0}{2+(1+\sin\varphi_0)}\frac{1}{\cos\varphi_t}, \tag{8-18}$$

When $q=\sigma_t$, the relation between limit loads p_t and the ratio of tensile strength to compressive strength of material is shown in Fig. 8.14.

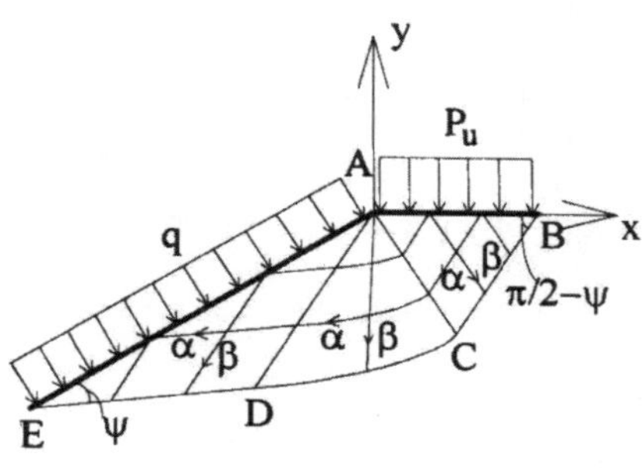

Fig. 8.13 Slip field of obtuse wedge

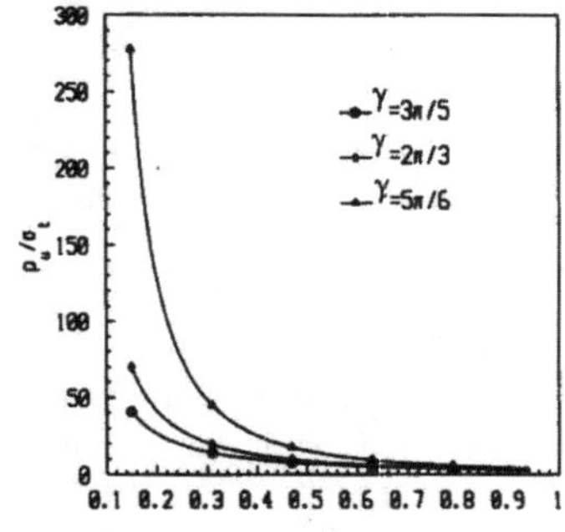

Fig. 8.14 Limit loads for different materials

When q=0, it is a slope problem. The twin-shear limit load is

$$p_t = (C_t \cdot ctg\,\varphi_t) \cdot tg^2(\frac{\pi}{4} + \frac{\varphi_t}{2}) \cdot \exp[(2\gamma - \pi) tg\varphi_t] - C_t \cdot ctg\varphi_t \qquad (8\text{-}52)$$

When $\varphi_t = \varphi_0$, it is the solution for the Mohr-Coulomb maerial.

When $\varphi_t = \varphi_0 = 0$, it is the solution for the twin-shear material in α=1. The limit load is

$$p_t = \frac{4}{3}(1 - \frac{\pi}{2} + \gamma)\sigma_s \qquad (8\text{-}53)$$

When $\varphi_t = \varphi_0 = 0$, and $C_t = C_0$, it is the conventional solution of the Tresca material.

$$p_t = (1 - \frac{\pi}{2} + \gamma)\sigma_s \qquad (8\text{-}54)$$

8.6.4. Example 4: Acute wedge

The twin-shear solution for an acute wedge was obtained by Zhang, Li and Yin in 1998. The acute wedge with angle $\angle BAD\ \gamma =< \pi/2$ is in a plane strain state, where $\angle BAC = \delta$, $\angle CAD = \nu$, and $\delta + \nu = \gamma$. Boundary surface AD is subjected to uniform pressure q, as shown in Fig. 8.15.

Determine the limit load p_t on the surface AB.

The slip lines field is illustrated in Fig. 8.15. The discontinuous line is denoted by AC, which divides the wedge into two regions. The angles between slip α lines and β lines are 2ψ. The external load p_t is assumed greater than pressure q, then the region ABC (Region I) is a active region.

The twin-shear limit load can be expressed as follows:

$$p_t = (q + C_t \cdot ctg\varphi_t) \cdot \frac{1 + \sin\varphi_t}{1 - \sin\varphi_t} \cdot \frac{\sin 2\nu}{\sin 2\delta} - C_t \cdot ctg\varphi_t \qquad (8\text{-}55)$$

where φ_t and C_t are the same as those of obtuse wedge (Example 3).

When $q = \sigma_t$, the relation between limit loads p_t and the ratio of tensile strength to compressive strength of material is shown in Fig. 8.16.

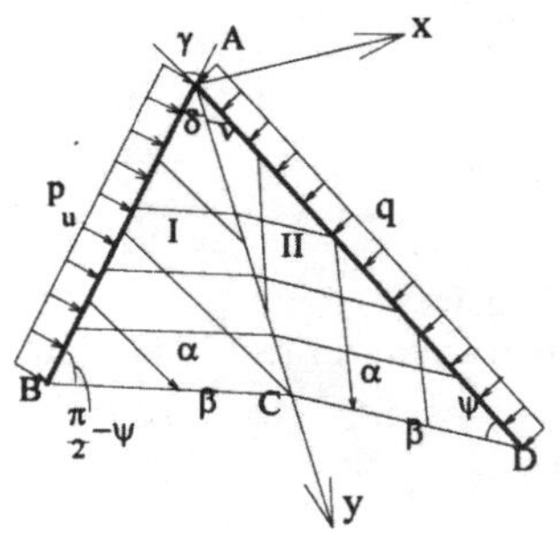

Fig. 8.15 Slip field of obtuse wedge

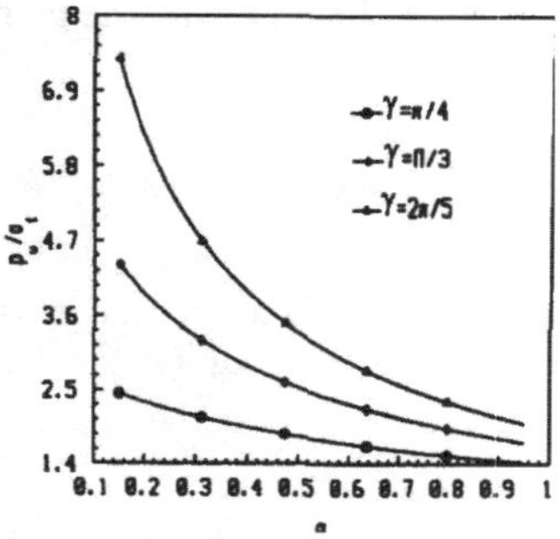

Fig. 8.16 Limit loads for different materials

When q=0, it is a slope problem. The twin-shear limit load is

$$p_t = (C_t \cdot ctg\varphi_t)\cdot\frac{1+\sin\varphi_t}{1-\sin\varphi_t}\cdot\frac{\sin 2v}{\sin 2\delta} - C_t \cdot ctg\varphi_t \tag{8-56}$$

When $\varphi_t=\varphi_0$, it is the solution for the Mohr-Coulomb material.

When $\varphi_t=\varphi_0=0$, it is the solution for the twin-shear material in $\alpha=1$. The limit load is

$$p_t = \frac{4}{3}(1-\cos\gamma)\sigma_s \tag{8-57}$$

When $\varphi_t=\varphi_0=0$, and $C_t=C_0$, it is the conventional solution of the Tresca material.

$$p_t = (1-\cos\gamma)\sigma_s \tag{8-58}$$

8.6.5. Example 5: Earth Pressure

Lateral earth pressure problem is one of important problems in soil mechanics, as shown in Figs. 8.17 and 8.18. Earth pressure problem deals with the magnitude and distribution of lateral earth pressure between a soil mass and an adjoining earth-retaining structure. The classical theories on earth pressure are those due to Coulomb (1773) and Rankine (1857).

Fig. 8.17 shows a passive earth pressure problem. Fig. 8.18 shows a active earth pressure problem.

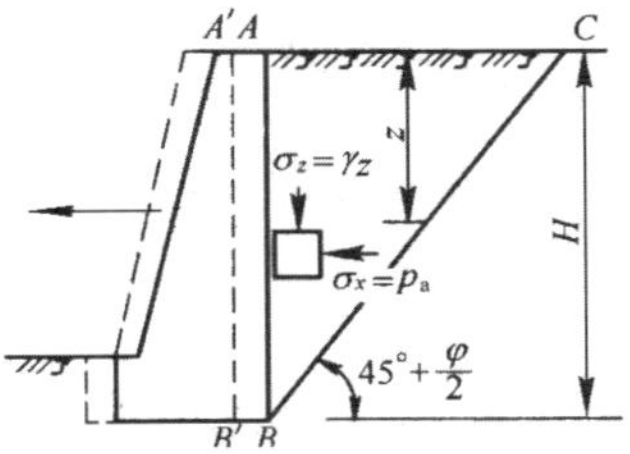

Fig. 8.17 Active earth pressure

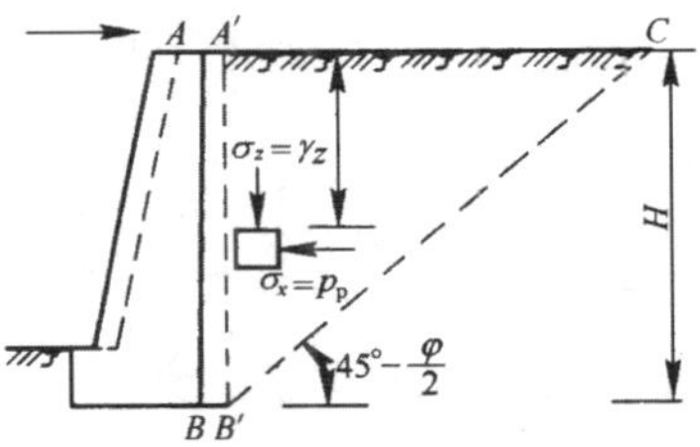

Fig. 8.18 Passive earth pressure

This problem is also considered as a limit load problem of a slope. The top and the slope side can be looked upon as the limit load problem of wedge, as show in Fig. 8.19a($\gamma \geq \pi/2$) and Fig. 8.19b($\gamma \leq \pi/2$). The weight of soil is not taken into account.

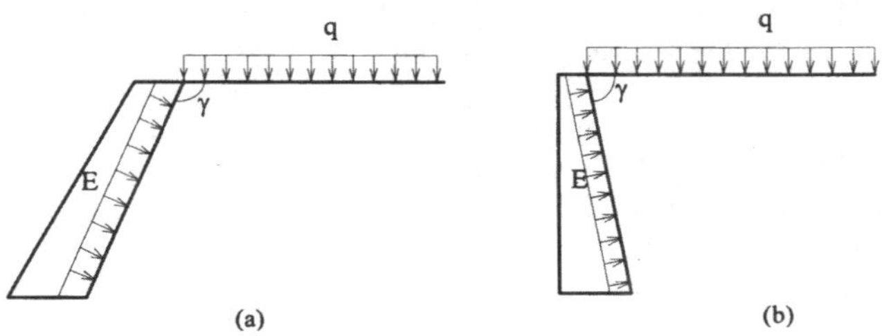

Fig. 8.19 Earth pressure (a) ($\gamma \geq \pi/2$) and (b) ($\gamma \geq \pi/2$)

The active earth pressure can be derived in the case of $q>E$ and $\gamma \geq \pi/2$ as follows:

$$E_{active} = (q + C_t \cdot ctg\varphi_t) \cdot tg^2(\frac{\pi}{4} + \frac{\varphi_t}{2}) \cdot \exp[(\pi - 2\gamma) \cdot tg\varphi_t] - C_t \cdot ctg\varphi_t \tag{8-59}$$

The passive earth pressure can be derived in the case of $E>q$ and $\gamma \geq \pi/2$ as follows:

$$p_t = (q + C_t \cdot ctg\varphi_t) \cdot tg^2(\frac{\pi}{4} + \frac{\varphi_t}{2}) \cdot \exp[(2\gamma - \pi) tg\varphi_t] - C_t \cdot ctg\varphi_t \tag{8-60}$$

The active earth pressure can be derived in the case of $q>E$ and $\gamma \leq \pi/2$ as follows:

$$p_t = (q + C_t \cdot ctg\varphi_t) \cdot \frac{1+\sin\varphi_t}{1-\sin\varphi_t} \cdot \frac{\sin 2\delta}{\sin 2\delta} - C_t \cdot ctg\varphi_t \tag{8-61}$$

The passive earth pressure can be derived in the case of $E>q$ and $\gamma \leq \pi/2$ as follows:

$$p_t = (q + C_t \cdot ctg\varphi_t) \cdot \frac{1+\sin\varphi_t}{1-\sin\varphi_t} \cdot \frac{\sin 2v}{\sin 2\delta} - C_t \cdot ctg\varphi_t \tag{8-62}$$

It is the same as Eq.8-55

Summary

The twin-shear slip line field theory for plane strain problems based on the twin-shear yield function is described in this chapter. The twin-shear slip line field theory for plane strain problems includes orthogonal and non-orthogonal slip line field. It is different from the used orthogonal or non-orthogonal slip line field theories based on single-shear strength theories (Tresca and Mohr-Coulomb materials). The new method can reflect the effect of intermediate principal stress of plane strain, where the intermediate principal stress is the principal stress in z direction.

The twin-shear slip theory can be used for those materials with obvious effect of the intermediate principal stress. Five examples are used to illustrate the applications of the twin-shear slip lines field.

Problems

Problem 8.1

The extension of a strip with a sufficiently large circular hole for non-SD material was studied by Prager and Hodge (1951) and Kachanov (1971) as shown in Fig. P8.1. The Tresca yield criterion was used in these studies. Can you obtain another study on this subject using the twin-shear slip line field for non-SD material ($\alpha = 1$)?

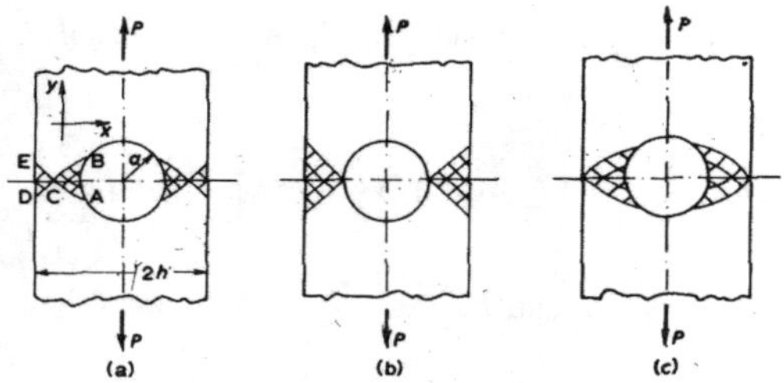

Fig. P8.1 Extension of a strip with a sufficiently large circular hole

Problem 8.2

Can you obtain a more complete study on the extension of a strip with a sufficiently large circular hole (Fig. P8.1) by using the twin-shear slip line field for SD materials ($\alpha \neq 1$)?

Problem 8.3.

The extension of a strip with ideal cuts (crack) for non-SD material was given by Kachanov (1971) as shown in Fig. P8.2. The Tresca yield criterion was used in these studies. Can you obtain a more complete study on this subject using the twin-shear slip line field for non-SD material ($\alpha = 1$)?

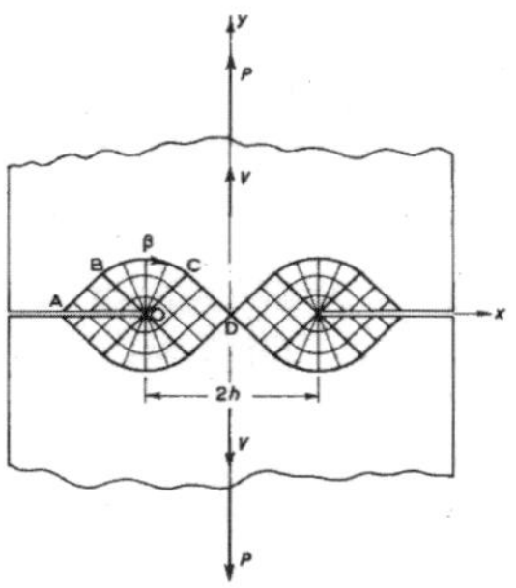

Fig. P8.2 A ideal cuts (crack)

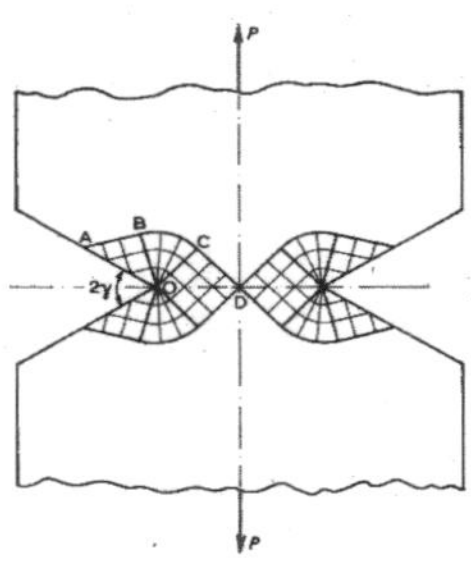

Fig. P8.3 An angular notches

Problem 8.4.

Can you obtain a more complete study on the extension of a strip with ideal cuts (Fig. P8.2) by using the twin-shear slip line field for SD materials ($\alpha \neq 1$)?

Problem 8.5.

The extension of a strip with angular notches for non-SD material was given by Kachanov (1971) as shown in Fig. P8.3. The Tresca yield criterion was used in the studies. Can you obtain a more complete study on this subject using the twin-shear slip line field ($\alpha = 1$)?

Problem 8.6.

Can you obtain a more complete study on the extension of a strip with angular notches (Fig. P8.3) by using the twin-shear slip line field for SD materials ($\alpha \neq 1$)?

Problem 8.7.

The extension of a strip with circular base for non-SD material was given by Kachanov (1971) as shown in Fig. P8.4. The Tresca yield criterion was used in the studies. Can you obtain a more complete study on this subject using the twin-shear slip line field ($\alpha = 1$)?

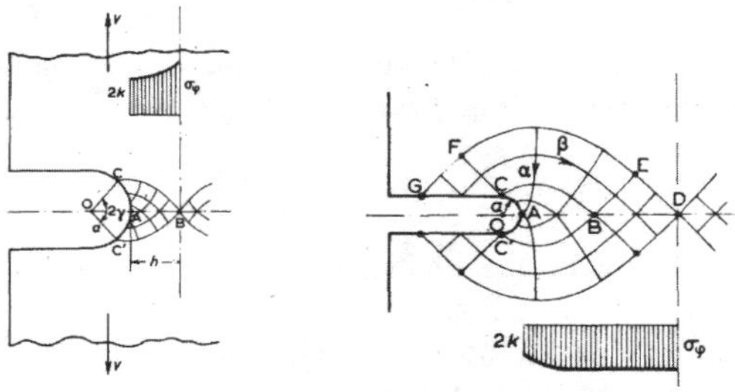

Fig. P8.4 Extension of a strip with a circular base

Problem 8.8.
Can you obtain a more complete study on the extension of a strip with circular base (Fig. P8.2) by using the twin-shear slip line field for SD materials ($\alpha \neq 1$)?

Problem 8.9.
Figure P8.5 shows the slip line field of indentation at the bottom of a flat trench. Find the limit load by using the twin-shear slip line field ($\alpha = 1$).

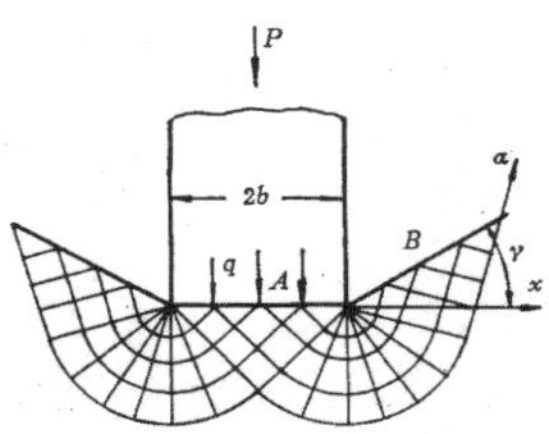

Fig. P8.5 Flat trench

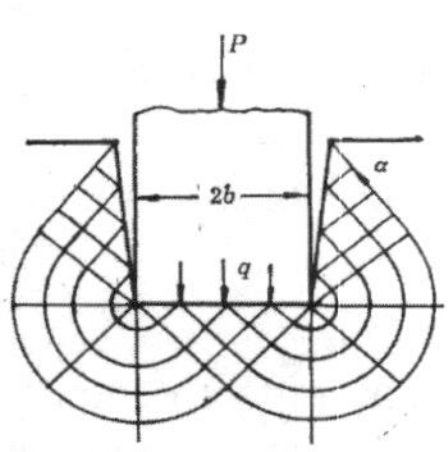

Fig. P8.6 Deep vertically sided

Problem 8.10.
Figure P8.5 shows the slip line field of indentation at the bottom of a flat trench. Find the limit load by using the twin-shear slip line field for SD materials ($\alpha \neq 1$).

Problem 8.11
Figure P8.6 shows the slip line field of indentation at the foot of a very deep vertically sided groove. Find the limit load by using the twin-shear slip line field ($\alpha = 1$).

Problem 8.12.
Figure P8.6 shows the slip line field of indentation at the foot of a very deep vertically sided groove. Find the limit load by using the twin-shear slip line field for SD materials ($\alpha \neq 1$).

Problem 8.13

Figure P8.7 shows the slip line field of indentation at the top of a flat punch. Find the limit load by using the twin-shear slip line field ($\alpha=1$).

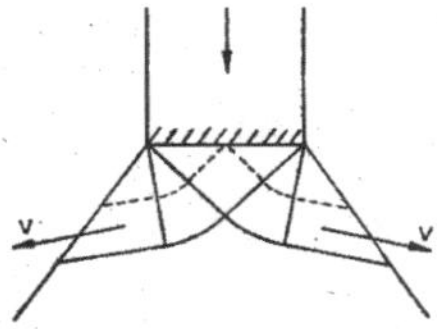

Fig. P8.7 Indentation at the top of a flat punch

Problem 8.14.

Figure P8.7 shows the slip line field of indentation at the top of a flat punch. Find the limit load by using the twin-shear slip line field for SD materials ($\alpha\neq 1$).

Problem 8.15.

Pure bending of a strip weakened by notches of various shapes was investigated by Green (1953) and Kachanov (1971). Fig. P8.8 shows the slip line field of a strip weakened by one-sided deep notch with a circular base. Find the limit load by using the twin-shear slip line field ($\alpha=1$).

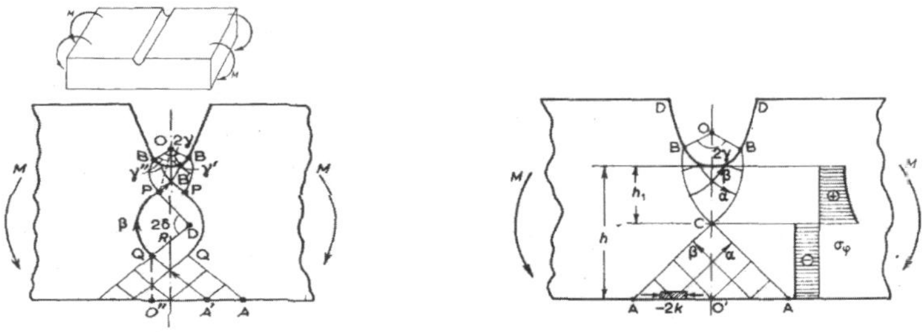

Fig. P8.8 Strip weakened by one-sided deep notch with a circular base

Problem 8.16.

Figure P8.8 shows the slip line field of a strip weakened by one-sided deep notch with a circular base. Find the limit load by using the twin-shear slip line field for SD materials ($\alpha\neq 1$).

Problem 8.17.

Figure P8.9 shows the slip line field of a strip weakened by two-sided deep notch with a circular base. Find the limit load by using the twin-shear slip line field for SD materials ($\alpha=1$).

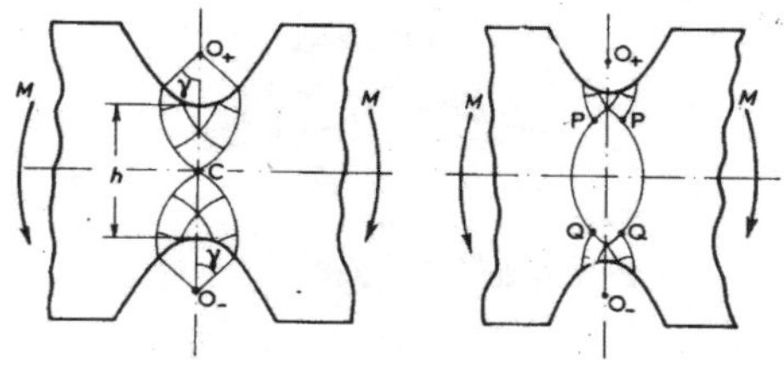

Fig. P8.9 Strip weakened by two-sided deep notch with a circular base

Problem 8.18.

Figure P8.9 shows the slip line field of a strip weakened by two-sided deep notch with a circular base. Find the limit load by using the twin-shear slip line field for SD materials ($\alpha \neq 1$).

Problem 8.19.

Figure P8.10 shows a slip line field of a strip weakened by center crack for non-SD materials. Find the slip line field for SD materials.

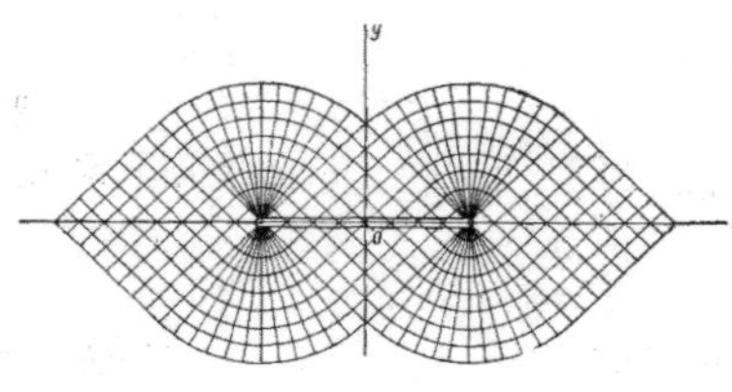

Fig. P8.10 shows a slip line field of a strip weakened by center crack

Problem 8.20.

Figure P8.11 shows the slip line field of a strip with a hole; find the limit pressure for non-SD materials and SD materials.

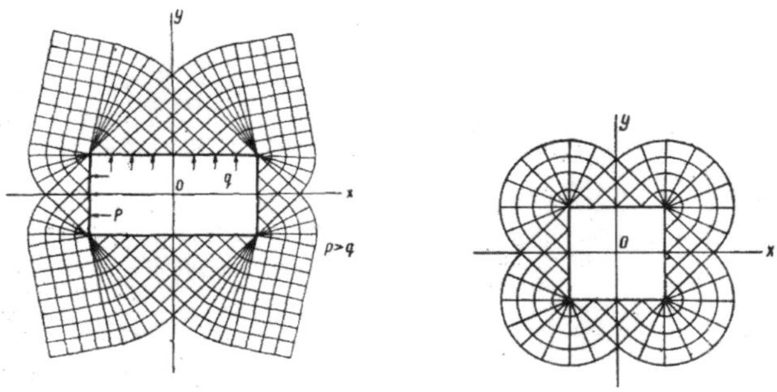

Fig. P8.11 Slip line field of a strip with a hole

Problem 8.21.
The slip line field and limit pressure were widely studied by some researchers. Figure P8.12 shows the slip line field around a circular hole loaded uniformly with a pressure, find the limit pressure for non-SD materials and SD materials by using the twin-shear slip line field theory.

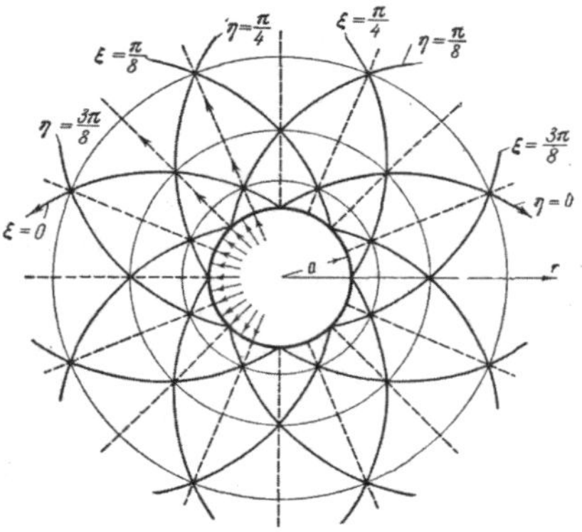

Fig. P8.12 Slip line field around a hole

References and Bibliography

Collins IF (1990) Plane Strain Characteristics Theory for Soils and Granular Materials with Density Dependent Yield Criteria, *J. Mech. Phys. Solids.*, **38**(1), 1–25.

Collins IE and Dewhurs, P (1993) A matrix technique for constructing slip-line field solutions to a class of plane-strain plasticity problems. *Int. J. Numer. Methods Eng.*, 7: 357–378.

Drucker DC, Greenberg HJ and Prager W (1951) The safety factor of an elastic-plastic body in plane strain. *J. Appl. Mech.*, 18: 371–378.

Ford H (1954) The Theory of Plasticity in Relation to Engineering Application. *J. AppL Math. Phys.*, 5: 1–35.

Geiringer H (1930) Beit zum Vollstiindigen ebenen Plastizitits-problem, Proc 3rd Intern. Congr. Appl. Mech, 2: 185–190.

Hencky H(1923) Ueber einige statisch bestimmte Faelle des Gleichgewichts in plastischen Koerpern, *Z. angew. Math Mech.*, 3: 245–251.

Hill R (1949) The Plastic Yielding of Notched Bars Under Tension, *Quart. Mech. AppL Math.*, 2: 40–52.

Hill R(1956), The Mechanics of Quasi-Static Plastic Deformation in Metals, In: *Survey in Mechanics*, G. K. Batcheler and R. M. Davis eds, Cambridge.

Johnson W and Mellor PB (1962), *Plasticity for Mechanical Engineers*. Van Nostrand, Princeton, N.J.

Johnson W, Sowerby R, Venter RD (1982), Plane-Strain Slip Line Fields for Metal Deformation Processes–A Source Book and Bibliography, Pergamon Press.

Matsuoka H (2000) Soil Mechanics. Nagoya Technical Institute Press: Nagoya.(in Japanese; Chinese translation, 2004 by Yao NP and Lou T)

Moore ID and and Rowe RK (1991) Objective solutions for bearing capacity of strain-softening soils. In: Beer G, Booker JR and Carter JP eds. *Computer Methods and Advances in Geomechanics.* Rotterdam: A.A.Balkema, Vol.2, 1183–1189.

Prager W (1955) *The Theory of Plasticity: A Survey of Recent Achievements*. Proc. Inst. Mech. Eng., 169: 41–57.

Prager W (1959) *Introduction to Plasticity*. Addison-Wesley, Reading, Mass.

Prager W and Hodge P G Jr (1951) *Theory of Perfectly Plastic Solids*. Wiley, New York.

Prandtl L (1920) l~ber die H~irte Plastischer Koerper, Goettinger Nachr., Math. Phys. KI., 74–85.

Thomsen EG, Yang CT and Kobayashi S (1965) *Mechanics of Plastic Deformation in Metal Processing.* Macmillan, New York.

Yu MH (1983), Twin Shear Stress Criterion, *Int. J. of Mech. Sci.*, **25**(1), 71–74.

Yu MH, He LN and Song, LY (1985), Twin Shear Stress Theory and Its Generalization, *Scientia Sinica (Science in China), Series A*, (English edn.), **28**(11), 1174–1183.

Yu MH and He LN (1991), A new model and theory on yield and failure of materials under complex stress state. In: *Mechanical Behaviour of Matterials-VI*, Pergamon, Oxford, Vol.3: 841–846.

Yu MH, Liu CY and He LN (1992), Generalized Twin Shear Stress Yield Criterion and Its Generalization, Chinese Science Bulletin (English edn.), **37**(24), 2085–2089.

Yu Mao-hong, et al (1992), Computer Image Analysis of the Plastic Zone of Structure, *Journal of Xi'an Jiaotong University*, **26**(2), 121–122 (in Chinese).

Yu MH, Liu JY and Ma Guo-wei (1994), Twin-shear slip line theory: orthogonal and non-orthogonal slip line fields. *Journal of Xi'an Jiaotong University*, 28(2):122–126.

Yu MH, Yang SY, Liu CY and Liu JY (1997) Unified plane-strain slip line theory. *China Civil Engrg. J.* **30**(2), 14–26 (in Chinese, English abstract).

Zhang YQ, Li JC and Yin ZN (1998) Application of twin-shear strength theory in wedge's limit analysis. In: Yu MH and Fan SC eds: *Strength Theory: Application, Development and Prospect for the 21st Century*, New York, Beijing: Science Press, 1998, 1155–1160.

9 Unified Slip-Line Field Theory for Plane Strain Problem

9.1 Introduction

The twin-shear slip line field theory based on the twin-shear yield criterion and generalized twin-shear criterion for plane strain problems has been described in Chapter 8.

Most available slip line field theories for plane strain problems are based on the one of single yield function such as the Tresca criterion (Maximum shear stress criterion), the Mohr-Coulomb strength theory (single-shear criterion), the Huber-von Mises criterion, the twin-shear yield criterion and the generalized twin-shear strength theory.

These kinds of slip line field theories can be used only for one kind of material, respectively, such as the Tresca material, the Mohr-Coulomb material, the Huber-von Mises material and the twin-shear material. The adaptability of various slip fields are illustrated in Table 9.1.

Table 9.1 The adaptability of various slip fields

Slip line fields		Applications of slip line fields theory	
Orthogonal slip field	Tresca slip line field	For $\varphi_t = \varphi_0 = 0$ and $\tau_s = 0{,}5\sigma_s$ materials	For $\sigma_t = \sigma_c$ materials Orthogonal slip field
	Mises slip line field	For $\varphi_t = \varphi_0 = 0$ and $\tau_s = 0{,}577\sigma_s$ materials	For $\sigma_t = \sigma_c$ materials Orthogonal slip field
	Twin-shear slip line field	For $\varphi_t = \varphi_0 = 0$ and $\tau_s = 0{,}667\sigma_s$ material	For $\sigma_t = \sigma_c$ materials Orthogonal slip field
Non-orthogonal slip field	Mohr-Coulomb slip field	For $\varphi_t \neq \varphi_0 \neq 0$, i.e. $\alpha \neq 1$ materials	For $\sigma_t \neq \sigma_c$ materials Non-orthogonal slip
	Generalized twin-shear slip field	For $\varphi_t > \varphi_0 \neq 0$, i.e. $\alpha \neq 1$ materials	For $\sigma_t \neq \sigma_c$ materials Non-orthogonal slip field

It is seen that the single slip field theory can not be adapted for other kinds of materials. No relations among these slip field theories are available.

In order to solve the above mentioned problems, the unified strength theory (Yu 1991) was used to extend the slip line field theory for plane strain problem by Yu-Yang-Liu in 1997. A new system of orthogonal and non-orthogonal unified slip line field theory for plane strain problems was proposed and developed. Various orthogonal and non-orthogonal plane strain slip line field theories based on different strength theories, such as the Tresca criterion, the Huber-von Mises criterion, the twin-shear yield criterion, the Mohr-Coulomb theory, and the generalized twin-shear strength theory, are special cases or linear approximation (Huber-von Mises) of the unified slip line field theory. Besides, a series of new unified slip line field can be introduced by different choice of strength criterion parameter b of the unified slip line field theory.

9.2 Unified Strength Theory in Plane Strain Condition

The stress state of plane strain structure is the same as the description in chapter 8. The main results are expressed as follows:

$$\varepsilon_z = \varepsilon_{xz} = \varepsilon_{yz} = \tau_{xz} = \tau_{yz} = v_z = 0\,, \tag{9-1}$$

z is a principal direction, and σ_z is a principal stress. The stress component in plane strain state have only four streses components σ_x, σ_y, τ_{xy} and σ_z.

The maximum principal stress σ_1 and minimum principal stress σ_3 ($\sigma_1 \geq \sigma_2 \geq \sigma_3$) of plane strain problem are

$$\begin{matrix} \sigma_1 = \sigma_{max} \\ \sigma_3 = \sigma_{min} \end{matrix} = \frac{\sigma_x + \sigma_y}{2} \pm \sqrt{\left(\frac{\sigma_x - \sigma_y}{2}\right)^2 + \tau_{xy}^2} \tag{9-2}$$

Introducing a parameter m, the intermediate principal stress σ_2 can be expressed as

$$\sigma_2 = \frac{m}{2}(\sigma_1 + \sigma_3), \tag{9-3}$$

where m is the intermediate principal stress state parameter, $0 < m \leq 1$ for plane strain problem. The intermediate principal stress state parameter m can be determined by both theoretical and experimental analysis. Generally, $m < 1$ for elastic region, i.e. $\sigma_z = m(\sigma_1 + \sigma_3)/2 < (\sigma_1 + \sigma_3)/2$; and $m \to 1$ for plastic region.

The stress state of plane strain can be expressed by two new variations R and p

$$p = \frac{1}{2}(\sigma_1 + \sigma_3) = \frac{1}{2}(\sigma_x + \sigma_y) = \frac{1}{m}\sigma_2 = \frac{1}{m}\sigma_z, \tag{9-4}$$

$$R = \frac{1}{2}(\sigma_1 - \sigma_3) = \sqrt{\left(\frac{\sigma_x - \sigma_y}{2}\right)^2 + \tau_{xy}^2}, \tag{9-5}$$

The relations between these stresses are

$$\begin{aligned} \sigma_x &= p + R\cos 2\theta \\ \sigma_y &= p - R\cos 2\theta \\ \tau_{xy} &= R\sin 2\theta \\ &= (p + C_0 \cdot \cot\varphi)\sin\varphi \cdot \sin 2\varphi \end{aligned} \tag{9-6}$$

where θ denote the angle from x coordinate to the direction of σ_1.

The mathematical formulae of the unified strength theory can be expressed in two equations as follows:

$$F = \sigma_1 - \frac{\alpha}{1+b}(b\sigma_2 + \sigma_3) = \sigma_t, \text{ when } \sigma_2 \le \frac{\sigma_1 + \alpha\sigma_3}{1+\alpha}, \tag{9-7a}$$

$$F' = \frac{1}{1+b}(\sigma_1 + b\sigma_2) - \alpha\sigma_3 = \sigma_t, \text{ when } \sigma_2 \ge \frac{\sigma_1 + \alpha\sigma_3}{1+\alpha}, \tag{9-7b}$$

Where σ_t and σ_c are uniaxial tensile and compressive strength of material respectively. Material parameter $\alpha = \sigma_t/\sigma_c$ is tensile-compressive strength ratio, which reflects material's Strength Difference (SD) effect.

It is seen from Eq (9-4) that b reflects the effect of intermediate principal stresses on the yield of materials.

The unified strength theory can also be expressed in terms of shear strength parameter C_0 and friction angle φ_0 as follows:

$$F = [\sigma_1 - \frac{1}{1+b}b(\sigma_2 + \sigma_3)] + [\sigma_1 + \frac{1}{1+b}(b\sigma_2 + \sigma_3)]\sin\varphi_0 = 2C_0 \cdot \cos\varphi_0$$

$$\text{when } \sigma_2 \le \frac{\sigma_1 + \sigma_3}{2} + \frac{\sigma_1 - \sigma_3}{2}\sin\varphi_0,$$

or

$$\text{when } \sigma_2 \le p + R\sin\varphi_0 \tag{9-8a}$$

$$F' = [\frac{1}{1+b}(\sigma_1 + b\sigma_2 - \sigma_3] + [\frac{1}{1+b}(\sigma_1 + b\sigma_2) + \sigma_3]\sin\varphi_0 = 2C_0 \cdot \cos\varphi_0$$

$$\text{when} \quad \sigma_2 \geq \frac{\sigma_1 + \sigma_3}{2} + \frac{\sigma_1 - \sigma_3}{2}\sin\varphi_0,$$

or

$$\text{when} \quad \sigma_2 \leq p + R\sin\varphi_0 \tag{9-8b}$$

The unified strength theory can consider the different effects of the intermediate principal stress for different materials. It establishes the relations among available strength theories, and creates a series of new yield criteria.

Because the intermediate principal stress equals

$$\sigma_2 = mp \leq p + R\sin\varphi_0 = \frac{1}{2}(\sigma_1 + \sigma_3) + \frac{1}{2}(\sigma_1 - \sigma_3)\sin\varphi_0, \tag{9-9}$$

it accords with the first expression of the unified strength theory Eq (9-8).

Combining Eqs (9-8a), (9-5) and (9-6), we get

$$R = \frac{2(1+b)C_0 \cdot \cos\varphi_0}{2 + b(1+\sin\varphi_0)} + \frac{b(1-m) + (2+b+bm)\sin\varphi_0}{2 + b(1+\sin\varphi_0)}p \tag{9-10}$$

where the compressive stress is usually defined positive.

We introduce two new parameters φ_{uni} and C_{uni} (Yu et al. 1997) as follows:

$$\sin\varphi_{uni} = \frac{b(1-m) + (2+b+bm)\sin\varphi_0}{2 + b(1+\sin\varphi_0)} \quad \text{when m} \neq 1, \tag{9-11a}$$

$$\sin\varphi_{uni} = \frac{2(b+1)\sin\varphi_0}{2 + b(1+\sin\varphi_0)} \quad \text{when m} = 1, \tag{9-11b}$$

$$C_{uni} = \frac{2(b+1)C_0 \cdot \cos\varphi_0}{2 + b(1+\sin\varphi_0)} \frac{1}{\cos\varphi_{uni}}, \tag{9-11c}$$

The parameters φ_{uni} and C_{uni} are referred to as the unified friction angle and the unified cohesion, respectively. The relations of $C_{\text{uni}}\sim b$ and $\Phi_{\text{uni}}\sim b$ for different φ_0 are shown in Fig. 9.1 and Fig. 9.2, respectively.

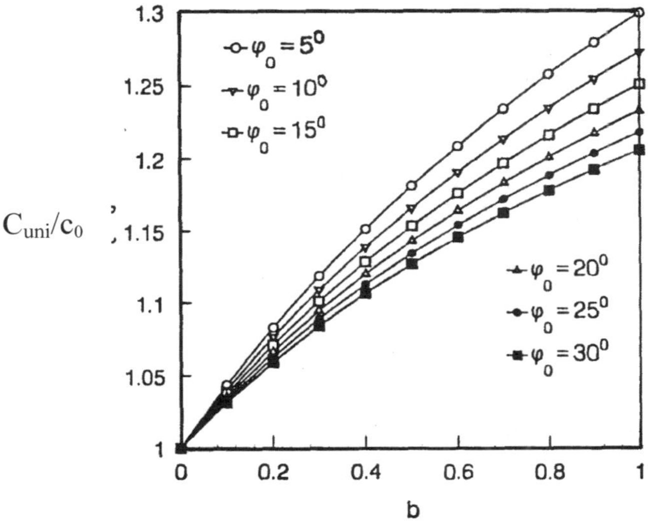

Fig. 9.1 Relation between unified cohesion C_{uni} and b

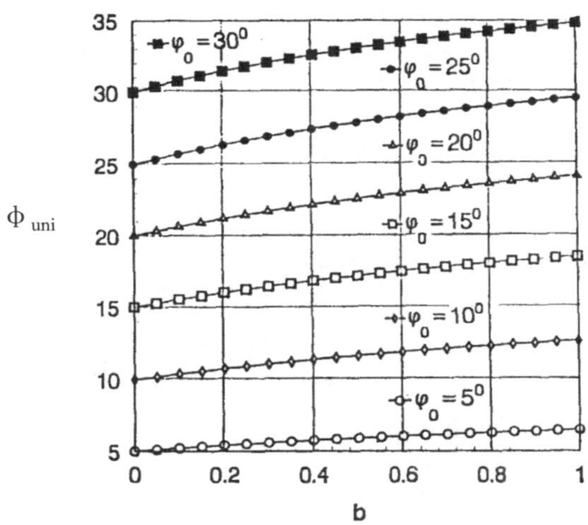

Fig. 9.2 Relation between unified friction angle φ_{uni} and b

Substituting Eqs (9-11) into Eq (9-10), the equation of the unified strength theory for plane strain problem can be rewritten as

$$R = p \sin \varphi_{uni} + C_{uni} \cos \varphi_{uni}, \tag{9-12}$$

It is the new expression of the unified strength theory for plane strain condition.

9.3 Unified Slip Line Field Theory for Plane Strain Problem (Statically Admissible Field)

Under plane strain condition, some restrictions as

$$\varepsilon_z = d\varepsilon_z = \gamma_{zx} = d\gamma_{zx} = \gamma_{zy} = d\gamma_{zy} = 0 \tag{9-13}$$

The plane strain equilibrium differential equations can be written as (see: Hill 1950 or Johnson and Mellor 1962)

$$\begin{aligned} \frac{\partial \sigma_x}{\partial x} + \frac{\partial \tau_{yx}}{\partial y} &= \gamma \sin \theta_0 \\ \frac{\partial \tau_{xy}}{\partial x} + \frac{\partial \sigma_y}{\partial y} &= -\gamma \cos \theta_0 \end{aligned} \tag{9-14}$$

in which γ is specific weight and θ_0 is the angle between the direction of gravity and the minus direction of y coordinate, as shown in Fig. 9.3.

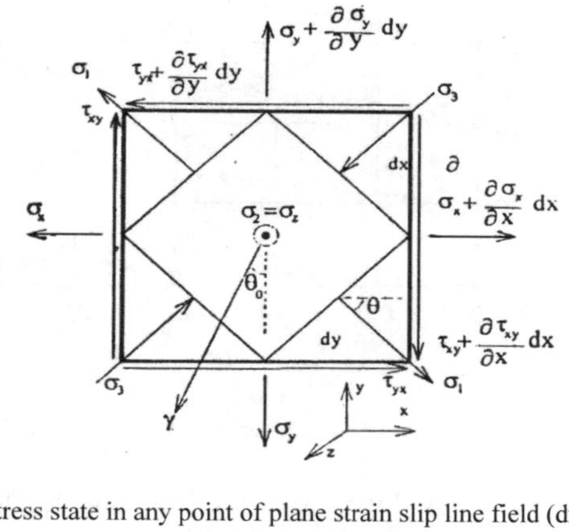

Fig. 9.3 Stress state in any point of plane strain slip line field (dx→0, dy→0)

Solving Eqs (9-9), (9-11), and (9-12), the stress governing equations are obtained as follows:

$$\frac{\partial p}{\partial x}(1+\frac{b(1-m)+(2+b+bm)\sin\varphi_0}{2+b(1+\sin\varphi_0)}\cos 2\theta)+\frac{\partial p}{\partial y}\frac{b(1-m)+(2+b+bm)\sin\varphi_0}{2+b(1+\sin\varphi_0)}\sin 2\theta+$$
$$+2R(\frac{\partial\theta}{\partial y}\cos 2\theta-\frac{\partial\theta}{\partial x}\sin 2\theta)=\gamma\sin\theta_0$$
$$\frac{\partial p}{\partial x}\frac{b(1-m)+(2+b+bm)\sin\varphi_0}{2+b(1+\sin\varphi_0)}\sin 2\theta+\frac{\partial p}{\partial y}(1-\frac{b(1-m)+(2+b+bm)\sin\varphi_0}{2+b(1+\sin\varphi_0)}\cos 2\theta)+$$
$$+2R(\frac{\partial\theta}{\partial x}\cos 2\theta+\frac{\partial\theta}{\partial y}\sin 2\theta)=-\gamma\cos\theta_0$$

(9-15)

Equation (9-15) satisfies the unified strength theory (Eq. 9-10), and plane strain static equilibrium equations (Eq. 9-14).

Equation (9-15) may be elliptic-type (two different real roots), parabolic-type (two same real roots) or hyperbolic-type (no real roots) pseudo-linear differential equations which can be solved by the method of characteristics

$$\alpha \text{ family:} \quad \frac{dy}{dx}=tg(\theta-\mu), \tag{9-16a}$$

$$\beta \text{ family:} \quad \frac{dy}{dx}=tg(\theta+\mu), \tag{9-16b}$$

in which

$$\mu=\frac{\pi}{4}-\frac{\varphi_{uni}}{2}.$$

We can see that the equations are elliptic-type pseudo-lineal differential equations. Equation (9-15) has not only two different real roots of the pseudo-linear equation (9-14), but also two characteristic lines at the angle of 2μ with each other.

Substituting Eq.9-11 into Eq.9-15, the governing equations (9-15) change to

$$\frac{\partial p}{\partial x}(1+\sin\varphi_{uni}\cdot\cos 2\theta)+\frac{\partial p}{\partial y}\sin\varphi_{uni}\cdot\sin 2\theta+2R(\frac{\partial\theta}{\partial y}\cos 2\theta-\frac{\partial\theta}{\partial x}\sin 2\theta)=\gamma\sin\theta_0$$
$$\frac{\partial p}{\partial x}\sin\varphi_{uni}\cdot\sin 2\theta+\frac{\partial p}{\partial y}(1-\sin\varphi_{uni}\cdot\cos 2\theta)+2R(\frac{\partial\theta}{\partial x}\cos 2\theta+\frac{\partial\theta}{\partial y}\sin 2\theta)=-\gamma\cos\theta_0 \tag{9-17}$$

It is interesting that Eq. (9-17) is similar to the conventional equation. It is only the substitution of C_{uni} and φ_{uni} for C_0 and φ_0. This substitution embodies the effect of yield criterion parameter b of the unified strength theory in an implicit way, builds the linkage of different slip line field theories and reflects the regular effect of intermediate principal stress in this new unified slip line field theory.

The unified slip lines field theory can be degenerated to orthogonal and non-orthogonal theories based on the Tresca critereion (α=1,b=0), the Huber-von Mises criterion (α=1, b=1/2), the twin Shear criterion (α=1, b=1), the Mohr-Coulomb criterion (b=0) and the generalized twin shear criterion (b=1). Moreover, when we choose other values of parameter b, a series new theories can be introduced. Hence, the unified strength criterion parameter b can also be regarding as a parameter of choosing for yield function.

In this section, the curvilinear coordinate system S_α and S_β coinciding with the slip lines α and β is choosed, as shown in Fig. 9.4.

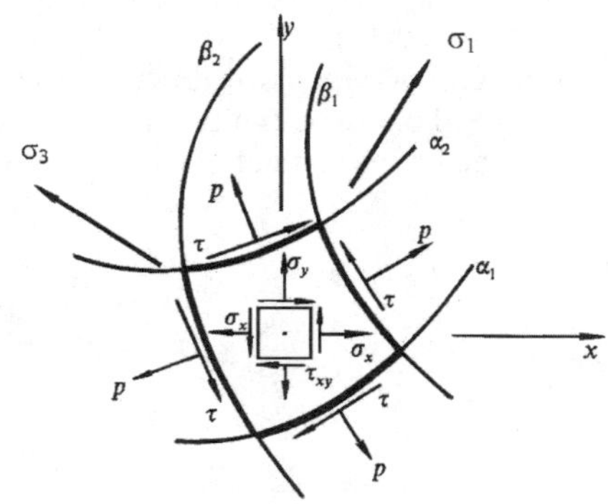

Fig. 9.4 Stress state in any point of plane strain slip line field

According to the rules of directional derivative, the stress governing equations, Eq. (9-17), can be expressed in curvilinear coordinates S_α and S_β

α family:

$$-\sin 2\mu \frac{\partial p}{\partial S_\alpha} + 2R\frac{\partial \theta}{\partial S_\alpha} + \gamma[\sin(\theta_0 + 2\mu)\frac{\partial x}{\partial S_\alpha} + \cos(\theta_0 + 2\mu)\frac{\partial y}{\partial S_\alpha}] = 0 \quad (9\text{-}18a)$$

β family:

$$\sin 2\mu \frac{\partial p}{\partial S_\beta} + 2R\frac{\partial \theta}{\partial S_\beta} + \gamma[\sin(\theta_0 - 2\mu)\frac{\partial x}{\partial S_\beta} + \cos(\theta_0 - 2\mu)\frac{\partial y}{\partial S_\beta}] = 0 \cdot \quad (9\text{-}18b)$$

Equations (9-17) and (9-18) are the stress governing equations (statically admissible) of the unified slip line field theory for plane strain problem. The equations differ from ordinary stress governing equations by substituting C_{uni} and φ_{uni} for C_0 and φ_0.

9.4. Unified Slip Line Field Theory for Plane Strain (Kinematically Admissible Field)

In this section, the associated flow rule is adopted, namely

$$d\varepsilon_{ij}{}^{p} = d\lambda \frac{\partial g}{\partial \sigma_{ij}} = d\lambda \frac{\partial f}{\partial \sigma_{ij}}, \tag{9-19}$$

where the potential function is the equation of unified strength theory, which can be expressed as

$$f = \frac{1}{2}(\sigma_x + \sigma_y)\sin\varphi_{uni} - \sqrt{\frac{1}{4}(\sigma_x - \sigma_y)^2 + \tau_{xy}{}^2} + C_{uni}\cdot\cos\varphi_{uni} = 0, \tag{9-20}$$

The plastic strain rates under small deformation and rigid plastic condition can be obtained by

$$\dot{\varepsilon}_x = \frac{\partial \dot{u}_x}{\partial x} = \frac{\partial v_x}{\partial x},$$

$$\dot{\varepsilon}_y = \frac{\partial \dot{u}_y}{\partial y} = \frac{\partial v_y}{\partial y},$$

$$\dot{\gamma}_{xy} = \frac{\partial \dot{u}_x}{\partial y} + \frac{\partial \dot{u}_y}{\partial x} = \frac{\partial v_x}{\partial y} + \frac{\partial v_y}{\partial x}, \tag{9-21}$$

Then we can get velocity governing equations of unified slip line field theory for plane strain problem

α family: $$dv_\alpha + [v_\alpha\, ctg(\frac{\pi}{2} - \varphi_{uni}) - v_\beta \cdot \csc(\frac{\pi}{2} - \varphi_{uni})]\,d\psi = 0, \tag{9-22a}$$

β family: $$dv_\beta + [v_\alpha \csc(\frac{\pi}{2} - \varphi_{uni}) - v_\beta \cdot ctg(\frac{\pi}{2} - \varphi_{uni})]\,d\psi = 0. \tag{9-22b}$$

where u_x and u_y are displacement components, $v_x = du_x/dt$ and $v_y = du_y/dt$ are velocity components along x and y directions, whereas v_α and v_β are velocity components along α and β slip lines, respectively. Superimposed dot means derivative with respect to time.

Substituting unified strength theory for plane strain into associated flue rule Eq. (9-19), and taking the rigid plastic postulate into consideration, we obtain

$$
\begin{aligned}
d\varepsilon_x &= -d\lambda \frac{\partial f}{\partial \sigma_x} = -\frac{d\lambda}{2}\left\{\sin\varphi_{uni} + \frac{1/2(\sigma_y-\sigma_x)}{\sqrt{1/4(\sigma_x-\sigma_y)^2+\tau_{xy}^2}}\right\} \\
d\varepsilon_y &= -d\lambda \frac{\partial f}{\partial \sigma_y} = -\frac{d\lambda}{2}\left\{\sin\varphi_{uni} - \frac{1/2(\sigma_y-\sigma_x)}{\sqrt{1/4(\sigma_x-\sigma_y)^2+\tau_{xy}^2}}\right\} \\
d\gamma_{xy} &= 2d\varepsilon_{xy} = -2d\lambda \frac{\partial f}{\partial \tau_{xy}} = d\lambda \frac{\tau_{xy}}{\sqrt{1/4(\sigma_x-\sigma_y)^2+\tau_{xy}^2}}
\end{aligned}
\tag{9-23}
$$

where the relation between shear strain ε_{xy} and engineering shear strain γ_{xy}, i.e., $\gamma_{xy}=2\varepsilon_{xy}$, is used. The compressive normal stress is defined positive.

Figure 9.5 gives the plane strain Mohr circle and the relation between the slip angles, were ψ is the angle between α slip line and x direction, i.e. $\psi=\theta-\mu=\theta-(\pi/4-\varphi_{uni}/2)$. Follow this definition, Eq. (9-23) can be simplified to

$$
\begin{aligned}
d\varepsilon_x &= -\frac{d\lambda}{2}[\sin\varphi_{uni}+\sin(2\psi-\varphi_{uni})] \\
d\varepsilon_y &= -\frac{d\lambda}{2}[\sin\varphi_{uni}-\sin(2\psi-\varphi_{uni})] \\
d\gamma_{xy} &= d\lambda\cos(2\psi-\varphi_{uni})
\end{aligned}
\tag{9-24}
$$

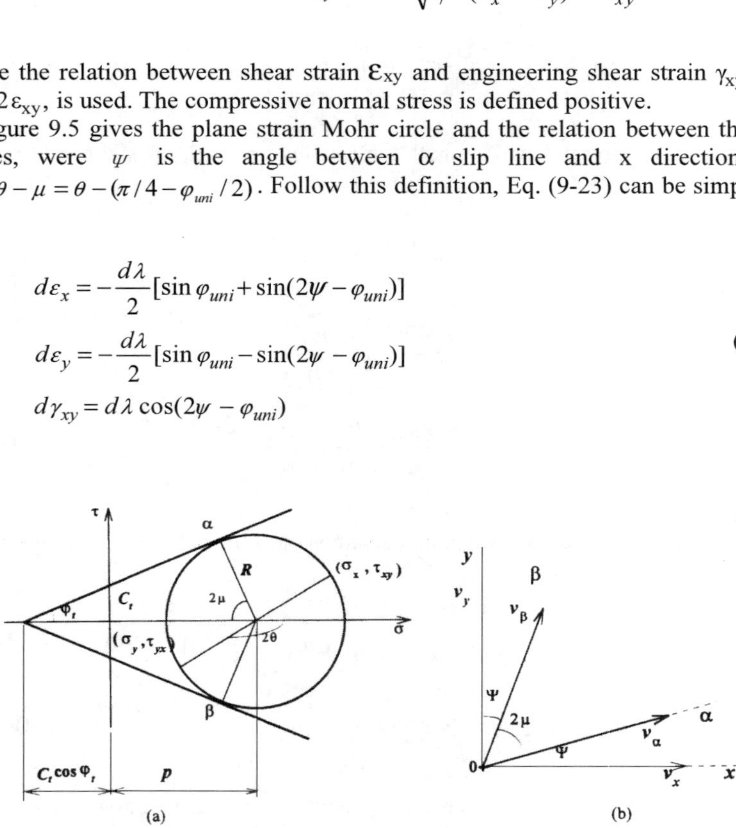

Fig. 9.5 Stress state and direction angle in any point of plane strain slip line field

When the direction of slip line is along x direction, *i.e.* $\Psi=0$ (α line) or $\psi = -(\pi/2 - \varphi_{uni})$ (β line), it follows from Eq. (9-24) that

$$d\varepsilon_x\Big|_{\psi=0} = d\varepsilon_x\Big|_{\psi=-(\pi/2-\varphi_{uni})} = 0 \tag{9-25}$$

This equation indicates that the rigid strain rate will vanishes along slip lines.

The relation in Fig. 9.5(b), *i.e.* the relation between velocity components along slip lines $v_\alpha = \dot{u}_\alpha$ ($v_\beta = \dot{u}_\beta$) and velocity components along *x* (*y*) axis v_x (v_y), can be obtained by

$$\begin{aligned} v_x = \dot{u}_x &= \frac{v_\alpha \sin(\psi + 2\mu) - v_\beta \sin\psi}{\sin 2\mu} \\ v_y = \dot{u}_y &= \frac{v_\alpha \cos(\psi + 2\mu) - v_\beta \cos\psi}{-\sin 2\mu} \end{aligned} \tag{9-26}$$

Combination of Eqs. (9-21), (9-25) and (9-26), gives,

$$\begin{aligned} d\varepsilon_x\Big|_{\psi=0} &= \dot{\varepsilon}_x\Big|_{\psi=0} dt = \frac{\partial v_x}{\partial x}\Big|_{\psi=0} dt \\ &= \{dv_\alpha + [v_\alpha ctg(\pi/2 - \varphi_{uni}) - v_\beta \csc(\pi/2 - \varphi_{uni})]d\psi\}dt = 0 \end{aligned} \tag{9-27}$$

In the above equation, we have used the relation $\mu = \pi/4 - \varphi_{uni}/2$ and definitions

$$dv_\alpha \equiv (\partial v_\alpha/\partial x)\Big|_{\psi=0} \text{ and } d\psi \equiv (\partial\psi/\partial x)\Big|_{\psi=0}.$$

When slip line field generates plastic flow, by integrating Eq. (9-21), we can work out the velocity field that satisfies the boundary conditions of the problem. The velocity field solution complies with the unified strength theory, associated flow rule and rigid plastic small deformation condition. When $\varphi_{uni}=\varphi_0$, which is the case of orthogonal slip line field, Eq. (9-22) can be simplified to the famous Geiringer velocity equations (Johnson and Mellor 1962)

$$\alpha \text{ family:} \quad dv_\alpha - v_\beta d\theta = 0, \tag{9-28a}$$

$$\beta \text{ family:} \quad dv_\beta + v_\alpha d\theta = 0. \tag{9-28b}$$

By now, we have extended the unified strength theory to slip line field theory for plane strain by means of the introduction of unified effective parameters: the unified internal cohesive C_{uni} and the unified friction angle φ_{uni} and got the governing equations of the unified slip line field theory for plane strain problem (Eqs 9-17, 9-18 and 9-22).

Unified slip line field theory can be used to materials with obvious intermediate principal stress effect and give the linkage of various slip line field theories without theoretical difficulty. Moreover, the unified slip field theory contains infinite number of orthogonal and non-orthogonal slip line field theories. With the different choice of parameter (α, β, γ) or C_{uni} and φ_{uni}, we can get a series of special cases of the unified slip field theory.

9.5 Special Cases of the Unified Slip Line Field Theory

9.5.1 Case 1: γ =0

This is the case when unified slip line field theory for plane strain problem is applied to materials where gravity is left out of consideration. In this case, velocity governing equations are the same as Eq. (9-22), whereas stress governing Eq. (9-18) can be simplified to

$$\alpha \text{ family:} \quad -\sin 2\mu \frac{\partial p}{\partial S_\alpha} + 2R\frac{\partial \theta}{\partial S_\alpha} = 0, \tag{9-29a}$$

$$\beta \text{ family:} \quad \sin 2\mu \frac{\partial p}{\partial S_\beta} + 2R\frac{\partial \theta}{\partial S_\beta} = 0 \tag{9-29b}$$

Integrating the above equations, after simplification, we obtain

$$\alpha \text{ family:} \quad p = C_\alpha \cdot \exp(2\theta \cdot ctg 2\mu) - C_{uni} \cdot ctg\varphi_{uni}, \tag{9-30a}$$

$$\beta \text{ family:} \quad p = C_\beta \cdot \exp(-2\theta \cdot ctg 2\mu) - C_{uni} \cdot ctg\varphi_{uni}, \tag{9-30b}$$

where C_α and C_β are integral constants, which will be determined by different boundary conditions. Parameters C_{uni} and φ_{uni} are defined in Eq. 9-11. This case is a new type of non-orthogonal slip line field.

As an example, a strip footing of width AA and semi-infinite length carrying a uniform load on the surface of a semi-infinite, homogeneous and isotropic soil is show in Fig.9.6. The material strength parameters for the soil are C_0 and φ_0 but the unit weight is assumed to be zero. When the pressure becomes equal to the ultimate bearing capacity, the footing is pushed downwards into the soil mass producing a state of plastic equilibrium below the footing. The surface AA is assumed to be smooth.

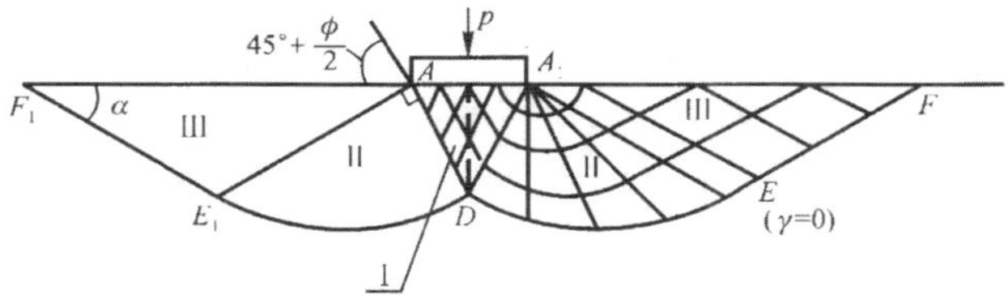

Fig. 9.6 Slip lines field of weight-less soil under a strip footing

The slip field can be indivited into three regions with the angles AAD being (45° + φ /2) as shown in Fig.9.6. The downward movement of the wedge AAD forces the adjoining soil sideways, producing outward lateral forces on both sides of the wedge AAD. Passive zones AEF and AE'F' therefore develop on both sides of the active wedge, with the angles EFA and E'F'A being (45° - φ /2). The transition between the downward movement of the active wedge and the lateral movement of the passive wedge takes place through zones of radial shear ADE and ADE', and the surfaces DE and DE' are logarithmic spirals to which the failure planes of the active and passive wedges are tangential. A state of plastic equilibrium thus exists above the surface EDF and DE'F', and the remainder of the soil is in a state of elastic equilibrium.

The following solution of limit load q is obtained by using the unified slip lines field theory. It is a new non-orthogonal slip line field system.

$$q = C_{uni} \cdot ctg\,\varphi_{uni} \left[\frac{1+\sin\varphi_{uni}}{1-\sin\varphi_{uni}} \exp(\pi\, tg\,\varphi_{uni}) - 1 \right] \tag{9-31}$$

This solution differs from the traditional solutions in the substitution of C_{uni} and φ_{uni} for C_0 and φ_0, and the effect of intermediate principal stress is taken into account and establishs the relation of different solutions, as shown in Fig. 9.7 (Yang and Yu 1996).

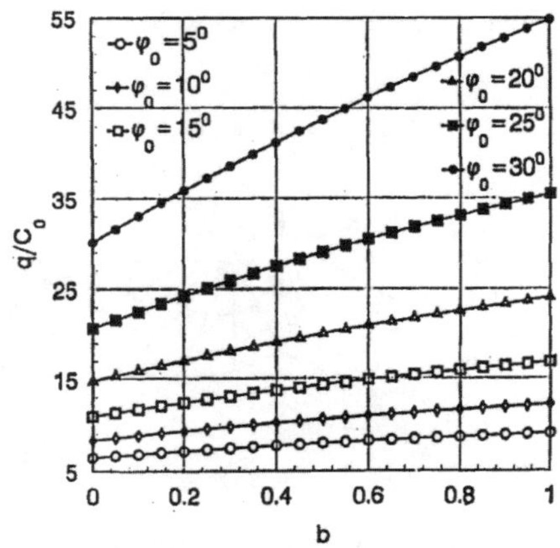

Fig. 9.7 Relation between limit load and unified strength parameter b of a strip footing

We can also draw the following conclusions:

1. when b=0, that is $\varphi_{uni}=\varphi_0$ and $C_{uni}=C_0$ (Eqs 9-16 and 9-17), the unified solution is simplified to the Tresca solution (α=1) or the Mohr-Coulomb solution ($\alpha \neq 1$).
2. when b ($b \neq 0$) is fixed, the larger the value of φ_0 is, the bigger the difference between the new solution and the Tresca solution (α=1) or the Mohr-Coulomb solution ($\alpha \neq 1$) is.
3. when φ_0 is fixed, the larger the value of b is, the bigger the difference of the two solutions is.

9.5.2 Case 2: φ_{uni}=0

This is the case of orthogonal unified slip line field theory for plane strain problems. Because of the perpendicularly of two families (α and β) of slip lines, $2\mu = \pi/2 - \varphi_{uni} = \pi/2$. In this case, it follows from Eqs (9-16) and (9-17) that

$$\varphi_{uni} = \varphi_0 = 0, \quad R = C_{uni} = \frac{2(1+b)}{2+b} C_0, \quad \alpha = 1. \tag{9-32}$$

Orthogonal unified slip line field theory for plane strain problem is used for non-SD materials, α=1. In another word, only this kind of materials can generate this case of orthogonal slip line field.

Substituting 2μ=π/2 into Eqs (9-18) and (9-22), we get the stress and velocity governing equations of this case

α family

$$-\frac{\partial p}{\partial S_\alpha}+2R\frac{\partial \theta}{\partial S_\alpha}+\gamma[\cos\theta_0\frac{\partial x}{\partial S_\alpha}-\sin\theta_0\frac{\partial y}{\partial S_\alpha}]=0$$
$$dv_\alpha-v_\beta d\theta=0 \qquad (9\text{-}33a)$$

β family

$$\frac{\partial p}{\partial S_\beta}+2R\frac{\partial \theta}{\partial S_\beta}+\gamma[-\cos\theta_0\frac{\partial x}{\partial S_\beta}+\sin\theta_0\frac{\partial y}{\partial S_\beta}]=0 \;.$$
$$dv_\beta+v_\alpha d\theta=0 \qquad (9\text{-}33b)$$

Integrating the stress governing Eq. (9-33), we obtain

α family: $p-\gamma\cos\theta_0 x+\gamma\sin\theta_0 y-2R\theta=C_\alpha,$ (9-34a)

β family: $p-\gamma\cos\theta_0 x+\gamma\sin\theta_0 y+2R\theta=C_\beta,$ (9-34b)

9.5.3
Case 3: $\varphi_{uni}=0$ and γ =0

This is the case of orthogonal unified slip line field theory for plane strain problem without consideration of gravity. From Eqs. (9-16) and (9-17), we can get the restrictions of this case

$$\varphi_{uni}=\varphi_0=0,\partial$$
$$R=C_{uni}=\frac{2(1+b)}{2+b}C_0, \qquad (9\text{-}35)$$
$$\gamma=0,\ \alpha=1$$

Substituting above equations into Eq. (9-35), we have the stress and velocity governing equations of this case

$$\frac{\partial}{\partial S_\alpha}(p-2R\theta)=0\;, \quad \alpha \text{ family} \qquad (9\text{-}36a)$$
$$dv_\alpha-v_\beta d\theta=0$$

$$\frac{\partial}{\partial S_\beta}(p+2R\theta)=0\;. \quad \beta \text{ family} \qquad (9\text{-}36b)$$
$$dv_\beta-v_\alpha d\theta=0$$

where the condition that R is constant with respect to S_α and S_β is used when material and strength theory is fixed

Integrating Eq. (9-36), we can get the solutions of this case

$$\alpha \text{ family:} \quad p - 2R\theta = C_\alpha, \tag{9-37a}$$

$$\beta \text{ family:} \quad p + 2R\theta = C_\beta. \tag{9-37b}$$

where C_α and C_β are integral constants. Substitute $\gamma=0$ into Eq. (9-30), we can also get above equations.

The form of Eqs (9-36) and (9-37) is similar to that of traditional solutions of orthogonal slip line field theories. But the definition of R is different. In this discussion, R is unified parameter and with different choice of yield criterion parameter b, the solution can be simplified to those of the Tresca, the Huber-von Mises and the twin-shear slip line field theories, they are:

1. $b=0$, $R = C_0 = \sigma_s/2 = 0.5\sigma_s$, it is the Tresca solution;
2. $b=1/2$, $R = 6/5\,C_0 = 3/5\sigma_s = 0.6\sigma_s$, it is the linear approximations to the Huber-von Mises solution (exact solution is $R = 1/\sqrt{3}\,\sigma_s = 0.58\sigma_s$);
3. $b=1$, $R = 4/3C_0 = 2/3\sigma_s = 0.667\sigma_s$ it is the solution of the twin-shear yield criterion.

9.6 Applications of the Unified Slip Line Field Theory

9.6.1 Example 1: Trapezoid structure

The trapezoid PVC specimen with a top angle 2ξ isconsidered. The uiform distributed load is applied on the top of the specimen, as shown in Fig. 9.8. The loading velocity is 1mm/min. The physical properties of the material are:

Young's modules: 367.88 kN/cm^2 Specific gravity: 1.35-1.45 g/cm^3

Poisson's Ratio: 0.27 Tensile strength: 5.886 kN/cm^2

Compressive strength: 7.575 kN/cm^2

Determine the limit load.

Solution

Friction angle and cohesion of this material can be determined by

$$\varphi_0 = \sin^{-1}\frac{\sigma_c - \sigma_t}{\sigma_c + \sigma_t} = 7.208^0$$
$$C_0 = \frac{\sigma_t(1+\sin\varphi_0)}{2\cos\varphi_0} = 3.339 \tag{9-38}$$

It is assumed that the surfaces AA' are smooth and there is no friction. It is also assumed that there is a constant pressure on the top. The rest of the boundary is stress-free.The slip line field is given in Fig.9.8.

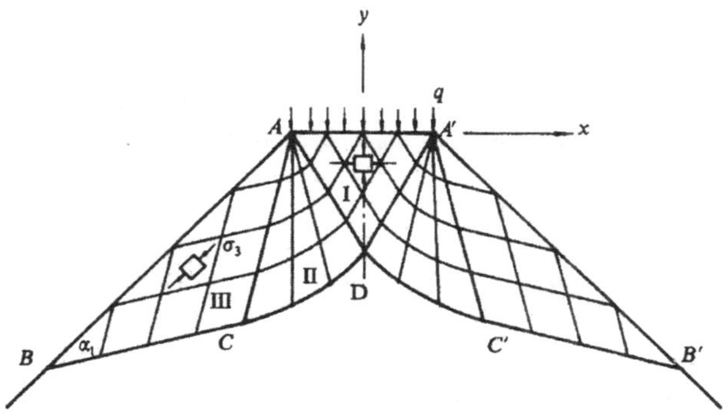

Fig. 9.8. Slip lines field of a trapeziod structure (Yu 1998)

The slip line field can be divited into three regions:I, II, and III, where I and III are uniform stress regions, and II is a centered fan region.

(1) The triangular region III (ABC)

It is formed by free boundary AB and the slip lines AC and BC. This is a constant stress region. The slip lines are straight with $\alpha_1=\pi/4-\varphi/2$. The mean stress is a constant and must satisfy condition throughout this region : $\sigma_n=\tau_n=0$, $\sigma_1=\sigma_t$, $\sigma_3=\sigma_n=0$. Along the β lines, we have

$$p = \frac{-C_{uni}\cdot\cos\varphi_{uni}}{1-\sin\varphi_{uni}} = C_\beta\cdot\exp(2\xi\cdot tg\varphi_{uni}) + C_{uni}\cdot ctg\,\varphi_{uni} \tag{9-39}$$

The integrate constant is determined by

$$C_\beta = \frac{-C_{uni}\cdot\cos\varphi_{uni}}{1-\sin\varphi_{uni}}\exp(-2\xi\cdot tg\,\varphi_{uni}) \tag{9-40}$$

(2) The triangular region I (AA'D)

It is formed by loading boundary AA' and the slip lines AD and A'D. This is a constant stress region. The slip lines are straight with angle $-(\frac{\pi}{4}+\frac{\varphi_{uni}}{2})$ (α lines) and $(\frac{\pi}{4}+\frac{\varphi_{uni}}{2})$ (β lines).

The integrate constant is determined by

$$C_{\beta}=\frac{-q-C_{uni}\cdot ctg\ \varphi_{uni}}{1+\sin\varphi_{uni}} \tag{9-41}$$

(3) Centered fan region II (ACD)

This is a slip region. The α lines are straight with different angle acrossed at point A. β lines are logarithmic spirals. The expression of the logarithmic spirals is

$$r_1=r_0\exp(2\xi\cdot tg\ \varphi_{uni})$$

The length AC of a side of center fan is

$$r_{AD}=\frac{\overline{AA'}}{2\cos(\frac{\pi}{4}+\frac{\varphi_{uni}}{2})}$$

The length AD of another side of center fan is

$$r_{AC}=\frac{\overline{AA'}\exp(2\xi\cdot tg\ \varphi_{uni})}{2\cos(\frac{\pi}{4}+\frac{\varphi_{uni}}{2})}$$

The limit loading of trapezoid specimen is obtained as follows

$$q=C_{uni}\cdot ctg\ \varphi_{uni}[\frac{1+\sin\varphi_{uni}}{1-\sin\varphi_{uni}}\exp(2\xi\cdot tg\ \varphi_{uni})-1] \tag{9-42}$$

With different choice of unified yield criterion parameter b, a series of limit loading are obtained as shown in Fig.9.9. The three curves are three results for three specimen, the top angles are: $2\xi=120^{\circ}$, $2\xi=80^{\circ}$ and $2\xi=60^{\circ}$. Figure 9.10 is the variation of the slip angles with the variation of unified yield criterion parameter b. From Fig. 9.10, we can see that the result of Unified Slip Line Field Theory ($2\mu=81.04^{\circ}$ for $b=0.8$) is much closer to the experimental result ($2\mu=79.5^{\circ}$) than that of the Mohr-Coulomb ($2\mu=82.8^{\circ}$).

It is worthy to mention that not only the limit loading q but also the slip angle 2μ are different. They are different for different material. The variation of slip angle 2μ with the unified yield criterion parameter b is shown in Fig.9.10.

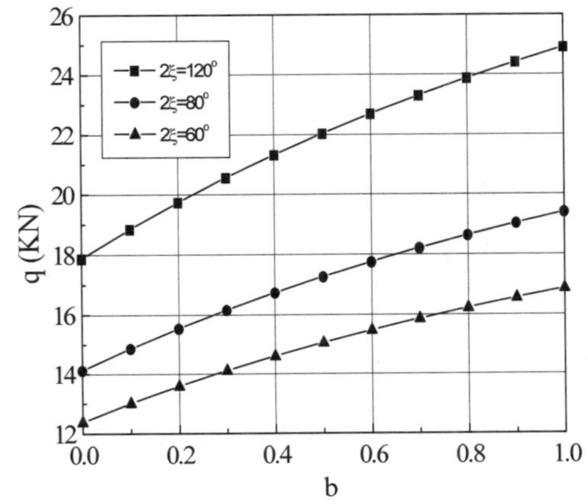

Fig. 9.9 Unified solutions of limit loading (Yang-Yu 1997)

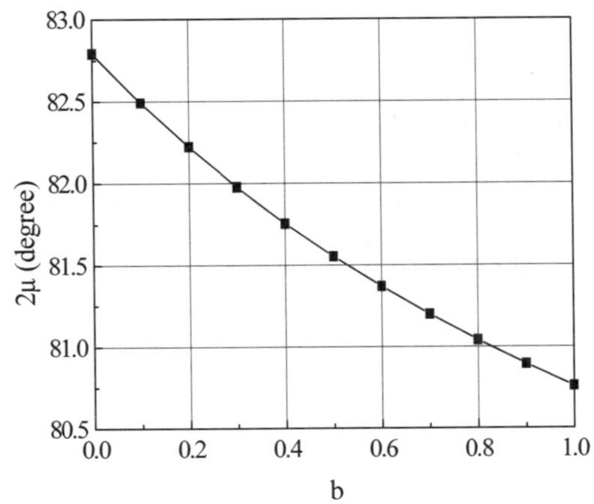

Fig. 9.10 Variation of slip angle 2μ with the unified yield criterion parameter b

9.6.2

Example 2: Strip footing

The slip lines field of strip footing under uniform punch pressure is similar to that of trapezoid structure as shown in Fig.9.11. The difference only in the angle π and 2ξ (2ξ=2θ_1 in Fig.9.11,b).

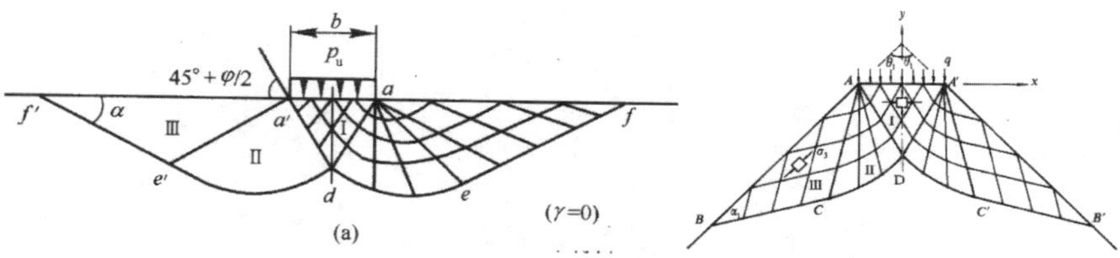

Fig. 9.11 Slip lines field of a semi-infinite surface (a) and a trapeziod structure (b)

The limit loading of strip footing can be obtained by 2ξ=π, which is given as follows

$$q = C_{uni} \cdot ctg\,\varphi_{uni} [\frac{1+\sin\varphi_{uni}}{1-\sin\varphi_{uni}} \exp(\pi \cdot tg\,\varphi_{uni}) - 1] \tag{9-43}$$

9.6.3

Example 3: Shallow strip footing

A shallow strip footing is shown in Fig.9.12, and the material parameters are: C_0= 9Kpa, $\varphi_0 = 14°$, $\gamma = 17.3KN/m^3$, width of strip B=3m, depth d=1.5m. The shear strength of the soil between the surface and depth d is neglected. This soil being considered only as a surcharge imposing a uniform pressure q=$\gamma_0 d$ on the horizonal plane at foundation level. Determine the limit load for strip footing.

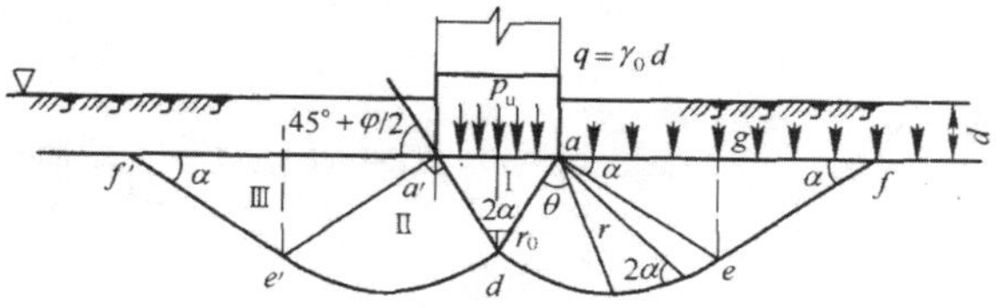

Fig. 9.12 Slip lines field of a falf space under shallow strip footing

Solution

Friction angle and cohesion of this material can be determined. The unified friction angle φ_{uni} and the unified cohesion C_{uni} can be determined by using Eqs.9-10 and 9-11. The relations of unified friction angle φ_{uni} and the unified cohesion C_{uni} with unified yield criterion parameter b, i.e. relations C_{uni}~b and Φ_{uni}~b are shown in Figs. 9.13 and 9.14, respectively.

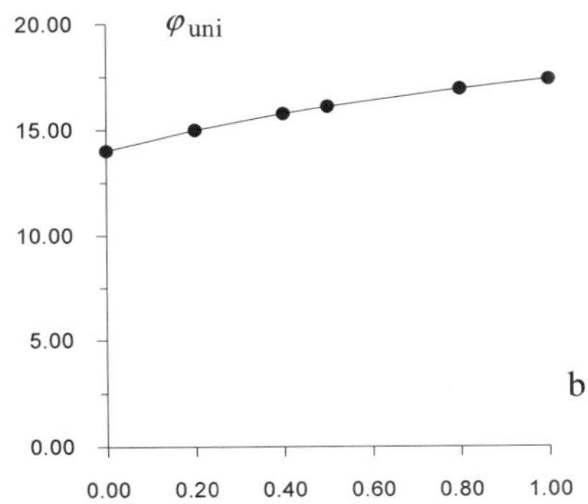

Fig. 9.13 Relations of unified friction angle φ_{uni} with unified yield criterion parameter b

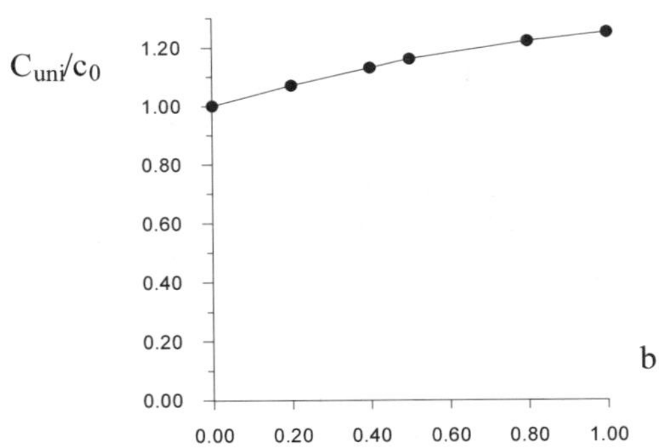

Fig. 9.14 Relations of unified cohesion C_{uni} with unified yield criterion parameter b

A series of slip line field and different results of limit loading were determined by Fan and Yu by using the unified slip field theory. The five slip line fields and relating limit loading are shown in Figs. 9.15 to 9.20.

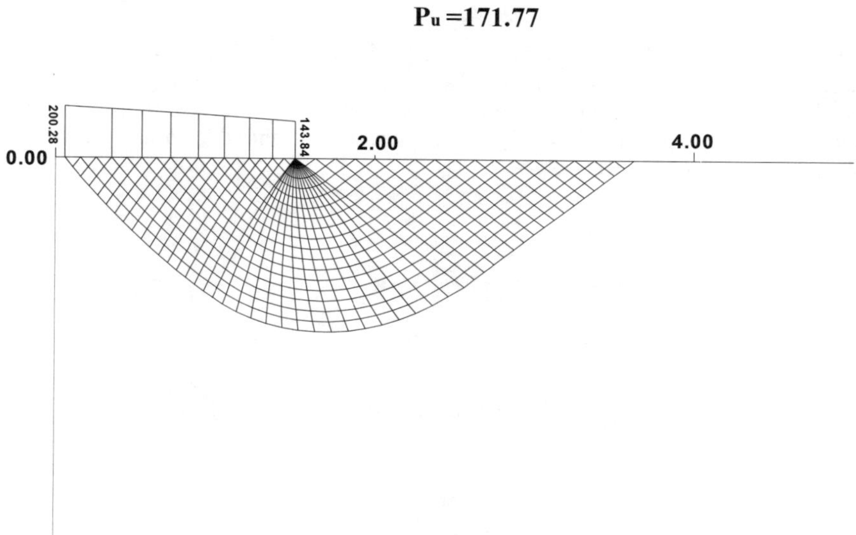

Fig. 9.15 Slip line field for b=0 material **(P_u =171.77)**

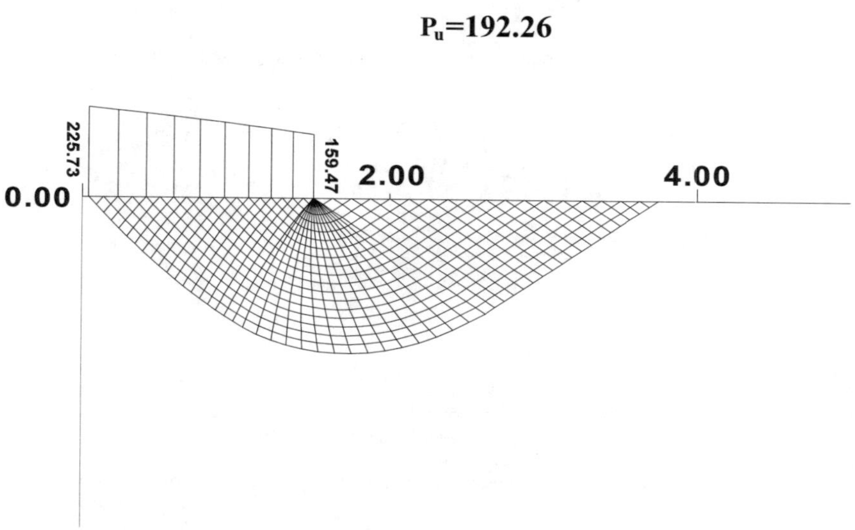

Fig. 9.16 Slip line field for b=0.2 material **(P_u=192.26)**

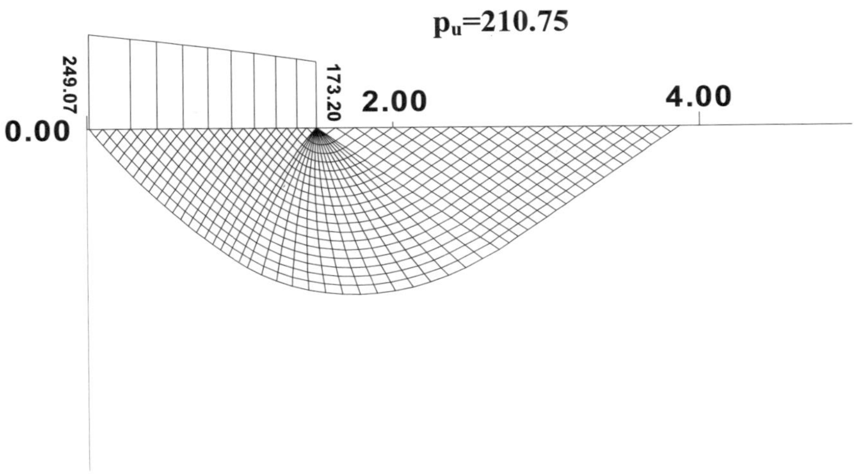

Fig. 9.17 Slip line field for b=0.4 material **(p_u=210.75)**

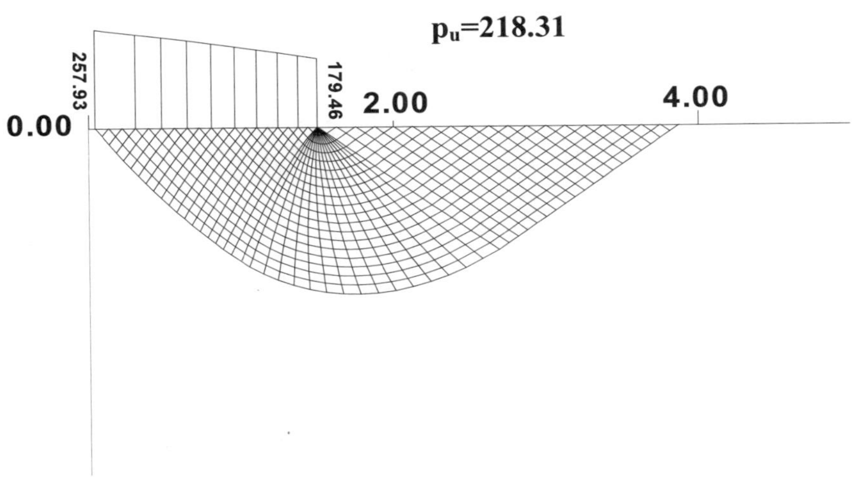

Fig. 9.18 Slip line field for *b*=0.5 material **(p_u=218.31)**

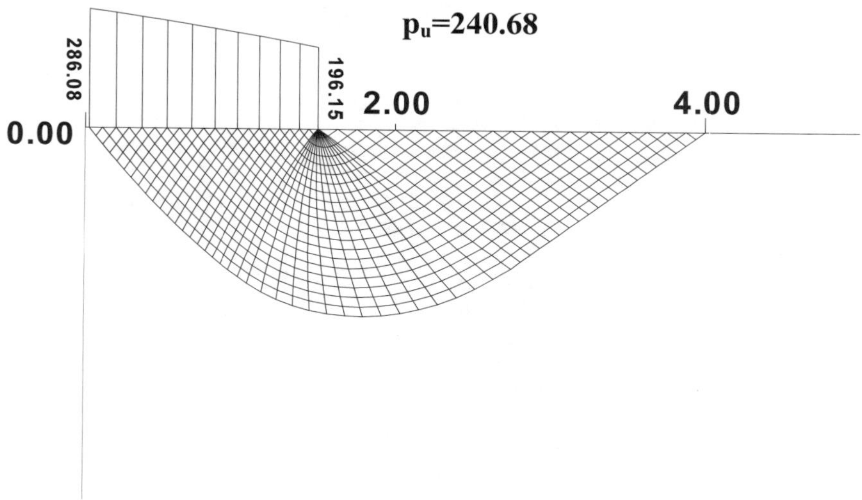

Fig. 9.19 Slip line field for b=0.8 material **(p_u=240.68)**

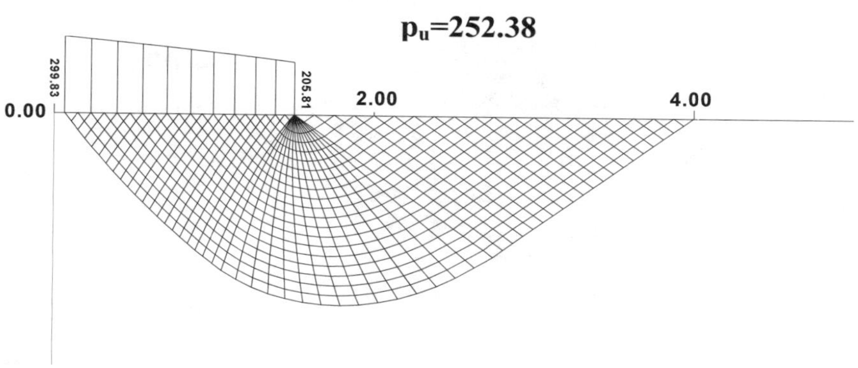

Fig. 9.20 Slip line field for b=1.0 material (Twin-shear theory **p_u=252.38**)

The limit loading of strip footing varies with the choice of the unified yield criterion parametere b. The relation of limit loading of strip footing with b is shown in Fig.9.21.

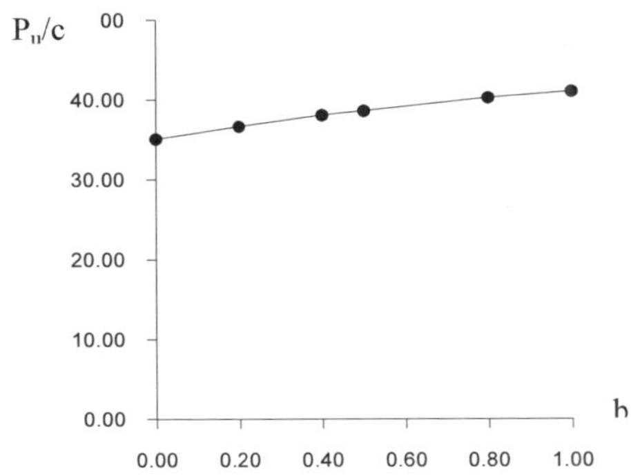

Fig. 9.21 Relations of limit loading with unified yield criterion parameter b

It is seen that a series of results can be obtained by using the unified slip lines field theory.

9.6.4 Example 4: Acute Wedge

An acute wedge is in a plane strain state. Its top angle $\gamma < \pi/2$, and surface AB is subjected to uniform pressure P_u, as shown in Fig. 9.22.

Determine the limit pressure P_u on surface AB.

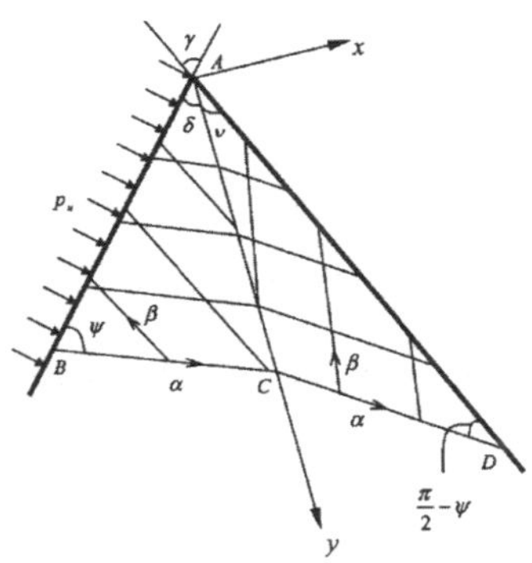

Fig. 9.22 Acute wedge under unilateral pressure

Solution:

The unified solution of acute wedge under unilateral pressure was given by Zhang, Hao and Yu in 2003.

When $0 < \gamma < \pi/2$, a stress discontinuous line in the wedge will appear. The characteristic field is shown in Fig.9.22, where $\angle BAC = \delta$, $\angle CAD = \nu$, $\delta + \nu = \gamma$. The regions ABC and ACD are regions of constant biaxial compression and uniaxial compression, respectively. The constant stress regions ABC and ACD are separated by the line of stress discontinuity AC which is inclined to AB at an angle δ to be determined. The different values a quantity may assume in the regions ABC and ACD will be distinguished by subscripts 1 and 2, respectively. The angle between characteristics α and β is 2ψ.

From the stress boundary condition of the wedge, we have $\sigma_3 = -p_u$ in region ABC , and $\sigma_1 = 0$ in region ACD. Thus it has

$$\varphi_1 = \frac{\pi}{2} - \delta, \quad \varphi_2 = \nu \tag{9-44}$$

It also has

$$\delta + \nu = \gamma \tag{9-45}$$

The unified limit load for acute wedge was given as follows

$$p_u = C_{uni} \cdot \cot \varphi_{uni} \left(\frac{1 + \sin \varphi_{uni}}{1 - \sin \varphi_{uni}} \frac{\sin 2\nu}{\sin 2\delta} - 1 \right) \tag{9-46}$$

where φ_{uni} and C_{uni} were introduced by Yu et al. in 1997 and 1998, they are defined as

$$\sin \varphi_{uni} = \frac{2\,(b+1)\,\sin\varphi}{2 + b\,(1+\sin)\varphi},$$

$$C_{uni} = \frac{2(b+1)C \cdot \cos\varphi}{2 + b(1 + \sin\varphi)} \cdot \frac{1}{\cos \varphi_{uni}} \tag{9-47}$$

φ_{uni} and C_{uni} are two parameters in the unified slip field theory.

For the case of $b = 0$ and $\alpha \neq 1$, the unified limit load reduces to

$$p_u' = c \cot\varphi \left(\frac{1 + \sin\varphi}{1 - \sin\varphi} \frac{\sin 2\nu}{\sin 2\delta} - 1 \right) \tag{9-48}$$

This is the solution on the basis of Mohr-Coulomb criterion (Shield 1954; Chen 1975).

For the case of $b=0$ and $\alpha=1$, the unified limit load reduces to

$$p_u'' = \lim_{\substack{\alpha\to 1\\(\varphi\to 0)}} p_u' = 2c(1-\cos\gamma) \tag{9-49}$$

This is the solution on the basis of Tresca criterion (Chen 1975).

When $\gamma=\pi/3$, the relation between the limit load p_u and the unified yield criterion b at different ratio of $\alpha=\sigma_t/\sigma_c$ are shown in Fig.9.23.

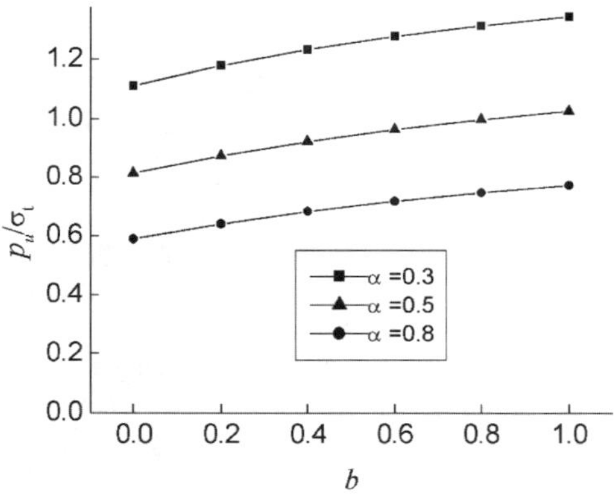

Fig. 9.23 The relation between limit load P_u and unified strength theory parameter b

It can be found that the SD effect of material and the influence of intermediate principal stress on the limit load are significant. Through the parameters α and b, the dependence of the result of the limit load on yield criterion is also reflected. As shown, at the same α, the Mohr-Coulomb criterion $(b=0)$ leads to the minimum value of p_u/σ_t while the twin-shear strengrh theory $(b=1)$ leads to the maximum value of p_u/σ_t.

9.6.5 Example 5: Experiments

A method of observing plastic zone in a structure under loading is applied. In order to obtain clear images of the development of structure's plastic zone, image analysis technique has been used.The plastic zone of structure under limit loading is obtained.

Three trapeziod specimens are the same as example 1, in which the top angle are: $2\xi=120^\circ$, $2\xi=80^\circ$ and $2\xi=60^\circ$. The three experimental results of limit loading are: 23.6kN for$2\xi=120^\circ$, 18.5kN for $2\xi=80^\circ$, and 16.9kN for $2\xi=60^\circ$. The theoretical results has been obtained as shown in Fig. 9.24.

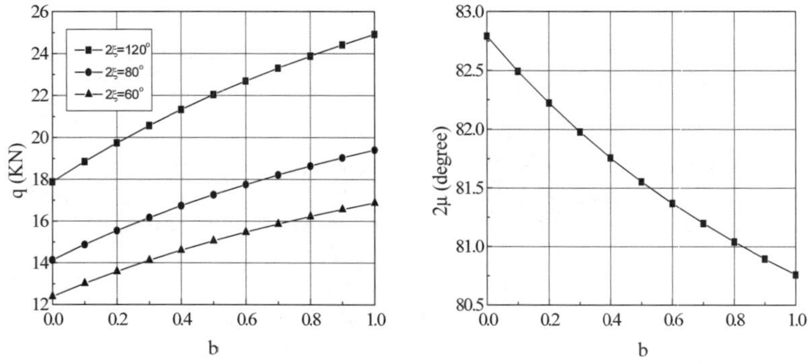

Fig. 9.24 Variations of limt loads and slip angle with the unified yield criterion parameter *b*

We can see from Fig. 9.24 that the three experimental results q_1=23.6kN, for$2\xi = 120°$, q_2=18.5kN for $2\xi = 80°$, and q_3=16.9kN for $2\xi = 60°$ are closed to the unified solution with $b \approx 0.8$. Obvious error will be produced when we use the traditional Mohr-Coulomb slip line field solution (*b*=0). We can also assume that this material can be simulated by b≈0.8, i.e. limit loads of cases with other top angles can be determined by Eq.9-42 with b≈0.8.

Furthermore, we can find from the experiment that the shape of slip line is non-orthogonal. The slip angle obtained by computer image analysis is 78° (for Mohr-Coulomb non-orthogonal slip line field theory, the angle is 82.8°). The test gives the change of 2μ, the angle between two slip lines, with different choice of unified yield criterion parameter *b*. From Fig. 9.24, we can see that the slip angle calculated through unified Slip Line Field Theory (2μ=81.04° when b=0.8) is much closer to the experiment result than the result based on the Mohr-Coulomb non-orthogonal slip line filed theory (2μ=82.8°).

9.7 Comparison of the Unified Slip Line Field Theory with Finite Element Method

Various finite element solutions for the bearing capacity of structures have been reported in the literature. Some are of very interesting and value. The development of plasticity in soil under a footing foundation were studied by Nayak and Zienkiewicz (1972) and Moore and Rowe (1991) et al. Fig. 9.25 to Fig. 9.28 show the results of plastic zones at footing displacement obtained by Moore and Rowe (1991). They used a 15 nodes element (as show at the right in Fig. 9.25) and a series of computational results were obtained. Each yielded integration point within the triangular elements has been marked with a small lines, where these lines are aligned with the direction of the shear deformation at those locations.The development of plastic zone at various footing penetrations are shown clearly.The final slip surface is shown in Fig. 9.28. It is observed that the soil materials collapses with a mechanism similar to that of Prandtl for perfectly plastic materials (Moore and Rowe 1991).

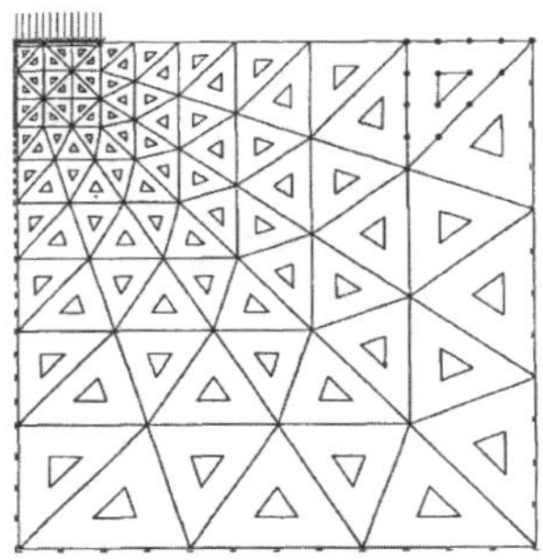

Fig. 9.25 Fine mesh (Moore and Rowe 1991)

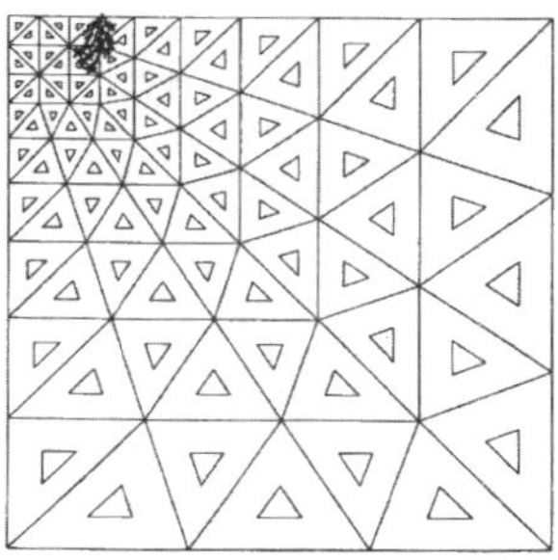

Fig. 9.26 Initial plastic zone

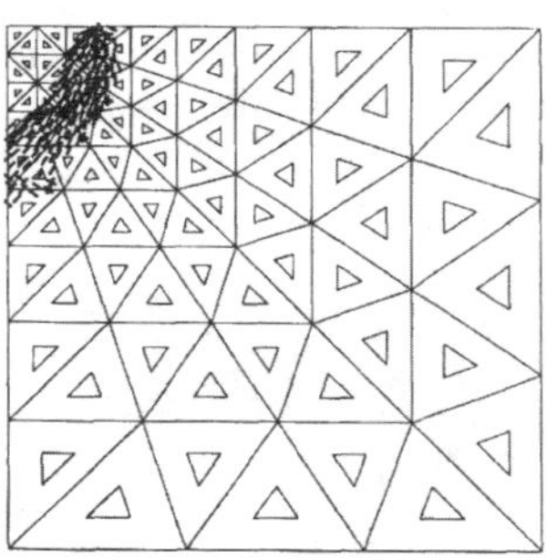

Fig. 9.27 Development of plastic zone

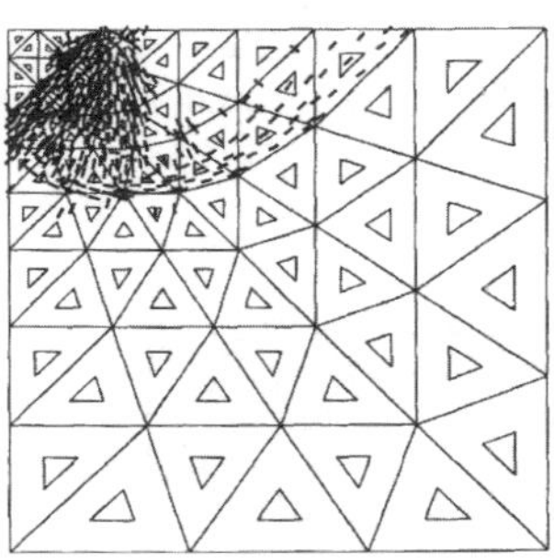

Fig. 9.28 Final plastic zone

The velocity fields at the final displasment is shown in Fig. 9.29. The shape roughtly resembles the Prandtl mechanism for perfectly plastic materials, although only part of the solid material in motion is responding plastically, with both the block of material directly under the rigid footing and much of the solid within the "passive motion" behaving elastically. The shape is also similar the results obtained by the unified slip field theory.

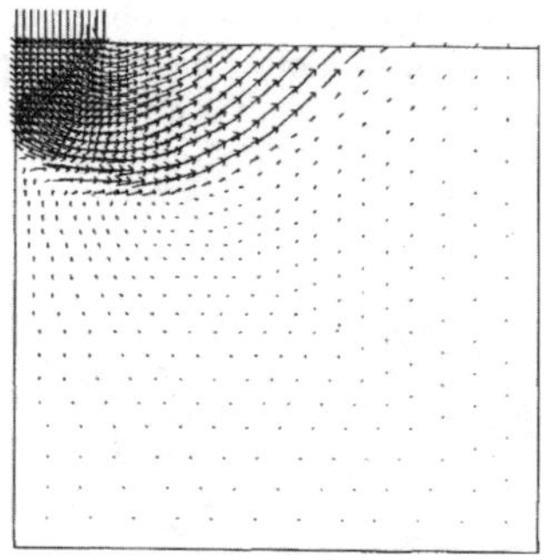

Fig. 9.29 Velocity fields at footing diaplacement (Moore and Rowe 1991)

The similar results were also obtained by other researcheres. A real-life problem in geomechanics were investigated by Wunderlich, Findeib and Cramer (Fig. 9.30) as well as Wunderlich, Findeib and Cramer Zimmermann and Commend et al.(Fig. 9.31). They showed the limit-load state obtained by the finite element analysis (2001). A localized failure in the shape of the well known Prandtle slip-lines which have been derived on a analytical basis was given. The results of the finite element analysis are in close agreement with observations in reality.

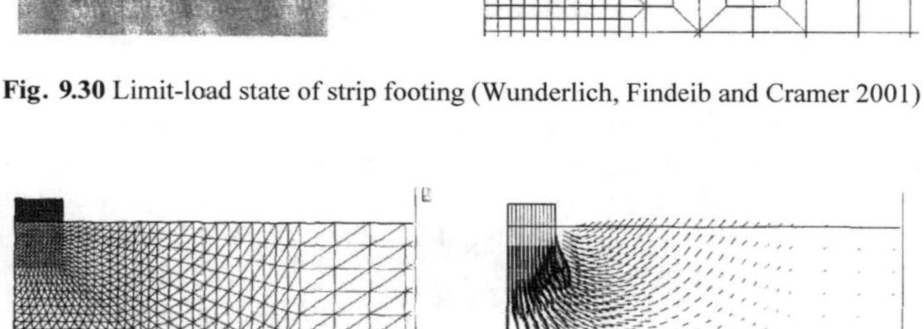

Fig. 9.30 Limit-load state of strip footing (Wunderlich, Findeib and Cramer 2001)

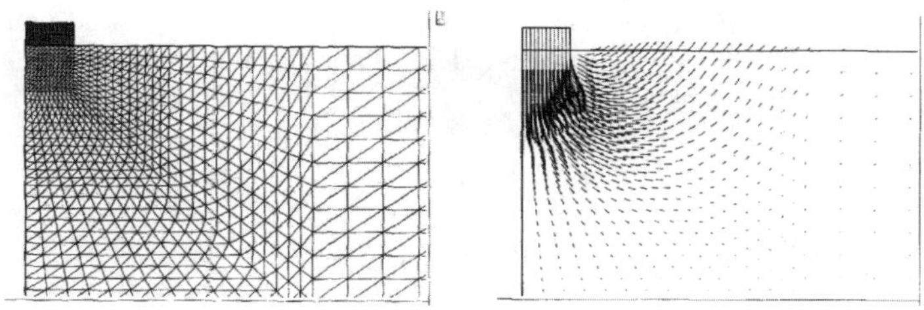

Fig. 9.31 Limit-load state of strip footing (Zimmermann and Commend 2001)

The numerical tests of a panel under compression were analyzed by de Borst et al (1995) and Chen and Baker (2001). The plastic strain develops in "X" shape. It is agreement with the experimental results shown in Fig. 9.32.

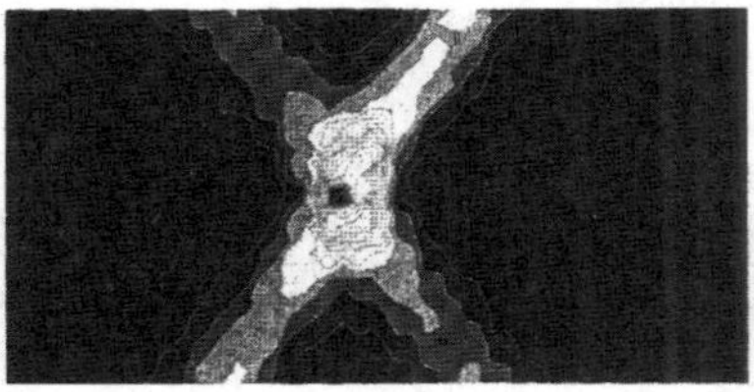

Fig. 9.32 The plastic strain develops in "X" shape (Chen and Baker 2001)

These kinds of results were obtained by Yu and Zeng in 1990. The calculation results by using two material models, i.e. the Mohr-Coulomb strength theory (b=0) and the twin-shear strength theory (b=1, Yu 1985) are illustrated in Fig.9.33 and Fig.9.34. Figure 9.33 is a result of thick plate with a hole under compression, and figure 9.34 is a result of thick plate with a hole under tension. It is seen that the four results are different.

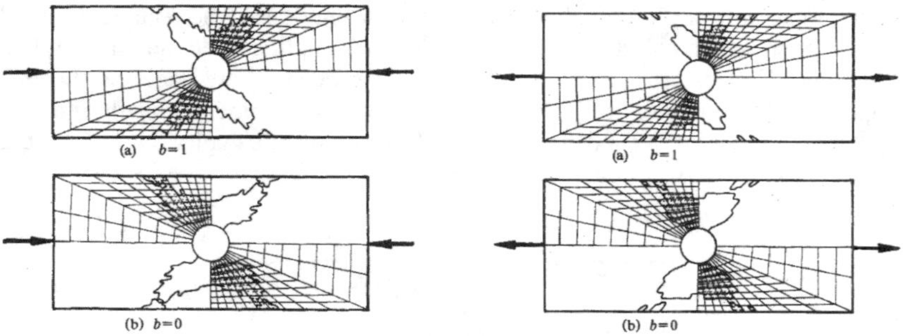

Fig. 9.33 Different results of two models in compression and in tension (Yu and Zeng 1993, see: Yu 1998)

It is worth noting that the result of the finit element method is always a single result by using a single yield criterion. The result of convenient slip line field is also a single result. It is also worthy to note that the shape and angle of the plastic zone are also different for different yield criterion.

9.8 Comparison of the Unified Slip Line Field Theory with Experimental Results

The experimental results of the plastic zones for a plate with different holes under tension are shown in Fig. 9.34 (see Yu 1998).

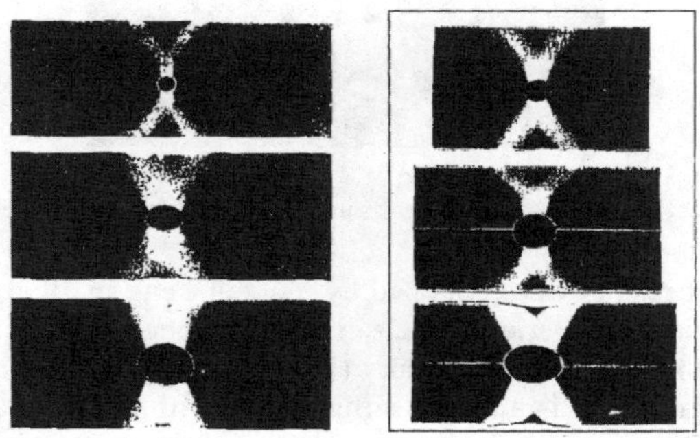

(a) Test results (Yu 1981) (b) Test results (Liu 1989)

Fig. 9.34 Plastic zones of tensile plate with different holes

The comparison of three methods are shown in Fig. 9.35. The rectangular thick plate specimen with a circular hole at the center was applied under an uniform load at the edge. The material and the experimental procedure are same as those in example 1 and example 5. In Fig. 9.35, curve 1 is the experimental result of plastic zone, curve 2 is that of unified elasto-plastic finite element calculation (b=0.8), and curve 3 is of Unified Slip Line Field Theory.

Fig. 9.35 Comparison of three methods (see Yu 1998)

As can be seen, the plastic zones are different in tension and in compression for SD (strength different) maerials. Fig. 9.33(a) shows the computational results by using the twin-shear strength theory and the Mohr-Coulomb strength theory for the same material in compression. Fig. 9.33(b) shows the computational results by using the twin-shear strength theory and the Mohr-Coulomb strength theory for the same material in tension.

The experimental results also present the difference of plastic zones in tension and in compression, as shown in Fig. 9.36.

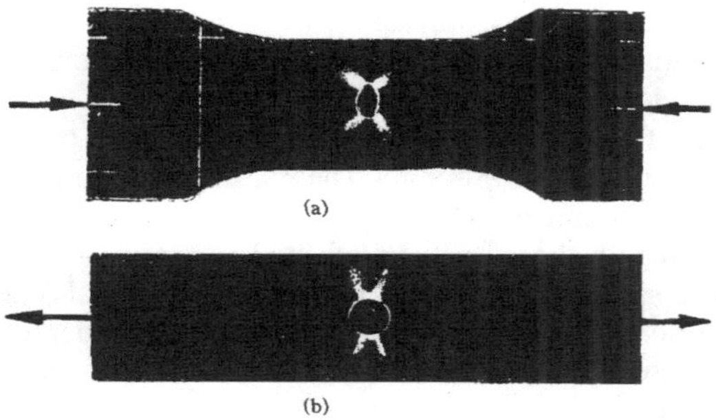

Fig. 9.36 Different results in compression and in tension (Yu and Liu 1989)

In order to confirm the results of the unified slip line theory, the elasto-plastic finite element program UEPP based on the unified strength theory (UST) is used to compute the limit loading q of the same structures. UEPP was established by Yu, Zeng, Ma,Yang, Wang et al. in the period of 1993 to 1998 (Yu and Zeng 1994; Yu, Yang, Fan et al. 1997).

From these comparisons, we can find that the result from the characteristics line theory are very close to those results from the experiments and finite element methods.

The detail of numerical analysis of generalized plasticity will be discussed in another book : "Computational Plasticity".

9.9 Discontinuous Bifurcations of Elasto-plastic Material for Plane Strain

During the process of the elastic-plastic deformation, the continuity of velocity vanishes when it is passing the certain characteristics surface, with the development of the deformation, namely, the phenomenon of discontinuous bifurcations is produced. Hill (1958), Marciniak and Kuczynski (1967), Storen and Rice (1975), Rudnicki and Rice (1975), Hutchinson and Tvergaard (1980, 1981), Tvergaard, Needleman and Lo (1981), Raniecki and Bruhns (1981), Bruhns (1984), Li (1987), Runesson and Mroz (1989), Hill JM and Wu YH (1993),

Zyczkowski (1999) et al. have done extensive researches for the discontinuous bifurcations.

Ottosen and Runesson (1991) put forward a general description of discontinuous bifurcations for the plane problem of isotropic and elastic-plastic body plane problem, obeying the Mohr-Coulomb yield. criterion. Hill JM and Wu YH (1993) used the Ashton-Warren Spring yield equation. It is a Mohr-Coulomb typed yield equation, in which the maximum principal stress σ_1 and the minimum principal stress σ_3 are taken into account. Zyczkowski (1999) studied the discontinuous bifurcations in the case of the Burzynski-Torere yield criterion.

The corresponding properties of discontinuous bifurcations have been obtained by using various yield criteria.

For selection of the yield function, the intermediate principal stress is not considered in Mohr-coulomb Strength Theory, and this theory can not match the experimental results of much materials.

The unified strength theory can be adapted for many kinds of materials, and make the former simple strength theories to be its special samples or its linear approach. Therefore, the unified strength theory is adopted to analyze the discontinuous bifurcations of materials for plane strain problem.

The mathematical expression of the unified strength theory for plane strain problems is

$$F=(1+b)(1+\sin\phi)\sigma_1-(1-\sin\phi)(b\sigma_2+\sigma_3)-2(1+b)c\cos\phi=0$$

$$\text{When } \sigma_2\le\frac{1}{2}(\sigma_1+\sigma_3)+\frac{\sin\phi}{2}(\sigma_1-\sigma_3) \tag{9-50a}$$

$$F'=(1+\sin\phi)(\sigma_1+b\sigma_2)-(1-\sin\phi)(1+b)\sigma_3-2(1+b)c\cos\phi=0$$

$$\text{When } \sigma_2\ge\frac{1}{2}(\sigma_1+\sigma_3)+\frac{\sin\phi}{2}(\sigma_1-\sigma_3) \tag{9-50b}$$

We chose the first equation of the unified strength theory (9-50a), because of the intermediate principal stress is

$$\sigma_2=\frac{m}{2}(\sigma_1+\sigma_3)\le\frac{1}{2}(\sigma_1+\sigma_3)+\frac{\sin\phi}{2}(\sigma_1-\sigma_3)\quad(0<m\le1) \tag{9-51}$$

Substituting Eq. (9-51) into Eq.(9-50), we have

$$F=(1+b)(1+\sin\phi)\sigma_1-\frac{1}{2}(1-\sin\phi)(bm\sigma_1+bm\sigma_3+2\sigma_3)-2(1+b)c\cos\phi=0 \tag{9-52}$$

definite $f=\frac{\partial F}{\partial \sigma}$, then we have

$$f_1=(1+b)(1+\sin\phi)\,\sigma_1-\frac{bm}{2}(1-\sin\phi)$$

$$f_2=-\frac{1}{2}(1-\sin\phi)(bm+2) \tag{9-52}$$

$$f_3=0$$

The angle of shear band is obtained as follows (Zhang 2000)

$$tg^2\theta=\frac{f_1}{f_2}=\frac{2(1+b)(1+\sin\varphi)-bm(1-\sin\varphi)}{(1-\sin\varphi)(2+bm)} \tag{9-53}$$

The direction angle of shear band is

$$\cos 2\theta=\frac{b(1-m)+(2+b+bm)\sin\varphi_0}{2+b(1+\sin\varphi_0)} \quad \text{when } m\neq 1, \tag{9-54}$$

Equation (9-54) for plastic plane strain problems can be simplified to

$$\cos 2\theta=\frac{2(1+b)\sin\varphi_0}{2+b(1+\sin\varphi_0)} \quad \text{when } m=1, \tag{9-55}$$

This result is similar to that of the unified slip field theory for plane strain problems. The direction angle of shear band is sensitive to the failure criterion. The unified slip field theory give us a effective method to study this problems.

The discontinuous bifurcations of elasto-plastic material for plane stress problems will be described in detail in Chapter 11.

Summary

A unified slip line field theory for plane strain problems based on the unified strength theory is described in this chapter. The unified slip line field theory for plane strain problems is composed of a group of orthogonal and non-orthogonal slip line fields. Many presently used orthogonal or non-orthogonal slip line field theories based on single shear strength theories are special cases (Tresca material slip field, Mohr-Coulomb material slip field and the twin-shear material slip field) or linear approximation (Huber-von Mises material) of the unified slip field theory system.

By introducing an intermediate principal stress parameter, m, the new method can reflect the intermediate principal stress of plane strain problem, where the intermediate principal stress is the principal stress in z direction. Experiments and Elasto-Plastic Finite Element calculations have been used to verified the validity of the unified slip theory. The unified strength theory, the unified slip theory can be used in many engineering fields, especially for materials with obvious intermediate principal stress effect.

The unified slip line field theory for plane strain problems has the following advantages:

1. It is a system of orthogonal and non-orthogonal slip line filed, which encompasses, in a piece-wise linear manner, other slip line field theories. All other slip line field theories are its special cases or linear approximations;
2. It can reflect the different effects of intermediate principal stress on different materials by introducing unified strength theory parameter b;
3. Giving a series of new slip line fields, and establishing the relations among available slip line field theories;
4. Introducing an intermediate principal stress parameter m, which can comply with different plane strain compressible and non-compressible materials.
5. It agrees with experimental results and can be easily used in many engineering fields.

Problems

Problem 9.1

The extension of a strip with a sufficiently large circular hole for non-SD material was shown in Fig. P8.1 of chapter 8. Can you obtain a more complete study on this subject using the unified slip field theory for non-SD material ($\alpha = 1$, and b=0, b=1/2, b=1).

Problem 9.2

The extension of a strip with a sufficiently large circular hole for non-SD material was shown in Fig. P8.1 of chapter 8. Can you obtain a more complete study on this subject using the unified slip line field theory for SD materials ($\alpha \neq 1$, and b=0,b=1/2, b=1).

Problem 9.3.
The extension of a strip with ideal (infinitely thin) cuts (crack) for non-SD material was shown in Fig. P8.2 of chapter 8. Can you obtain a more complete study on this subject using the unified slip line field theory ($\alpha = 1$, and b=0, b=1/2, b=1).

Problem 9.4.
The extension of a strip with ideal (infinitely thin) cuts (crack) for non-SD material was shown in Fig. P8.2 of chapter 8. Can you obtain a more complete study on this subject using the unified slip line field theory for SD materials ($\alpha \neq 1$, and b=0,b=1/2, b=1).

Problem 9.5.
The extension of a strip with angular notches for non-SD material was shown in Fig. P8.3 last chapter. Can you obtain a more complete study on this subject using the unified slip line field theory (α=1, and b=0, b=1/2, b=1).

Problem 9.6.
The extension of a strip with angular notches for non-SD material was shown in Fig. P8.3 of chapter 8. Can you obtain a more complete study on this subject using the unified slip line field theory for SD materials ($\alpha \neq 1$, and b=0,b=1/2, b=1).

Problem 9.7.
The extension of a strip with circular base for non-SD material was shown in Fig. P8.4 of chapter 8. Can you obtain a study on this subject using the unified slip line field theory for non-SD materials (α=1, and b=0, b=1/2, b=1).

Problem 9.8.
The extension of a strip with circular base for non-SD material was shown in Fig. P8.4 of chapter 8. Can you obtain a more complete study on this subject using the unified slip field theory for SD materials ($\alpha \neq 1$, and b=0, b=1/2, b=1).

Problem 9.9.
The slip line field of indentation at the bottom of a flat trench was shown in Fig.P8.5 of chapter 8. Find the limit load by using the unified slip line field theory for non-SD materials (α=1, and b=0, b=1/2, b=1).

Problem 9.10.
The slip line field of indentation at the bottom of a flat trench was shown in Fig.P8.5 last chapter. Find the limit load by using the unified slip field theory for SD materials ($\alpha \neq 1$, and b=0, b=1/2, b=1).

Problem 9.11.
The slip line field of indentation at the foot of a very deep vertically sided groove was shown in Fig.P8.6of chapter 8. Find the limit load by using the unified slip line field theory for non-SD materials (α=1, and b=0, b=1/2, b=1).

Problem 9.12.
The slip line field of indentation at the foot of a very deep vertically sided groove was shown in Fig.P8.6 of chapter 8. Find the limit load by using the unified slip field theory for SD materials ($\alpha \neq 1$, and b=0, b=1/2, b=1).

Problem 9.13.

The slip line field of indentation at the top by a flat punch was shown in Fig.P8.7 of chapter 8. Find the limit load by using the unified slip line field theory for non-SD materials (α=1, and b=0, b=1/2, b=1).

Problem 9.14.

The slip line field of indentation at the top by a flat punch was shown in Fig.P8.7 of chapter 8. Find the limit load by using the unified slip field theory for SD materials (α ≠ 1, and b=0, b=1/2, b=1).

Problem 9.15.

The slip line field of a strip weakened by one-sided deep notch with a circular base was shown in Fig.P8.8 last chapter. Find the limit load by using the unified slip line field theory for non-SD materials (α=1, and b=0, b=1/2, b=1).

Problem 9.16.

The slip line field of a strip weakened by one-sided deep notch with a circular base was shown in Fig.P8.8 of chapter 8. Find the limit load by using the unified slip field theory for SD materials (α ≠ 1, and b=0, b=1/2, b=1).

Problem 9.17.

The slip line field of a strip weakened by two-sided deep notch with a circular base was shown in Fig.P8.9 of chapter 8. Find the limit load by using the unified slip line field theory for non-SD materials (α=1, and b=0, b=1/2, b=1).

Problem 9.18.

The slip line field of a strip weakened by two-sided deep notch with a circular base was shown in Fig.P8.9 last chapter. Find the limit load by using the unified slip field theory for SD materials (α ≠ 1, and b=0, b=1/2, b=1).

Problem 9.19.

Figure P9.1 shows the slip line field of a pure bending strip weakened by two-sided angular notch. The limit moment for the Tresca material is

$$M = \left(1 + \frac{\pi}{2} - \gamma\right) h^2 \sigma_s, \quad \text{for } \gamma \geq \frac{\pi}{4} + \frac{1}{2}$$

Find the limit load by using the unified slip line field theory for non-SD materials (α=1, and b=0, b=1/2, b=1).

Problem 9.20.

Find the limit moment of a pure bending strip weakened by two-sided angular notch (Fig. P9.1) by using the unified slip line field theory for SD materials (α ≠ 1, and b=0, b=1/2, b=1).

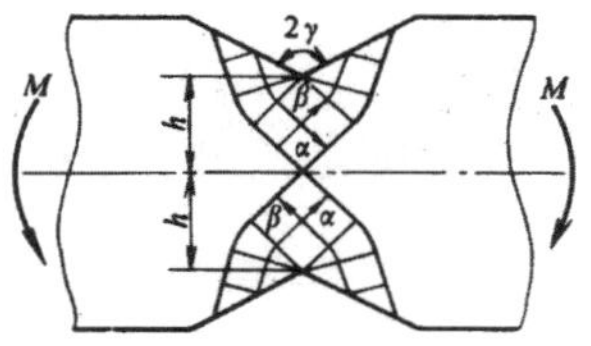

Fig. P9.1 Two-sided angular notch

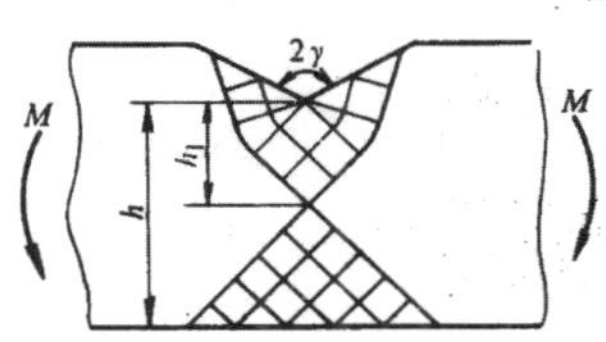

Fig. P9.2 One-sided angular notch

Problem 9.21.
Figure P9.2 shows the slip line field of a pure bending strip weakened by one-sided angular notch. The limit moment for the Tresca material is

$$M = \left(\frac{\pi + \pi - 2\gamma}{4 + \pi - 2\gamma} \right) \frac{h^2 \sigma_s}{2}, \quad \text{for } \gamma \geq 1$$

Find the limit moment by using the unified slip line field theory for non-SD materials (α=1, and b=0, b=1/2, b=1).

Problem 9.22.
Find the limit moment of a pure bending strip weakened by one-sided angular notch by using the unified slip line field theory for SD materials (α ≠ 1, and b=0, b=1/2, b=1).

Problem 9.23.
Figure P9.3 shows the slip line field of a pure bending strip weakened by one-sided angular notch with $\gamma < 1$. Find the limit moment by using the unified slip line field theory for non-SD materials and SD materials

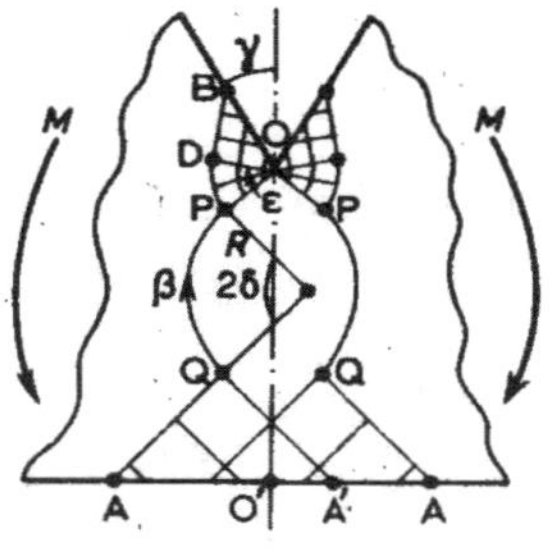

Fig. P9.3 Bending strip with angular notch

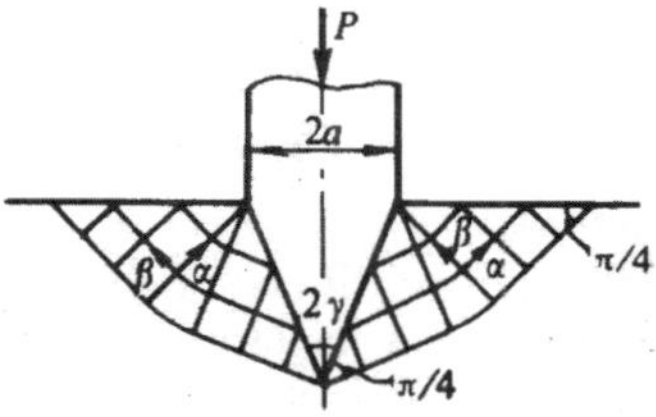

Fig. P9.4 Indentation of a semi-infinite mass

Problem 9.24.
Figure P9.4 shows the slip line field of an indentation of a semi-infinite mass of rigid-perfectly plastic material by a rigid straight-sided, acute-angle indenter (Hill 1950; Johnson and Mellor 1961). The limit load for the Tresca material is

$$P = 2a(1+\gamma)\sigma_s$$

Find the limit load by using the unified slip line field theory for non-SD materials (α=1, and b=0, b=1/2, b=1).

Problem 9.25.
Find the limit load of an indentation of a semi-infinite mass of rigid-perfectly plastic material by a rigid straight-sided, acute-angle indenter by using the unified slip line field theory for SD materials ($\alpha \neq 1$, and b=0, b=1/2, b=1).

Problem 9.26.
The slip line field of a thick-walled cylinder is shown in Fig.P9.5. The limit pressure for the Tresca material is

$$P = \sigma_s \ln \frac{b}{a}$$

Find the limit pressure by using the unified slip line field theory for non-SD materials (α=1, and b=0, b=1/2, b=1).

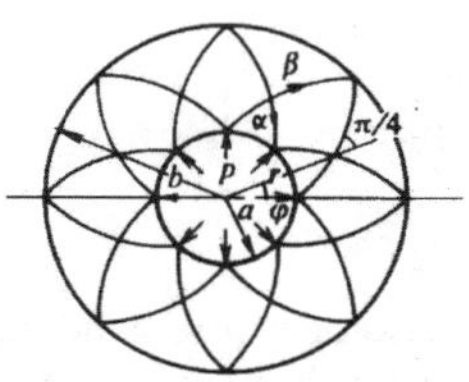

Fig. P9.5 Slip line field of a thick-walled cylinder (Tresca material)

Problem 9.27.
Find the limit pressure for a thick-walled cylinder by using the unified slip line field theory for SD materials ($\alpha \neq 1$, and b=0, b=1/2, b=1).

Problem 9.28.
Comparing the results of the slip line field theory for a thick-walled cylinder with the elasto-plastic analysis in Chapter 15.

Problem 9.29.
Figure P9.6 shows a slip line field of a strip weakened by center crack for non-SD materials. Determine thelimit load by using the unified slip line field theory for SD materials ($\alpha \neq 1$, and b=0, b=1/2, b=1).

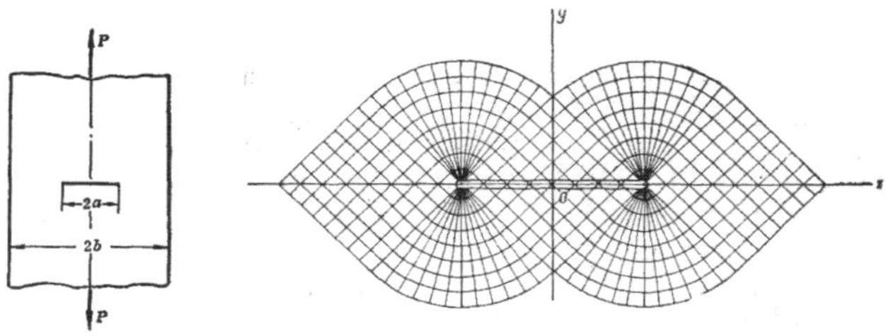

Fig. P9.6 A slip line field of a strip weakened by center crack

Problem 9.30.
The slip line field of a strip with a hole under uniform pressure was shown in Fig. P8.9 last chapter. Find the limit pressurefor non-SD materials and SD materials by using the unified slip field theory.

Problem 9.31.
The slip line field of extrusion through a die over a smooth container wall was studied by Johnson et al (1962). Figure P9.7 shows the fan shaped slip line field The limit pressure were given by using the Tresca condition as follows

$$P = \left(1 + \frac{\pi}{2}\right) h \sigma_s$$

Find the limit pressure for SD materials by using the unified slip line field theory ($\alpha \neq 1$, and b=0, b=1/2, b=1).

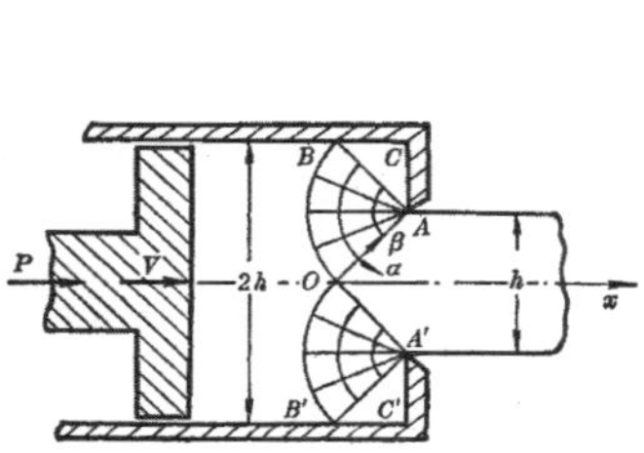

Fig. P9.7 Extrusion through a die over a smooth container wall

Problem 9.32.
Find the limit pressureof extrusion through a die over a smooth container wall for SD materials by using the unified slip line field theory ($\alpha \neq 1$, and b=0, b=1/2, b=1).

Problem 9.33.
The slip line field of drawing of a strip was studied by Johnson et al (1962). Figure P9.8 shows the fan shaped slip line field. Limit drawing force for the Tresca material is

$$P=\frac{2h\sigma_s(1+\gamma)\sin\gamma}{1+2\sin\gamma} \quad \text{when } \gamma \geq 42^\circ 27', \ H=2h$$

Find the drawing force for non-SD materials by using the unified slip line field theory (α=1, and b=0, b=1/2, b=1).

Problem 9.34.
Find the drawing force for SD materials by using the unified slip line field theory (α ≠ 1, and b=0, b=1/2, b=1).

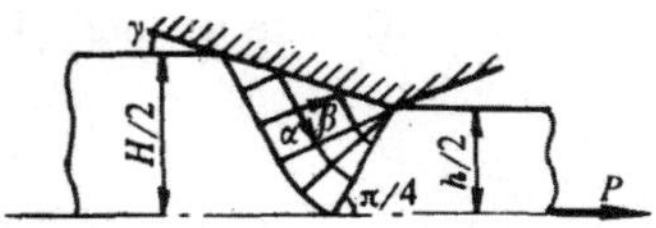

Fig. P9.8 Extrusion of a strip

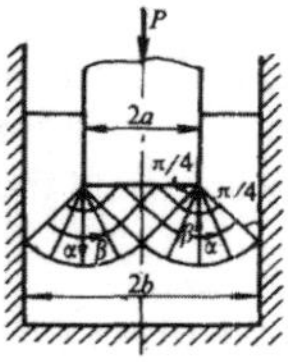

Fig. P9.9

Problem 9.35.
The slip line field and limit pressure for extrusion problems were studied by Johnson et al. as shown in Fig. P9.9. The limit extrusion force for the Tresca material is

$$P=2\left(1+\frac{\pi}{2}\right)a\sigma_s \quad \text{when } b=2a$$

Find the limit pressure of extrusion by using the unified slip line field theory for non-SD materials (α=1, and b=0, b=1/2, b=1).

Problem 9.36.
Find the limit pressure of extrusion by using the unified slip line field theory for SD materials (α ≠ 1, and b=0, b=1/2, b=1).

Problem 9.37.
Determine the limit load of a deep beam shown in Fig. P9.10

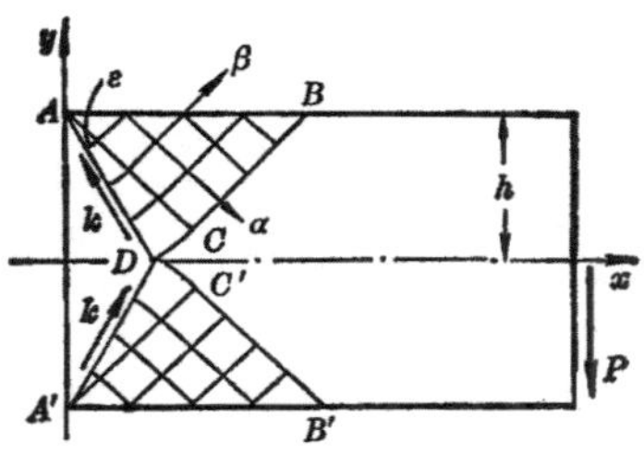

Fig. P9.10

Problem 9.38.
A series of results can be obtained by using the unified slip line field theory for various plane strain problems. All the solutions of the problems in this chapter have not been studied before. Can you try to give a series of new results for these problems, and try to use in practice.

Problem 9.39.
The slip line field and limit pressure for slop stability problems were given by Соколовский ВВ (1960) for the Mohr-Coulomb material as shown in Fig. P9.11. The limit pressure is

$$q = C \cdot \cot\varphi \left\{ \frac{1+\sin\varphi}{1-\sin\varphi} \exp\left[(\pi - 2\chi)\tan\varphi\right] - 1 \right\}$$

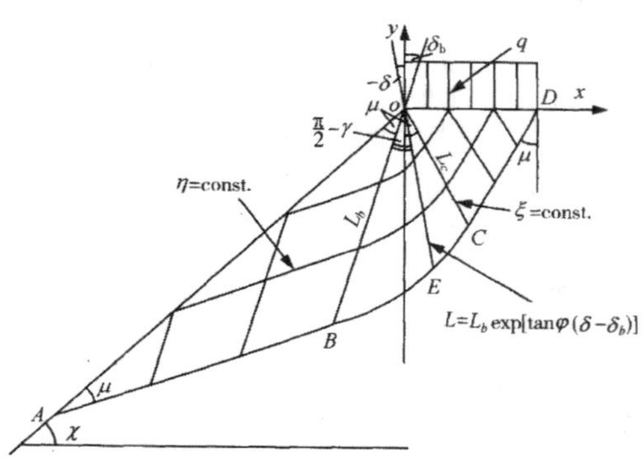

Fig. P9.11

Find the limit pressure of slop by using the unified slip line field theory for non-SD materials (α=1, and b=0, b=1/2, b=1) and SD materials (α ≠ 1, and b=0, b=1/2, b=1).

References and Bibliography

Chen G. and Baker G (2001). Gradient-dependent plasticity with a 3-node incompatible element for the plastic multiplier. In: Valliappan S and Khalili N eds., *Computational Mechanics : New Frontiers for the New Millennium,* Amsterdam: Elsevier, 1129–1134.

de Borst R., Pamin J and Sluys LJ (1995) Computation issues in gradient plasticity, In: *Continuum Models for Materials with Micro-structure*, H.B. Mfihlhaus (ed.), Wiley: Chichester, 159–200.

Drucker DC, Greenberg HJ and Prager W (1951) The safety factor of an elastic-plastic body in plane strain, *J. Appl. Mech.*, 18: 371–378.

Ford H (1954) The Theory of Plasticity in Relation to Engineering Application, *J. AppL Math. Phys*., 5: 1–35.

Geiringer H (1930) Beit zum Vollstiindigen ebenen Plastizitits-problem, Proc 3rd Intern. Congr. Appl. Mech, 2: 185–190.

Hencky H(1923) Ueber einige statisch bestimmte Faelle des Gleichgewichts in plastischen Koerpern Z. angew. Math Mech., 3: 245–251.

Hill R (1949) The Plastic Yielding of Notched Bars Under Tension, *Quart. Mech. AppL Math.*, 2: 40–52.

Hill R(1956), The Mechanics of Quasi-Static Plastic Deformation in Metals, In: Survey in Mechanics, G. K. Batcheler and R. M. Davis eds, Cambridge,

Johnson W and Mellor PB (1962), *Plasticity for Mechanical Engineers*, Van Nostrand, Princeton, N.J.

Johnson W, Sowerby R, Venter RD (1982), *Plane-Strain Slip Line Fields for Metal Deformation Processes--A Source Book and Bibliography*, Pergamon Press, Oxford.

Moore ID and and Rowe RK (1991) Objective solutions for bearing capacity of strain-softening soils. In: Beer G, Booker JR and Carter JP eds. *Computer Methods and Advances in Geomechanics*. Rotterdam: A.A.Balkema, Vol.2,1183–1189.

Panayotounakos D (1996) Ad Hoc Exact Solutions for the Stress and Velocity Fields in Rigid Perfectly Plastic Materials Under Plane-Strain Conditions. *Int. J. Plast.*, 12:1069–1190.

Prager W (1955) The Theory of Plasticity: A Survey of Recent Achievements, *Proc. Inst. Mech. Eng.*, 169: 41–57.

Prager W (1959) *Introduction to Plasticity,* Addison-Wesley, Reading, Mass.

Prager W and Hodge P G Jr (1951) *Theory of Perfectly Plastic Solids,* Wiley New York.

Prandtl L (1920) l~ber die H~irte Plastischer Koerper, Goettinger Nachr., *Math. Phys.* KI.,74–85.

Thomsen EG, Yang CT and Kobayashi S (1965) *Mechanics of Plastic Deformation in Metal Processing,* Macmillan, New York.

Wunderlich, Findeib and Cramer in (2001) Finite element analysis of bearing capacities in geomechanics considering dilatant and contractant constitutive laws. In: Valliappan S and Khalili N eds., *Computational Mechanics : New Frontiers for the New Millennium*, Amsterdam: Elsevier, 2001, 533–538.

Yu Mao-hong (1983), Twin Shear Stress Criterion, *Int. J. of Mech. Sci.*, **25**(1), 71–74,

Yu Mao-hong, He, Li-nan, Song, Ling-yu (1985), Twin Shear Stress Theory and Its Generalization, *Scientia Sinica (Science in China), Series A*, (English edn.), **28**(11), 1174–1183.

Yu Mao-hong and He Li-nan (1991), A new model and theory on yield and failure of materials under complex stress state. In: *Mechanical Behaviour of Matterials-VI.*

Yu Mao-hong, He Li-nan (1992), Generalized Twin Shear Stress Yield Criterion and Its Generalization, *Chinese Science Bulletin* (English edn.), **37**(24), 2085–2089.

Yu Mao-hong, et al (1992), Computer Image Analysis of the Plastic Zone of Structure, *Journal of Xi'an Jiaotong University*, **26**(2), 121–122 (in Chinese).

Yu Mao-Hong, Liu Jian-Yu and Ma Guo-wei (1994), Twin-shear slip line theory: orthogonal and non-orthogonal slip line fields. *Journal of Xi'an Jiaotong University*, 28(2):122–126.

Yu Mao-hong. (1998) *Twin-shear Theory and Its Applications*. Beijing: Science Press (in Chinese).

Yu MH and Fan SC eds (1998) *Strength Theory: Applications, Developments and Prospects for the 21st Century*. New York, Beijing: Science Press.

Yu Mao-hong. (2004) *Unified Strength Theory and Its Applications*. Berlin: Springer.

Zenkiewicz OC(1992), Computational Mechanics Today, *Int. J. Numer. Methods in Engrg.*, **34**, 9–33.

Zhang YQ (2000) Analysis of damage and Discontinuous bifurcations for Elarto-Plastic Geomaterials. Dissertation of Doctor at Xi'an Jiaotong University, Xi'an, China (in Chinese)

Zhang YQ and Yu MH (2001) Discontinuous bifurcations of metallic materials for plane stress. Chinese *J. of Mechanical Engineering*, 37(4): 62–85 (in Chinese, English Abstract).

Zhang YQ and Yu MH (2001) Discontinuous bifurcations of elasto-plastic materials for plane stress. *Acta Mechanica Sinica*, 33(5): 706–713 (in Chinese, English Abstract).

Zimmermann Th and Commend S (2001) Stabilized finite element applications in geomechanics. In: Valliappan S and Khalili N eds., *Computational Mechanics : New Frontiers for the New Millennium,* Amsterdam: Elsevier, 2001, 533–538.

Березанлев ВГ (1952) Осесимметричная Задача Теории Предельного Равновесия Сыпучей Среды. Москва: Гос. Изд. Тех. Тео. Литературы (in Russia)

Соколовский ВВ (1942) Статика Сыпучей Среды. Изд. АН СССР, Москва (in Russia)

Соколовский ВВ (1960) Статика Сыпуей Среды (Third ed.).Гос. Изд. Физ-Мат. Литературы. Москва (in Russia)

10 Twin-Shear Characteristics Field for Plane Stress Problem

10.1 Introduction

The twin-shear slip field and the unified slip line field theory for plane strain problems have been described in Chapters 8 and 9. This chapter deals with the twin-shear characteristics filed for plane stress problems.

For plane stress problems, characteristic methods can be used to solve the quasi-linear differential equation systems of stress and velocity fields. Judgments on the types of these differential equation systems can be made using the theory of characteristics. They may be elliptic or hyperbolic, depending on the considered stress state. The methods of characteristics based on the Huber-von Mises and the Tresca criteria can be found in the literature of Kachanov (1971), Martin (1973), Yan (1988) and Panoyotomakos (1999) for plane stress problems. A new characteristics method for plane stress problems was established based on the twin-shear yield criterion (Yu 1961, 1983) by Yan and Bu in 1993 and 1996.

The methods of characteristics based on the Huber-von Mises, the Tresca and the twin-shear criteria can be applied to the limit analysis of the plane stress problems. However, they are only adapted for the non-SD (strength differential) materials with $\tau_0 \approx 0.58\sigma_t$, $\tau_0 = 0.5\sigma_t$ and $\tau_0 = 2\sigma_t/3$, respectively, but fail for the SD materials. Although the method of characteristics based on the Mohr-Coulomb criterion takes account of SD effect, it is only adapted for the material with $\tau_0=\sigma_t\,\sigma_c/\,(\sigma_t + \sigma_c)$.

10.2 Characteristics Method based on the Tresca Criterion and the Huber-von Mises criterion

10.2.1 Characteristics Method based on the Tresca Criterion

The Tresca criterion in plane stress state is shown in Fig. 10.1. If χ and λ are introduced for the cases of $\sigma_x\sigma_y \leq \tau^2_{xy}$ and $\sigma_x\sigma_y \geq \tau^2_{xy}$ respectively, the following expressions of stresses satisfy automatically the Tresca criterion in plane stress state,

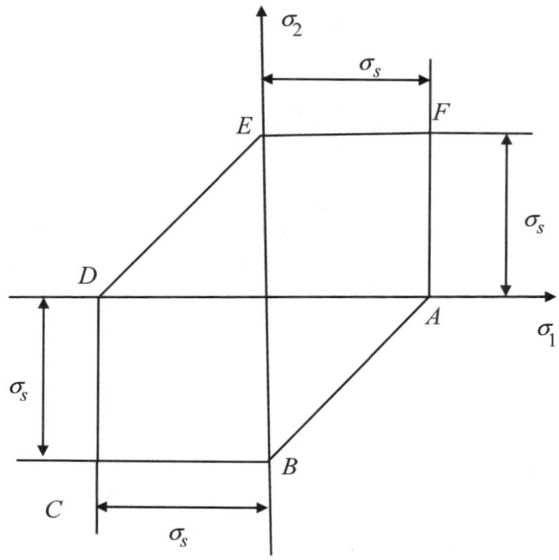

Fig. 10.1 Tresca criterion in plane stress state

$$\sigma_x = 2k\chi + k\cos 2\varphi \tag{10-1a}$$

$$\sigma_y = 2k\chi - k\cos 2\varphi \tag{10-1b}$$

$$\tau_{xy} = k\sin 2\varphi \tag{10-1c}$$

on lines *AB* and *DE* in Fig. 10.1.

$$\sigma_x = \sigma_s[s(1-\lambda) + \lambda\cos 2\varphi] \tag{10-2a}$$

$$\sigma_y = \sigma_s[s(1-\lambda) - \lambda\cos 2\varphi] \tag{10-2b}$$

$$\tau_{xy} = k\sin 2\varphi \tag{10-2c}$$

on lines *BC*, *CD*, *EF* and *FA* in Fig. 10.1, where

$$s = \begin{cases} 1 & when \;\; \sigma_1 > 0, \sigma_2 > 0 \\ \\ -1 & when \;\; \sigma_1 < 0, \sigma_2 < 0 \end{cases} \tag{10-3}$$

The equations of equilibrium determined by the stress components in Eq. (10-1) and Eq. (10-2) are of the hyperbolic type and the parabolic type, respectively. The equations of the characteristics are

$$\frac{dy}{dx} = \tan\left(\varphi \pm \frac{\pi}{4}\right) \tag{10-4}$$

$$\frac{dy}{dx} = -\frac{s + \cos 2\varphi}{\sin 2\varphi} = \begin{cases} -\cot\varphi, \\ \tan\varphi, \end{cases} \tag{10-5}$$

It can be seen that Eq. (10-4) has two orthogonal families of characteristics, which differentiate π/4 with the direction of principal stresses σ_1 and σ_2. It coincides with the definition of slip line. The two families of characteristics have the following forms,

$$\chi + \varphi = C_1 \tag{10-6a}$$

$$\chi - \varphi = C_2 \tag{10-6b}$$

Eq. (10-6) is the same as Hencky equation for the plane strain problem, when assuming $\chi = \sigma/2k$. Eq. (10-5) has only one family of characteristics, which is parallel to the intermediate principal stress by considering that it is a three-dimensional problem. The characteristic line is the intersecting line between xy-plane and the plane which offsets an angle of π/4 from the directions of the maximum and minimum stresses. This agrees with the explanation of the slip line equation (17-4), as long as considering the difference of the cases $\sigma_x\sigma_y \le \tau^2_{xy}$ and $\sigma_x\sigma_y \ge \tau^2_{xy}$.

From the above analysis, if the Tresca criterion is applied, the characteristics with respect to the lines *AB* and *DE* ($\sigma_x\sigma_y > \tau^2_{xy}$) can be solved by the hyperbolic type equations. On other lines ($\sigma_x \sigma_y > \tau^2_{xy}$), it can be solved by integrating the differential equation directly.

10.2.2
Characteristics Method based on the Huber-von Mises Criterion

The Huber-von Mises criterion in plane stress state is given in Fig. 10.2. If a variable ω is introduced, the stress components satisfy Huber-von Mises criterion can be obtained by

$$\sigma_x = k\left(\sqrt{3}\cos\omega + \sin\omega\cos 2\varphi\right) \tag{10-7a}$$

$$\sigma_y = k\left(\sqrt{3}\cos\omega - \sin\omega\cos 2\varphi\right) \tag{10-7b}$$

$$\tau_{xy} = k\sin\omega\sin 2\varphi \tag{10-7c}$$

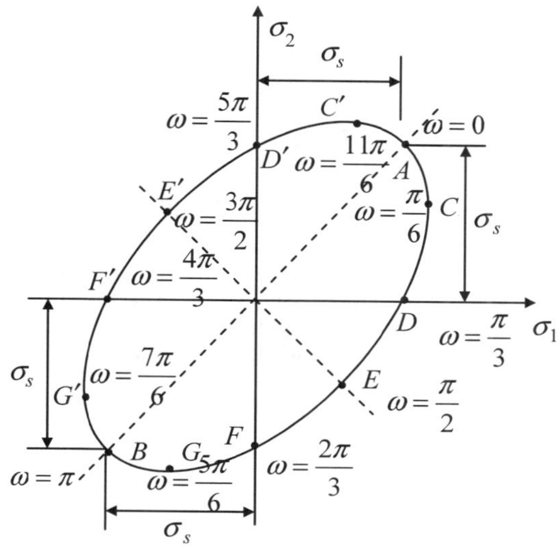

Fig. 10.2 Huber-von Mises criterion in plane stress state

The characteristics of the equilibrium equation determined by Eq. (10-7) is

$$\frac{dy}{dx} = \frac{\sqrt{3}\sin\omega\sin 2\varphi \pm \sqrt{3-4\cos^2\omega}}{\sqrt{3}\sin\omega\cos 2\varphi - \cos\omega} \tag{10-8}$$

It is obvious that there exist two families of characteristics only when $3-4\cos^2\omega > 0$ ($\pi/6 < \omega < 5\pi/6$ or $7\pi/6 < \omega < 11\pi/6$) (lines CG and $C'G'$ in Fig. 10.2). Otherwise, there is no characteristics ($3-4\cos^2\omega < 0$) (lines CC' and GG' in Fig. 10.2) or there is only one family of the characteristics ($3-4\cos^2\omega = 0$) (points C, C', G, and G'). For the hyperbolic type, two supplementary incremental expressions are needed

$$\psi = \frac{\pi}{2} - \frac{1}{2}\cos^{-1}\left(\frac{\cot\omega}{\sqrt{3}}\right) \tag{10-9a}$$

$$\chi = -\frac{1}{2}\int_{\frac{\pi}{6}}^{\omega} \frac{\sqrt{3-4\cos^2\omega}}{\sin\omega} d\omega \tag{10-9b}$$

The equation of characteristics can then be obtained as

$$\frac{dy}{dx} = \tan(\varphi \pm \psi) \tag{10-10}$$

Along these two families of the characteristics, it has

$$\chi + \varphi = C_1 \tag{10-11a}$$

$$\chi - \varphi = C_2 \tag{10-11b}$$

which has the same form with Eq. (10-6), however, χ is much more complex than that in Eq. (10-6). Meanwhile, it can be found that the two families of characteristics are not orthogonal to each other. The angle between the two families of the characteristics is 2ψ, which varies with locations.

From the above analysis, for Huber-von Mises criterion, the problem can be solved by the characteristics method of the hyperbolic type when $\pi/6 < \omega < 5\pi/6$, or $7\pi/6 < \omega < 11\pi/6$, however, the process will be very tedious due to the non-orthogonal feature. On other segments it will be more difficult to solve the characteristics.

10.3 Characteristics Method Based on the Twin-Shear Yield Criterion

Characteristics method based on the twin-shear yield criterion was proposed by Yan and Bu in 1993.

In(σ_x, σ_y, τ_{xy}) plane stress states, three principal stresses are

$$\sigma_1 = \frac{\sigma_x + \sigma_y}{2} + \sqrt{\left(\frac{\sigma_x - \sigma_y}{2}\right)^2 + \tau^2_{xy}} \tag{10-12a}$$

$$\sigma_2 = \frac{\sigma_x + \sigma_y}{2} - \sqrt{\left(\frac{\sigma_x - \sigma_y}{2}\right)^2 + \tau^2_{xy}} \tag{10-12b}$$

$$\sigma_3 = 0 \tag{10-12c}$$

The twin-shear yield criterion has the following form,

$$f = \sigma_1 - \frac{1}{2}(\sigma_2 + \sigma_3) = \sigma_s = \frac{3}{2}k, \text{ when } \sigma_2 \le \frac{1}{2}(\sigma_1 + \sigma_3) \tag{10-13a}$$

$$f = \frac{1}{2}(\sigma_2 + \sigma_3) - \sigma_3 = \sigma_s = \frac{3}{2}k, \text{ when } \sigma_2 \ge \frac{1}{2}(\sigma_1 + \sigma_3) \tag{10-13b}$$

The yield loci of the twin-shear yield criterion in the plane stress state are shown in Fig. 10.3.

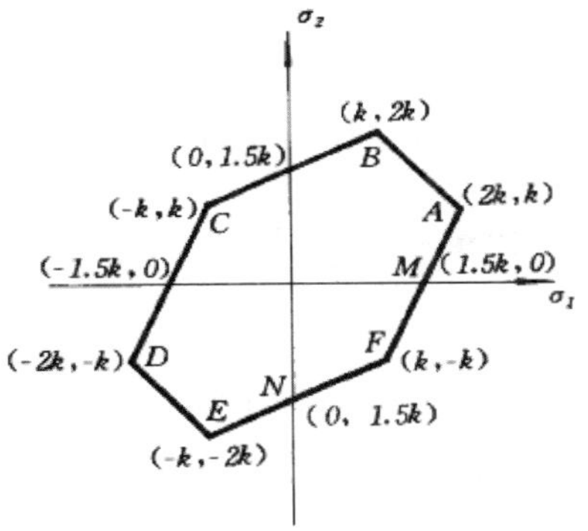

Fig. 10.3 Twin-shear yield criterion in plane stress state

The functions of the six segments in Fig. 10.3 can be expressed as follows,

$$f_1 = 2\sigma_1 - \sigma_2 = 3k \tag{10-14a}$$

$$f_2 = 2\sigma_1 - \sigma_2 = -3k \tag{10-14b}$$

$$f_3 = 2\sigma_2 - \sigma_1 = 3k \tag{10-14c}$$

$$f_4 = 2\sigma_2 - \sigma_1 = -3k \tag{10-14d}$$

$$f_5 = \sigma_1 - \sigma_2 = -3k \tag{10-14e}$$

$$f_6 = \sigma_1 - \sigma_2 = 3k \tag{10-14f}$$

Substituting Eq. (10-10) into Eq. (10-14), we derive the twin-shear yield criterion function expressed in terms of the three principal stresses as follows,

$$\frac{1}{2}\left|\sigma_x + \sigma_y\right| + 3\sqrt{\frac{1}{4}(\sigma_x - \sigma_y)^2 + \tau^2_{xy}} = 3k\ ,$$

$$\text{when } \left|\sigma_x + \sigma_y\right| \le 6\sqrt{\frac{1}{4}(\sigma_x - \sigma_y)^2 + \tau^2_{xy}} \tag{10-15a}$$

$$\left|\sigma_x + \sigma_y\right| = 3k, \ \text{ when } \left|\sigma_x + \sigma_y\right| \ge 6\sqrt{\frac{1}{4}(\sigma_x - \sigma_y)^2 + \tau^2_{xy}} \tag{10-15b}$$

Neglecting body force, the equations of equilibrium for plane stress problems are

$$\frac{\partial \sigma_x}{\partial x} + \frac{\partial \tau_{xy}}{\partial y} = 0, \qquad \frac{\partial \tau_{xy}}{\partial x} + \frac{\partial \sigma_y}{\partial y} = 0 \tag{10-16}$$

10.3.1 Hyperbolic Type

For the yield condition of Eq. (10-15a), assuming

$$\frac{\sigma_1-\sigma_2}{2}=\lambda k, \quad \frac{\sigma_1+\sigma_2}{2}=3(1-\lambda)sk \tag{10-17}$$

where

$$\begin{aligned} &S=1, \text{ when } \sigma_1+\sigma_2>0\,; \\ &S=0, \text{ when } \sigma_1+\sigma_2=0\,; \\ &S=-1, \text{ when } \sigma_1+\sigma_2<0 \end{aligned} \tag{10-18}$$

Combining Eqs. (10-10) and (10-17) gives

$$\sqrt{\left(\frac{\sigma_x-\sigma_y}{2}\right)^2+\tau^2_{xy}}=\lambda k\,, \quad \frac{\sigma_x+\sigma_y}{2}=3(1-\lambda)sk \tag{10-19}$$

The corresponding stress components are

$$\sigma_x=k[3s(1-\lambda)+\lambda\cos 2\varphi] \tag{10-20a}$$

$$\sigma_y=k[3s(1-\lambda)-\lambda\cos 2\varphi] \tag{10-20b}$$

$$\tau_{xy}=k\lambda\sin 2\varphi \tag{10-20c}$$

The above stresses satisfy Eq. (10-15a) automatically, in which λ is in the range of

$$\frac{1}{2}\le\lambda\le 1 \tag{10-21}$$

Combining Eq. (10-20) and Eq. (10-16) derives

$$-3s\frac{\partial\lambda}{\partial x}+\frac{\partial\lambda}{\partial x}\cos 2\varphi-2\lambda\sin 2\varphi\frac{\partial\lambda}{\partial x}+\frac{\partial\lambda}{\partial y}\sin 2\varphi+2\lambda sos2\varphi\frac{\partial\lambda}{\partial y}=0 \tag{10-22a}$$

$$\frac{\partial\lambda}{\partial x}\sin 2\varphi+2\lambda sos2\varphi\frac{\partial\varphi}{\partial x}-3s\frac{\partial\lambda}{\partial y}-\frac{\partial\lambda}{\partial y}sos2\varphi+2\lambda\sin 2\varphi\frac{\partial\varphi}{\partial y}=0 \tag{10-22b}$$

For any line in xy-plane, increments of λ and φ can be written as

$$d\lambda=\frac{\partial\lambda}{\partial x}dx+\frac{\partial\lambda}{\partial y}dy \tag{10-23a}$$

$$d\varphi = \frac{\partial \varphi}{\partial x}dx + \frac{\partial \varphi}{\partial y}dy \tag{10-23b}$$

Eq. (10-22) together with Eq. (10-23) makes an algebraic equation system with $\partial\lambda/\partial x$, $\partial\lambda/\partial y$, $\partial\varphi/\partial x$, $\partial\varphi/\partial y$ as unknowns. Let the determinant of coefficients vanish, i.e.

$$\Delta = \begin{vmatrix} -3s+\cos 2\varphi & \sin 2\varphi & -2\lambda\sin 2\varphi & 2\lambda\cos 2\varphi \\ \sin 2\varphi & -3s-\cos 2\varphi & 2\lambda\cos 2\varphi & 2\lambda\sin 2\varphi \\ dx & dy & 0 & 0 \\ 0 & 0 & dx & dy \end{vmatrix} = 0 \tag{10-24}$$

It is then obtained,

$$\frac{dy}{dx} = \frac{-3s\sin 2\varphi \pm 2\sqrt{2s}}{1-3s\cos 2\varphi} \tag{10-25}$$

Assuming $s = -3\cos 2\varphi$, Eq. (10-25) can be rewritten as

$$\frac{dy}{dx} = \tan(\varphi \mp \psi), \psi = \frac{1}{2}\cos^{-1}\left(-\frac{s}{3}\right) \tag{10-26}$$

Eq. (10-26) determines two families of characteristics. As can be seen, the two families of characteristics differ angles $\pm\psi$ from the direction of the principal stress σ_1. Here those corresponding to the plus sign are assigned as family α and those to the minus sign as family β. In order to derive the relationship between λ and φ, from Eqs. (10-22), (10-23) and (10-24), it has

$$\varphi + \sqrt{2s}\ln\lambda = C_1 \quad \text{(along } \alpha \text{ line)} \tag{10-27a}$$

$$\varphi - \sqrt{2s}\ln\lambda = C_2 \quad \text{(along } \beta \text{ line)} \tag{10-27b}$$

Thus, based on the twin-shear yield criterion, it can be solved by the characteristics method for stress state on lines of *FA*, *BC*, *CD*, *EF* (see Fig. (10.3)).

10.3.2 Elliptic Type

For the yield condition Eq. (10-15b) (AB and DE in Fig. (10.3)), it has

$$\frac{\partial\lambda}{\partial x}\cos 2\varphi - 2\lambda\sin 2\varphi\frac{\partial\varphi}{\partial x} + \frac{\partial\lambda}{\partial y}\sin 2\varphi + 2\lambda\cos 2\varphi\frac{\partial\varphi}{\partial y} = 0 \tag{10-28a}$$

$$\frac{\partial\lambda}{\partial x}\sin 2\varphi + 2\lambda\cos 2\varphi\frac{\partial\varphi}{\partial x} - \frac{\partial\lambda}{\partial y}\cos 2\varphi + 2\lambda\sin 2\varphi\frac{\partial\varphi}{\partial y} = 0 \tag{10-28b}$$

where $\lambda \le 1/2$.

Let the determinant of the coefficients of Eqs. (10-28) and (10-23) vanish, it has

$$\Delta = \begin{vmatrix} \cos 2\varphi & \sin 2\varphi & -2\lambda \sin 2\varphi & 2\lambda \cos 2\varphi \\ \sin 2\varphi & -\cos 2\varphi & 2\lambda \cos 2\varphi & 2\lambda \sin 2\varphi \\ dx & dy & 0 & 0 \\ 0 & 0 & dx & dy \end{vmatrix} = 0 \tag{10-29}$$

Then it is obtained

$$(dx)^2 + (dy)^2 = 0 \tag{10-30}$$

There is no real root for dy/dx. The equilibrium differential equation determined by the stress components which satisfy Eq. (10-15b) is an elliptic type, thus the characteristics do not exist.

Based on the above analysis, it can be found that the problem of the stress state on the lines of *FA*, *BC*, *CD*, *EF* (see Fig. (10.3)) is of the hyperbolic type and can be solved by the characteristic method. However, the problem of the stress state on the other two lines *AB* and *DE* (see Fig. (10.3)) is of the elliptic type which can not be solved by the characteristic method. Nevertheless, the yield condition of Eq.(10-15b) is very simple and can be solved directly by using equilibrium equation and the yield function.

10.4 Twin-shear Characteristics Field for Plane Stress Problems (Velocity Field)

10.4.1 Velocity Field corresponding to Eq. (10-15b)

Applying associate flow rule, the strain rate can be obtained from Eq. (10-15b) as

$$\xi_x = \xi_y, \zeta_{xy} = 0 \tag{10-31}$$

The velocity components are thus

$$\frac{\partial V_x}{\partial x} - \frac{\partial V_y}{\partial y} = 0 \tag{10-32a}$$

$$\frac{\partial V_x}{\partial x} + \frac{\partial V_y}{\partial y} = 0 \tag{10-32b}$$

The increments of the velocity components are

$$\frac{\partial V_x}{\partial x} dx + \frac{\partial V_x}{\partial y} dy = dV_x \tag{10-33a}$$

$$\frac{\partial V_y}{\partial x}dx+\frac{\partial V_y}{\partial y}dy=dV_y \tag{10-33b}$$

The determinant of coefficients of $\frac{\partial V_x}{\partial x},\frac{\partial V_x}{\partial y},\frac{\partial V_y}{\partial x},\frac{\partial V_y}{\partial y}$ in the above four equations equal to 0, which gives

$$\Delta=\begin{vmatrix}1 & 0 & 0 & -1\\ 0 & 1 & 1 & 0\\ dx & dy & 0 & 0\\ 0 & 0 & dx & dy\end{vmatrix}=0 \tag{10-34}$$

Then

$$(dx)^2+(dy)^2=0 \tag{10-35}$$

There is no real root for *dy/dx*. The equation is of the elliptic type and the characteristics do not exist.

10.4.2 Velocity Field corresponding Eq. (10-15a)

Corresponding to the yield condition of Eq. (10-15), it has

$$\zeta_x=\alpha\frac{\partial f}{\partial \sigma_x}=\alpha\left[\frac{s}{6}+\frac{1}{2}\frac{\sigma_x-\sigma_y}{\sqrt{(\sigma_x-\sigma_y)^2+4\tau^2{}_{xy}}}\right] \tag{10-36a}$$

$$\zeta_y=\alpha\frac{\partial f}{\partial \sigma_y}=\alpha\left[\frac{s}{6}-\frac{1}{2}\frac{\sigma_x-\sigma_y}{\sqrt{(\sigma_x-\sigma_y)^2+4\tau^2{}_{xy}}}\right] \tag{10-36b}$$

$$\zeta_{xy}=\alpha\frac{\partial f}{\partial \tau_{xy}}=\alpha-\frac{2\tau_{xy}}{\sqrt{(\sigma_x-\sigma_y)^2+4\tau^2{}_{xy}}} \tag{10-36c}$$

where f is the twin-shear yield function, α (>0) is a scalar factor. Combining Eq. (10-20) and Eq. (10-36) derives

$$\frac{\dfrac{\partial V_x}{\partial x}}{\dfrac{s}{6}+\dfrac{1}{2}\cos 2\varphi}=\frac{\dfrac{\partial V_y}{\partial y}}{\dfrac{s}{6}-\dfrac{1}{2}\cos 2\varphi}=\frac{\dfrac{\partial V_x}{\partial y}+\dfrac{\partial V_y}{\partial z}}{\sin 2\varphi}=\frac{\dfrac{\partial V_x}{\partial x}-\dfrac{\partial V_y}{\partial y}}{\cos 2\varphi} \tag{10-37}$$

Eq. (10-37) can be rewritten as

$$\left(\frac{s}{3}-\cos 2\varphi\right)\frac{\partial V_x}{\partial x}-\left(\frac{s}{3}+\cos 2\varphi\right)\frac{\partial V_y}{\partial y}=0 \tag{10-38a}$$

$$\cos 2\varphi\left(\frac{\partial V_x}{\partial y}+\frac{\partial V_y}{\partial x}\right)-\sin 2\varphi\left(\frac{\partial V_x}{\partial x}-\frac{\partial V_y}{\partial y}\right)=0 \tag{10-38b}$$

The determinant of the coefficient must vanish, which gives

$$\Delta=\begin{vmatrix} \frac{s}{3}-\cos 2\varphi & 0 & 0 & -\frac{s}{3}-\cos 2\varphi \\ -\sin 2\varphi & \cos 2\varphi & \cos 2\varphi & \sin 2\varphi \\ dx & dy & 0 & 0 \\ 0 & 0 & dx & dy \end{vmatrix}=0 \tag{10-39}$$

Then the following equation is obtained

$$\frac{dy}{dx}=\frac{-3\sin 2\varphi \pm 2\sqrt{2}}{s-3\cos 2\varphi}=\frac{-3\sin 2\varphi \pm 2\sqrt{2}s}{1-3s\cos 2\varphi} \tag{10-40}$$

Eq. (10-40) is the same with the characteristics of the stress field. Thus, the characteristics of the velocity field just coincide with those of the stress field for the twin-shear yield criterion, just like the Tresca criterion and the Huber-von Mises criterion.

Combining Eqs. (10-33), (10-38) and (10-39) gives

$$dV_x+dV_y\frac{dy}{dx}=0 \tag{10-41}$$

Corresponding velocity field can then be obtained through Eqs. (10-26) and (10-41)

$$dV_x+dV_y\tan(\varphi-\psi)=0\text{, along } \alpha \text{ line} \tag{10-42a}$$

$$dV_x+dV_y\tan(\varphi+\psi)=0\text{, along } \beta \text{ line} \tag{10-42b}$$

Considering an arbitrary point P at an plane curve L on the xy plane, if local coordinate system (t, n) is applied, the first equation of Eq. (10-38) can be rewritten as

$$\left(\frac{s}{3}-\cos 2\varphi\right)\frac{\partial V_t}{\partial t}-\left(\frac{s}{3}+\cos 2\varphi\right)\frac{\partial V_n}{\partial n} \tag{10-43}$$

where t and n are respectively in the tangential and normal directions of the curve L. Coefficients of $\partial V_n/\partial n$ of Eq. (10-43) equals to 0, then

$$\cos 2\varphi=-s/3 \tag{10-44}$$

From Eq. (10-43), it has

$$\zeta_t = \partial V_t / \partial t = 0 \tag{10-45}$$

10.5 Applications of the Twin-shear Characteristics Method

10.5.1 Velocity Discontinuous Line in Uniaxial Tension

In the plane stress problem, it is possible that velocity discontinuities exist not only in the tangential, but also in the normal direction of the slip line. A 'necking' deformation zone may be formed. Assuming that the neck has a small width b (in the limiting case b→0) and by a simple derivation, it can be seen that the line of velocity discontinuity is just one of the velocity characteristics, and the relative velocity v is perpendicular to the velocity characteristics of the other family (Kachanov 1971). In this example, $\sigma_y = \sigma_1$, $\sigma_x = 0$, $\tau_{xy} = 0$, so $s = 1$. There are two families of velocity characteristics, and the value of ψ (the angle between the velocity characteristics and the principal stress σ_1) is given by

$$\psi = \frac{1}{2}\cos^{-1}\left(-\frac{1}{3}\right) = 54^{\circ}44' \tag{10-46}$$

Therefore, the angle γ between the relative velocity v and principal stress σ_1 is

$$\gamma = 90^{\circ} - \psi = 35^{\circ}16' \tag{10-47}$$

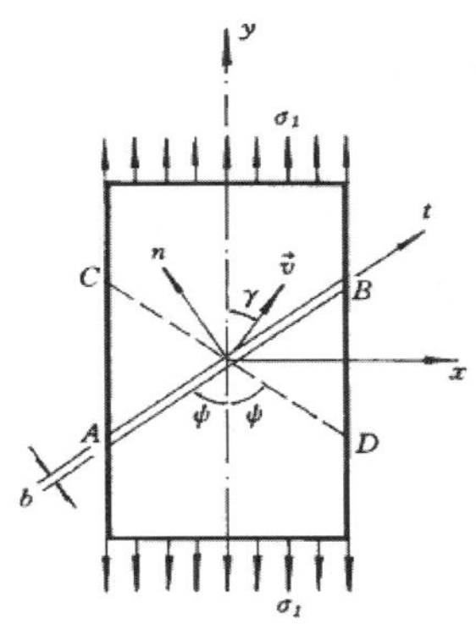

Fig. 10.4 Plate specimen in uniaxial tension

It can be seen that by using the twin-shear yield criterion, the solution of this problem is simple and straightforward.

The same value of ψ can also be obtained by using the Huber-von Mises yield criterion, but the process will be more cumbersome (Kachanov 1971). This value

($54^\circ 44'$) has been verified by experiment (Nadai 1950) and coincide with the result of Huber-von Mises (Yan 1988). The line of velocity discontinuity is known as Lüder's line.

Based on the Tresca yield criterion, ψ can take an arbitrary value between 45° and 90°; obviously, it is unreasonable.

10.5.2 Limit Load of an Infinite Thin Plate with a Circular Hole

An infinite thin plate, having a circular hole with radius a (Fig. 10.5), is subjected to a two-directional uniform tension q at infinity. Find the limit load q_s and the corresponding stress distribution.

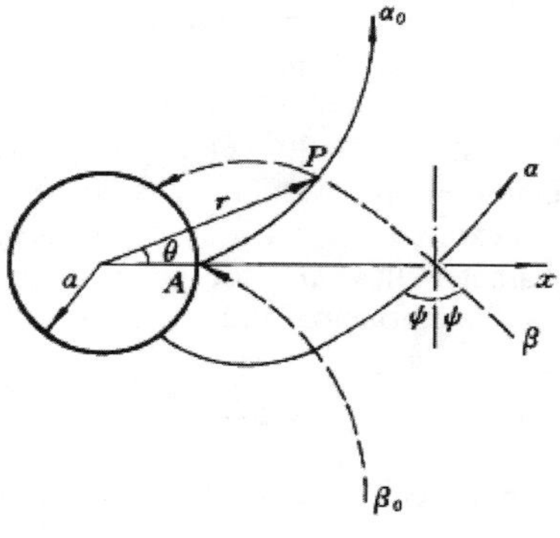

Fig. 10.5 An infinite plate with a circular hole

It is obvious that this is a plane stress problem. As the hole is free and the plate experiences two-directional uniform tension at infinity, on the edge of the hole there will be $\sigma_\theta > 0$ and $\sigma_r = 0$ and at infinity there will be $\sigma_\theta = \sigma_r > 0$. Thus $\sigma_\theta \geq \sigma_r \geq 0$ holds in the whole plate, and then s = +1. Near the hole, the stress satisfies Eq. (10-15a). And equations of the characteristics passing through the point A ($r = a$, $\theta = 0$) will be

$$\theta = \pm \frac{1}{\sqrt{2}} \ln \frac{r}{a}, \text{ '+' for } \alpha \text{ line, '−' for } \beta \text{ line} \tag{10-48}$$

Along AP, it has

$$\varphi + \sqrt{2} \ln \lambda = C_1 \tag{10-49}$$

From $\varphi = \theta + \pi/2$ and $\sigma_r \big|_{r=a} = 0$, we know $\lambda_A = 3/4$, thus, it has

$$\lambda = \frac{3}{4} \left(\frac{a}{r} \right)^{1/2} \tag{10-50}$$

The stress components are derived as follows

$$\sigma_\theta = 3k\left[1-\frac{1}{2}\left(\frac{a}{r}\right)^{1/2}\right] \tag{10-51a}$$

$$\sigma_r = 3k\left[1-\left(\frac{a}{r}\right)^{1/2}\right] \tag{10-51b}$$

With r increasing, λ will decrease. When r = 9a/4, λ = 1/2, which is the critical value for the hyperbola. When r > 9a/4, the equation is ellipse. Stress components can be derived from differential equation and yield criterion.

$$\sigma_r = \frac{3k}{2}\left[1-\frac{27}{16}\left(\frac{a}{r}\right)^{2}\right] \tag{10-52a}$$

$$\sigma_\theta = \frac{3k}{2}\left[1+\frac{27}{16}\left(\frac{a}{r}\right)^{2}\right] \tag{10-52b}$$

Thus, limit load is

$$q_s = \sigma_r\big|_{r\to\infty} = \frac{3k}{2} = \sigma_s \tag{10-53}$$

Expressions of corresponding stresses are

$$\sigma_r = \begin{cases} 2\sigma_s\left[1-\left(\frac{a}{r}\right)^{1/2}\right], a \le r \le \frac{9}{4}a \\ \sigma_s\left[1-\frac{27}{16}\left(\frac{a}{r}\right)^{2}\right], r > \frac{9}{4}a \end{cases} \tag{10-54a}$$

$$\sigma_\theta = \begin{cases} 2\sigma_s\left[1-\frac{1}{2}\left(\frac{a}{r}\right)^{1/2}\right], a \le r \le \frac{9}{4}a \\ \sigma_s\left[1+\frac{27}{16}\left(\frac{a}{r}\right)^{2}\right], r > \frac{9}{4}a \end{cases} \tag{10-54b}$$

For comparison, characteristics and stress distribution under the three yield criterion are shown in Figs. 10.6, 10.7 and 10.8. The same value of limit load $q_s = \sigma_s$ is obtained based on each of the aforementioned three yield criterion (Kachanov 1971).

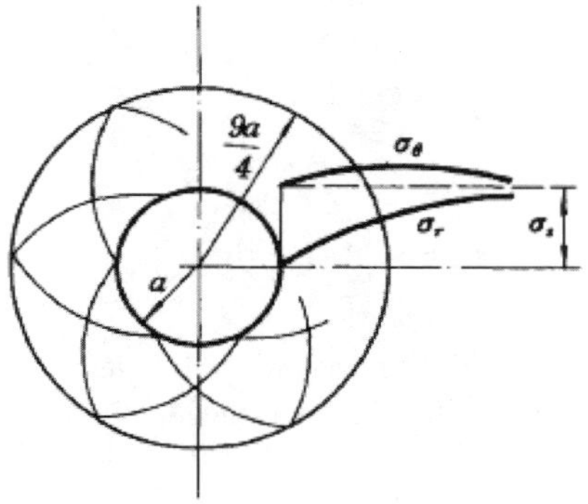

Fig. 10.6 Results of example 2: Twin-shear criterion

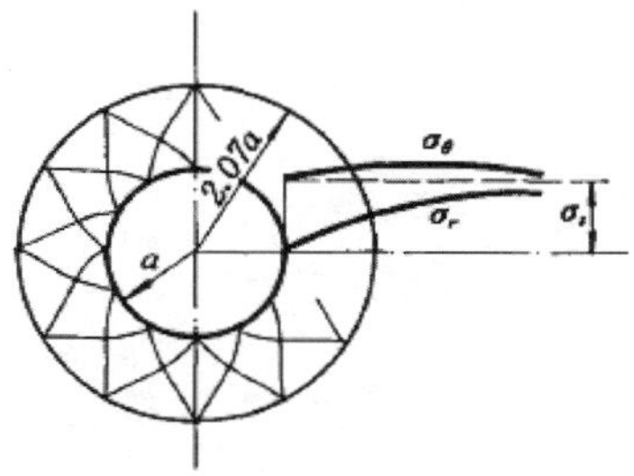

Fig. 10.7 Result of example 2: von-Mises criterion

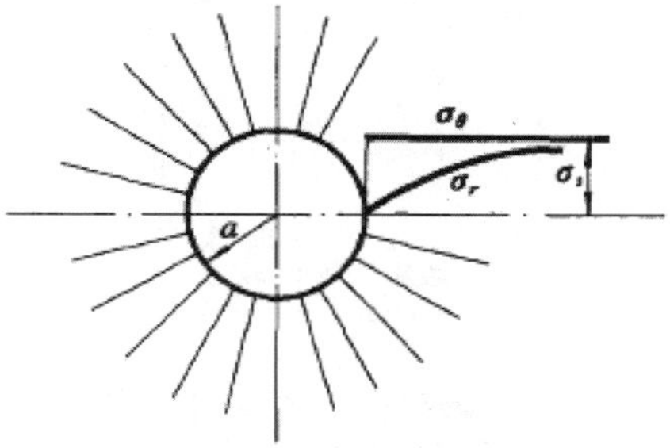

Fig. 10.8 Results of example 2: Tresca criterion

10.5.3 Thin Circular Ring Plate subjected to a Uniform Internal Pressure

A thin circular ring plate with inner radius a and outer $b = 2a$, subjected to a uniform internal pressure q (Fig. 10.9). Find the limit load and the corresponding stress distribution.

Through analysis similar to that in example 2, near the outer edge of the ring plate

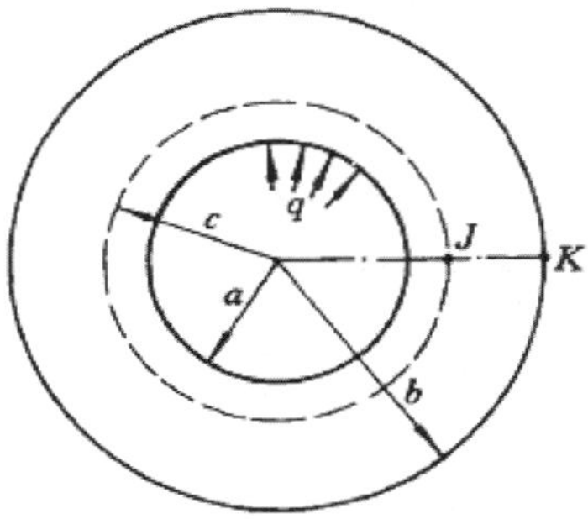

Fig. 10.9 Example 3

$$\lambda = \frac{3}{4}\left(\frac{2a}{r}\right)^{(1/2)} \tag{10-55}$$

$$\sigma_\theta = 3k\left[1 - \frac{1}{2}\left(\frac{2a}{r}\right)^{(1/2)}\right] \quad (c \le r \le b) \tag{10-56a}$$

$$\sigma_r = 3k\left[1 - \frac{1}{2}\left(\frac{2a}{r}\right)^{(1/2)}\right] \quad (c \le r \le b) \tag{10-56b}$$

With r decreasing, λ will increase, and $\lambda = 1$ is the limit of availability of Eq. (10-13). Take c to denote this radius; thus, from Eq. (10-10)

$$c = \frac{9}{8}a \tag{10-57}$$

In the region $a \le r \le (9/8)a$, $s = -1$, thus

$$\sigma_\theta = -(3 - 4\lambda)k \quad (a \le r \le c) \tag{10-58a}$$

$$\sigma_r = -(3 - 2\lambda)k \quad (a \le r \le c) \tag{10-58b}$$

$$\psi = \frac{1}{2}\cos^{-1}\frac{1}{3} = 35^\circ 16' \tag{10-59}$$

The equations of characteristics passing through the point $r = a,\ \theta = 0$, are

$$\theta = \pm\sqrt{2}\ln\frac{r}{a} \text{ ('+' for } \alpha \text{ line, '–' for } \beta \text{ line)} \tag{10-60}$$

Then,

$$-\frac{\pi}{2}-\sqrt{2}\ln 1=-\frac{\pi}{2}+\theta-\sqrt{2}\ln\lambda=-\frac{\pi}{2}+\sqrt{2}\ln\frac{r}{c}-\sqrt{2}\ln\lambda \tag{10-61}$$

Hence,

$$\lambda=\frac{r}{c}=\frac{8r}{9a} \tag{10-62}$$

Substituting Eq. (10-19) into Eq. (10-18)

$$\sigma_\theta=-3k\left(1-\frac{32}{27}\frac{r}{a}\right) \tag{10-63a}$$

$$\sigma_r=-3k\left(1-\frac{16}{27}\frac{r}{a}\right) \tag{10-63b}$$

Since $\left|\sigma_\theta+\sigma_r\right|=3k\left|2-(48/27)(r/a)\right|\le 3k$ in the region $a \le r \le (9/8)\,a$, the yield condition Eq. (10-15a) is always applicable for this region. Thus, the limit load is

$$q_s=\sigma_r\big|_{r=a}=1.222k=0.815\sigma_s$$

The characteristics and stress distribution are shown in Fig. 10.10.

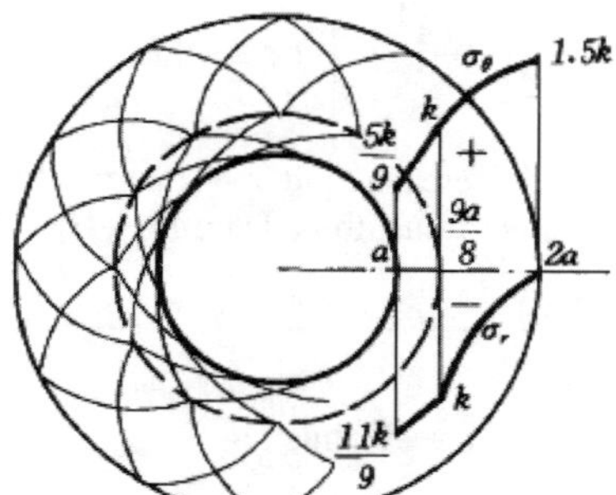

Fig. 10.10 Results of example 3

Based on the Huber-von Mises and Tresca yield criterion, the values of limit loads are $q_s = 1.330k = 0.768\sigma_s$ for Huber-von Mises yield criterion; and $q_s = 1.386k = 0.693\sigma_s$ for the Tresca yield criterion.

10.6 Comparison of These Different Methods

Yan and Bu (1993) have compared the three different characteristic methods based respectively on the twin-shear yield criterion, the Tresca criterion, and the Huber-von Mises criterion. The advantages and disadvantages of these methods are shown in Table 10.1 and Table 10.2.

Through the comparison, the following conclusions can be drawn:

1. It would be simpler to derive the characteristics based on the Tresca criterion and the twin-shear criterion than the Huber-von Mises criterion.
2. The twin-shear yield criterion has the advantage in describing the velocity discontinuity for the unixial tension example.
3. It is suggested to use the Tresca criterion if the experimental data of a certain material offset the Huber-von Mises ellipse toward the Tresca polygon. If the experimental data locates outside the Huber-von Mises ellipse, the twin-shear yield criterion is suggested. Thus, application of the Huber-von Mises criterion can be avoided.

Table 10.1. Comparison of the three yield criterion

Criterion	Range of hyperbolic type	Solution for other ranges	Example of uniaxial tension	
			Equation type	Result
Tresca	$\frac{\sigma_1}{\sigma_2} < 0$	Directly, parabolic characteristics method	Parabolic	$\psi = 45°$
von Mises	$\frac{\sigma_1}{\sigma_2} < \frac{1}{2}, \frac{\sigma_1}{\sigma_2} > 2$	Cumbersome	Hyperbolic	$\psi = 54°44'$
Twin-shear	$\frac{\sigma_1}{\sigma_2} < \frac{1}{2}, \frac{\sigma_1}{\sigma_2} > 2$	Directly	Hyperbolic	$\psi = 54°44'$

Table 10.2. Comparison of the infinite plate with a hole

Criterion	Uniform tension		Uniform compression	
	Equation type	Complexity	Equation type	Complexity
Tresca	Parabolic	Very simple	Hyperbolic	Simple
Huber-von Mises	Hyperbolic, elliptic	Cumbersome	Hyperbolic	Cumbersome
Twin-shear	Hyperbolic, elliptic	Simple	Hyperbolic	Simple

Summary

The methods of characteristics for plane stress problems based on the Huber-von Mises and the Tresca criteria can be found in the literature of Kachanov (1971), Martin (1973), Yan (1988) and Panoyotomakos (1999). A new characteristics method for plane stress problems was established based on the twin-shear yield criterion by Yan and Bu in 1993 and 1996. The three different characteristic methods based respectively on the twin-shear yield criterion, the Tresca criterion, and the Huber-von Mises criterion are described in this chapter. The advantages and disadvantages of these methods are reviewed.

These three methods can be unified to a unified characteristics field theory for plane stress problems by using the unified strength theory (Yu, Zhang, Li 1999). The unified characteristics field theory has a unified method to obtain a series of solutions. The results of the three characteristics methods based on the Tresca criterion, the Huber-von Mises criterion and the twin-shear yield criterion are special cases of the unified characteristics field theory. The unified characteristics field theory for plane stress problems will be described in next chapter.

Problems

Problem 10.1.
Compare the characteristics method for plane strain and plane stress problems if the Tresca criterion is used.

Problem 10.2.
Compare the characteristics method for plane strain and plane stress problems if the Huber-von Mises criterion is used.

Problem 10.3.
Compare the characteristics method for plane strain and plane stress problems if the twin-shear yield criterion is used.

Problem 10.4.
The extension of a strip with a sufficiently large circular hole for non-SD material was shown in Fig. P10.1 (chapter 10). Can you obtain another study on this subject in plane stress state using the twin-shear characteristics line field for non-SD material ($\alpha = 1$). Comparing the results between plane strain and plane stress.

Problem 10.5.
The extension of a strip with a sufficiently large circular hole for non-SD material was shown in Fig. P10.1. Can you obtain a more complete study on this subject in plane stress state using the twin-shear characteristics line field for SD materials ($\alpha \neq 1$). Comparing the results between plane strain and plane stress.

Problem 10.6.
The extension of a strip with ideal (infinitely thin) cuts for non-SD material was shown in Fig. P10.2 last chapter. Can you obtain a more complete study on this subject in plane stress state using the twin-shear characteristics line field for non-SD material ($\alpha = 1$). Comparing the results between plane strain and plane stress.

Problem 10.7.
Can you obtain a more complete study on the extension of a strip with ideal cuts (see: Fig. P10.2 in chapter 10) in plane stress state using the twin-shear characteristics line field for SD materials ($\alpha \neq 1$). Comparing the results between plane strain and plane stress.

Problem 10.8.
The extension of a strip with angular notches for non-SD material was shown in Fig. P10.3 (chapter 10). Can you obtain a more complete study on this subject in plane stress state using the twin-shear characteristics line field ($\alpha = 1$). Comparing the results between plane strain and plane stress.

Problem 10.9.
Can you obtain a more complete study on this subject mentioned above in plane stress state using the twin-shear characteristics line field for SD materials ($\alpha \neq 1$). Comparing the results between plane strain and plane stress.

Problem 10.10.
The extension of a strip with circular base for non-SD material was shown in Fig. P10.4 (in Chapter 10). Can you obtain a more complete study on this subject in plane stress state using the twin-shear characteristics line field ($\alpha = 1$). Comparing the results between plane strain and plane stress. Comparing the results between plane strain and plane stress.

Problem 10.11.
The extension of a strip with circular base for non-SD material was shown in Fig. P10.4 (in chapter 10). Can you obtain a more complete study on this subject using the twin-shear characteristics line field for SD materials ($\alpha \neq 1$). Comparing the results between plane strain and plane stress.

Problem 10.12.

The bending of a strip with a one-sided notch for non-SD material was given by Kachanov (1971) as shown in Fig. P10.1. The Tresca yield criterion was used in the studies. Can you obtain a more complete study on this subject in plane stress state using the twin-shear characteristics line field ($\alpha = 1$).

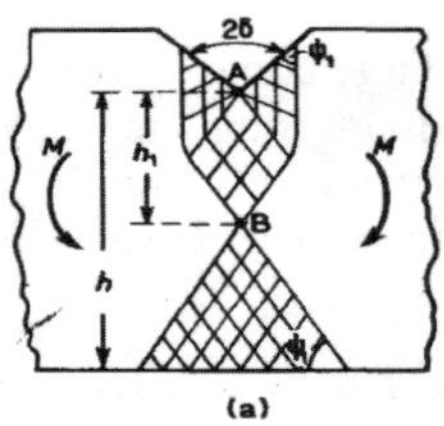

P10.1. Bending of a strip with a one-sided notch

Problem 10.13.

Can you obtain a more complete study on the above problem using the twin-shear characteristics line field for SD materials ($\alpha \neq 1$).

Problem 10.14.

The indentation of a strip of finite thickness with flat indenter for non-SD material was given by Hill (1950) and exactly by Johnson and Woo (1956) as shown in Fig. P10.2. The Tresca yield criterion was used in the studies. Can you obtain a more complete study on this subject in plane stress state using the twin-shear characteristics line field ($\alpha = 1$).

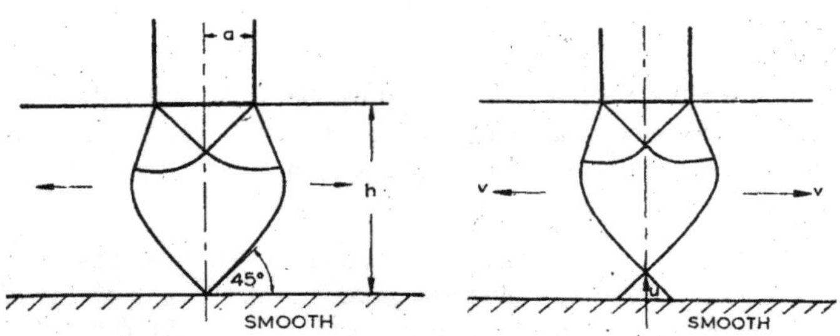

Fig. P10.2. Indentation of a strip of finite thickness with flat indenter

Problem 10.15.

Can you obtain a more complete study on above problem using the twin-shear characteristics line field for SD materials ($\alpha \neq 1$).

Problem 10.16.

The slip line field of indenting with two dies and by three equal size dies spaced at 100 to each other were discussed by Johnson and Mellor (1962) for non-SD material as shown in Fig. P10.3. The Tresca yield criterion was used in the studies.

Can you obtain a more complete study on this subject in plane stress state using the twin-shear characteristics line field ($\alpha = 1$).

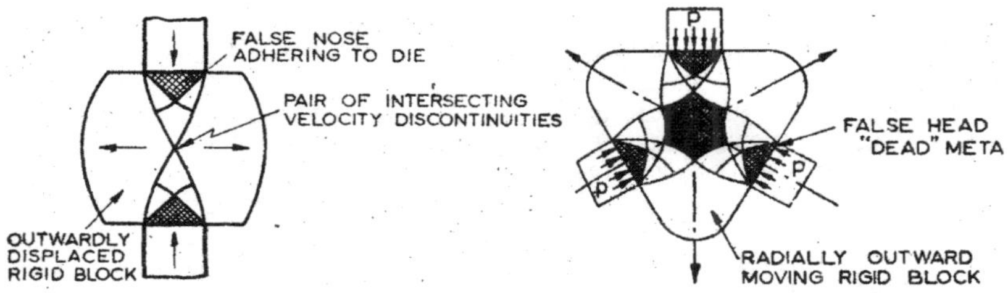

Fig. P10.3. Indentation by two and three dies

Problem 10.17.
The slip line field of indenting with two dies and by three equal size dies spaced at 100 to each other were discussed by Johnson and Mellor (1962) for non-SD material as shown in Fig. P10.3. The Tresca yield criterion was used in the studies. Can you obtain a more complete study on above subject using the twin-shear characteristics line field for SD materials ($\alpha \neq 1$).

Problem 10.18.
The cutting with opposed wedge-shaped indenters was discussed by Johnson and Mellor (1962) for non-SD material as shown in Fig. P10.4. The Tresca yield criterion was used in the studies. Can you obtain a more complete study on this subject in plane stress state using the twin-shear characteristics line field ($\alpha = 1$).

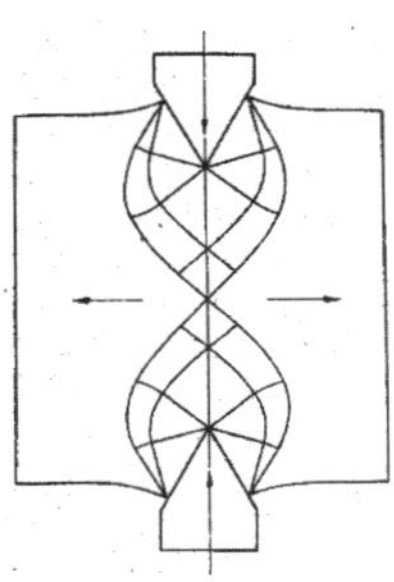

Fig. P10.4. Cutting with opposed wedge-shaped indenters

Problem 10.19.
Can you obtain a more complete study on this subject using the twin-shear strength theory for SD materials ($\alpha \neq 1$) using the twin-shear characteristics line field. Comparing the different results between plane strain and plane stress.

References and Bibliography

Kachanov LM (1971) Foundations of the Theory of Plasticity. North-Holland Pub., Amsterdam, London.

Martin JB (1973) Plasticity: Fundamentals and General Results, The MIT Press.

Nadai A (1950) Theory of flow and fracture of solids. McGraw-Hill Book Co., Inc., New York

Panayotounakos D(1999) Ad hoc exact solutions for the stress and velocity fields in rigid perfectly plastic materials under plane-stress conditions, Int. J. Non-Linear Mech. 34: 71–84.

Runesson K, Ottosen N and Peric D (1991) Discontinuous bifurcations of elastic-plastic solutions at plane stress and plane strain, *Int. J. Plasticity* 7: 99–101.

Shield RT (1954), Stress and velocity fields in soil mechanics. *J. Math. Phys*., 33(2) 144–156.

Yan ZD (1988) *Plasticity*. Tijing University Press, Tijing (in Chinese).

Yan ZD and Bu XM (1993) The method of characteristics for solving the plane stress problem of ideal rigid-plastic body on the basis of Twin shear stress yield criterion. In: *Advances in Engineering Plasticity and its Applications,* Lee WB ed., Elsevier Science Publishers, 295–302.

Yan ZD and Bu XM (1993) On the characterist9ics line method for plane stress problem based on the three kinds of yield criteria. Engineering Mechanics (in Chinese), 10 (Suppl.):89–96.

Yan ZD and Bu XM (1996) An Effective Characteristics Method for Plastic Plane Stress Problems, *J. of Eng. Mech.* ASCE, 102, 502–506.

Yu MH (1961) General behaviour of isotropic yield function. Xi'an: Research Report of Xi'an Jiaotong University.

Yu MH (1983) Twin shear stress yield criterion. Int. J. Mech. Sci. 25, 71–74.

Yu MH, He LN and Song LY (1985) Twin shear strength theory and its generalization. Scientia Sinica (Science in China), A28, 1174–1183.

Yu MH, Zhang YQ, Yang SY (1998) Another important generalization of the unified strength theory. *J. of Xi'an Jiaotong University.* 32, 108–110 (in Chinese, English Abstract).

Yu MH, Zhang YQ, Li JC (1999) The unified characteristics theory for plastic plane stress problems. *J. of Xi'an Jiaotong University.* 33(4): 1–4 (in Chinese, English Abstract).

11 Unified Characteristics Field Theory for Plane Stress Problem

11.1 Introduction

The characteristics field for plane stress problems based on the twin-shear yield criterion and the twin-shear strength criterion has been described in Chapter 10. They are corresponding to the special cases of b=1 in the Yu unified strength theory.

The unified characteristics field theory for solving the plastic plane stress problem has been developed for ideal rigid-plastic bodies based on the Yu unified strength theory by Yu and Zhang in 1998 and 1999. The characteristic methods in terms of the Tresca criterion, the Huber-von Mises criterion, the Mohr-Coulomb criterion, twin-shear yield criterion (Yu 1961) and the generalized twin-shear criterion (Yu 1985) are the special cases or linear approximation (Huber-von Mises) of the proposed theory. Besides, a series of new characteristics methods can be obtained if the unified strength theory parameter b in the Yu unified strength theory takes different value.

The unified characteristics field theory can consider the strength-differential effect (SD effect) and the effect of intermediate principal stress. It can be used for a wide variety of materials. This chapter will give detail derivation of the unified characteristics lines field theory for plane stress problem. The theory can be used conveniently in all sorts of plane stress problems for idea-plastic materials.

11.2 Unified Yield Function in Plane Stress State

The Yu unified strength theory (unified yield function) has been described in Chap. 4 in detail. It can be expressed in terms of the principal stresses as follows

$$F = \sigma_1 - \frac{\alpha}{1+b}(b\sigma_2 + \sigma_3) = \sigma_t \text{ , when } \sigma_2 \le \frac{\sigma_1 + \alpha\sigma_3}{1+\alpha} \tag{11-1a}$$

$$F' = \frac{1}{1+b}(\sigma_1 + b\sigma_2) - \alpha\sigma_3 = \sigma_t \text{ , when } \sigma_2 \ge \frac{\sigma_1 + \alpha\sigma_3}{1+\alpha} \tag{11-1b}$$

It should be noted that the parameter b plays an important role in the unified yield function. It builds a bridge among different strength criteria. It is this parameter that distinguishes one criterion from another. On the other hand, the scope of application of each criterion is also represented by this parameter. Hence, the Yu unified strength theory is not a single strength criterion but a theoretical system including a series of regular strength criteria, and it can be applied to more than one kind of material. In practice, when basic material parameters are obtained by experiments. Whenever parameter b is obtained, the yield criterion for this sort of material is determined and the application is possible. Consequently, b can be regarded as a parameter by which the suitable yield criterion for material of interest can be determined.

The Yu unified strength theory (UST) is a series of piecewise linear yield criteria on the π-plane as shown in Fig. 11.1. The exact form of expression depends on the choice of parameter b. With different choices of parameter b, the UST can be simplified to the Tresca ($\alpha = 1$ and $b = 0$), the linear approximations of the Huber-von Mises ($\alpha = 1$ and $b = 1/2$), the Mohr-Coulomb ($0 < \alpha < 1$ and $b = 0$), the TS ($\alpha = 1$ and $b = 1$), the GTS ($0 < \alpha < 1$ and $b = 1$) and a series of new strength criteria. In the stress space, the lower and upper bounds of the yield surfaces on the π-plane are special cases of the UST, i.e., $b = 0$ ($\alpha = 1$ for the Tresca or $0 < \alpha < 1$ for the Mohr-Coulomb) and $b = 1$ ($\alpha = 1$ for the TS or $0 < \alpha < 1$ for the GTS), respectively. When the parameter b varies between 0 and 1, a series of yield surfaces between the two limiting surfaces can be obtained. Various limit loci of the unified strength theory in the plane stress state are shown in Fig. 11.1. The unified yield criterion, the Mohr–Coulomb strength theory, the twin-shear strength theory and a series of new failure criteria can be obtained from the Yu unified strength theory.

The Yu unified strength theory in plane stress states can be expressed in terms of the three principal stresses as follows:

$$\begin{aligned}
&\sigma_1 - \frac{\alpha b}{1+b}\sigma_2 = \sigma_t; \qquad \frac{1}{1+b}(\sigma_1 + b\sigma_2) = \sigma_t \\
&\sigma_1 - \frac{\alpha}{1+b}\sigma_2 = \sigma_t; \qquad \frac{1}{1+b}\sigma_1 - \alpha\sigma_2 = \sigma_t \\
&\sigma_2 - \frac{\alpha}{1+b}\sigma_1 = \sigma_t; \qquad \frac{1}{1+b}(\sigma_2 + b\sigma_1) = \sigma_t \\
&\sigma_2 - \frac{\alpha}{1+b}\sigma_1 = \sigma_t; \qquad \frac{1}{1+b}\sigma_2 - \alpha\sigma_1 = \sigma_t \\
&\sigma_2 - \frac{\alpha}{1+b}\sigma_1 = \sigma_t; \qquad \frac{1}{1+b}\sigma_2 - \alpha\sigma_1 = \sigma_t \\
&-\frac{\alpha}{1+b}(b\sigma_2 + \sigma_1) = \sigma_t; \qquad \frac{1}{1+b}b\sigma_2 - \alpha\sigma_1 = \sigma_t
\end{aligned} \tag{11-2}$$

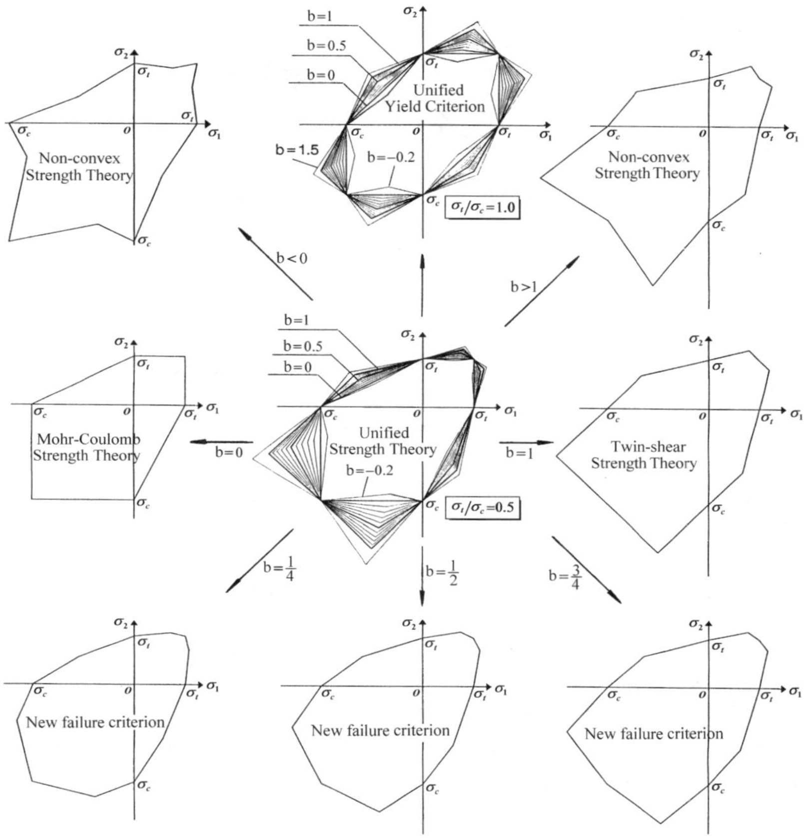

Fig. 11.1 Variation of the Yu unified strength theory in the plane stress state

There are three stress components σ_x, σ_y, and τ_{xy} in plane stress states: Assuming

$$A = \frac{1}{2}\left(\sigma_x + \sigma_y\right), \tag{11-3}$$

$$B = \sqrt{\left(\frac{\sigma_x - \sigma_y}{2}\right)^2 + \tau_{xy}{}^2}, \tag{11-4}$$

the unified characteristics for plane stress state can thus be derived in a unified form.

The yield loci on the plane stresses (σ_1, σ_2) can be illustrated as Fig.11.2.

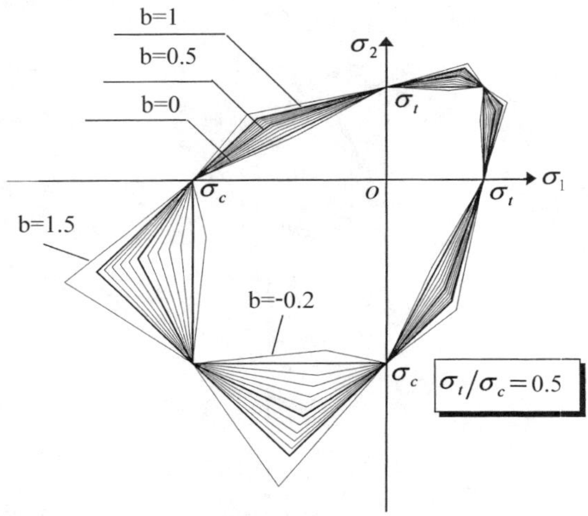

Fig. 11.2 Variation of the yield loci of the Yu unified strength theory in plane stress state

11.3 Characteristics Field for Plane Stress problems

For the cases of plane stress and plane strain, σ_{I} and σ_{III} are assumed to be two principal stresses in the xy-plane and $\sigma_{\mathrm{I}} \geq \sigma_{\mathrm{III}}$, and σ_{II} is assumed to be the out-of-plane principal stress. Assuming $A=(\sigma_x+\sigma_y)/2=(\sigma_{\mathrm{I}}+\sigma_{\mathrm{III}})/2$ and $B=\sqrt{\left[\left(\sigma_x-\sigma_y\right)/2\right]^2+\tau_{xy}{}^2}=\left(\sigma_{\mathrm{I}}-\sigma_{\mathrm{III}}\right)/2$, the Yu unified strength theory in plane state can be expressed as

$$F=mA+nB=\sigma_t \tag{11-5}$$

where m and n are material parameters.

In the case of plane stress, the out-of-plane principal stress σ_{II} vanishes. Then, there are three cases to be distinguished in the state of plane stress.

Case A. When $\sigma_{\mathrm{I}} \geq \sigma_{\mathrm{III}} \geq 0$, it has $\sigma_1=\sigma_{\mathrm{I}}$, $\sigma_2=\sigma_{\mathrm{III}}$ and $\sigma_3=0$. From Eq. (11-2), we can obtain

$$m=\frac{1+b-\alpha b}{1+b}, n=\frac{1+b+\alpha b}{1+b}, \text{ when } B \leq A \leq \frac{2+\alpha}{\alpha} B$$
$$m=1, n=\frac{1-b}{1+b}, \text{ when } A \geq \frac{2+\alpha}{\alpha} B \qquad (11\text{-}6a)$$

Case B. When $\sigma_{\mathrm{I}} \geq 0 \geq \sigma_{\mathrm{III}}$, it has $\sigma_1 = \sigma_{\mathrm{I}}$, $\sigma_2 = 0$ and $\sigma_3 = \sigma_{\mathrm{III}}$. From Eq. (11-2), it can be derived that

$$m=\frac{1+b-\alpha b}{1+b}, n=\frac{1+b+\alpha b}{1+b}, \text{ when } B \leq A \leq \frac{2+\alpha}{\alpha} B$$
$$m=\frac{1-\alpha-\alpha b}{1+b}, n=\frac{1+\alpha+\alpha b}{1+b}, \text{ when } -B \leq A \leq \frac{\alpha-1}{1+\alpha} B \qquad (11\text{-}6b)$$

Case C. When $0 \geq \sigma_{\mathrm{I}} \geq \sigma_{\mathrm{III}}$, it has $\sigma_1 = 0$, $\sigma_2 = \sigma_{\mathrm{I}}$ and $\sigma_3 = \sigma_{\mathrm{III}}$. From Eq. (11-2), we can obtain that

$$m=-\alpha, n=\alpha\frac{1-b}{1+b}, \text{ when } A \leq -(1+2\alpha)B$$
$$m=\frac{b-\alpha-\alpha b}{1+b}, n=\frac{b+\alpha+\alpha b}{1+b}, \text{ when } -(1+2\alpha)B \leq A \leq -B \qquad (11\text{-}6c)$$

11.3.1 Characteristics of Stress Field

Neglecting body force, the equations of equilibrium for plane stress problems are

$$\frac{\partial \sigma_x}{\partial x}+\frac{\partial \tau_{xy}}{\partial y}=0, \qquad \frac{\partial \tau_{xy}}{\partial x}+\frac{\partial \sigma_y}{\partial y}=0 \qquad (11\text{-}7)$$

The characteristics for the plane stress problems based on the Yu unified strength theory can be discussed in the following cases.

1. $mn \neq 0$

If a variable λ is introduced, the stress components satisfying the Yu unified strength theory can be expressed as

$$\sigma_x = \frac{1}{nm}\sigma_t\left[n(1-\lambda)+m\lambda\cos 2\varphi]\right] \tag{11-8a}$$

$$\sigma_y = \frac{1}{nm}\sigma_t\left[n(1-\lambda)-m\lambda\cos 2\varphi]\right] \tag{11-8b}$$

$$\tau_{xy} = \frac{\sigma_t}{n}\lambda\sin 2\varphi \tag{11-8c}$$

where φ is angle of the larger one of the principal stresses σ_1 and σ_2 from the x-axis with counterclockwise positive. When $\lambda \geq 0$, the angle can be determined with respect to different yield conditions as given in Eq. (11-2)

Substituting Eq. (11-8) into the equilibrium equation Eq. (11-7) gives,

$$n\frac{\partial\lambda}{\partial x}-m\cos 2\varphi\frac{\partial\lambda}{\partial x}+2m\lambda\sin 2\varphi\frac{\partial\varphi}{\partial x}-m\sin 2\varphi\frac{\partial\lambda}{\partial y}-2m\lambda\cos 2\varphi\frac{\partial\varphi}{\partial y}=0 \tag{11-9}$$

$$m\sin 2\varphi\frac{\partial\lambda}{\partial x}+2m\lambda\cos 2\varphi\frac{\partial\varphi}{\partial x}-n\frac{\partial\lambda}{\partial y}-m\lambda\cos 2\varphi\frac{\partial\lambda}{\partial y}+2m\lambda\sin 2\varphi\frac{\partial\varphi}{\partial y}=0 \tag{11-10}$$

With the following two supplementary incremental equations,

$$\frac{\partial\lambda}{\partial x}dx+\frac{\partial\lambda}{\partial y}dy=d\lambda;\qquad \frac{\partial\varphi}{\partial x}dx+\frac{\partial\varphi}{\partial y}dy=d\varphi \tag{11-11}$$

the characteristics of the plane stress problem is then derived as,

$$\frac{dy}{dx}=tg(\varphi\mp\psi) \tag{11-12}$$

From Eq. (11-12), the two families of the characteristics have angles $\mp\psi$ with the larger one of the principal stresses σ_1 and σ_2. The positive sign corresponds to the α family of the characteristic lines, vice versa; the negative sign represents the β family. The angle ψ is determined by $\cos 2\psi = -m/n$. It can be further derived that

$$2m\varphi+(n^2-m^2)^{1/2}\ln\lambda=\text{Constant}\quad(\text{along }\alpha\text{ line}) \tag{11-13a}$$

$$2m\varphi-(n^2-m^2)^{1/2}\ln\lambda=\text{Constant}\quad(\text{along }\beta\text{ line}) \tag{11-13b}$$

2. $mn=0$

When $m=0, n\neq 0$, Eq. (11 -5) can be rewritten as

$$F = nB = n\left[\left(\frac{\sigma_x - \sigma_y}{2}\right)^2 + \tau_{xy}^2\right]^{1/2} \tag{11-14}$$

Introducing another variable ω, the stresses satisfying the yield condition Eq. (11-14) are

$$\sigma_x = \sigma_t\omega + \frac{\sigma_t}{n}\cos 2\varphi \tag{11-15a}$$

$$\sigma_y = \sigma_t\omega - \frac{\sigma_t}{n}\cos 2\varphi \tag{11-15b}$$

$$\tau_{xy} = \frac{\sigma_t}{n}\sin 2\varphi \tag{11-15c}$$

Substituting Eq. (11-15) into Eq. (11.7), it is derived,

$$n\frac{\partial\omega}{\partial x} - 2\sin 2\varphi\frac{\partial\varphi}{\partial y} + 2\cos 2\varphi\frac{\partial\varphi}{\partial y} = 0 \tag{11-16a}$$

$$n\frac{\partial\omega}{\partial y} - 2\cos 2\varphi\frac{\partial\varphi}{\partial x} + 2\sin 2\varphi\frac{\partial\varphi}{\partial y} = 0 \tag{11-16b}$$

Through similar derivation of the case $nm\neq 0$, the characteristics can be given by

$$\frac{dy}{dx} = tg\left(\varphi \pm \frac{\pi}{4}\right) \tag{11-17b}$$

And

$$n\omega + 2\varphi = \text{Constant (along } \alpha \text{ line)} \tag{11-18a}$$

$$n\omega - 2\varphi = \text{Constant (along } \beta \text{ line)} \tag{11-18b}$$

It can be seen from Eq. (11-17), the two family characteristic lines are orthogonal, and both have an angle $4/\pi$ with the principal stresses.

When $m\neq 0$, $n=0$, the yield condition can be written as,

$$F = mA = \frac{m}{2}(\sigma_x + \sigma_y) = \sigma_t \tag{11-19}$$

Introducing an variable λ, the stress components satisfying the yield condition Eq. (11-19) are

$$\sigma_x = \frac{1}{m}\sigma_t + \lambda\sigma_t \cos 2\varphi \tag{11-20a}$$

$$\sigma_y = \frac{1}{m}\sigma_t - \lambda\sigma_t \cos 2\varphi \tag{11-20b}$$

$$\tau_{xy} = \lambda\sigma_t \sin 2\varphi \tag{11-20c}$$

which derives,

$$(dx)^2 + (dy)^2 = 0 \tag{11-21}$$

It is known that dy/dx in Eq. (11-21) has no real root. However, due to the simple expression of Eq. (11-19), the plane stress problem can be solved based on the equilibrium equation (11-7) and the yield condition (11-19) directly.

11.3.2 Characteristics of Velocity Field

Again the analysis is categorized into two cases, i.e. $mn \neq 0$ and $mn = 0$.

1. $mn \neq 0$

Based on the yield condition Eq. (11-5) and its associated flow rule, it has

$$\zeta_x = \eta\frac{\partial F}{\partial \sigma_x} = \frac{\eta}{2}\left[m + n\frac{\sigma_x - \sigma_y}{((\sigma_x - \sigma_y)^2 + 4\tau_{xy}^2)^{1/2}}\right] \tag{11-22a}$$

$$\zeta_y = \eta\frac{\partial F}{\partial \sigma_y} = \frac{\eta}{2}\left[m - n\frac{\sigma_x - \sigma_y}{((\sigma_x - \sigma_y)^2 + 4\tau_{xy}^2)^{1/2}}\right] \tag{11-22b}$$

$$\zeta_{xy} = \eta\frac{\partial F}{\partial \tau_{xy}} = 2\eta n\frac{\tau_{xy}}{((\sigma_x - \sigma_y)^2 + 4\tau_{xy}^2)^{1/2}} \tag{11-22c}$$

in which η is a non-negative ratio.

Substituting Eq. (11-8) into Eq. (11-22), and representing the left-hand-side strain components by velocity variables V_x and V_y, it is derived,

$$(m - n\cos 2\varphi)\frac{\partial V_x}{\partial x} - (m + n\cos 2\varphi)\frac{\partial V_y}{\partial y} = 0 \tag{11-23a}$$

$$\cos 2\varphi\left(\frac{\partial V_x}{\partial y}+\frac{\partial V_y}{\partial x}\right)-\sin 2\varphi\left(\frac{\partial V_x}{\partial y}-\frac{\partial V_y}{\partial x}\right)=0 \tag{11-23b}$$

Besides, the differentiation of the velocity variables gives,

$$\frac{\partial V_x}{\partial x}dx+\frac{\partial V_x}{\partial y}dy=dV_x \tag{11-24a}$$

$$\frac{\partial V_y}{\partial x}dx+\frac{\partial V_y}{\partial y}dy=dV_y \tag{11-24b}$$

The characteristics can then be deduced as

$$\frac{dy}{dx}=\frac{-n\sin 2\varphi \pm (n^2-m^2)^{1/2}}{m-n\cos 2\varphi} \tag{11-25}$$

Eq. (11-25) has exactly same form with Eq. (11-12). It can be concluded that the characteristics of the velocity field are consistent with that of the stress field. The two family velocity characteristics are further derived as

$$dV_x+dV_y\tan(\varphi-\psi)=0 \quad \text{(along } \alpha \text{ line)} \tag{11-26a}$$

$$dV_x+dV_y\tan(\varphi+\psi)=0 \quad \text{(along } \beta \text{ line)} \tag{11-26b}$$

2. $mn=0$

It can be categorized into the following two cases:

(1) when m=0, n≠0, based on the yield condition Eq. (11-14) and the associated flow rule, the plastic strain components are given by

$$\zeta_x=\eta\frac{\partial F}{\partial \sigma_x}=\frac{\eta n}{2}\cos 2\varphi \tag{11-27a}$$

$$\zeta_y=\eta\frac{\partial F}{\partial \sigma_y}=\frac{-\eta n}{2}\cos 2\varphi \tag{11-27b}$$

$$\zeta_{xy}=\eta n\sin 2\varphi \tag{11-27c}$$

Substituting the left-hand-side plastic strains by the velocity variables V_x and V_y, the characteristic equations are derived as,

$$\frac{dy}{dx}=tg\left(\varphi\pm\frac{\pi}{4}\right) \tag{11-27c}$$

Eq. (11-28) is consistent with the characteristic equations derived for the stress field. The two family characteristic lines are further deduced as

$$dV_x + dV_y \tan(\varphi + \pi/4) = 0 \quad \text{(along } \alpha \text{ line)} \tag{11-28a}$$

$$dV_x + dV_y \tan(\varphi - \pi/4) = 0 \quad \text{(along } \beta \text{ line)} \tag{11-28b}$$

(2) when $n = 0$, $m \neq 0$, the differential equation is an elliptic type, which has no real solution for characteristic lines.

The above developed characteristics for plane stress problems are suitable to different materials. They degrade to the twin-shear characteristics derived in Chap. 10 if the parameter b is set to 1. The traditional characteristic theory based on the Tresca criterion, the Mohr-Coulomb criterion, the von Mises criterion can be approximated. If different parameter b is adopted, a series of the characteristic lines can be obtained corresponding to materials with or without the SD effect.

11.4 Applications of the Unified Characteristics Field for Plane Stress Problems

On the basis of the unified characteristics of stress and velocity fields, many plastic plane problems can be studied.

Example 1

An infinite thin plate, having a circular hole with radius a (Fig. 11.3), is subjected to a two-directional uniform tension q at infinity. Find the limit load q_s and the corresponding stress distribution.

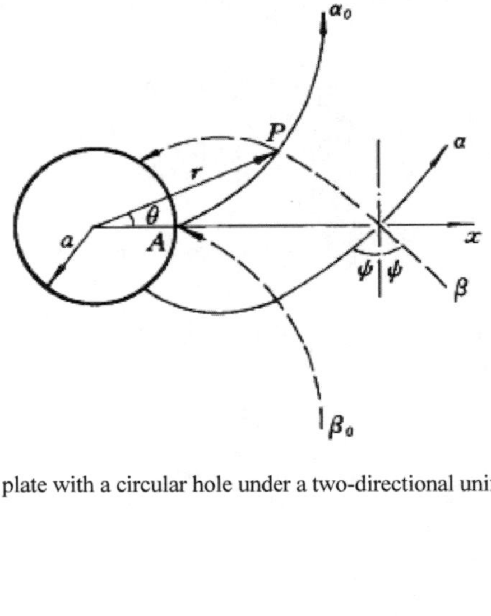

Fig. 11.3 An infinite thin plate with a circular hole under a two-directional uniform tension at infinity

The present example differs from that discussed in Chap. 10. The unified characteristic field theory is a straightforward extension if the parameter b is incorporated into the twin-shear yield and strength criteria. The current result is thus applicable to various engineering materials. Similarly, as the hole is free and the plate experiences two-directional uniform tension at infinity, on the edge of the hole there will be $\sigma_\theta > 0$ and $\sigma_r = 0$ and at infinity there will be $\sigma_\theta = \sigma_r > 0$. Thus $\sigma_\theta \geq \sigma_r \geq 0$ holds in the whole plate.

For the stress state, the Yu unified strength theory has the following form,

$$F = mA + nB = \sigma_t, m = \frac{1+b-\alpha b}{1+b}, n = \frac{1+b+\alpha b}{1=b}, \text{ when } B \leq A \leq \frac{2+\alpha}{\alpha} B \quad (11\text{-}29a)$$

$$F = mA + nB = \sigma_t, m = 1, n = \frac{1+b}{1=b}, \text{ when } B \leq A \leq \frac{2+\alpha}{\alpha} B \quad (11\text{-}29b)$$

in which $A = (\sigma_x + \sigma_y)/2$, $B = \left[\left[(\sigma_x - \sigma_y)/2\right]^2 + \tau_{xy}^2\right]^{1/2}$. Near the inner edge, combing the above yield condition and Eq. (11-7) gives

$$\sigma_\theta = \frac{1}{mn} \sigma_t [n(1-\lambda) + m\lambda] \quad (11\text{-}30a)$$

$$\sigma_r = \frac{1}{mn} \sigma_t [n(1-\lambda) - m\lambda] \quad (11\text{-}30b)$$

where $\alpha n/2(m+\alpha) \leq \lambda \leq n/2$, because

$$\psi = \arccos(-m/n) \quad (11\text{-}31)$$

ψ is the angle between the characteristic lines and σ_θ, then the differential equation of the characteristics is derived as

$$\frac{dr}{rd\theta} = \pm tg\psi = \pm \frac{2^{1/2}}{(n-m)^{1/2}} \quad (11\text{-}32)$$

The characteristic lines passing through A are then obtained,

$$\theta = \pm \frac{(n-m)^{1/2}}{2^{1/2}} \ln \frac{r}{\alpha} \quad (11\text{-}33)$$

in which the α family may take the positive sign, while the β family takes the negative sign. Along the path AP (α line), it has

$$2m\varphi + (2(n-m))^{1/2}\ln\lambda = \text{Constant} \tag{11-34}$$

Due to that $\varphi = \theta + \frac{\pi}{2}$, and $\sigma_r\big|_{r=a} = 0$, we know $\lambda_A = \frac{n}{2}$. Thus,

$$\lambda = \frac{n}{2}(\frac{a}{r})^m \tag{11-35}$$

Substituting Eq. (11-35) into Eq. (11-30) gives,

$$\sigma_\theta = \frac{\sigma_t}{m}\left[1 + \frac{m-n}{2}(\frac{a}{r})^m\right] \quad \text{for } a \le r \le c \tag{11-36a}$$

$$\sigma_r = \frac{\sigma_t}{m}\left[1 - (\frac{a}{r})^m\right] \quad \text{for } a \le r \le c \tag{11-36b}$$

With the increase of r, the difference of σ_r and σ_θ decreases until $\lambda = \alpha n/(2(m+\alpha))$, when

$$c = (1 + m/\alpha)^{1/m} a \tag{11-37}$$

Substituting Eq. (11-37) into Eq. (11-36), we have

$$(\sigma_\theta)_c = \frac{\sigma_t}{m}\left[1 + \frac{m-n}{2}\frac{\alpha}{m+\alpha}\right] = \frac{1+\alpha}{m+\alpha} \tag{11-38a}$$

$$(\sigma_r)_c = \sigma_t/(m+\alpha) \tag{11-38b}$$

Fig. 11.4 and Fig. 11.5 plot respectively $q = (\sigma_\theta)_c/\sigma_t$ and $q = (\sigma_r)_c/\sigma_t$ versus the parameter b.

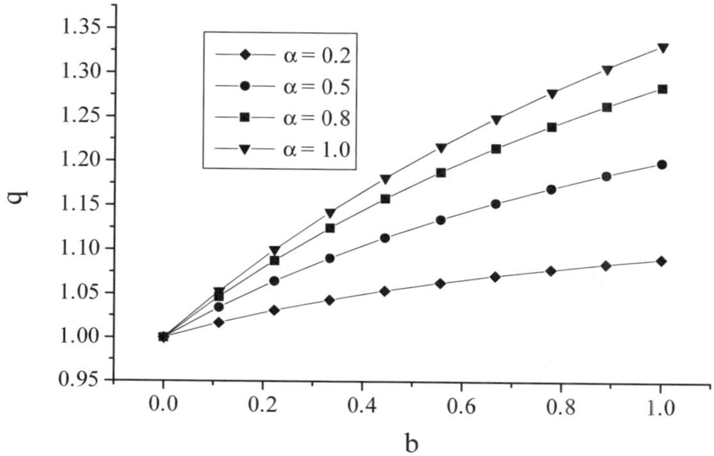

Fig.11.4 Relation between $q=(\sigma_\theta)_c/\sigma_t$ and unified strength theory parameter b.

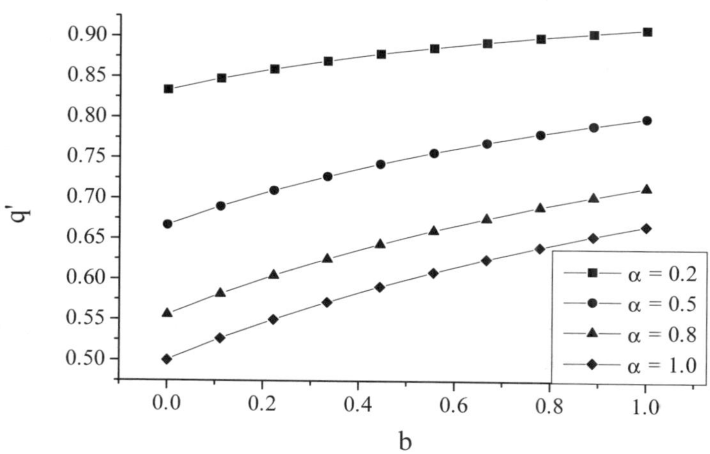

Fig. 11.5 Relation between $q'=(\sigma_r)_c/\sigma_t$ and unified strength theory parameter b.

When $r \geq c$, the stress states on the plate satisfy the yield condition Eq. (11-29b). The characteristics do not exist. The stresses can be solved directly from the dynamic equilibrium equation

$$\frac{d\sigma_r}{dr} + \frac{\sigma_r - \sigma_\theta}{r} = 0 \tag{11-39}$$

and the yield condition (11-29b) as follows,

$$\sigma_r = (p - \sigma_t)(c/r)^{\frac{2}{1+n}} + \sigma_t \quad \text{for } r > c \tag{11-40a}$$

$$\sigma_\theta = \frac{n-1}{1+n}(p - \sigma_t)(c/r)^{\frac{2}{1+n}} + \sigma_t \quad \text{for } r > c \tag{11-40b}$$

in which $p = (\sigma_r)_c$, and the limit load is obtained as

$$q_s = \sigma_r \big|_{r=\infty} = \sigma_t \tag{11-41}$$

For the special case of b=0, and m=n=1, the yield condition becomes,

$$F = A + B = \sigma_t \quad (B \leq A) \tag{11-42}$$

Due to $\psi = (1/2)\arccos(-1) = \pi/2$, the two family characteristic lines overlap each other and they are parallel to the principal stress σ_r, namely the characteristic lines are a family of radial lines and the stress components are,

$$\sigma_\theta = \sigma_t \tag{11-43a}$$

$$\sigma_r = \sigma_t(1 - a/r) \tag{11-43b}$$

Example 2 A trapezoidal plate under compression

Fig. 11.6 shows a symmetrical trapezoidal plate under compression. The angle between the two extended sides is 2ξ.

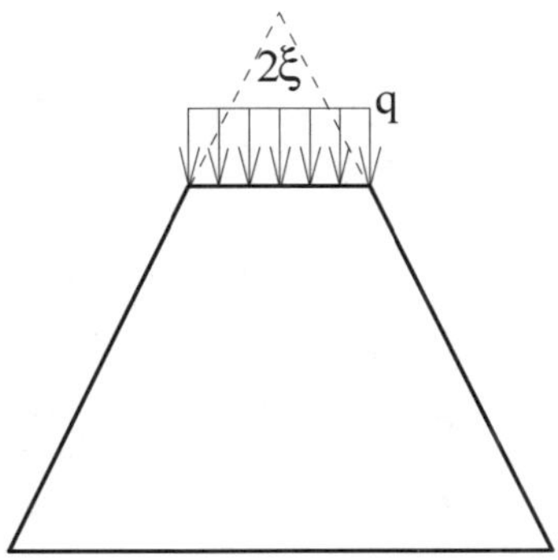

Fig. 11.6 A trapezoidal plate under compression

If the plate is subjected to a uniformly distributed pressure at the top, the limit load, based on the unified characteristic theory, can be derived as

$$q = \frac{D(n+m)-(n-m)}{m(n-m)}\sigma_t \quad (11\text{-}44)$$

in which

$$D = \exp(2m\zeta /(n^2 - m^2)^{1/2}) \quad (11\text{-}45)$$

$$m = (b-\alpha-\alpha b)/(1+b) \quad (11\text{-}46)$$

$$n = (b+\alpha+\alpha b)/(1+b) \quad (11\text{-}47)$$

σ_t and σ_c are respectively the uniaxial tensile and compressive strength of the material, $\alpha = \sigma_t / \sigma_c$. It is seen that the limit load is derived when $m \neq 0$, $n \neq \pm m$. Obviously $m \neq n$. For the cased when m=0 corresponding to b=$\alpha/(1-\alpha)$ and n=-m corresponding to $b = 0$, the limit load can be approximated by,

$$\lim_{m\to 0} q = 2(1+\xi)\sigma_t / n = (1+\xi)\sigma_t / \alpha \quad (11\text{-}48)$$

$$\lim_{m\to 0} q = \sigma_t / \alpha \quad (11\text{-}49)$$

respectively.

Through the above analysis, the limit load with respect to the Mohr-Coulomb criterion ($b = 0$) can be approximated by the unified characteristic field theory. For different parameter b, the limit load differs. Fig. 11.7 plots the relationship between the limit load q and the parameter b.

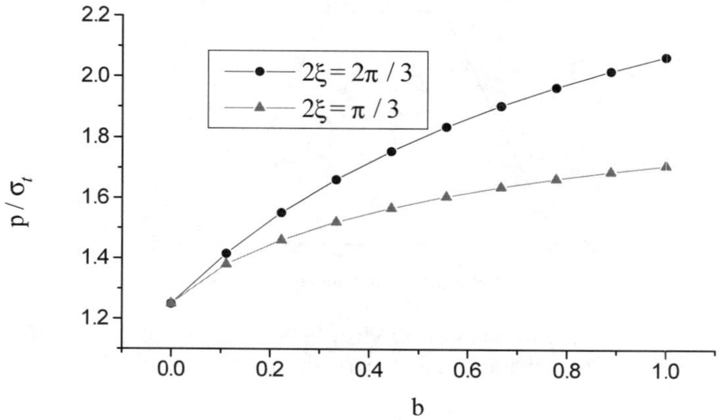

Fig. 11.7 Limit load q versus unified strength theory parameter b when $\alpha = 0.8$

11.5 Discontinuous Bifurcations of Elasto-plastic Material for Plane Stress

During the process of the elastic-plastic deformation, the continuity of velocity vanishes when it is passing the certain characteristics surface, with the development of the deformation, namely, the phenomenon of discontinuous bifurcations is produced. Hill (1958), Marciniak and Kuczynski (1967), Storen and Rice (1975), Rudnicki and Rice (1975), Hutchinson and Tvergaard (1980, 1981), Tvergaard, Needleman and Lo (1981), Raniecki and Bruhns (1981), Bruhns (1984), Li (1987), Runesson and Mroz (1989), Hill and Wu (1993), Zyczkowski (1999) et al. have done extensive researches for the discontinuous bifurcations.

In 1991, Ottosen and Runesson (1991) put forward a general description of discontinuous bifurcations for the plane problem of isotropic and elastic-plastic body plane problem, obeying the Mohr-Coulomb yield. criterion. Hill and Wu (1993) used the Ashton-Warren Spring yield equation. It is a non-linear yield equation, in which the maximum principal stress σ_1 and the minimum principal stress σ_3 are taken into account. Zyczkowski (1999) studied the discontinuous bifurcations in the case of the Burzynski-Torere yield criterion.

The corresponding properties of discontinuous bifurcations have been obtained for the specific yield criterion and plastic potential function. But the maximum critical hardening modulus is not unique.

For selection of the yield function, the intermediate principal stress is not considered in Mohr-coulomb Strength Theory, and this theory can not match the experimental results of much materials. Since the differences between compressive meridian and tensile meridian of limit surface can not be reflected in the Drucker-Prager criterion, there are some deficiencies in using the criterion in practice. The twin-shear strength theory (1985) takes the influence of the intermediate principal stress into consideration, but it can only be used for the materials that the shear strength limitation τ_0, tensile stress limitation σ_t and compressive strength σ_c satisfying the equation $\tau_0 = 2\sigma_t\sigma_c./2(2\sigma_c+\sigma_t)$. Furthermore, some corner models are proposed, but there is no clear physical conception for mathematical data-fitting of testing point for certain materials. The unified strength theory (Yu and He 1991; Yu 1994; Yu 1998) has a unified mechanical model, gets over the disadvantages of the former simplex strength theories. It can be adapted for many kinds of materials, and make the former simplex strength theories to be its special samples or its linear approach. Therefore, the Yu unified strength theory is adopted to analyze the discontinuous bifurcations of materials for plane stress problem.

The discontinuous bifurcations can be described as follows. Take account of a macro-symmetrical solid subjected to uniform load, with the increasing of the load, when the discontinuous bifurcation occurs, the strain rate is not continuous any longer as passing the local banded area. The discontinuous of strain rate must satisfying Maxwell compatibility condition:

$$[\dot{\varepsilon}] = \frac{1}{2}(m \otimes n + n \otimes m) \tag{11-50}$$

Where, [ ] denotes the difference produced by discontinuity, ⊗ denotes the Dyad tensor, *m* is a vector describing the mode of discontinuity of strain rate, *n* is the unit normal vector of local band area.

According to the different directions of *m* and *n*, two kinds of discontinuous bifurcations will occur, named split mode and shear band mode, represent the bifurcation mode when *m* and *n* are parallel or not parallel with each other, respectively.

For the standing out of the key points, Ottosen and Runesson (1991) assumed the deformation is small, uniform temperature is independent with the strain rate, described the properties of discontinuous bifurcations of elastic-plastic materials under plane stress, proposed the maximum critical hardening modulus when the discontinuous bifurcations occur. Because the hardening modulus is not unique, unify them and get the unique maximum critical hardening modulus

$$H_{db} = \max H_{cr}(n) \tag{11-51}$$

Where, the critical hardening modulus

$$H_{cr} = -P_{ij}D^e_{ijkl}Q_{kl} + n_j D^e_{ijst}Q_{st}R^e_{il}P_{mn}D^e_{mnkl}n_k \tag{11-52}$$

The Einstein sum criterion is satisfied (in this chapter, all the Latin letter of subscripts is 1 or 2). H_{db} being larger, equal or smaller than 0 represent the cases that the material is at the plastic hardening state, ideal plastic state and plastic softening state respectively, as the discontinuous bifurcations occur. D^e_{ijkl} is the symmetric elastic stiffness tensor. For the plane stress problem of anisotropic materials,

$$D^e_{ijkl} = \mu\left[\left(\delta_{ik}\delta jl + \delta_{il}\delta_{kl}\right) + \frac{2v}{1-v}\delta_{ij}\delta_{kl}\right] \tag{11-53}$$

where, μ is Lamé constant, ν is Poisson's ratio, and generally, $0 \le \nu \le 0.5$.

Introducing new expressions,

$$P_{ij} = \frac{\partial F}{\partial \sigma_{ij}} \bigg/ \left\| \frac{\partial F}{\partial \sigma_{ij}} \right\|, \quad Q_{ij} = \frac{\partial G}{\partial \sigma_{ij}} \bigg/ \left\| \frac{\partial G}{\partial \sigma_{ij}} \right\|, \quad R^e_{il} = \frac{1}{\mu}\left[-\frac{1+v}{2}n_i n_l + \delta_{il}\right] \tag{11-54}$$

where F, G are yield function and plastic potential function, and the symbol $\|\ \|$ is the norm of the tensor.

Apparently, from Eq. (11-51), the initial azimuth angle and its corresponding plastic hardening modulus can be determined. If the vector is known, the discontinuous vector of strain rate m can be expressed as

$$m = 2n\cdot g - (1+v)(n\cdot g\cdot n)n + v(trg)n \tag{11-55}$$

where, the symbol 'tr' denotes the trace of the tensor.

11.6 Discontinuous Bifurcations of Non-associated Flow Elasto-plastic Materials based on Yu Unified Strength Theory

Select the Yu unified strength theory as the strength theory model. The Yu unified strength theory can be expressed as follows,

$$F=(1+b)(1+\sin\phi)\,\sigma_1-(1-\sin\phi)(b\sigma_2+\sigma_3)-2(1+b)\,c\cos\phi=0$$

$$\text{When } \sigma_2\le\frac{1}{2}(\sigma_1+\sigma_3)+\frac{\sin\phi}{2}(\sigma_1-\sigma_3) \quad \text{(11-56a)}$$

$$F'=(1+\sin\phi)(\sigma_1+b\sigma_2)-(1-\sin\phi)(1+b)\sigma_3-2(1+b)c\cos\phi=0 \quad \text{When}$$

$$\sigma_2\ge\frac{1}{2}(\sigma_1+\sigma_3)+\frac{\sin\phi}{2}(\sigma_1-\sigma_3) \quad \text{(11-56b)}$$

where, c and ϕ are cohesion and friction angle, respectively. The parameter in the Yu unified strength theory b is referred as the unified yield criterion parameter; it is also the parameter of effect of intermediate principal stress

$$b=\frac{(1+\alpha)\tau_0-\sigma_t}{\sigma_t-\tau_0} \quad \text{(11-57)}$$

Generally, $0\le b\le 1$. b can also be the selection parameter of strength theory. For a certain material, σ_t, σ_t and τ_0 are determined from the experiments and from Eq. (11-57), the corresponding value of b can be calculated. Substituting this value for b in Eq.(11-56), the strength theory fit for this material can be obtained.

For the case of non-associated flow, the plastic potential function and yield function have the same mode and the different friction angle generally. Therefore, let

$$m_1=(1+b)(1+\sin\phi),\quad m_2=b(\sin\phi-1),\quad m_3=\sin\phi-1 \quad \text{(11-58)}$$

$$l_1=1+\sin\phi,\quad l_2=b(1+\sin\phi),\quad l_3=(\sin\phi-1)(1+b) \quad \text{(11-59)}$$

$$m_1'=(1+b)(1+\sin\psi),\quad m_2'=b(\sin\psi-1),\quad m_3'=\sin\psi-1 \quad \text{(11-60)}$$

$$l_1'=1+\sin\psi,\quad l_2'=b(1+\sin\psi),\quad l_3'=(\sin\psi-1)(1+b) \quad \text{(11-61)}$$

Incorporating Eq. (11-46), then

$$F=m_1\sigma_{\mathrm{I}}+m_2\sigma_{\mathrm{II}}+m_3\sigma_{\mathrm{III}}-2(1+b)c\cos\phi=0,$$

$$G=m'_1\sigma_{\mathrm{I}}+m'_2\sigma_{\mathrm{II}}+m'_3\sigma_{\mathrm{III}}$$

$$\sigma_{\mathrm{II}}\le\frac{1}{2}(\sigma_{\mathrm{I}}+\sigma_{\mathrm{III}})+\frac{\sin\phi}{2}(\sigma_{\mathrm{I}}-\sigma_{\mathrm{III}}) \quad \text{(11-62a)}$$

$$F = l_1\sigma_{\mathrm{I}} + l_2\sigma_{\mathrm{II}} + l_3\sigma_{\mathrm{III}} - 2(1+b)c\cos\phi = 0\,,\ G = l'_1\sigma_{\mathrm{I}} + l'_2\sigma_{\mathrm{II}} + l'_3\sigma_{\mathrm{III}}$$

$$\sigma_2 \geq \frac{1}{2}(\sigma_{\mathrm{I}} + \sigma_{\mathrm{III}}) + \frac{\sin\phi}{2}(\sigma_{\mathrm{I}} - \sigma_{\mathrm{III}}) \tag{11-62b}$$

where, $\sigma_{\mathrm{I}} \geq \sigma_{\mathrm{II}} \geq \sigma_{\mathrm{III}}$ are the principal stresses. For the plane stress problem, let the direction of coordinates coincide with the principal direction and $\sigma_1 \geq \sigma_2$, and assume tensor *P, Q* have the same direction, with the satisfying of $P_1 \geq P_2$, $Q_1 \geq Q_2$. Therefore, three cases can be discussed respectively:

1. the case of $\sigma_1 \geq \sigma_2 \geq 0$, from Eq. (11-62), we can obtain,

$$F = m_1\sigma_1 + m_2\sigma_2 - 2(1+b)c\cos\phi = 0\,,\quad G = m'_1\sigma_1 + m'_2\sigma_2\,, \tag{11-63a}$$

$$\text{when}\quad \sigma_2 \leq \frac{1}{2}(1+\sin\phi)\sigma_1$$

$$F = l_1\sigma_1 + l_2\sigma_2 - 2(1+b)c\cos\phi = 0\,,\quad G = l'_1\sigma_1 + l'_2\sigma_2\,,\ \text{when} \tag{11-63b}$$

$$\sigma_2 \leq \frac{1}{2}(1+\sin\phi)\,\sigma_1$$

Therefore, when $\sigma_2 \leq (1+\sin\phi)\sigma_1/2$,

$$P_1 = \frac{m_1}{R}\,,\quad P_2 = \frac{m_2}{R}\,,\quad Q_1 = \frac{m'_1}{R'}\,,\quad Q_2 = \frac{m'_2}{R'} \tag{11-64}$$

where,

$$R = \sqrt{m_1^2 + m_2^2}\,,\quad R' = \sqrt{m_1'^2 + m_2'^2} \tag{11-65}$$

Since $P_1 > P_2$, $Q_1 > Q_2$, and

$$2m_1m'_1 - m_1m'_2 - m_2m'_1 \geq 0\,,\quad m_1m'_2 + m_2m'_1 - 2m_2m'_2 \leq 0 \tag{11-66}$$

Therefore, from Eq. (11-51), we can obtain,

$$\tan^2\theta = \frac{2m_1m'_1 - m_1m'_2 - m_2m'_1}{2m_2m'_2 - m_1m'_2 - m_2m'_1}\,,\quad \frac{H_{db}}{2\mu} = \frac{1+\nu}{4RR'}\frac{(m_2m'_1 - m_1m'_2)^2}{(m_1 - m_2)(m'_1 - m'_2)} \tag{11-67}$$

Apparently, the hardening modulus when the discontinuous bifurcations occur $H_{\mathrm{db}} \geq 0$, namely, the bifurcations occur at plastic hardening or ideal plastic state, and the bifurcation mode is shear band mode. With ignoring the different tensile and compressive property ($\phi = \psi = 0$) of materials, the relationship between the initial bifurcation angle θ and the parameter of the unified strength theory b can be shown by Fig. 11.8. The parameter b is also a influence parameter of intermediate principal stress.

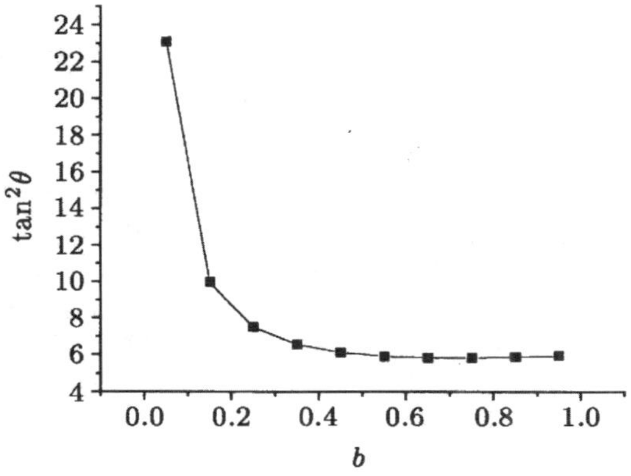

Fig. 11.8 Influence of the intermediate principal stress on the initial direction of discontinuous bifurcation

When $\sigma_2 \geq (1+\sin\phi)\sigma_1/2$,

$$P_1 = \frac{l_1}{S}, \quad P_2 = \frac{l_2}{S}, \quad Q_1 = \frac{l'_1}{S'}, \quad Q_2 = \frac{l'_2}{S'} \tag{11-68}$$

where,

$$S = \sqrt{l_1^2 + l_2^2}, \quad S' = \sqrt{l'^2_1 + l'^2_2} \tag{11-69}$$

When $0 < b < 1$, and since $P_1 > P_2$, $Q_1 > Q_2$, and

$$l_1 l'_2 + l_2 l'_1 - 2 l_2 l'_2 > 0 \tag{11-70}$$

Then, from Eq. (11-51)

$$\theta = 90^\circ, \quad \frac{H_{db}}{2\mu} = -(1+\nu)\frac{l_2 l'_2}{SS'} \tag{11-71}$$

Due to $H_{db} \leq 0$, the discontinuous bifurcations don't occur at the plastic hardening state, and the bifurcation mode is split mode. When b=1, $P_1=P_2$, $Q_1=Q_2$, the direction of bifurcation is arbitrary. When b=0, as the intermediary of the two modes, the mode of discontinuous bifurcations belongs to either split mode or shear band mode.

2. case of $\sigma_1 \geq 0 \geq \sigma_2$, when $\sigma_2 \geq (1+\sin\phi)\sigma_1/(\sin\phi-1)$, through analysis, we can obtain

$$\tan^2\theta = \frac{2m_1m'_1 - m_1m'_3 - m_3m'_1}{2m_3m'_3 - m_1m'_3 - m_3m'_1}, \quad \frac{H_{db}}{2\mu} = \frac{1+\nu}{4RR'}\frac{(m_3m'_1 - m_1m'_3)^2}{(m_1 - m_3)(m'_1 - m'_3)} \tag{11-72}$$

where, $R=\sqrt{m_1^2+m_3^2}$, $R'=\sqrt{m'^2_1+m'^2_3}$, from Eq. (11-72), $H_{db} \geq 0$, the bifurcations occur at plastic hardening or ideal plastic state, and the bifurcation mode is shear band mode. With ignoring the different tensile and compressive property ($\phi = \psi = 0$) of materials, the relationship between the initial bifurcation angle θ and the parameter of the unified strength theory b can be shown by Fig. 11.9.

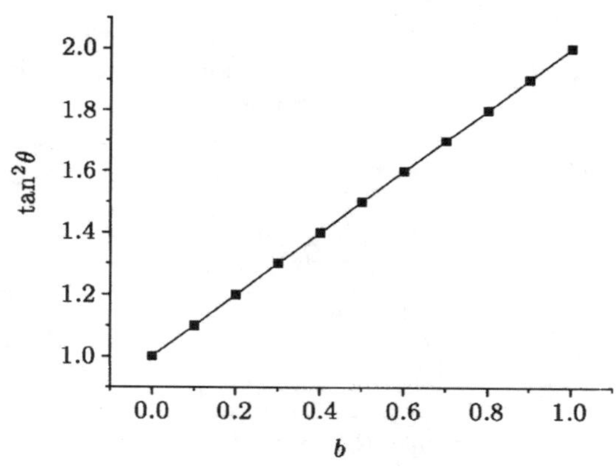

Fig. 11.9 Relation of the unified strength theory parameter b with initial direction of discontinuous bifurcation

When $\sigma_2 \leq (1+\sin\phi)\sigma_1/(\sin\phi-1)$,

$$\tan^2\theta = \frac{2l_1l'_1 - l_1l'_3 - l_3l'_1}{2l_3l'_3 - l_1l'_3 - l_3l'_1}, \quad \frac{H_{db}}{2\mu} = \frac{1+\nu}{4SS'}\frac{(l_3l'_1 - l_1l'_3)^2}{(l_1 - l_3)(l'_1 - l'_3)} \tag{11-73}$$

where, $S=\sqrt{l_1^2+l_3^2}$, $S'=\sqrt{l'^2_1+l'^2_3}$, from Eq. (11-73), $H_{db}\geq 0$, the bifurcation mode is shear band mode. With ignoring the different tensile and compressive property ($\phi=\psi=0$) of materials, the relationship between the initial bifurcation angle θ and the parameter of the unified strength theory b can be shown by Fig. 11.10.

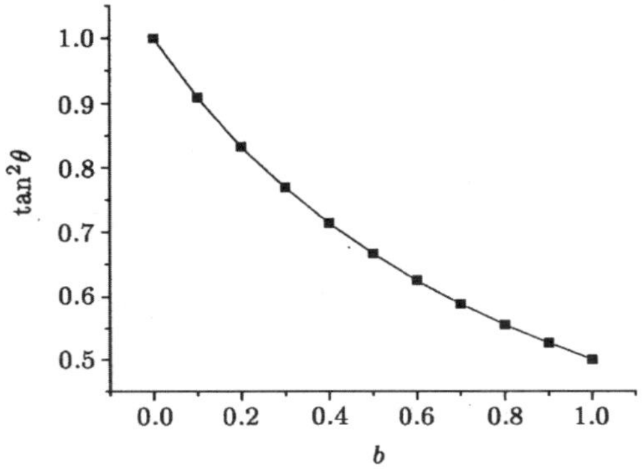

Fig. 11.10 Relation of the unified strength theory parameter b with initial direction of discontinuous bifurcation

3. case of $0\geq\sigma_1\geq\sigma_2$, when $\sigma_2\geq(1+\sin\phi)\sigma_1/2$, by calculation, when $0\leq b\leq 1$,

$$\theta=0^\circ,\quad \frac{H_{db}}{2\mu}=-(1+\nu)\frac{m_2 m'_2}{RR'} \tag{11-74}$$

where, $R=\sqrt{m_2^2+m_3^2}$, $R'=\sqrt{m'^2_2+m'^2_3}$, from Eq. (11-74), $H_{db}\leq 0$, the bifurcations occur at plastic softening or ideal plastic state, and the bifurcation mode is split mode (also shear band mode, as $b=0$). When $b=1$, an arbitrary direction can be the initial direction of the bifurcation, belonging to split mode, and the corresponding hardening modulus is

$$\frac{H_{db}}{2\mu}=-(1+\nu)\frac{m_2 m'_2}{RR'}=-(1+\nu)\frac{m_3 m'_3}{RR'} \tag{11-75}$$

When $\sigma_2 \geq (1+\sin\phi)\sigma_1 / 2$, by calculation,

$$\left.\begin{aligned} \tan^2\theta &= \frac{2l_2l'_2 - l_2l'_3 - l_3l'_2}{2l_3l'_3 - l_2l'_3 - l_3l'_2} \\ \frac{H_{db}}{2\mu} &= \frac{1+\nu}{4SS'} \frac{(l_3l'_2 - l_2l'_3)^2}{(l_2 - l_3)(l'_2 - l'_3)} \end{aligned}\right\} \quad (11\text{-}76)$$

where, $S=\sqrt{l_2^2+l_3^2}$, $S'=\sqrt{l'^2_2+l'^2_3}$, from Eq. (11-76), $H_{db} \geq 0$, the bifurcations occur at plastic hardening or ideal plastic state, and the bifurcation mode is shear band mode. With ignoring the different tensile and compressive property ($\phi = \psi = 0$) of materials, the relationship between the initial bifurcation angle θ and the parameter of strength theory b can be shown by Fig. 11.11.

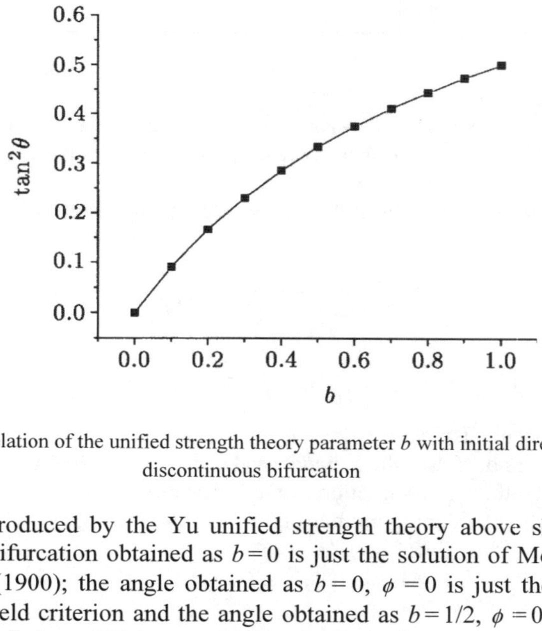

Fig. 11.11 Relation of the unified strength theory parameter b with initial direction of discontinuous bifurcation

The results produced by the Yu unified strength theory above show that the initial angle of bifurcation obtained as $b=0$ is just the solution of Mohr-Coulomb strength theory (1900); the angle obtained as $b=0$, $\phi=0$ is just the solution of Tresca (1991) yield criterion and the angle obtained as $b=1/2$, $\phi=0$ is the linear approach of the solution of Huber-von Mises yield criterion.

11.7 Discussion and Experimental Verification

11.7.1 Discussion

Assume the material satisfy the associate flow law. A special case is discussed in this part and compared with the existing unified plane stress characteristics field theory (Yu et al. 1998, 1999).

Define A = $(\sigma_1+\sigma_2)$ / 2, B = $(\sigma_1-\sigma_2)$ / 2. For plane stress problem, materials are assumed satisfy the associate flow law, following conclusions are drawn:

For $\sigma_1 \geq \sigma_2 \geq 0$ (viz. A ≥ B), when $\sigma_2 \leq (1+\sin\phi)\sigma_1/2$, it is equal to A ≤ (2+α)B / α (α is the tension-compression ratio), discontinuous bifurcations are shear band mode. From Eq. (11-67)

$$\tan^2\theta = -\frac{m_1}{m_2} \tag{11-77}$$

then

$$\cos 2\theta = -\frac{m_1+m_2}{m_1-m_2} = -\frac{1+b-\alpha b}{1+b+\alpha b} \tag{11-78}$$

Here, sinϕ = (1−α)/(1+α) is used. In such a stress state, the angle of characteristics is given by unified characteristics theory as follows

$$\cos 2\theta = -\frac{m}{n} \tag{11-79}$$

where, m = (1+b−αb) / (1+b), n = (1+b+αb) / (1+b). It is obviously that Eq. (11-78) and Eq. (11-79) have given the same result.

When $\sigma_2 \geq (1+\sin\phi)\sigma_1/2$ (viz. A ≥ (2+α)B / α), if 0 < b ≤ 1, discontinuous bifurcations are of split mode. In such a stress state, when 0 < b ≤ 1, from the unified plane stress characteristics field theory, we know that there is no characteristics, because there is no root for the characteristics angle cos2θ = −(1+b) / (1−b). When b = 0, these two will give the same results.

The same conclusion will be drawn for the two cases of $\sigma_1 \geq 0 \geq \sigma_2$ (A ≤ B), and $0 \geq \sigma_1 \geq \sigma_2$ (A ≤ −B).

From the above analysis, following conclusions can be obtained: In the plane stress state, shear band mode bifurcations is the same with the characteristics in essence. Angle of the unified characteristics given by Yu et al. (1999) can be obtained by the degeneration of the angle of shear band. So, the bifurcation theory can be used to analyze the characteristics.

11.7.2 Experimental Verification

A rectangular thin plate, made of PVC hard plastic, is subjected to uniform tension on two short sides as shown in Fig. 11.12. Unixial tensile strength of the PVC material: $\sigma_t = 5.886\text{kN/cm}^2$; unixial compressive strength: $\sigma_c = 7.575\text{kN/cm}^2$; tension-compression ratio: $\alpha = \sigma_t/\sigma_c = 0.777$.

When subjected to tension, local slip band, which has the angle $\theta=61°$ with the tensional direction, appeared in the specimen. Based on the present unified characteristic field theory, with different b, the result of the characteristics will be different and a series of solutions can be obtained. On the other words that different strength theory will lead to different azimuth angle of shear band.

Localized shear band on the plate is observed after the stretching. The angle θ of the shear band from the force direction is 60°. From a back analysis, the angle θ is determined as 60.97° when b = 0.675. It implies that the commonly used Mohr-Coulomb criterion may give very different prediction (48.60° for this example) of the shear band direction. It can not coincide with the experimental result. So, $b = 0.675$ is used to simulate the PVC-like material, viz. the strength theory of $b = 0.675$ can be used for shear band analysis of this kind of material.

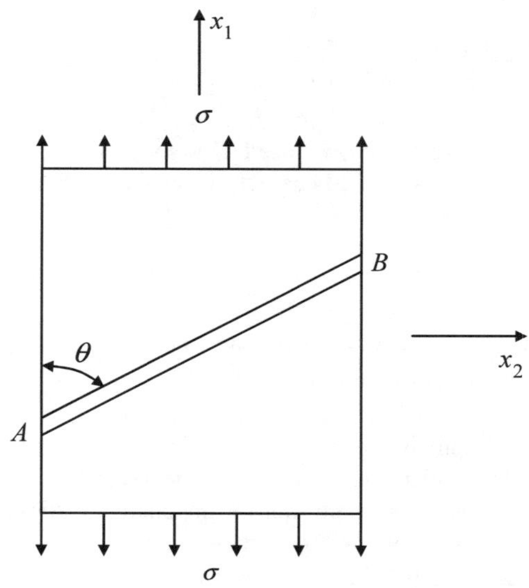

Fig. 11.12 Rectangular thin plate subjected to tension

In this section, unification of the maximum critical hardening modulus has been done for the discontinuous bifurcations of elastic-plastic materials. Based on this unification, the phenomenon of discontinuous bifurcations has been investigated and obtained the initial azimuth angle and unified analytical model of the corresponding hardening modulus of discontinuous bifurcations under the non-associated plastic flow condition. This solution can be adapted for many kinds of materials and make the former simplex strength theories to be its special samples or its linear approach.

Through the analysis of the results we can know that the isomerism of compression and tension and the intermediate principal stress have so great influence on the discontinuous bifurcations that govern the stress state. Therefore, these two factors above can't be ignored during the analysis of the bifurcation, or else the results obtained can't accord with the results of actual practice.

On the other hand, the influence of different strength criterion to the discontinuous bifurcations of elastic-plastic materials can be established by giving different value of parameter b. A conclusion can be drawn that the influence of strength criterion is relative to the stress state of material. Actually, the value of parameter b can be calculated from Eq. (11-57). Substituting it in the unified solution, the properties of bifurcation fit for the material can be obtained.

Comparing with the theory of characteristics field of plane stress, we can know that the discontinuous bifurcations of shear band mode of plane stress coincide with the characteristics. Thus, the theory of bifurcation can be adopted to analyze the characteristics. Similar discontinuous bifurcations for plane strain problems can be seen in Chapter 9.

Summary

Based on the unified strength theory the unified characteristics field theory for plastic plane stress problem is described in this chapter. It is the unification of the plastic characteristics methods based on the Tresca criterion, the Huber-von Mises criterion and the twin-shear yield criterion. A series of new results can be also obtained by using the unified characteristics field theory. The strength-differential effect (SD effect and the effect of intermediate principal stress are taken into account in the unified characteristics field theory.

The unified solutions of two examples, i.e. an infinite thin plate with a hole and a trapezoidal plate under compression, are given. It concludes a series of new results.

The discontinuous bifurcations problems under plane stress are also studied by using the unified strength theory in this chapter.

Problems

Problem 11.1
Derive the characteristics field for plane stress problem by using the Mohr-Coulomb strength criterion or the Yu unified strength theory with b = 0.

Problem 11.2
Derive the characteristics field for plane stress problem by using the twin-shear strength or the Yu unified strength theory with b=1.

Problem 11.3
Derive the characteristics field for plane stress problem by using the Yu unified strength theory with b = 0.5.

Problem 11.4
Discuss the discontinuous bifurcation based on the Mohr-Coulomb strength criterion or the Yu unified strength theory with b = 0.

Problem 11.5
Discuss the discontinuous bifurcation based on the twin-shear strength or the Yu unified strength theory with b = 1.

Problem 11.6
Discuss the discontinuous bifurcation based on the Yu unified strength theory with b = 0.5.

Problem 11.7
Discuss the discontinuous bifurcation for plane strain problems (see chapters 8 and 9) based on the Yu unified strength theory with b = 0.5.

References and Bibliography

Bruhns OT (1984) Bifurcation problems in plasticity. In: *The Constitutive Law in Thermoplasticity*, Lehmann Thed. Springer-Verlag, Wien, New York.

Hill RA (1958) General theory of uniqueness and stability in elastic-plastic solids. *J. Mech. and Phys. of Solids*, 6: 236–249.

Hill JM and Wu YH (1993) Plastic flows of granular materials of shear index n–1, (1) yield functions; (2) Plane and axially symmetric problems for $n = 2$, *J. Mech. and Phys. of Solids*, 40(1): 77–93; 95–115.

Hutchinson JW and Tvergaard V (1980) Surface instabilities on statically strained plastic solids. *Int. J. Mech. Sci.*, 22: 339–358.

Hutchinson JW and Tvergaard V (1981) Shear band formation in plane strain. *Int. J. Solids Struct.*, 17: 451–466.

Li GC (1987) Shear band bifurcations of ductile materials. *Acta Mechanics Sinica*, 19, 61–68.

Marciniak Z and Kuczynski K (1967) Limit strains in the process of stretch forming sheet metal. *Int. J. Mech. Sci.*, 9: 609–619.

Ranniecki B, Bruhns OT (1981) Bounds to bifurcation stresses in solids with non-associated flow law at finite strain. *J. Mech. and Phys. of Solids.*, 29: 153–172.

Rudnicki JW and Rice JR (1975) Conditions for the localization of deformation in pressure-sensitive dilatant materials. J. Mech. Phys. Solids, 23: 231–241.

Runesson K, Mroz Z (1989) A note on non-associated plastic flow rules. *International Journal of Plasticity*, 5: 639–658.

Runesson K, Ottosen N, Peric D (1991) Discontinuous bifurcations of elastic-plastic solutions at plane stress and plane strain. *International Journal of Plasticity*, 7: 99–121.

Storen S and Rice JR (1975) Localized necking in thin sheets. *J. Mech. and Phys. of Solids*, 23: 421–433.

Thomas TY (1961) *Plastic Flow and Frature in Solids.*, Academic Press, New York.

Tvergaard V, Needleman A and Lo KK (1981) Flow localization in the plane strain tensile test. *J. Mech. and Phys. of Solids*. 29: 115–127.

Yu MH (1961) General behaviour of isotropic yield function (in Chinese), Res. Report of Xi'an Jiaotong University. Xi'an.

Yu MH (1983) Twin shear stress yield criterion. *Int. J. Mech. Sci*. 25: 71–74.

Yu MH, He LN, S LY (1985) Twin shear stress theory and its generalization, *Scientia Sinica (Science in China)*, Series A, 28(12): 1174–1183.

Yu MH, He LN (1991) A New Model and Theory on Yiwld and Failure of Materials under Complex Stress State. *Mechanical behaviour of materials-6*, Pergamon Press, 3: 841–846.

Yu MH (1994) Unified Strength Theory for geomaterials and its applications. Chinese *Journal of Geothechnical Engineering,* 16(2): 1–10 (in Chinese).

Yu MH (1998) *Twin Shear Theory and Its Application*. Berjing: Science Press (in Chinese).

Yu MH, Zhang YQ, Yang SY (1998) Another important generalization of the unified strength theory. *J. of Xi'an Jiaotong University*. 32: 108–110 (in Chinese, English Abstract).

Yu MH, Zhang YQ, Li JC (1999) The unified characteristics theory for plastic plane stress problems. *J. of Xi'an Jiaotong University*. 33(4): 1–4 (in Chinese, English Abstract).

Zhang YQ and Yu MH (1999) Solutions to plastic plane stress problems. *J. of Xi'an Jiaotong University*, 33(6): 60–63 (in Chinese, English Abstract).

Zhang YQ and Yu MH (2001) Discontinuous bifurcations of metallic materials for plane stress. Chinese *J. of Mechanical Engineering*, 37(4): 62–85 (in Chinese, English Abstract).

Zhang YQ and Yu MH (2001) Discontinuous bifurcations of elasto-plastic materials for plane stress. *Acta Mechanica Sinica*, 33(5): 706–713 (in Chinese, English Abstract).

Zhang YQ, Hao H and Yu MH (2003) A unified characteristics theory for plastic plane stress and strain problems. *J. of Applied Mechanics*, 70: 649–654.

Zyczkowski M (1999) Discontinuous bifurcations in the case of the Burzynski-Torere yield criterion. Acta Mechanica, 132: 19–35.

12 Unified Characteristics Line Theory for Spatial Axisymmetric Problem

12.1 Introduction

The twin-shear slipe-line field and the unified slipe-line field theory for plane strain problem and the twin-shear characteristics field and the unified characteristics field theory for plane stress problem have been described in Chapters 8, 9, 10 and 11. It can be applied to metal plastic forming, the limit analysis of structure in civil engineering and mechanical engineering. The partial differential equations of the axisymmetric problem are not exactly hyperbolic, they may be elliptical and parabolic. The two families of characteristics lines of the parabolic functions become one family, and there are no characteristics lines for the elliptical equations. The characteristics line field theories of the spatial axisymmetric problem is more complex. It is difficult to solve analytically, and still remains open.

A lots of simplified methods were used to discuss the characteristics line theory of spatial axisymmetric problems. Levin (1953), Shield (1955) and Kachanov (1971) et al. used Haar von-Karman complete plasticity condition and the Tresca criterion. They studied the indentation problem of a punch, and assumed that circumferential stress σ_θ is equal to one of the other two principal stresses, namely $\sigma_\theta \approx \sigma_1$ or $\sigma_\theta \approx \sigma_3$. The results of the characteristics line theory, however, did not agree well with the reality (Hill 1950). Szczepinski (1979) used the hypotheses $\sigma_\theta \approx \sigma_r$ and the Tresca criterion to study the characteristics line theory of spatial axisymmetric problems. After that, the infinite differential method was also used for spatial axisymmetric problems (Shield 1955).

Based on the same method, Cox analyzed the limit loading capability of cohesive circular smooth soil foundations ($c-\varphi$ soil and $\rho=0$). Larkin analyzed circular shallow soil foundations ($\rho=0$) when $c=0$. The above analyzed results were summarized by Chen (1975). Shield and Cox used the single-sheer strength theory (Tresca-Mohr-Coulomb theory) and the hypotheses $\sigma_\theta = \sigma_2 < \sigma_1$ of Haar and von-Karnam.

Some axisymmetric characteristics problems were solved by Ishlinsky (1944), and Sokolovsky (1942) using the Mohr-Coulomb theory. Figs. 12.1 and 12.2 show the two examples of axisymmetric characteristics field (from Березанлев 1960). Figure 12.3 is an example of the axisymmetric characteristics for a long-rod penetration in concrete target (Yu, Li, Wei and Chen, see Yu 2002a).

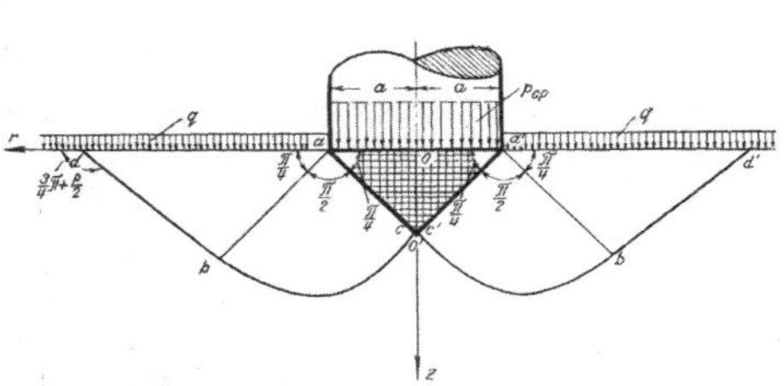

Fig. 12.1. Characteristics field of an indentation of a circular head punch (non-SD material)

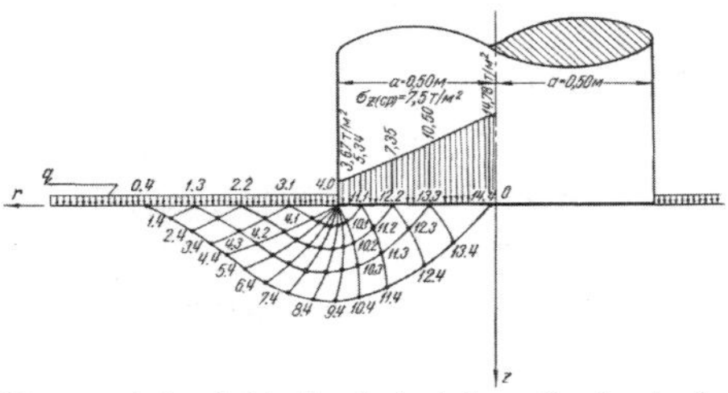

Fig. 12.2. Characteristics field of an indentation of a circular head punch (SD material)

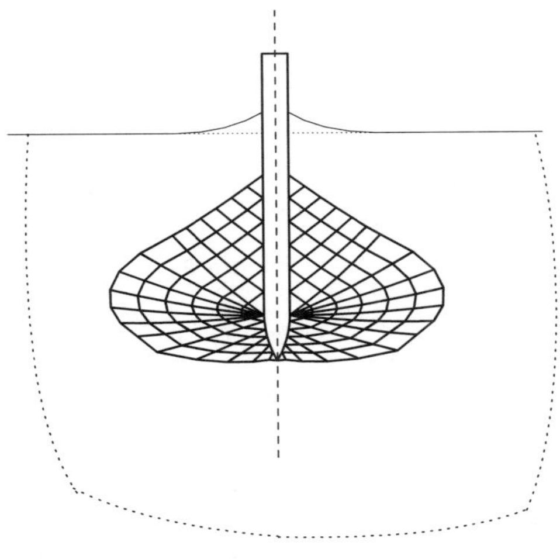

Fig. 12.3 Axisymmetric characteristics field of a long-rod in normal penetration

Most solutions can be adapted only for the Tresca material and the Mohr-Coulomb material

A unified characteristics line field theory for axisymmetric problems which can be adapted for more materials was introduced by Yu et al. in 2001 (Yu, Li and Zhang 2001). It is based on the peculiarity of spatial axisymmetric plastic problems and the unified strength theory.

The stress field and velocity field of the unified axisymmetric characteristics line field theory are described in this chapter, and two examples by using the new characteristics theory for axisymmetric plastic problem are given. The unified axisymmetric characteristics line field theory is also used to solve the high velocity impact problem. The normal penetration of a long-rod under high velocity is described in this chapter.

12.2 The Unified Strength Theory

The Yu unified strength theory has been described in Chapter 4. The mathematical expression of the unified strength theory is

$$f=\sigma_1-\frac{\alpha}{1+b}(b\sigma_2+\sigma_3)=\sigma_t \qquad \text{when } \sigma_2\le\frac{\sigma_1+\alpha\sigma_3}{1+\alpha} \tag{12-1a}$$

$$f'=\frac{1}{1+b}(\sigma_1+b\sigma_2)-\alpha\sigma_3=\sigma_t \qquad \text{when } \sigma_2\ge\frac{\sigma_1+\alpha\sigma_3}{1+\alpha} \tag{12-1b}$$

The unified strength theory can be expressed in terms of internal cohesion c_0 and the friction angle φ_0 as follows

$$f=[\sigma_1-\frac{1}{1+b}(b\sigma_2+\sigma_3)]+[\sigma_1+\frac{1}{1+b}(b\sigma_2+\sigma_3)]\sin\varphi_0=2c_0\cos\varphi_0,$$

$$\text{when} \quad \sigma_2\le\frac{\sigma_1+\sigma_3}{2}+\frac{\sigma_1-\sigma_3}{2}\sin\varphi_0$$

or

$$\text{when} \quad \sigma_2\le p+R\sin\varphi_0 \tag{12-2a}$$

$$f'=[\frac{1}{1+b}(\sigma_1+b\sigma_2)-\sigma_3]+[\frac{1}{1+b}(\sigma_1+b\sigma_2)+\sigma_3]\sin\varphi_0=2c_0\cos\varphi_0,$$

$$\text{when} \quad \sigma_2\ge\frac{\sigma_1+\sigma_3}{2}+\frac{\sigma_1-\sigma_3}{2}\sin\varphi_0$$

or

$$\text{when} \quad \sigma_2\le p+R\sin\varphi_0 \tag{12-2b}$$

The relations among c_0, φ_0 and the other parameters of material are shown as

$$\alpha=\frac{1-\sin\varphi_0}{1+\sin\varphi_0} \qquad \sigma_t=\frac{2c_0\cos\varphi_0}{1+\sin\varphi_0} \tag{12-3}$$

It is shown that the unified strength theory consists of a series of strength theories from the single shear strength theory (lower limit, Mohr-Coulomb 1900) to the twin shear strength theory (upper limit, Yu 1985). The variaties of yield loci of the unified strength theory on the deviatoric plane have been illustrated in Figure 2 in Preface and Fig. 4.7, 4.8 in Chapter 4.

12.3 Unified Characteristics Line Field Theory for Spatial Axisymmetric Problems (Stress Field)

There are four non-zero stress components σ_r, σ_θ, σ_z and τ_{rz} in the axisymmetric problem, the circumferential stress σ_θ must be one of principal stresses.

According to the stress state of a spatial axisymmetric problems, we introduce a parameter m for circumferential stress σ_θ $(0 \le m \le 1)$

$$\sigma_\theta = \sigma_3 + m(\frac{\sigma_1 + \sigma_3}{2} - \sigma_3) = \sigma_3 + m\frac{\sigma_1 - \sigma_3}{2} \tag{12-4}$$

For the axisymmetric problem, we define

$$P = \frac{\sigma_1 + \sigma_3}{2}, \ R = \frac{\sigma_1 - \sigma_3}{2} \tag{12-5}$$

The following relations can be derived

$$\sigma_1 = P + R \qquad \sigma_2 = \sigma_\theta = P + (m-1)R \qquad \sigma_3 = P - R \tag{12-6}$$

The stresses of the problem can be expressed as

$$\left.\begin{aligned} \sigma_r &= P + R\cos 2\theta \\ \sigma_z &= P - R\cos 2\theta \\ \tau_{rz} &= R\sin 2\theta \\ \sigma_\theta &= P + (m-1)R \end{aligned}\right\} \tag{12-7}$$

where θ is the angle between the directions of the maximum principal stress and axis r.

Using the first expression of the unified strength theory Eq.(12-2a) because the intermediate principal stress is $\sigma_2 = \sigma_\theta = P + (m-1)R \le P + R \sin \varphi_0$.

Substituting Eq. (12-6) into Eq. (12-2), the unified strength theory for spatial axisymmetric problem is obtained as follows

$$R = -\frac{2(1+b)\sin\varphi_0}{2(1+b) + mb(\sin\varphi_0 - 1)} P + \frac{2(1+b)c_0\cos\varphi_0}{2(1+b) + mb(\sin\varphi_0 - 1)} \tag{12-8}$$

The expression of the unified strength theory for spatial axisymmetric problem can be simplified to the simple expression as follows

$$R = -P\sin\varphi_{uni} + C_{uni}\cos\varphi_{uni} \tag{12-9}$$

where $\sin\varphi_{uni}$ and C_{uni} are the unified material parameters introduced from the unified strength theory by Yu et al. (2001) for spatial axisymmetric problem. The parameter φ_{uni} and C_{uni} are referred to as the unified friction angle φ_{uni} and the unified internal cohesion C_{uni}, respectively.

$$\sin\varphi_{uni} = \frac{2(1+b)\sin\varphi_0}{2(1+b) + mb(\sin\varphi_0 - 1)}, \tag{12-10}$$

$$C_{uni}=\frac{2(1+b)c_0\cos\varphi_0}{2(1+b)+mb(\sin\varphi_0-1)}\cdot\frac{1}{\cos\varphi_{uni}} \tag{12-11}$$

When $m=0.5$, $\sin\varphi_{uni}$ and C_{uni} can be expressed as

$$\sin\varphi_{uni}=\frac{2(1+b)\sin\varphi_0}{2(1+b)+0.5b(\sin\varphi_0-1)}, \tag{12-12}$$

$$C_{uni}=\frac{2(1+b)c_0\cos\varphi_0}{2(1+b)+0.5b(\sin\varphi_0-1)}\cdot\frac{1}{\cos\varphi_{uni}} \tag{12-13}$$

The relations between the unified material parameters φ_{uni} and C_{uni} and the conventional material parameter φ_0 and C_0 are shown in Fig. 12.4 and Fig. 12.5.

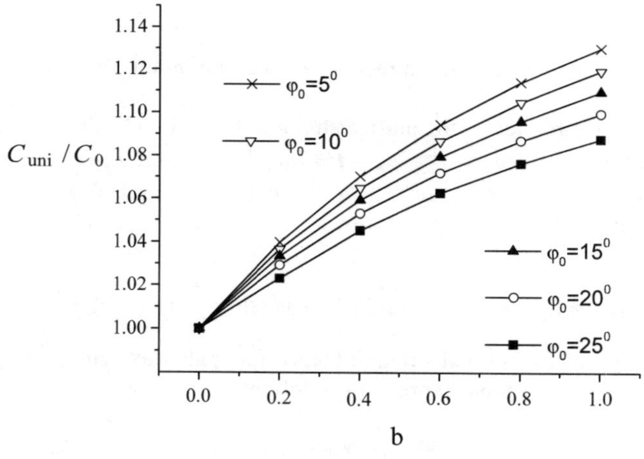

Fig. 12.4. Relation between unified cohesion C_{uni} and b

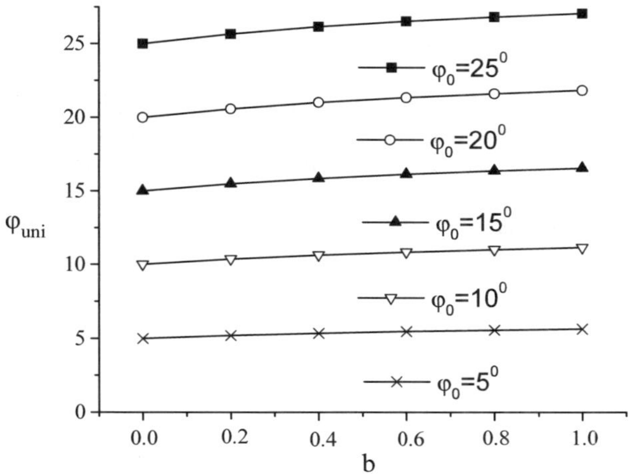

Fig. 12.5. Relation between unified internal friction angle φ_{uni} and *b*

When stress P presents compressive stress, the Eq.(12-9) changes to

$$R = P\sin\varphi_{uni} + c_{uni}\cos\varphi_{uni} \tag{12-14}$$

The equilibrium differential equations of the axisymmetric problem are shown as

$$\left.\begin{aligned} &\frac{\partial\sigma_r}{\partial r} + \frac{\partial\tau_{rz}}{\partial z} + \frac{\sigma_r - \sigma_z}{r} = 0 \\ &\frac{\partial\sigma_z}{\partial z} + \frac{\partial\tau_{rz}}{\partial r} + \frac{\tau_{rz}}{r} = 0 \end{aligned}\right\} \tag{12-15}$$

Substituting Eq. (12-7) and Eq. (12-14) into Eq.(12-15), the governing differential equations are expressed as

$$\frac{\partial P}{\partial r}(1+\sin\varphi_{uni}\cos 2\theta) + \frac{\partial P}{\partial z}\sin\varphi_{tuni}\sin 2\theta + 2R\left(\frac{\partial\theta}{\partial z}\cos 2\theta - \frac{\partial\theta}{\partial r}\sin 2\theta\right) = \frac{R}{r}(m-1-\cos 2\theta)$$

$$\frac{\partial P}{\partial r}\sin\varphi_{uni}\sin 2\theta + \frac{\partial P}{\partial z}(1-\sin\varphi_{uni}\cos 2\theta) + 2R\left(\frac{\partial\theta}{\partial z}\cos 2\theta + \frac{\partial\theta}{\partial r}\sin 2\theta\right) = -\frac{R}{r}\sin 2\theta$$

(12-16)

Assuming that functions P and θ are given respectively along some curve $z = z(r)$ in the roz-plane, we get

$$dP=\frac{\partial P}{\partial r}dr+\frac{\partial P}{\partial z}dz\,,\ d\theta=\frac{\partial \theta}{\partial r}dr+\frac{\partial \theta}{\partial z}dz \tag{12-17}$$

Eqs.(12-12) and (12-13) constitute a system of four linear inhomogeneous algebraic equations for the partial derivatives. The characteristics line equations of the solutions can be given, let the determinant of the algebraic and the relavant numerators are zero.

Letting the determinant of the system equal to zero, we obtain the differential equations of the characteristics lines as follows

$$\alpha \text{ line group: } \frac{dz}{dr}=\tan(\theta-\mu) \tag{12-18a}$$

$$\beta \text{ line group: } \frac{dz}{dr}=\tan(\theta+\mu) \tag{12-18b}$$

There exist two families of characteristics lines, where the parameter 2μ is the angle between the characteristics lines α and β and the relation between parameters μ and φ_{uni} is

$$\mu=\frac{\pi}{4}-\frac{\varphi_{uni}}{2}.$$

According to the direction derivation formulas

$$\frac{\partial}{\partial r}=[\sin(\theta+\mu)\frac{\partial}{\partial S_\alpha}-\sin(\theta-\mu)\frac{\partial}{\partial S_\beta}]\Big/\sin 2\mu \tag{12-19a}$$

$$\frac{\partial}{\partial z}=-[\cos(\theta+\mu)\frac{\partial}{\partial S_\alpha}-\cos(\theta-\mu)\frac{\partial}{\partial S_\beta}]\Big/\sin 2\mu \tag{12-19b}$$

where S_α and S_β are the convected coordinate system of the characteristics lines α and β, respectively. Therefore, the unified characteristics line equations are introduced as follows

$$\sin 2\mu\frac{\partial P}{\partial S_\alpha}-2R\frac{\partial\theta}{\partial S_\alpha}=\frac{R}{r}(m-1)(\cos 2\mu\frac{\partial z}{\partial S_\alpha}+\sin 2\mu\frac{\partial r}{\partial S_\alpha})+\frac{R}{r}\frac{\partial z}{\partial S_\alpha}$$

$$\alpha \text{ line group} \tag{12-20a}$$

$$\sin 2\mu\frac{\partial P}{\partial S_\beta}-2R\frac{\partial\theta}{\partial S_\beta}=-\frac{R}{r}(m-1)(\cos 2\mu\frac{\partial z}{\partial S_\beta}-\sin 2\mu\frac{\partial r}{\partial S_\beta})-\frac{R}{r}\frac{\partial z}{\partial S_\beta}$$

$$\beta \text{ line group} \tag{12-20b}$$

It is the stress field of the unified characteristics line theory for axisymmetric problem.

12.4 Unified Characteristics Line Field Theory for Spatial Axisymmetric Problems (Velocity field)

The associated flow rule is

$$d\varepsilon_{ij}^{p} = d\lambda \frac{\partial f}{\partial \sigma_{ij}} \tag{12-21}$$

The unified strength theory of the spatial axisymmetric problem can be derived from the Eq. (12-3) and Eq. (12-8)

$$f = \frac{1}{2}(\sigma_r + \sigma_\theta)\sin\varphi_{uni} - \sqrt{\left(\frac{\sigma_r - \sigma_\theta}{2}\right)^2 + \tau_{rz}^2} + C_{uni}\cos\varphi_{uni} = 0 \tag{12-22}$$

Under the small deformation condition, the strain rates can be expressed as

$$\begin{aligned}
\dot{\varepsilon}_r &= \frac{\partial u}{\partial r} = -\lambda \frac{\partial f}{\partial \sigma_r} \\
\dot{\varepsilon}_\theta &= \frac{\dot{u}}{r} = -\lambda \frac{\partial f}{\partial \sigma_\theta} \\
\dot{\varepsilon}_z &= \frac{\partial v}{\partial z} = -\lambda \frac{\partial f}{\partial \sigma_z} \\
\dot{\gamma}_{rz} &= \frac{\partial u}{\partial z} + \frac{\partial v}{\partial r} = -\lambda \frac{\partial f}{\partial \tau_{rz}}
\end{aligned} \tag{12-23}$$

From Eqs. (12-21), (12-22), (12-23) and Eq (12-10), we can get

$$\dot{\varepsilon}_r = -\frac{\lambda}{2}(\sin\varphi_{uni} - \cos 2\theta) \tag{12-24a}$$

$$\dot{\varepsilon}_z = -\frac{\lambda}{2}(\sin\varphi_{uni} + \cos 2\theta) \tag{12-24b}$$

$$\dot{\lambda}_{rz} = \lambda \sin 2\theta \tag{12-24c}$$

where θ is the angle from the direction of principle stress σ_r to the direction of axis r.

When the angle between the characteristics line α and axis r is denoted by ψ, the relation between φ and θ is shown as

$$\theta = \psi + \mu = \psi + \left(\frac{\pi}{4} - \frac{\varphi_{uni}}{2}\right) \tag{12-25}$$

Substituting the above relation into Eq. (12-20) derives

$$\dot{\varepsilon}_r = -\frac{\lambda}{2}[\sin\varphi_{uni} + \sin(2\psi - \varphi_{uni})] \tag{12-26a}$$

$$\dot{\varepsilon}_z = -\frac{\lambda}{2}[\sin\varphi_{uni} - \sin(2\psi - \varphi_{uni})] \tag{12-26b}$$

$$\dot{\lambda}_{rz} = \lambda\cos(2\psi - \varphi_{uni}) \tag{12-26c}$$

When one of the characteristics lines α or β overlaps with axis r, namely $\psi = 0$ or $\psi = -(\frac{\pi}{2} - \varphi_{uni})$, Eq. (12-21) can be simplified to

$$\dot{\varepsilon}_r = (\frac{\partial\dot{u}}{\partial r})_{\psi=0} = 0 \text{ or } \dot{\varepsilon}_r = (\frac{\partial\dot{u}}{\partial r})_{\psi=-(\frac{\pi}{2}-\varphi_{uni})} = 0 \tag{12-27}$$

It shows that the strain rate is zero along the direction of the characteristics lines.

If the velocities along the characteristics lines α and β are denoted by V_α and V_β, the velocities along axes r and z are denoted by u and w, respectively, the relations among them can be expressed as

$$u = \frac{V_\alpha\sin(\psi + 2\mu) - V_\beta\sin\psi}{\sin 2\mu} \tag{12-28a}$$

$$w = \frac{V_\alpha\cos(\psi + 2\mu) - V_\beta\cos\psi}{-\sin 2\mu} \tag{12-28b}$$

It can derive the following equations by getting differential with respect to r and combining with Eq.(12-22)

$$\left(\frac{\partial u}{\partial r}\right)_{\psi=0} = \left(\frac{\partial V_\alpha}{\partial r}\right)_{\psi=0} + V_\alpha\cot 2\mu\left(\frac{\partial\psi}{\partial r}\right)_{\psi=0} - V_\beta\csc 2\mu\left(\frac{\partial\psi}{\partial r}\right)_{\psi=0} = 0 \tag{12-29}$$

From the above equations, the velocity equation along the characteristics line α and line β are obtained as follows

α line group: $$dV_\alpha + \left[V_\alpha\cot\left(\frac{\pi}{2} - \varphi_{uni}\right) - V_\beta\csc\left(\frac{\pi}{2} - \varphi_{uni}\right)\right]d\psi = 0 \tag{12-30a}$$

β line group: $$dV_\beta + \left[V_\alpha\csc\left(\frac{\pi}{2} - \varphi_{uni}\right) - V_\beta\cot\left(\frac{\pi}{2} - \varphi_{uni}\right)\right]d\psi = 0 \tag{12-30b}$$

It is the velocity field of the unified characteristics line theory for axisymmetric problem.

12.5 Applications of the Unified Characteristics Field Theory

Example 1. Circular Cone under Compression
A circular cone under compression is shown in Fig. 12.6. The top radius of the circular cone is 20mm and the side face at rest has the oblique angle 45^0 from axis r. The material parameters are: E = 210 GPa, C_0 = 120MPa, and $\varphi_0 = 0$. Assuming that the contact surface is smooth, determine the characteristics line field and the limit loading of an circular cone during compression based on the unified characteristics line field theory.

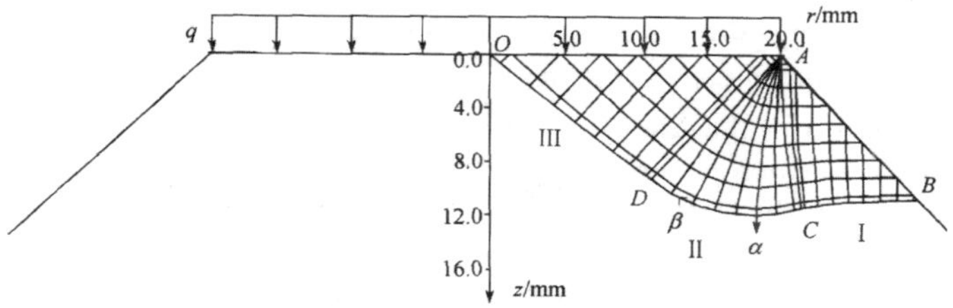

Fig. 12.6. Characteristics line field of circular cone under compression

Only the right part of the cone should be considered because of axisymmetry. Due to $\varphi_0 = 0$ it can be obtained that $\varphi_t = 0$ and $\mu = \dfrac{\pi}{4}$, which shows that the characteristics lines are orthogonal with each other. According to this condition , the characteristics lines can be expressed as

$$\alpha \text{ line group: } \frac{dz}{dr} = \tan(\theta - \frac{\pi}{4}) \tag{12-31a}$$

$$\beta \text{ line group: } \frac{dz}{dr} = \tan(\theta + \frac{\pi}{4}) \tag{12-31b}$$

It can derive the governing differential equations

$$\alpha \text{ line group } \frac{\partial P}{\partial S_\alpha} - 2R\frac{\partial \theta}{\partial S_\alpha} = \frac{R}{r}[(m-1)\frac{\partial r}{\partial S_\alpha} + \frac{\partial z}{\partial S_\alpha}] \tag{12-32a}$$

$$\beta \text{ line group: } \frac{\partial P}{\partial S_\beta} + 2R\frac{\partial \theta}{\partial S_\beta} = -\frac{R}{r}[(1-m)\frac{\partial r}{\partial S_\beta} + \frac{\partial z}{\partial S_\beta}] \tag{12-32b}$$

If the parameter m is 0.5, the differential terms in Eq. (12-32) are substituted by differential coefficient terms and Eq. (12-32) can be solved by numerical integral. The stress field is divided into three regions I, II and III, which correspond to the sides AB and AC, AC and AD, AD and AO, resperctively.

Firstly, we begin to construct the characteristics line field of the region I from no-stress side AB and get the values P and θ of the region. There are $P=-R$, $\sigma_1=\sigma_2=0$, $\sigma_3=2R$ and $\theta=\dfrac{3\pi}{4}$ on the side AB.

Secondly, for the singularity of point A, the characteristics lines field of the region II is constructed with lines α and lines β which is orthogonal to lines α. The two families of lines form a center fan of characteristics line at point A.

Finally, we can find that the characteristics line field of the region III is dependent on the sides AO and AD, and the value θ is zero along the side AO. Therefore, the values P and θ can be derived in the region II. If the total loading acted on the top surface is divided by the area of the surface, the ultimate pressure q can be obtained. The unified characteristics line field with $b=1$ is shown in Fig. 12.6 for the circular cone of top angle 45^0.

In order to confirm the results of the theory, the elasto-plastic finite element program UEPP based on the unified strength theory (UST) is used to compute the limit loading q of the same cone. The relation between limit loading q and parameter b is shown in Fig. 12.7 with the two methods. Curve 1 is the solution of the characteristics line theory for spatial axisymmetric problem, curve 2 is the result from finite element method; curve 3 is the solution of the slip line theory for plane strain problem.

From the Fig. 12.7, we can find that the result from the characteristics line theory are very chose to those from the two computer programs . When $b=0.8$, the limit loading with the new method is close to the experiment data in Suh, Lee and Rogers (1968).

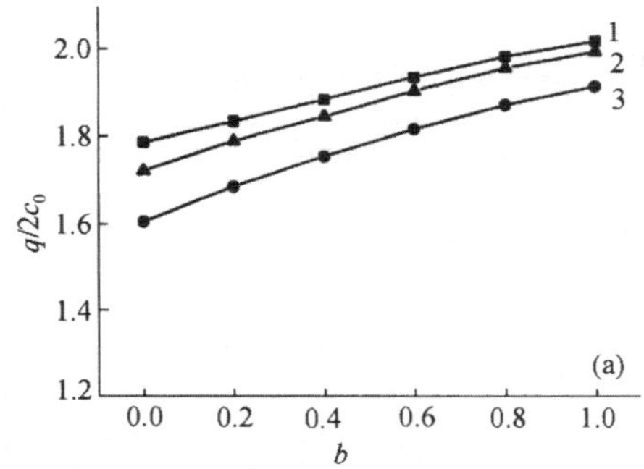

Fig. 12.7. Relation curves between limit loading q and strength criterion parameter b

The relation between loading q_A and displacement δ at the top point A is calculated with the finite element program UEPP. The two curves in Fig. 12.8 show that the limit loading of spatial axisymmetric problem (curve 1) is little bigger than that of plane strain problem (curve 2), which are close to the results in Fig. 12.7.

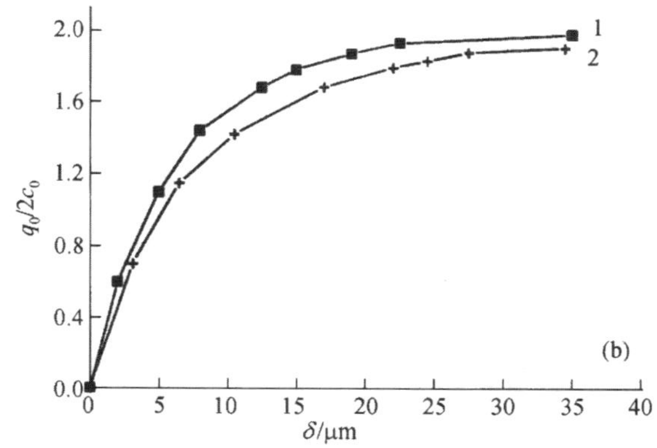

Fig. 12.8. Relation curves between displacement δ and loading q_A

Example 12.2. Semi-Infinite Body under the Circular Punch Pressure q

A semi-infinite body under the circular punch pressure q on the free surface is shown in Fig.12.9. The radium of the pressure region is 2m. The parameters are $c_0 = 0.3$MPa and $\varphi_0 = 15^0$.

The right part of the characteristics line field is shown in Fig.12.9 when $b = 1$ (the twin-shear strength theory, Yu 1985). The relations between q/c_0 and b is shown in Fig. 12.10. They are obtained by using the unified characteristics line field theory (curve 1) and the unified slip field theory (curve 2). The solution at $b = 0$ ($q/c_0 = 13.9$) in Fig.12.10 is the complete solution of Cox (1961), which is based on the Haar von-Karman condition and the Mohr-Coulomb strength theory .

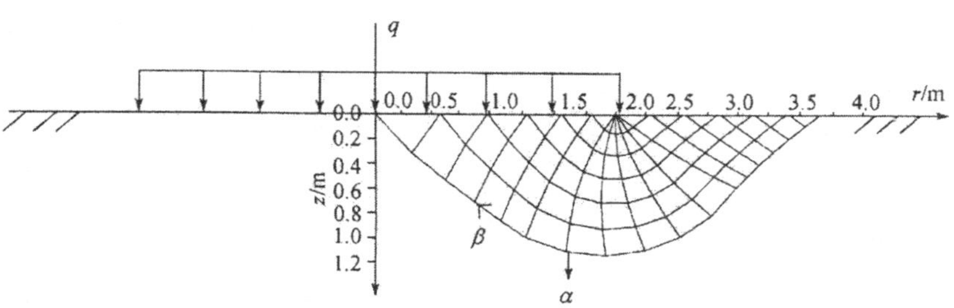

Fig. 12.9. Semi-infinite body under the compression and the characteristics line field

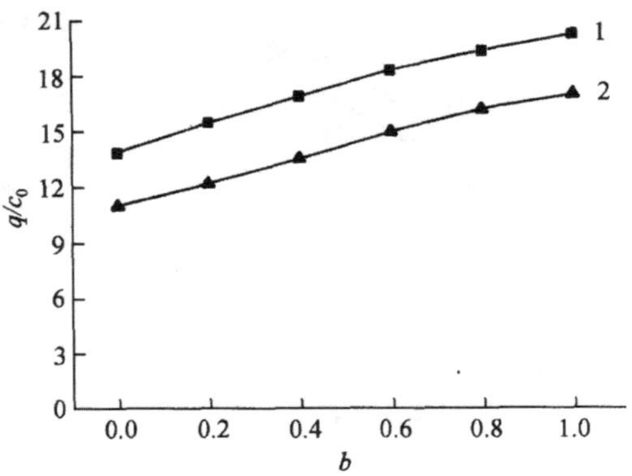

Fig. 12.10. The relation between the limit loading q and the parameter b

A series of results can be introduced for the SD materials with the different tensile-compression strength ($\varphi_0 \neq 0$) and the non-SD materials with same tensile-compression strength ($\varphi_0 = 0$) by the unified characteristics line field theory. The solutions of the Mohr-Coulomb material and Tresca material are special cases of the solutions of the unified characteristics line theory.

12.6 Penetration of High Velocity Rod to Target

12.6.1 Penetration problems of long rod

Much research has been reported on impact and penetration problems because of their importance in modern engineering applications and protective engineering. Due to the experimental, analytical and computational complexities involved in general cases, attention has been paid to the penetration of a long-rod, with different impact velocities, into a target. Studies on this subject are mainly based on tests and supplemented with analytical or numerical methods. The analytical models began with the work of Bishop et al. (1945), who developed equations for the quasi-static expansion of cylindrical and spherical cavities and used these equations to estimate forces on conical nose punches pushed slowly into metal targets. Later, Hill (1950) and Hopkins (1960) derived and discussed the dynamic, spherically symmetric, cavity-expansion equations for an incompressible target material. The cavity expansion theory was further developed by Luk and Forrestal (1987), Forrestal and Tzou (1997), Xu et al. (1997) and Mastilovic and Krajcinovic (1999) to model the penetration of projectiles through soil, porous rock, ceramic and concrete targets. Based on the characteristics line theory, Simmons et al. (1962) investigated the deformation of thin plates subjected to high-impact loads. Slip line field theory is used by Wijk (1999) to find the penetration resistance on the rod. An overview on projectiles penetrating into geological targets was given by Heuze (1990).

The dynamic failure mechanism of concrete is intricate since discontinuities such as cleavage cracks and defects with different shapes and orientations are commonly encountered in concrete and they have significant influence on the deformation and failure characteristics of concrete. The initially existing cracks and defects will be nucleated, and will evolve until material loses strength, when subjected to dynamic loading. The damage theory has been considered to be more suitable for the cleavage analysis of the concrete material (Krajcinovic, 1996; Lorrain and Loland, 1983). In the previous studies of the impact and penetration to concrete target, the commonly used constitutive models are elastic-plastic and/or brittle damage models (Luk and Forrestal, 1987; Forrestal and Tzou, 1997; Xu et al., 1997; Mastilovic and Krajcinovic, 1999), where elastoplasticity and damage are treated separately.

How to define the failure criterion of the targets is critical for better analyzing penetration problems (Zukas et al., 1978). The failure criteria such as the Mohr-Coulomb strength theory and the Tresca yield criterion were often applied to penetration problems, as can be seen from the above references. These criteria do not consider all of the stress components in stress space. The effect of intermediate principal stress is not taken into account in the Tresca criterion and the Mohr-Coulomb strength theory, which are not consistent with the test results of many materials. A unified strength theory, which was suggested by Yu (1991), considers all of the components in stress space. It covers a series of strength theories, such as the Mohr-Coulomb's single-shear strength theory and the twin-shear strength theory when the tension and compression strengths of materials are different, as well as the Tresca criterion, the Huber-von Mises criterion and the twin-shear criterion when the tension and compression strengths of materials are the same.

The unified strength theory was applied to model penetration problem by Yu, Li and Wei (see Yu 2002a), and a unified plasticity-damage penetration model related to the crack density is proposed. The relation between radial traction and velocity at cavity-surface can be obtained by analyzing the distributions of stress and velocity of the target material. Based on the cylindrical cavity expansion theory and spatial axisymmetrical unified characteristics line theory, the attacking capability of a long-rod can be assessed from the derived relation as the rod impacts and penetrates the target with initial velocities of $300 \sim 1100 m/s$. The results are compared with those of the experiments available in the open literatures.

12.6.2
Damage modeling of target material

For the present problems, the cracks of the target are mainly caused by tensile stress waves and developed along the radial orientation (Forrestal and Tzou, 1997; Xu et al., 1997; Mastilovic and Krajcinovic, 1999). Therefore, a damaged loading surface is defined based on an equivalent tensile strain. When concrete material is subjected to tension, it will not fail unless the value of the stress is larger than its tensile strength, assuming the damage is caused by the activation and growth of initially existing micro cracks. Although the damage evolution is rate-dependent in general (Chen et al., 2001), it is assumed that the damage evolution in this paper is rate-independent and isotropic for the purpose of simplicity. The definition of effective stress σ^* by Kachanov (1986) is used and mode I cracking is considered.

Since the penetration damage is accumulated as a function of time and applied stress, crack density is expressed as a function of the equivalent tensile strain and time which is similar to the model proposed by Liu and Katsabanis (1997)

$$c_f = m_1 < \bar{\varepsilon} - \varepsilon_{cr} >^{m_2} t \tag{12-33}$$

where the angular bracket $< \cdot >$ denotes the function is valid only when the value inside the bracket is larger than zero, c_f is the total number of cracks per unit volume, m_1 and m_2 are material constants, ε_{cr} is the static critical tensile strain which can be easily determined from uniaxial static tensile test results ($\varepsilon_{cr} = \sigma_{st}/E$), and t represents the accumulated time in the evolution of damage. According to the definition suggested by Whittaker et al. (1992), the scalar damage ω is related to the crack density c_f and expressed as

$$\omega = 1 - e^{-c_f V_l} \tag{12-34}$$

where V_l is the damaged volume. If the maximum principal strain is less than the critical value ε_{cr}, there is no damage in the target and the probability of fracture is equal to zero. On the other hand, if the maximum principal strain exceeds ε_{cr}, the material stiffness will degrade, based on the research by Huang et al. (1994).

12.6.3 Fundamental equations of the target

A cylindrical symmetric cavity is expanded from an initial radius of zero at velocity $\mathrm{v}_\mathrm{r}^0 = \mathrm{v}_\mathrm{r}^0(t)$, as a rod penetrates a target. The expansion of cavity produces four regions of response as shown in Fig.12.11. The four regions are: (1) plastic region (the material reaches unified strength theory, $\mathrm{v}_\mathrm{r}^0 t \le r \le ct$); (2) damaged region (the material reaches its tensile strength, $ct < r \le c_1 t$); (3) elastic region ($c_1 t < r \le c_d t$) and (4) an undisturbed region ($r > c_d t$).

v_r^0 is the cavity-expansion rate, which is rate-dependent and decreased with time t. c and c_1 are interface velocities. c_d is the elastic, dilatation velocity. r is radial Eulerian coordinate.

There are three principal stresses, σ_r, σ_θ and σ_z, which are radial, circumferential and axial stress components (positive in compression) and $\sigma_r \ge \sigma_z \ge \sigma_\theta$ for the problem.

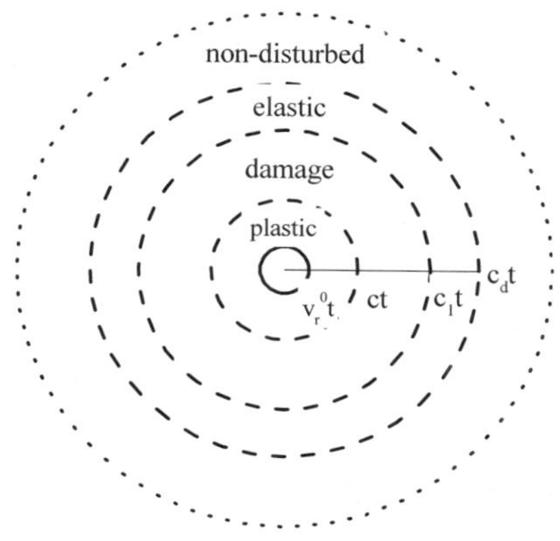

Fig. 12.11. Cylindrical cavity-expansion

The target material is considered to be compressive and is assumed to satisfy linear pressure-volumetric strain relation

$$\sigma_m = K\varepsilon_v = K(1-\rho_0/\rho) \tag{12-35}$$

where ρ_0 and ρ are densities of the original and deformed material, respectively, ε_v is the volumetric strain, K is the bulk modulus and σ_m is given by

$$\sigma_m = \frac{\sigma_r + \sigma_\theta + \sigma_z}{3} \tag{12-36}$$

According to the material characteristics of a cylindrical cavity-expansion, the strength theory can be deduced when the material comes into a plastic yielding state, namely

$$\sigma_r - \sigma_\theta = A_t \sigma_r + B_t \tag{12-37}$$

with

$$A_t = \frac{2\sin\varphi_t}{1+\sin\varphi_t}, \quad B_t = \frac{2c_t \cos\varphi_t}{1+\sin\varphi_t} \tag{12-38}$$

The equations of mass and momentum conservation in cylindrical coordinates take the forms of

$$\frac{\partial v}{\partial r} + \frac{v}{r} = -\frac{1}{\rho}\frac{d\rho}{dt}, \tag{12-39a}$$

$$\frac{\partial \sigma_r}{\partial r} + \frac{1}{r}(\sigma_r - \sigma_\theta) = -\rho\frac{dv}{dt} \tag{12-39b}$$

At the two interfaces $r = c_1 t$ and $r = ct$, the target material satisfies the Hugoniot jump condition that mass and momentum across the interface remain conservative,

$$[\rho(\mathrm{v} - \mathrm{v}_J)] = 0, \tag{12-40a}$$

$$[\sigma_r + \rho \mathrm{v}(\mathrm{v} - \mathrm{v}_J)] = 0 \tag{12-40b}$$

where the brackets [...] stand for the magnitude of the discontinuity of the bracketed variable across the wave front (interface) that propagates with the velocity v_J.

12.7 Elastic-Damage-Plastic Analysis of the Target

Define E_0, K_0, ν_0 and E, K, ν to be the Young's modulus, bulk modulus and Poission's ratio of the original and damaged materials, respectively. Because an isotropic model is used, only two of three moduli are independent. $\bar{\sigma}_{r1}$ and $\bar{\sigma}_{r2}$, $\bar{\mathrm{v}}_1$ and $\bar{\mathrm{v}}_2$, $\bar{\rho}_1$ and $\bar{\rho}_2$ are the dimensionless radial stresses, velocities and densities of the damaged and plastic regions at $r = ct$, respectively. $\bar{\sigma}_{r3}$ and $\bar{\sigma}_{r4}$, $\bar{\mathrm{v}}_3$ and $\bar{\mathrm{v}}_4$ are the dimensionless radial stresses and velocities of the elastic and damaged regions at $r = c_1 t$, respectively. $\bar{\sigma}_{ri} = \sigma_{ri}/K_0$, $\bar{\mathrm{v}}_i = \mathrm{v}_i/c$ and $\bar{\rho}_j = \rho_j/\rho_0$ ($i = 1 \sim 4$, $j = 1 \sim 2$) are used for the dimensionless variables.

12.7.1 Elastic region ($c_1 t \le r \le c_d t$)

According to the linear elastic stress-strain relations,

$$\begin{cases} \sigma_r = -\dfrac{E_0}{(1-2\nu_0)\,(1+\nu_0)}\left[(1-\nu_0)\dfrac{\partial u}{\partial r} + \nu_0 \dfrac{u}{r}\right] \\ \sigma_\theta = -\dfrac{E_0}{(1-2\nu_0)\,(1+\nu_0)}\left[(1-\nu_0)\dfrac{u}{r} + \nu_0 \dfrac{\partial u}{\partial r}\right] \end{cases} \tag{12-41}$$

and Eq. (12-39), the differential equation of wave can be derived, with the dimensionless variables $\xi = r/ct$ and $\bar{u} = u/ct$, as follows:

$$(1 - \lambda^2 \xi^2)\frac{d^2\bar{u}}{d\xi^2} + \frac{1}{\xi}\frac{d\bar{u}}{d\xi} - \frac{1}{\xi^2}\bar{u} = 0 \tag{12-42}$$

where $\lambda = c/c_d$ and u is radial particle displacement.

The boundary conditions are

$$\bar{u}(r = c_d t) = 0 \,,\ \sigma_\theta(r = c_1 t) = -\sigma_f \tag{12-43}$$

Solving Eqs. (12-42) and (12-43), the radial velocity and stress in the region can be found to be

$$\bar{v}(\xi) = \frac{1}{c}\frac{du}{dt} = -\left[\frac{\sqrt{1-\lambda^2\xi^2}}{2\lambda\xi} + \frac{\lambda\xi}{2}\ln\frac{1+\sqrt{1-\lambda^2\xi^2}}{\lambda\xi}\right]G \tag{12-44a}$$

$$\bar{\sigma}_r(\xi) = \frac{\sigma_r}{K_0} = -\frac{3}{1+\nu_0}[(1-2\nu_0)\frac{\sqrt{1-\lambda^2\xi^2}}{2\lambda\xi^2} + \frac{\lambda}{2}\ln\frac{1+\sqrt{1-\lambda^2\xi^2}}{\lambda\xi}]G \tag{12-44b}$$

where $\eta = c/c_p$, $\eta_1 = c_1/c_p$, $c_p = \sqrt{K_0/\rho_0}$, σ_f is the static tensile strength of the material, and

$$G = -\frac{\sigma_f}{K_0}\left[\frac{3(1-2\nu_0)}{1+\nu_0}\cdot\frac{\eta\sqrt{\eta^2-\lambda^2\eta_1^2}}{2\lambda\eta_1^2} - \frac{3\lambda}{2(1+\nu_0)}\ln\frac{\eta+\sqrt{\eta^2-\lambda^2\eta_1^2}}{\lambda\eta_1}\right]^{-1} \tag{12-45}$$

12.7.2 Damaged region ($ct \le r \le c_1 t$)

As the cracks appear and develop along the radial direction, the circumferential stress is zero, that is, $\sigma_\theta = 0$. The density ρ of the deformed material is a function of the scalar damage ω. At $r = c_1 t$, the material comes into plastic yielding, the radial stress can be expressed as

$$\bar{\sigma}_{r1} = \frac{\bar{B}_t}{1-A_t} \tag{12-46}$$

where $\bar{B}_t = B_t/K_0$. The distributions of the dimensionless stress and velocity, and the relationship between η_1 and η can be derived from Eqs. (12-35) and (12-36) and Eq. (12-39), as ε_v is very small and $1-\varepsilon_v \approx 1$.

$$\bar{\sigma}_r(\xi) = -t_1/\xi + t_2 \tag{12-47}$$

$$\bar{v}(\xi) = -t_2/(\eta^2\xi) + (1+\nu)t_1/(3\bar{K}) \tag{12-48}$$

$$\frac{B_t}{(1-A_t)K_0}=\frac{\eta_1^{2}(1+\nu)\bar{\sigma}_{r4}+3\bar{K}\eta_1\cdot\eta\bar{v}_4}{\eta^2(1+\nu)-3\bar{K}}-3\bar{K}(\frac{\eta_1}{\eta})\cdot\frac{\eta_1\cdot\eta\bar{v}_4+\bar{\sigma}_{r4}}{\eta^2(1+\nu)-3\bar{K}} \quad (12\text{-}49)$$

where

$$t_1=3\bar{K}(\frac{\eta_1}{\eta})\cdot\frac{\eta_1\cdot\eta\bar{v}_4+\bar{\sigma}_{r4}}{\eta^2(1+\nu)-3\bar{K}} \quad (12\text{-}50a)$$

$$t_2=\frac{\eta_1^{2}(1+\nu)\bar{\sigma}_{r4}+3\bar{K}\eta_1\cdot\eta\bar{v}_4}{\eta^2(1+\nu)-3\bar{K}} \quad (12\text{-}50b)$$

$$\bar{K}=K/K_0 \quad (12\text{-}50c)$$

12.7.3
Elastic-damaged and plastic-damaged interface ($r=c_1t$ and $r=ct$)

The relations between radial stresses and velocities at $r=c_1t$ can be derived from Eq. (12-34) and Eq. (12-39),

$$\bar{\sigma}_{r4}=\bar{\sigma}_{r3}+\frac{3(\eta_1-\eta\bar{v}_3)(\eta\bar{v}_4-\eta\bar{v}_3)}{3-(1+\nu_0)(\bar{\sigma}_{r3}-\bar{\sigma}_f)} \quad (12\text{-}51a)$$

$$\eta\bar{v}_4=\eta\bar{v}_3+\frac{(\eta_1-\eta\bar{v}_3)[(1+\nu_0)(\bar{\sigma}_{r3}-\bar{\sigma}_f)-\frac{1+\nu}{\bar{K}}\bar{\sigma}_{r4}]}{3-(1+\nu_0)(\bar{\sigma}_{r3}-\bar{\sigma}_f)} \quad (12\text{-}51b)$$

From Eqs. (12-35)~(12-37) and Eq. (12-40), the radial stress, velocity and density at $r=ct$ can be obtained as below

$$\bar{\sigma}_{r2}=\frac{2\bar{K}}{2\bar{K}+2-A_t}\bar{\sigma}_{r1}+\frac{[2\bar{K}-\bar{\rho}_1(2\bar{K}+\bar{B}_t)]\bar{\rho}_1\eta^2(1-\bar{v}_1)^2}{2\bar{K}+2-A_t} \quad (12\text{-}52)$$

$$\bar{v}_2=(\bar{v}_1-1)^2\cdot\frac{\bar{\rho}_1\bar{\sigma}_{r1}(2-A_t)+\bar{\rho}_1^2\eta^2(\bar{v}_1-1)^2(2-A_t)-\bar{\rho}_1(2\bar{K}+\bar{B}_t)}{\bar{\rho}_1^2\eta^2(\bar{v}_1-1)^2(2-A_t)-2\bar{K}}+1 \quad (12\text{-}53)$$

$$\bar{\rho}_2=\frac{\bar{\rho}_1\eta^2(1-\bar{v}_1)^2(2-A_t)-2\bar{K}}{\bar{\sigma}_{r1}(2-A_t)+\bar{\rho}_1\eta^2(1-\bar{v}_1)^2(2-A_t)-(2\bar{K}+\bar{B}_t)} \quad (12\text{-}54)$$

12.7.4
Plastic region ($v_r^0t\leqslant r\leqslant ct$)

The boundary conditions are

$$\bar{v}(\xi=v_r^0/c)=v_r^0/c,\ \bar{v}(r=ct)=\bar{v}_2,\ \bar{\sigma}_r(r=ct)=\bar{\sigma}_{r2} \quad (12\text{-}55)$$

The differential forms of radial stress and velocity can be derived, from Eqs. (12-40), (12-35) and (12-37), to be

$$\frac{d\overline{\mathrm{v}}}{d\xi}=\frac{2\overline{K}(1-\varepsilon_v)^2\overline{\mathrm{v}}+(2-A_t)(A_t\overline{\sigma}_r+\overline{B}_t)(\xi-\overline{\mathrm{v}})(1-\varepsilon_v)}{\xi[(2-A_t)\eta^2(\xi-\overline{\mathrm{v}})^2-2\overline{K}(1-\varepsilon_v)^2]} \tag{12-56}$$

$$\frac{d\overline{\sigma}_r}{d\xi}=\frac{2\overline{K}(1-\varepsilon_v)[(A_t\overline{\sigma}_r+\overline{B}_t)(1-\varepsilon_v)+\eta^2(\xi-\overline{\mathrm{v}})\overline{\mathrm{v}}]}{\xi[(2-A_t)\eta^2(\xi-\overline{\mathrm{v}})^2-2\overline{K}(1-\varepsilon_v)^2]} \tag{12-57}$$

where $\varepsilon_v=[(2-A_t)\overline{\sigma}_r-\overline{B}_t]/(2\overline{K})$. The numerical solutions of Eqs. (12-55) and (12-56) for the plastic region can be got by using Runge-Kutta method. The radial stress and velocity of the cavity surface can be obtained when the condition $\overline{\mathrm{v}}(\xi=\mathrm{v}_r^0/c)=\mathrm{v}_r^0/c$ is satisfied.

12.7.5 Solution procedure and the normal penetration dept

For the purpose of simplicity, the solution is obtained with an inverse procedure from the elastic region to the plastic region, because the boundary condition of the cavity surface is unknown. The procedure is summarized as follows:

(1). The mathematical expressions of $\overline{\mathrm{v}}_3$ and $\overline{\sigma}_{r3}$ in elastic region can be found from Eqs. (12-44a) and (12-44b). The mathematical expressions of $\overline{\mathrm{v}}_4$ and $\overline{\sigma}_{r4}$ in damaged region can be found from Eqs. (12-50a) and (12-50b). The relationship between η_1 and η can be got when the above expressions $\overline{\mathrm{v}}_3$, $\overline{\sigma}_{r3}$, $\overline{\mathrm{v}}_4$ and $\overline{\sigma}_{r4}$ are inserted into Eq. (12-48).

(2). Choose a value for η and calculate η_1 from the relationship between η_1 and η.

(3). The values of $\overline{\mathrm{v}}_3$, $\overline{\sigma}_{r3}$, $\overline{\mathrm{v}}_4$ and $\overline{\sigma}_{r4}$ can be obtained by substituting η_1 and η into Eqs.(12-44a), (12-44b), (12-50a) and (12-50b). $\overline{\mathrm{v}}_2$ and $\overline{\sigma}_{r2}$ can be derived by substituting $\overline{\mathrm{v}}_1$ and $\overline{\sigma}_{r1}$ into Eq.(12-51) to (12-53).

(4). The calculations of Eqs. (12-55) and (12-56) are proceeded from the plasticity-damage interface $\xi=1$ to the cavity surface $\xi=\mathrm{v}_r^0/c$. The particle velocity and radial stress in the plastic region can then be obtained. When the boundary condition $\overline{\mathrm{v}}\big|_{\xi=\mathrm{v}_r^0/c}=\mathrm{v}_r^0/c$ is satisfied, the value of v_r^0 corresponding to the chosen value of η is determined. Since $\eta\overline{\mathrm{v}}_r^0=\mathrm{v}_r^0/c_p$, v_r^0/c_p can be obtained. An inverse procedure is used to calculate the interface velocity and the cavity-expansion velocity. The relation between the radial stress σ_r^0 and the cavity-expansion velocity v_r^0 can then be obtained.

(5). After the numerical relationship between v_r^0 and σ_r^0 are obtained, a mathematical penetration expression of σ_r^0 versus v_r^0 can be deduced by curve fitting methods.

12.7.6 Final depth of normal penetration

When a long-rod with an ogival nose penetrates vertically into a semi-infinite concrete target, f_n and f_τ are normal force and tangential force. Tangential force $f_\tau = \mu \cdot f_n$, where μ is the friction parameter. The ogival nose is shown in Fig. 12.12, where s is the radius of the rod, and θ_0 and θ_1 are the center angles of the top and the side surface of the rod, respectively.

Following Forrestal and Tzou (1997) it is assumed that the normal force f_n on the ogive nose is actually equal to the force $\sigma_r^0(v_r^0)$ needed to expand the cavity at the rate v_r^0. Letting V_z to be the penetration velocity of the rod, we have

$$v_r^0 = V_z \cdot \cos\theta_1 \tag{12-58}$$

If the rod is assumed to be rigid and the shank side resistance is neglected, the final normal depth after the rod completely penetrates into the target can be derived from v_r^0 versus σ_r^0 and the Newton's second law, namely

$$z_{max} = \frac{M}{2n_2} \ln\left|1 + \frac{n_2}{m_2 - Mg} \cdot V_0^2\right| \tag{12-59}$$

where $m_2 = 2\pi s^2 A_1 K_0 \int_{\theta_0}^{90^0} (\sin\theta_1 - \sin\theta_0)(\cos\theta_1 + \mu\sin\theta_1) d\theta_1$,

$n_2 = 2\pi s^2 A_3 \rho_0 \int_{\theta_0}^{90^0} (\sin\theta_1 - \sin\theta_0)(\cos\theta_1 + \mu\sin\theta_0)\cos^2\theta_1 d\theta_1$, A_1 and A_3 are the coefficients of curve-fit v_r^0 versus σ_r^0, M is the mass of the rod, and V_0 is the impact velocity of the rod.

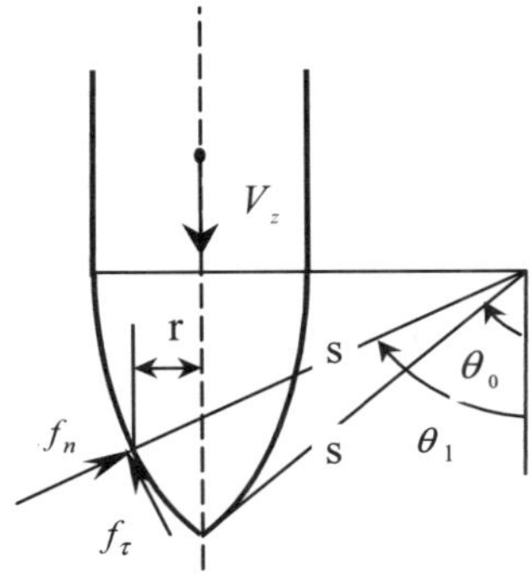

Fig. 12.12. Geometry of rod nose and traction components

12.8 Comparison and Verification

A special concrete considered by Mastilovic and Krajcinovic (1999) is defined with the following parameters, $\nu_0 = 0.22$, $E_0 = 11.3GPa$, $\rho_0 = 2260Kg/m^3$, $\sigma_f = 13MPa$ and $\mu = 0.1$. The parameters of a long-rod with an ogival nose are $M = 1.6Kg$, $s = 0.0915m$ and the radius of the rod shank $a = 0.0153\ m$.

According to the test data (Forrestal and Tzou, 1997) and some of the numerical investigations of the damage evolution pattern (Mastilovic and Krajcinovic, 1999), the damage value is about 0.23 when the dynamic tensile stress reaches the dynamic failure stress, implying $c_f V_l$ can be derived. The degraded moduli due to damage can be obtained from the analysis of Huang et al. (1994).

The relations of the plastic-damaged interface velocity c versus cavity-expansion velocity v_r^0 and v_r^0 versus σ_r^0 can be derived by using the methods from steps (1) to (5), which are described in 12.7.5. The relations are shown in Fig. 12.13 and Fig. 12.14.

It is shown that the plastic-damaged interface velocity c is increased with the increase of expansion velocity v_r^0, as shown in Fig. 12.13.

The radial stress σ_r^0 is also incresed with the expansion velocity v_r^0 as shown in Fig. 12.14. For the same cavity-expansion velocity v_r^0, the radial stress at cavity surface is changing for the different values of strength criterion parameter b. The radial stress σ_r^0 is increased with the increase of parameter b.

The dimensionless radial stress $\overline{\sigma}_r^0$ ($\overline{\sigma}_r^0 = \sigma_r^0 / K_0$) can be expressed as a second order polynomial of $\mathrm{v}_\mathrm{r}^0 / \sqrt{K_0/\rho_0}$, that is,

$$\bar{\sigma}_r{}^0 = A_1 + A_2(\mathrm{v}_\mathrm{r}^0/\sqrt{K_0/\rho_0}) + A_3(\mathrm{v}_\mathrm{r}^0/\sqrt{K_0/\rho_0})^2 \tag{12-60}$$

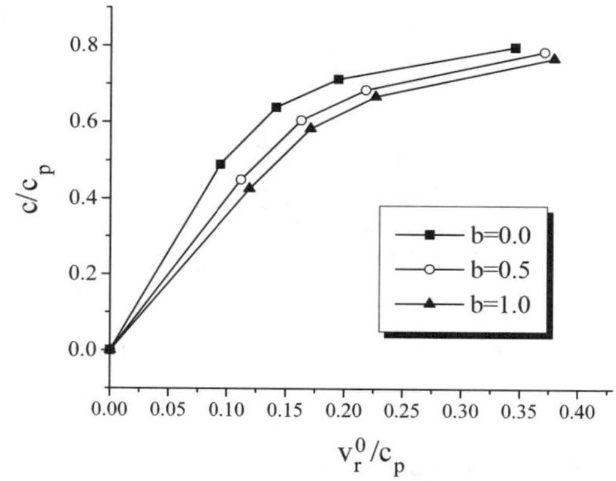

Fig.12-13. Plastic-damage velocity vs cavity-expansion velocity

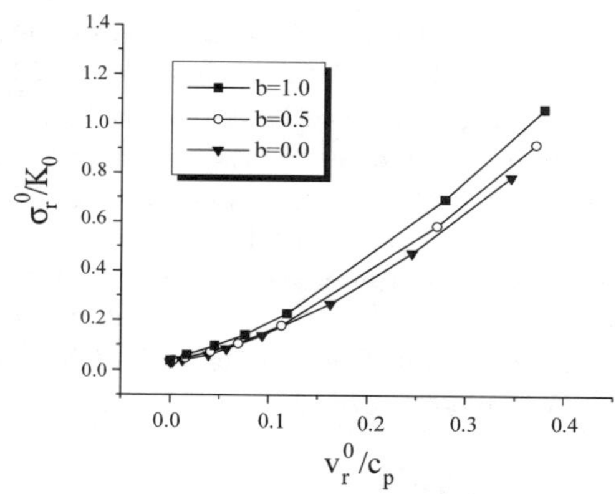

Fig.12.14. Radial stress vs velocity of cavity-expansion

According to the conditions

$$\sigma_r^0\Big|_{\mathrm{v}_\mathrm{r}^0=0} = \sigma_{rst}^0 \text{ and } \frac{d\sigma_r^0}{d\mathrm{v}_\mathrm{r}^0}\Big|_{\mathrm{v}_\mathrm{r}^0=0} = 0, \tag{12-61}$$

where σ_{rst}^0 is the dimensionless radial traction of quasi-static cavity-expansion. It can be found that $A_1 = \sigma_{rst}^0$ and $A_2 = 0$. The value of σ_{rst}^0 can be derived from the spatial axisymmetric unified characteristics line theory (Yu et al., 2001). The spatial axisymmetric characteristics lines at the contact surface is shown as

Fig. 12.3, in which the effect of surface friction is not taken into account. Based on the spatial axisymmetric characteristics line theory, we can get:

(1) $A_1 = 0.03377$ when $b = 1$;
(2) $A_1 = 0.03203$ when $b = 0.5$, which is close to the quasi-static cavity-expansion radial limit stress $\sigma_{rst}^0 = 0.0323$ of Mastilovic and Krajcinovic (1999);
(3) $A_1 = 0.03072$ when $b = 0$.

The value A_3 can be obtained from the curve of V_r^0 vs σ_r^0, that is, $A_3 = 6.8{\sim}7.0$ (when $b = 1$), $A_3 = 4.5{\sim}4.7$ (when $b = 0.5$) and $A_3 = 2.2{\sim}2.5$ (when $b = 0$).

The final depths for the different values of strength criterion parameter b can be derived from Eqs. 12-58, 12-59 and 12-60. They are shown in Fig. 12.15.

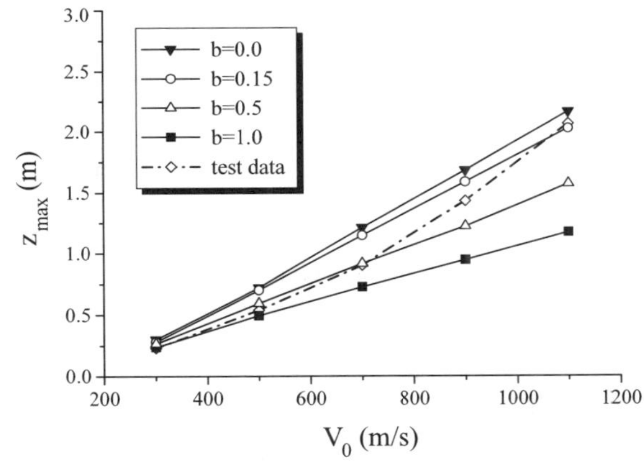

Fig. 12.15. Final depth vs initial impact velocity of long rod

When the impact velocity is $V_0 = 300m/s$, the final depths $Z_{max} = 0.236$m for $b = 1$ are close to the test data $Z_{max} = 0.23$m (Forrestal and Tzou, 1997). When the impact velocity is $V_0 = 700$m/s, $Z_{max} = 0.915$m for $b = 0.5$ are close to the test data $Z_{max} = 0.9$m. When the initial impact velocity is $V_0 = 1100$m/s, $Z_{max} = 2.02$m for $b = 0.15$ are close to the test data $Z_{max} = 2.06$m. The penetration depths of the rod are situated within the area of $0 < b < 1$.

Fig. 12.16 shows the relationship between the rod mass and the final penetration depths when $b = 1.0$. The rod penetrates more deeply if its mass is heavier. The effect of the mass on the finial depth is more obvious when the initial impact velocity of the rod is higher. One reason why tungsten alloy rods (Rosenberg, 1997) are usually applied in modern arms is that their densities are higher.

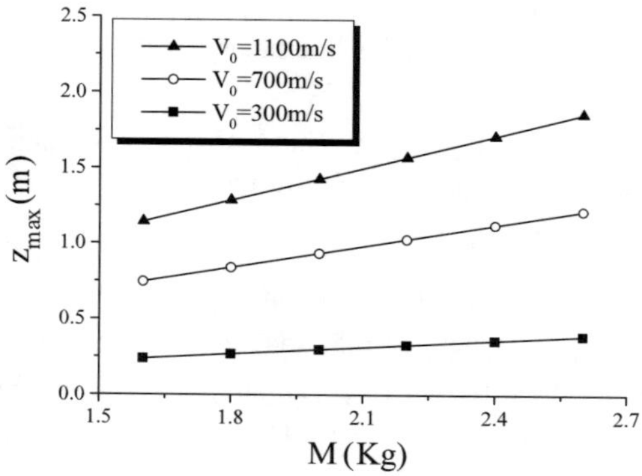

Fig. 12.16. Final depth vs mass of the rod

Summary

The characteristics line theory for spatial axisymmetric plastic problem is very important in the plasticity and engineering.

The unified characteristics line field theory of spatial axisymmetric problem and its applications are described in this chapter. The effects of the intermediate principle stress σ_2 and the SD effect of materials are taken into account in the unified characteristics field theory. A series of characteristics line field for spatial axisymmetric problem suitable for different kinds of materials can be derived from the new theory, the previous theories are special cases or linear approximation of the unified characteristics line theory. The new theory can be applied to the limit analysis of spatial axisymmetric plastic problems in plasticity and engineering.

Based on the cylindrical cavity-expansion theory and the unified strength theory, a unified plastic-damage model is proposed for penetration problems. The proposed model is used to simulate penetration of a long-rod into a concrete target. The spatial axisymmetric characteristics line theory is used for the analysis of quasi-static normal penetration of a long-rod. The results show that: (1) The rod mass has obvious affect on the final penetration depth; (2) By comparison with the available test date, it appears that the proposed procedure is effective for penetration analysis. The test results are situated within the analysis results $(0 < b < 1)$; (3) When the initial impact velocity of a rod is higher than $1500m/s$, the material behaviour and penetration process of rods and targets will change significantly.

Problems

Problem 12.1

The characteristyics fields for indentation of a circular head punch are shown in Fig. P12.1. The limit pressures was obtained by using the Tresca yield criterion as follows.

$$p_b = \sigma_s(1+\frac{\pi}{4})$$

Find the limit pressure of indentation of a circular head punch by using the unified characteristics line field theory of spatial axisymmetric problem for non-SD materials ($\alpha = 1$, and $b = 0$, $b = 1/2$, $b = 1$).

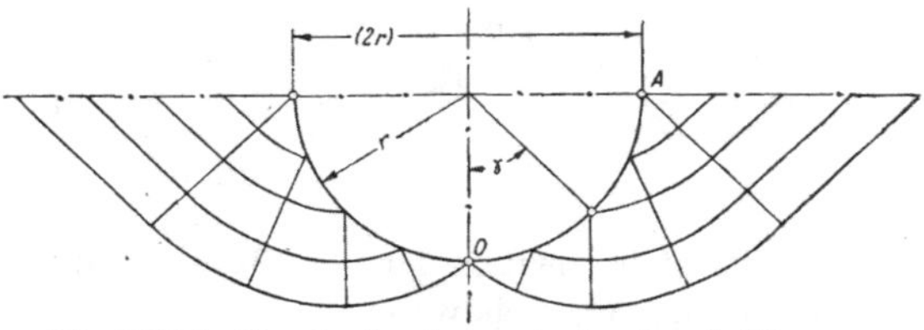

Fig. P12.1. Circular head penestration into half space

Problem 12.2

Find the limit pressure of indentation of a circular head punch by using the the unified characteristics line field theory of spatial axisymmetric problem for SD materials ($\alpha \neq 1$, and $b = 0$, $b = 1/2$, $b = 1$).

Problem 12.3

The characteristyics fields for indentation of a circular head punch are shown in Fig. P12.2. This problem was studied by Hill (1950) Prager-Hodge (1951) and Levin (1953) by using the Tresca yield criterion. A is a initial situation; b is the indentation of the circular head; c is the deep indentation of the circular head and bar. The limit pressures are:

$$p_b = \sigma_s(1+\frac{\pi}{4}) \text{ and } p_c = \sigma_s(1+\frac{3\pi}{4}).$$

Find the limit pressure of deep indentation of a circular head punch by using the the unified characteristics line field theory of spatial axisymmetric problem for non-SD materials ($\alpha = 1$, and $b = 0$, $b = 1/2$, $b = 1$).

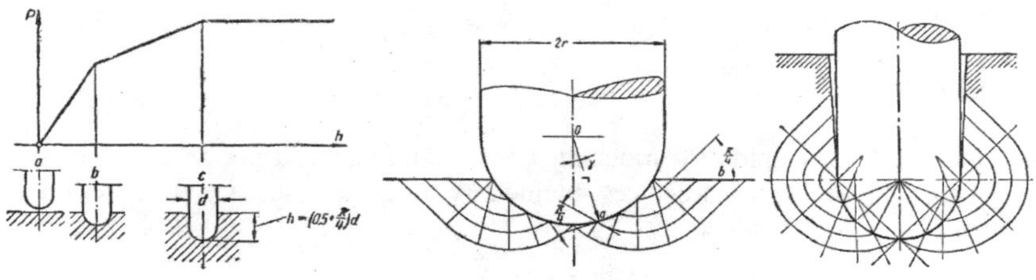

Fig. P12.2. Circular head penestration into infinit surface

Problem 12.4

Find the limit pressure of deep indentation of a circular head punch by using the the unified characteristics line field theory of spatial axisymmetric problem for SD materials (α ≠ 1, and b = 0, b = 1/2, b = 1).

Problem 12.5

The characteristyics fields of indentation for non-SD materials under a circular head punch with a uniform pressure *q* shown in Fig. P12.3. The limit pressures was obtained by using the Tresca yield criterion. Find the limit pressure of indentation of a circular head punch by using the the unified characteristics line field theory of spatial axisymmetric problem for non-SD materials (α = 1, and b = 0, b = 1/2, b = 1).

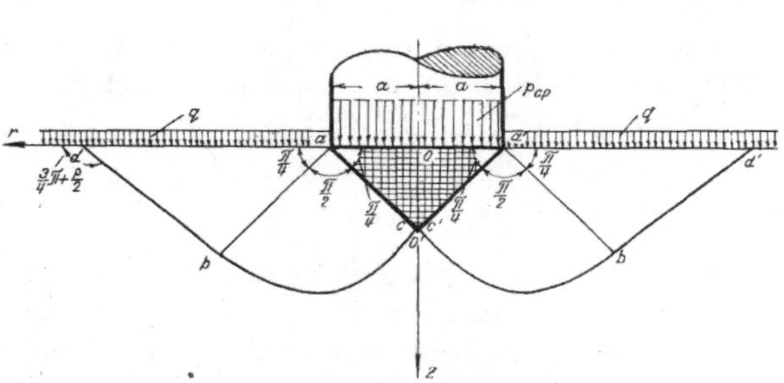

Fig. P12.3. Slip field under circular foundation for non-SD materials

Problem 12.6

Find the limit pressure of deep indentation of a circular head punch by using the the unified characteristics line field theory of spatial axisymmetric problem for SD materials (α ≠ 1, and b = 0, b = 1/2, b = 1).

References and Bibliography

Bishop RF, Hill R and Mott NF (1945) The theory of indentation and hardness tests. The Proc. of the Phys. Soc. 57(3): 147–159.

Chen WF (1975) Limit Analysis and Soil Plasticity. Elsevier, Amsterdam, New York.

Chen Z, Deng M and Chen EP (2001) Rate-dependent transition from the tensile damage to discrete fracture in dynamic brittle failure. *Theoretical and Applied Fracture Mechanics*. 35(3): 229–235.

Collins IF, Dewhurst P (1975) A slip line field analysis of asymmetrical hot rolling. International Journal of Mechanical Science, (17): 643·

Forrestal, M. J., Tzou, D. Y., 1997. A spherical cavity-expansion penetration model for concrete targets. Int. J. Solids Structures 34, 4127-4146.

Heuze FE (1990). An overview of projectile penetration into geological materials, with emphasis on rock. Int. J. Rock Mech. Min. Sci. and Geomech. Abstr. 27 (1), 1–14.

Hill R (1950). The mathematical theory of plasticity. Clarendon Press, London.

Hopkins HG (1960). Dynamic expansion of spherical cavities in metal. In: Progress in Solid Mechanics, Vol. 1, Chapter III, ed. I. N. Sneddon and R. Hill. North-Holland Publishing Company, Amsterdam, New York.

Ishlinsky AYu (1944) Axisymmetrical plasticity problem and Brinell test. *Appl. Math. Mech.* **8** (in Russian).

Johnson W, Sowerby R, Venter RD (1982) Plane strain slip line fields for metal deformation Processes-A source book and Bibliography. Oxford: Pergamon Press,.

Kachanov LM (1975) Foundations of the Theory of Plasticity. North-Holland, London,.

Kachanov LM (1986) Introduction to continuum damage mechanics. Martinus Nijhoff Publishers, Dordrecht, The Netherlands.

Krajcinovic D (1996). Damage Mechanics. Elsevier, Amsterdam.

Liu L, Katsabanis PD (1997). Development of a continuum damage model for blasting analysis. Int. Journal of Rock Mech. Min. Sci. 34, 217–231.

Lorrain M and Loland KE (1983) Damage theory applied to concrete. Fracture Mechanics of Concrete, Edited by Writtmann.

Luk VK and Forrestal MJ (1987) Penetration into semi-infinite reinforced-concrete targets with spherical and ogival nose projectiles. Int. J. Impact Eng. 6 (4): 291–301.

Levin E (1953) Indentation pressure of a smooth circular punch. *Quarterly. Appl. Math.*, (13), 133.

Mastilovic S and Krajcinovic D (1999). High-velocity expansion of a cavity within a brittle material. Journal of the Mechanics and Physics of Solids 47, 577-610.

Rosenberg Z, Dekal E, Hohler V, et al. (1997). Hypervelocity penetration of tungsten alloy rods into ceramic tiles: experiments and 2-D simulations. Int. J. of Impact Engineering 20, 675-683.

Shield RT (1955) The Plastic Indentation of a Layer by a Flat Punch. *Quarterly. Appl. Math.*, (13), 27.

Simmons J, Hauser F, Dorn JE (1962) *Mathematical Theories of Plastic Deformations under Impulsive Loading*. Cambridge University Press, London, England.

Sokolovsky VV (1960) *Statics of Cohesionless Medium*. Publ. Books on Phys. Math. Moscow (in Russian).

Suh, NP., Lee RS., Rogers CR (1968).. The Yielding of Truncated Solid Cones under Quasi-Static and Dynamic Loading, *J. Mech. Phys. Solids* 16, 357.

Szczepinski W (1979).. Introduction to the Mechanics of Plastic Forming of Metals. Sijthoff and Noordhoff, Netherlands,.

Whittaker KT, Singh RH and Sun D (1992). Rock fracture mechanics principles, design and applications. Elsevier, Amsterdam.

Wijk AG (1999). High-velocity projectile penetration into thick armor targets. Int. J. Impact Eng. 22, 45-54.

Xu Y, Keer LM, Luk VK (1997) Elastic-cracked model for penetration into unreinforced concrete targets with ogival nose projectiles. Int. J. Solids Structures 34(12), 1479–1491.

Yu MH, He LN, Song LY (1985) Twin shear stress theory and its generalization, *Scientia Sinica* (Science in China), English Edition, Series A, 28(11): 1113–1120.

Yu MH, He LN (1991) A new Model and Theory on Yield and Failure of Materials Under Complex Stress State, Mechanical Behaviors of Materials~6, Pergamon Press, Vol. 3: 841–846.

Yu MH (1992) *New System of Strength Theory*. Xi'an Jiaotong Universitry Press, Xi'an (in Chinese).

Yu MH (1994) Unified strength theory for geomaterials and its applications. Chinese J. Geot. Eng. 16(2): 1–10, (in Chinese, English Abstract).

Yu MH, Yang SY, et al (1997). Unified Plane-strain slip Line field theory system. *Chinese J. Civil Engineering*, 30(2): 14～26 (in Chinese, English Abstract).

Yu MH (1998) *Twin Shear Theory and its Application*. Science Press, Beijing 892 pp. (in Chinese).

Yu MH, Li JC and Zhang YQ (2001) Unified characteristics line theory of spatial axisymmetric plastic problem. *Science in China* (Series E), English edn. 44(2), 207–215; Chinese edn. 44(4), 323–331.

Yu MH (2002a) *Concrete Strength Theory and Its Applications*. Higher Education Press, Beijing (in Chinese).

Yu MH (2002b) Advances of strength theories of materials under complex stress state in the 20th Century. *Applied Mechanics Reviews*, 55(3): 169-258.

Zhao DW, Xu JZ, Yang H., et al (1998) Application of Twin Shear Stress Yield criterion in axisymmetric Indentation of a Semi-Infinite Medium. In : *Strength Theory* Science Press, New York, Beijing, 1079-1084.

Zukas JA, Nicholas T, Swift HF, et al. (1982). Impact Dynamics. Wiley, New York.

Березанлев ВГ (1952) Осесимметричная Задача Теории Предельного Равновесия Сыпучей Среды. Москва: Гос. Изд. Тех. Тео. Литературы (in Russia)

Соколовский ВВ (1942) Статика Сыпучей Среды. Изд. АН СССР, Москва (in Russia)

Соколовский ВВ (1960) Статика Сыпуей Среды (Third ed.).Гос. Изд. Физ-Мат. Литературы. Москва (in Russia)

13 Unified Solution of Plastic Zones at Crack Tip under Small Scale Yielding

13.1 Introduction

Estimation of the crack-tip plastic zone under small-scale yielding (SSY) is one of the important topics in elastic-plastic fracture mechanics. Literatures show that researchers have endeavored in finding the shape and size of mode-I, -II and -III crack-tip plastic zone under SSY. Dugdale (1960) and Barenblatt (1962) independently introduced their model representing a yield-strip or cohesive-strip zone around a crack-tip. It is called Dugdale yield-strip model or cohesive model. On the other hand, Irwin (1960) suggested an approximation method by increasing the effective length of a crack based on the anti-plane elastic-perfectly-plastic solution as well as Hult and McClintock (1957). It is known as the Irwin's plastic-zone correction model. The shape and size of yielding zone were derived from the Huber-von Mises criterion. Irwin (1968) further extended his model for plane-strain condition by enhancing the yield stress by a factor of $\sqrt{3}$. Broek (1982) presented his estimation using the Tresca criterion. The above-mentioned models are all based on the SSY condition and are constructed with the assumptions in the linear elastic fracture mechanics (LEFM). They yield reasonably good estimations for certain materials such as metals having equal tensile and compressive strength.

Recent investigations revealed that the Tresca criterion do not take into account the effect of the intermediate principal stress properly, while the Huber-von Mises criterion only puts in place the average effect of the three principal stresses. In other words, Tresca criterion is only applicable for those materials governed by a shear-stress limit ($\tau_s = 0.5\sigma_s$), so is the Huber-von Mises criterion for materials governed by another shear-stress limit ($\tau_s = 0.58\sigma_s$). The twin-shear yield criterion, which is equivalent to a different shear-stress limit ($\tau_s = 0.67\sigma_s$). It agrees well with the results ($\tau_s = 0.7\sigma_s$) reported by the British Royal Air Force who conducted multi-axial tests for four groups of nickel alloy at 750° Celsius (Winston, 1984).

Later, a unified strength theory (UST) was proposed (Yu et al. 1991, 1992). An influence parameter '*b*' for the effect of intermediate principal stress and the effect of intermediate principal shear-stress was put in place. The parameter '*b*' normally varies between 0 and 1. The earlier twin-shear yield criterion is embraced as a special case (*b=1*). The UST unifies prevailing strength criteria through the variable parameter '*b*', which actually bears a physical meaning as a material parameter. Adopting specific values of '*b*' can lead to results equivalent to those through Tresca criterion, Huber-von Mises criterion or its varieties. In addition, stretching beyond the normally admissible range (*b<0* or *b>1*) can generate a new series of strength criteria for unconventional materials, which strength envelope may not necessarily be convex.

Against this background, estimations of crack-tip plastic zone under SSY are re-investigated using the UST by Qiang in 1998. It provides unified solutions for shapes and sizes of mode-I, mode-II and mode-III crack-tip plastic zone in the K-dominant region. The minimum and maximum crack-tip plastic zones can be deduced through the normally admissible lower-bound (b = 0) and upper-bound (b=1) respectively. Adopting a proper value of the material parameter 'b' can lead to accurate solutions. Estimations of the crack-tip plastic zones using different values of 'b' are illustrated and compared, including those deductions equivalent to the Tresca and the Huber-von Mises criteria. The influences on the resulting plastic zone due to strength-differential (SD) effect and Poisson's ratio are also studied. The investigation of the SD effect has practical significance because most materials have unequal tensile and compressive strength.

The present approach is also extended to include some non-conventional materials under unsteady state. The definition of 'conventional material under steady state' refers to Drucker's postulate that its failure/yield strength envelope is of convex shape in the stress space (i.e. the π-space). The convexity condition is satisfied when $0 \leq b \leq 1$. Beyond this conventionally admissible range (i.e. $b < 0$ or $b > 1$), the UST is ready to be applied to a whole spectrum of non-conventional materials, which strength envelope is not necessarily of convex shape. The shape and size of crack-tip plastic zone of those unconventional materials are also presented.

13.2 Unified Strength Theory

The UST has been described in detail in Chapter 4. The mathematical modeling and the mathematical formulation are expressed as follows.

$$F = \tau_{13} + b\tau_{12} + \beta(\sigma_{13} + b\sigma_{12}) = C \quad \text{when} \quad \tau_{12} + \beta\sigma_{12} \geq \tau_{23} + \beta\sigma_{23}$$
$$F' = \tau_{13} + b\tau_{23} + \beta(\sigma_{13} + b\sigma_{23}) = C \quad \text{when} \quad \tau_{12} + \beta\sigma_{12} \leq \tau_{23} + \beta\sigma_{23} \qquad (13\text{-}1)$$

where $\tau_{ij} = (\sigma_i - \sigma_j)/2$ are the principal shear stresses and $\sigma_{ij} = (\sigma_i + \sigma_j)/2$ are the associated normal stresses. σ_i (i=1,2,3) are the principal stresses. β is a coefficient which accounts for the effect of the corresponding normal stresses. Both β and C are strength characteristics of a material, which values can be determined from standard experimental procedures for uniaxial tensile and uniaxial compressive tests. The relations are as follows:

where σ_t and σ_c are the uniaxial tensile and compressive strength respectively, $\alpha = \sigma_t / \sigma_c$ is the strength ratio. This ratio is a measure of the strength differential (SD) effect.

Substituting Eq. (13-2) into Eq. (13-1) leads to the alternative expressions in terms of the principal stresses σ_i, the tensile strength σ_t and the strength ratio α as follows.

$$\sigma_1 - \frac{\alpha}{1+b}(b\sigma_2 + \sigma_3) = \sigma_t \text{ when } \sigma_2 \le \frac{\sigma_1 + \alpha\sigma_3}{1+\alpha}$$
$$\frac{1}{1+b}(\sigma_1 + b\sigma_2) - \alpha\sigma_3 = \sigma_t \text{ when } \sigma_2 \ge \frac{\sigma_1 + \alpha\sigma_3}{1+\alpha} \tag{13-3}$$

Equation (13-3) shows that the UST not only takes into account the effect of the intermediate principal shear stress but also the SD effect. It is a function of the strength ratio α. It can be seen from the envelope plotted in the diviatoric plane (see Fig. 13.1). It should be noted that the envelopes have symmetrical characteristics at 60°-intervals. For the sake of clarity, envelopes corresponding to different values of 'b' are plotted only in the diagonally opposite sector bounded by $\theta = 0°$ and $\theta = 60°$. Along the radiation line r_t (at $\theta = 0°$), the limiting value is the tensile strength. Along the radiation line r_c (at $\theta = 60°$), the limiting value is the compressive strength. The UST criterion is a bi-linear surface approximation for the envelope between the radiation lines r_t and r_c. For values of 'b' ranging between 0 and 1, the shape of the whole envelope ($0° \le \theta < 360°$) is convex. Outside this range (i.e. $b < 0$ or $b > 1$), the shape of the whole envelope become corrugated, and hence it represents those material not obeying the Drucker's convexity postulate. In this sense, the UST unified the prevailing strength criteria. Its main features are further elaborated below.

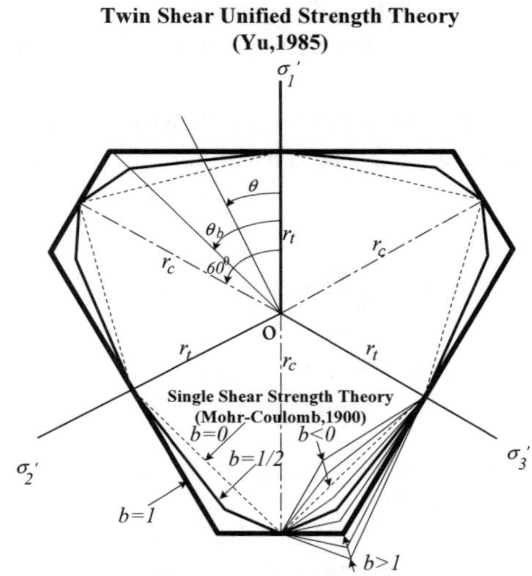

Fig. 13.1 Loci of the UST in π-plane having different values of 'b'

a) UST consists of two material parameters - b and β (or its equivalent form α), which enables it to represent the strength envelope in the deviatoric plane for all conventional materials. In other words, it embraces all traditional strength criteria. For example, setting $b = 0$ and $\alpha = 1$ leads to the well-known Tresca criterion, which neglects totally the effect of intermediate principal shear stress and is only meant for materials having equal tensile and compressive strength. Again, setting $\alpha = 1$ and $b = 1/2$ yields segmental linear envelope very close to the smooth envelope by Mises criterion. It is worth noting that setting $b = 1$ leads to Yu's twin-shear criterion – an earlier version of the UST.

b) The UST provides admissible lower and upper bounds of the strength envelope for materials obeying the Drucker's convexity postulate. When $b = 0$, it sets the lower bounds. Amongst them, Tresca envelope (α = 1) is the one for materials having equal tensile and compressive strength; Mohr-Coulomb envelope is another one but it is for materials having different tensile and compressive strengths (α ≠ 1). On the other hand when $b = 1$, it sets the upper bounds. Yu's twin-shear criterion is one of it. Hence, the UST can be used to find the minimum and maximum yield/failure zone.

c) When the value of 'b' is chosen outside the conventionally admissible range (i.e. $b < 0$ or $b > 1$), the corresponding strength envelope becomes non-convex shape. It violates the Drucker's convexity postulate. In this range, the UST yields strength envelopes for a whole spectrum of non-conventional materials, which do not obey the Drucker's postulate.

13.3 Stress Fields Around Crack-Tip

By virtue of linear elastic fracture mechanics, the 2D stress field in the vicinity of a mode-I crack-tip can be described as follows (Kanninen, 1985).

$$\begin{Bmatrix} \sigma_x \\ \sigma_y \\ \tau_{xy} \end{Bmatrix} = \frac{K_I}{(2\pi r)^{1/2}} \cos(\theta/2) \begin{Bmatrix} 1-\sin(\theta/2)\sin(3\theta/2) \\ 1+\sin(\theta/2)\sin(3\theta/2) \\ \sin(\theta/2)\cos(3\theta/2) \end{Bmatrix} \tag{13-4}$$

For mode-II crack-tip,

$$\begin{Bmatrix} \sigma_x \\ \sigma_y \\ \tau_{xy} \end{Bmatrix} = \frac{K_{II}}{(2\pi r)^{1/2}} \cos(\theta/2) \begin{Bmatrix} -\sin(\theta/2)[2+\cos(\theta/2)\cos(3\theta/2)] \\ \sin(\theta/2)\cos(\theta/2)\sin(3\theta/2) \\ \cos(\theta/2)[1-\sin(\theta/2)\sin(3\theta/2)] \end{Bmatrix} \tag{13-5}$$

For mode-III crack-tip,

$$\begin{Bmatrix} \tau_{zx} \\ \tau_{zy} \end{Bmatrix} = \frac{K_{III}}{(2\pi r)^{1/2}} \begin{Bmatrix} -\sin(\theta/2) \\ \cos(\theta/2) \end{Bmatrix} \tag{13-6}$$

in which K_I, K_{II} and K_{III} are the stress-intensity factors for mode-I, mode-II and mode-III respectively. σx, σy are the normal stresses in x- and y-direction respectively, and τ_{xy}, τ_{zx}, τ_{zy} are the shear stresses in the xy-, zx- and zy-plane respectively. Position vector is expressed in polar coordinates (r,θ) measured from the crack-tip.

The stress components in Eqs. (13-4)-(13-6) can be expressed in terms of the corresponding principal stresses σ_i. For 2-dimenional plane problems, the principal stresses are:

$$\left.\begin{matrix} \sigma_1 \\ \sigma_2 \end{matrix}\right\} = \frac{\sigma_x+\sigma_y}{2} \pm \sqrt{\left(\frac{\sigma_x-\sigma_y}{2}\right)^2 + \tau_{xy}^2} \tag{13-7a}$$

$$\sigma_3 = \begin{cases} 0 & \text{for plane stress} \\ \nu(\sigma_1+\sigma_2) & \text{for plane strain} \end{cases} \tag{13-7b}$$

in which ν is Poisson's ratio. Due to symmetry, the following derivations are presented only for the half-plane ($\theta \in [0,\pi]$). By substituting Eqs. (13-4)-(13-6) into Eqs (13-7a)-(13-7b) and arranging the resulting stresses in the order such that $\sigma_1 \geq \sigma_2 \geq \sigma_3$, we have

For mode-I (plane-stress case)

$$\left.\begin{matrix}\sigma_1 \\ \sigma_2\end{matrix}\right\} = \frac{K_I}{\sqrt{2\pi r}} \cos\frac{\theta}{2}(1 \pm \sin\frac{\theta}{2}) \qquad \sigma_3 = 0 \tag{13-8}$$

For mode-I (plane-strain case)

$$\left.\begin{matrix}\sigma_1 \\ \sigma_2\end{matrix}\right\} = \frac{K_I}{\sqrt{2\pi r}} \cos\frac{\theta}{2}(1 \pm \sin\frac{\theta}{2}) \qquad \sigma_3 = \frac{2\nu K_I}{\sqrt{2\pi r}} \cos\frac{\theta}{2} \qquad \text{for } 0 \leq \theta < 2\sin^{-1}(1-2\nu) \tag{13-9}$$

and

$$\sigma_1 = \frac{K_I}{\sqrt{2\pi r}} \cos\frac{\theta}{2}(1 + \sin\frac{\theta}{2}) \qquad \sigma_2 = \frac{2\nu K_I}{\sqrt{2\pi r}} \cos\frac{\theta}{2} \qquad \sigma_3 = \frac{K_I}{\sqrt{2\pi r}} \cos\frac{\theta}{2}(1 - \sin\frac{\theta}{2}) \qquad \text{for } 2\sin^{-1}(1-2\nu) \leq \theta < \pi \tag{13-10}$$

For mode-II (plane-stress case)

$$\sigma_1 = \frac{K_{II}}{\sqrt{2\pi r}}(-\sin\frac{\theta}{2} + \frac{1}{2}\sqrt{4 - 3\sin^2\theta}) \qquad \sigma_2 = 0 \qquad \sigma_3 = \frac{K_{II}}{\sqrt{2\pi r}}(-\sin\frac{\theta}{2} - \frac{1}{2}\sqrt{4 - 3\sin^2\theta}) \qquad \text{for } 0 \leq \theta < 2\sin^{-1}\frac{1}{\sqrt{3}} \tag{13-11}$$

and

$$\sigma_1 = 0 \qquad \left.\begin{matrix}\sigma_2 \\ \sigma_3\end{matrix}\right\} = \frac{K_{II}}{\sqrt{2\pi r}}(-\sin\frac{\theta}{2} \pm \frac{1}{2}\sqrt{4 - 3\sin^2\theta}) \qquad \text{for } 2\sin^{-1}\frac{1}{\sqrt{3}} \leq \theta < \pi \tag{13-12}$$

For mode-II (plane-strain case)

$$
\begin{aligned}
\sigma_1 &= \frac{K_{II}}{\sqrt{2\pi r}}\left(-\sin\frac{\theta}{2}+\frac{1}{2}\sqrt{4-3\sin^2\theta}\right)\\
\sigma_2 &= \frac{K_{II}}{\sqrt{2\pi r}}\left(-\sin\frac{\theta}{2}\right)\\
\sigma_3 &= \frac{K_{II}}{\sqrt{2\pi r}}\left(-\sin\frac{\theta}{2}-\frac{1}{2}\sqrt{4-3\sin^2\theta}\right)
\end{aligned}
\qquad (13\text{-}13)
$$

For mode-III (plane-stress and plane-strain cases)

$$
\begin{aligned}
\sigma_1 &= \frac{K_{III}}{\sqrt{2\pi r}}\\
\sigma_2 &= 0\\
\sigma_3 &= -\frac{K_{III}}{\sqrt{2\pi r}}
\end{aligned}
\qquad (13\text{-}14)
$$

13.4 Shape and Size of Plastic Zone for Mode-I Crack Tip

In the preceding session, the equations are all based on linear elastic fracture mechanics. Beyond the elastic limit, the material is assumed perfectly plastic. The UST is employed to determine these limits. The loci of the limiting boundary are obtained by substituting the principal-stress equations in preceding session into the UST equation. By symmetry, the loci are only derived in the half plane ($0 \le \theta \le \pi$) in terms of polar coordinates (r_p, θ) as follows.

13.4.1 Plane stress case

Substituting Eq. (13-8) into Eq. (13-3), the formulae of shape and size of plastic zone for mode-I rack tip (plane-stress case) can be obtained as follows:

$$
r_p = \frac{1}{2\pi}\left(\frac{K_I}{\sigma_t}\right)^2\left\{\cos\frac{\theta}{2}\left[1-\frac{\alpha b}{1+b}+\sin\frac{\theta}{2}\left(1+\frac{\alpha b}{1+b}\right)\right]\right\}^2, \quad when \quad \theta \ge 2\arcsin\frac{\alpha}{2+\alpha}
$$

$$
r_p = \frac{1}{2\pi}\left(\frac{K_I}{\sigma_t}\right)^2\left[\cos\frac{\theta}{2}\left(1+\frac{1-b}{1+b}\sin\frac{\theta}{2}\right)\right]^2, \quad when \quad \theta \le 2\arcsin\frac{\alpha}{2+\alpha}
$$

(13-15)

The shape and size of plastic zones for a series of 'b' values are plotted in Fig. 13.2.

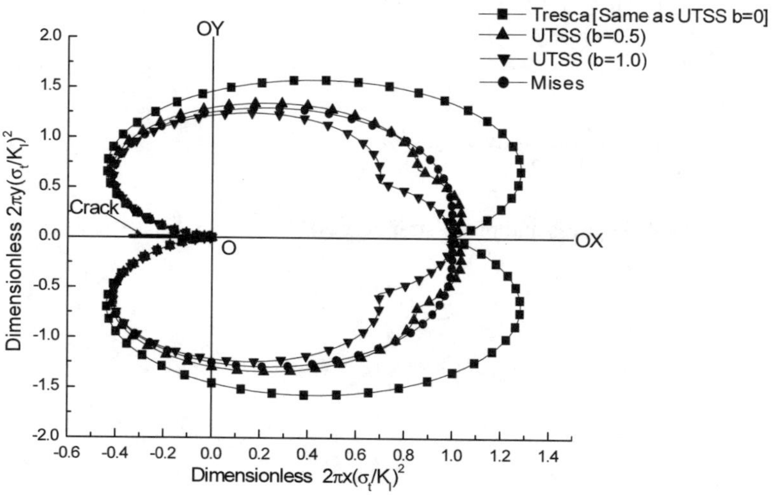

Fig. 13.2 Mode-I crack-tip plastic zone by different criteria in plane-stress state ($\alpha = 1,\ \nu = 1/3$)

13.4.2 Plane strain case

Substituting Eqs. (13-9)-(13-10) into Eq. (13-3), the formulae of shape and size of plastic zone for mode-I rack tip (plane-strain case) can be obtained as follows:

(a) when $0 \le \theta < 2\sin^{-1}(1-2\nu)$

$$r_p = \frac{1}{2\pi}\left(\frac{K_I}{\sigma_t}\right)^2 \left\{\cos\frac{\theta}{2}\left[1-\frac{\alpha b}{1+b}-\frac{2\alpha\nu}{1+b}+\sin\frac{\theta}{2}\left(1+\frac{\alpha b}{1+b}\right)\right]\right\}^2,\ when\ \theta \ge 2\sin^{-1}\left(\frac{\alpha(1-2\nu)}{2+\alpha}\right)$$
$$r_p = \frac{1}{2\pi}\left(\frac{K_I}{\sigma_t}\right)^2 \left[\cos\frac{\theta}{2}\left(1-2\alpha\nu+\frac{1-b}{1+b}\sin\frac{\theta}{2}\right)\right]^2,\ when\ \theta \le 2\sin^{-1}\left(\frac{\alpha(1-2\nu)}{2+\alpha}\right)$$

(13-16)

(b) when $2\sin^{-1}(1-2\nu) \le \theta < \pi$

$$r_p = \frac{1}{2\pi}\left(\frac{K_I}{\sigma_t}\right)^2 \left\{\cos\frac{\theta}{2}\left[1-\frac{a(1+2b\nu)}{1+b}+\sin\frac{\theta}{2}\left(1+\frac{\alpha}{1+b}\right)\right]\right\}^2,\ when\ \theta \ge 2\sin^{-1}\left(\frac{(1+\alpha)(2\nu-1)}{1-\alpha}\right)$$
$$r_p = \frac{1}{2\pi}\left(\frac{K_I}{\sigma_t}\right)^2 \left[\cos\frac{\theta}{2}\left(\frac{1+2b\nu}{1+b}-\alpha+\frac{1+a+\alpha b}{1+b}\sin\frac{\theta}{2}\right)\right]^2,\ when\ \theta \le 2\sin^{-1}\left(\frac{(1+\alpha)(2\nu-1)}{1-\alpha}\right)$$

(13-17)

The shape and size of plastic zones for a series of 'b' values are plotted in Fig. 13.3.

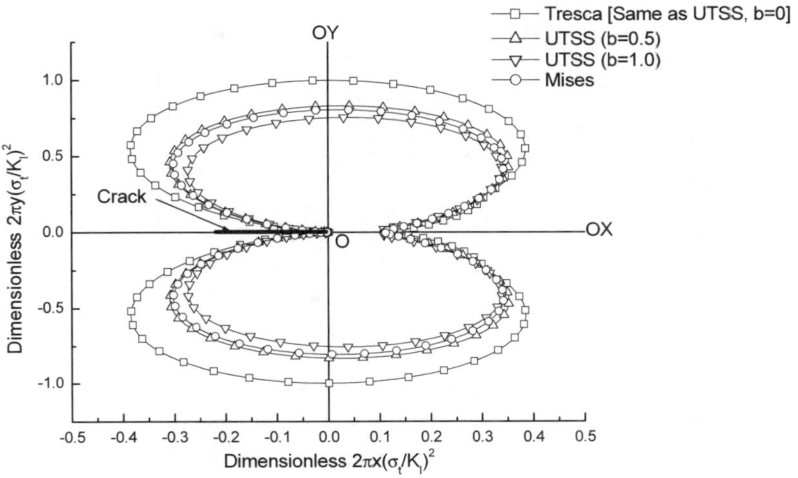

Fig. 13.3 Mode-I crack-tip plastic zone by different criteria in plane-strain state ($\alpha = 1$, $\nu = 1/3$)

13.4.3
Shape and size of plastic zone of mode-I crack tip for SD materials

The plotting in Fig. 13.3 is for materials of equal tensile and compressive strength, i.e. $\alpha=1$. To visualize the effect in strength-differential (SD) materials, the shapes and sizes of plastic zones corresponding to different α-values are plotted in the same figure, including both plane-stress and plane-strain cases. Figures 13.4-13.6 are for mode-I crack-tips with b=0, b=0.5 and b=1 respectively.

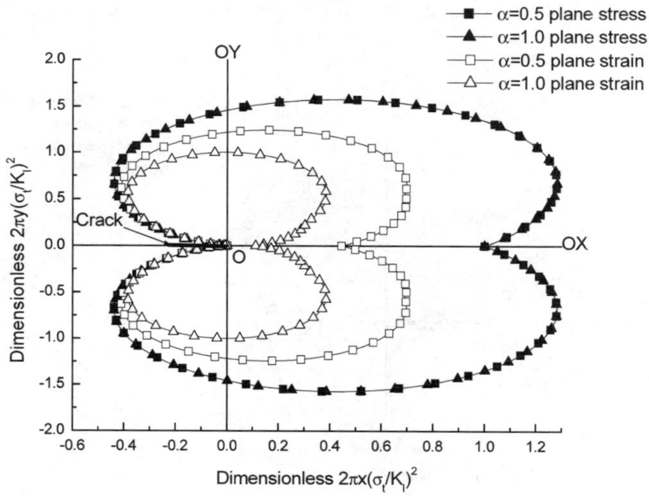

Fig. 13.4 Effect of SD on mode-I crack-tip plastic zone ($b = 0, \nu = 1/3$)

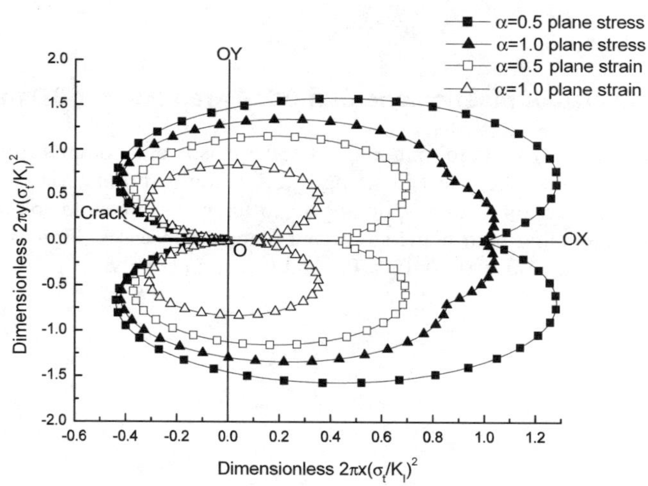

Fig. 13.5 Effect of SD on mode-I crack-tip plastic zone ($b = 0.5, \nu = 1/3$)

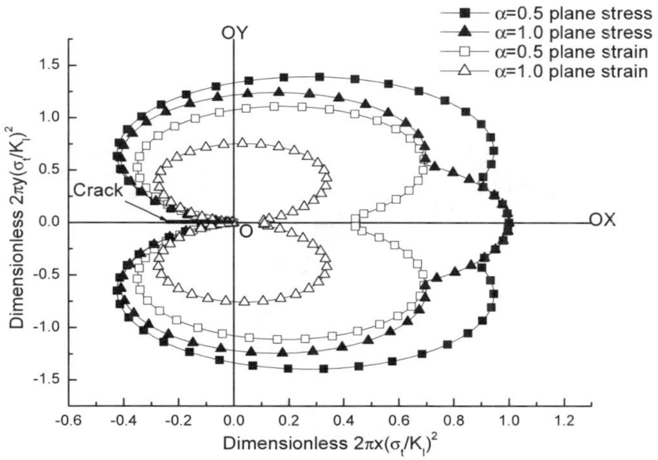

Fig. 13.6 Effect of SD on mode-I crack-tip plastic zone ($b = 1.0, v = 1/3$)

13.5 Shape and Size of Plastic Zone for Mode-II Crack Tip

13.5.1 Plane stress case

Substituting Eqs. (13-11)-(13-12) into Eq. (13-3), the shape and size of plastic zone for mode-II crack tip (plane-stress case) can be obtained as follows:

(a) when $0 \le \theta < 2\sin^{-1}\left(\frac{1}{\sqrt{3}}\right)$

$$r_p = \frac{1}{2\pi}\left(\frac{K_{II}}{\sigma_t}\right)^2\left[\left(\frac{\alpha}{1+b}-1\right)\sin\frac{\theta}{2}+\frac{\alpha+b+1}{2(1+b)}\sqrt{4-3\sin^2\theta}\right]^2, \text{ when } \sigma_1+a\sigma_3 \ge 0$$

$$r_p = \frac{1}{2\pi}\left(\frac{K_{II}}{\sigma_t}\right)^2\left[\left(\alpha-\frac{1}{1+b}\right)\sin\frac{\theta}{2}+\frac{\alpha+\alpha b+1}{2(1+b)}\sqrt{4-3\sin^2\theta}\right]^2, \text{ when } \sigma_1+a\sigma_3 \le 0$$

(13-18)

(b) when $2\sin^{-1}\left(\frac{1}{\sqrt{3}}\right) \le \theta < \pi$

$$r_p = \frac{1}{2\pi}\left(\frac{K_{II}}{\sigma_t}\right)^2\left[\alpha\sin\frac{\theta}{2}+\frac{\alpha(1-b)}{2(1+b)}\sqrt{4-3\sin^2\theta}\right]^2, \text{ when } \sigma_2 \le \frac{\alpha\sigma_3}{1+\alpha}$$

$$r_p = \frac{1}{2\pi}\left(\frac{K_{II}}{\sigma_t}\right)^2\left[\left(\alpha-\frac{b}{1+b}\right)\sin\frac{\theta}{2}+\frac{\alpha+\alpha b+b}{2(1+b)}\sqrt{4-3\sin^2\theta}\right]^2, \text{ when } \sigma_2 \ge \frac{\alpha\sigma_3}{1+\alpha}$$

(13-19)

The shape and size of plastic zones for a series of 'b' values are plotted in Fig. 13.7.

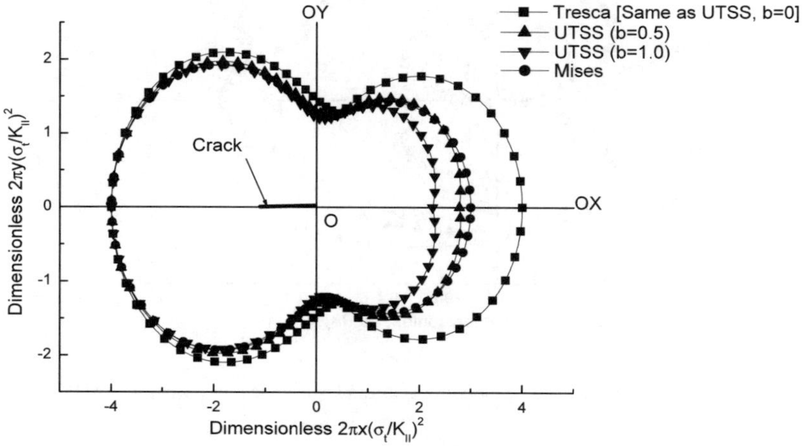

Fig. 13.7 Mode-II crack-tip plastic zone by different criteria in plane-stress state ($\alpha = 1$, $v = 1/3$)

13.5.2 Plane strain case

For mode-II (plane-strain case)
Substituting Eq. (13-13) into Eq. (13-3) leads to

$$r_p = \frac{1}{2\pi}\left(\frac{K_{II}}{\sigma_t}\right)^2\left[\left(\frac{\alpha(1+2vb)}{1+b}-1\right)\sin\frac{\theta}{2}+\frac{\alpha+b+1}{2(1+b)}\sqrt{4-3\sin^2\theta}\right]^2, \ when\ \sigma_2 \le \frac{\sigma_1+\alpha\sigma_3}{1+\alpha}$$
$$r_p = \frac{1}{2\pi}\left(\frac{K_{II}}{\sigma_t}\right)^2\left[\left(\alpha-\frac{1+2vb}{1+b}\right)\sin\frac{\theta}{2}+\frac{\alpha+\alpha b+b}{2(1+b)}\sqrt{4-3\sin^2\theta}\right]^2, \ when\ \sigma_2 \ge \frac{\sigma_1+\alpha\sigma_3}{1+\alpha}$$
(13-20)

The shape and size of plastic zones of mode-II crack-tip for a series of 'b' values are plotted in Fig. 13.8.

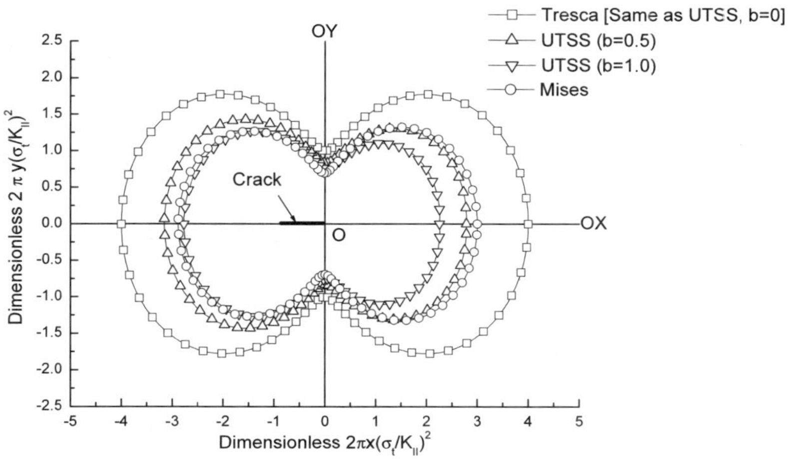

Fig. 13.8 Mode-II crack-tip plastic zone by different criteria in plane-strain state ($\alpha = 1$, $\nu = 1/3$)

13.5.3 Shape and size of plastic zone of mode-II crack tip for SD materials

The plotting in Fig. 13.7 and 13.8 is for materials of equal tensile and compressive strength, i.e. $\alpha = 1$. The shapes and sizes of plastic zones for SD materials (corresponding to different α-values), including both plane-stress and plane-strain cases are plotted. Figures 13.9-13.11 are for mode-II crack-tips with $b = 0$, $b = 0.5$ and $b = 1$ respectively.

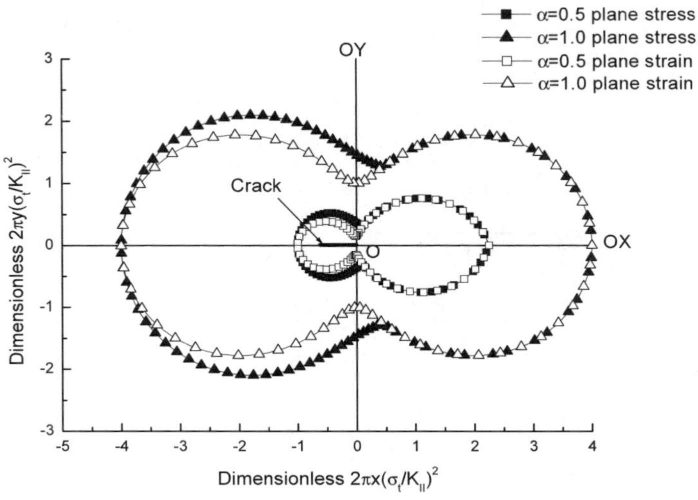

Fig. 13.9 Effect of SD on mode-II crack-tip plastic zone ($b = 0, \nu = 1/3$)

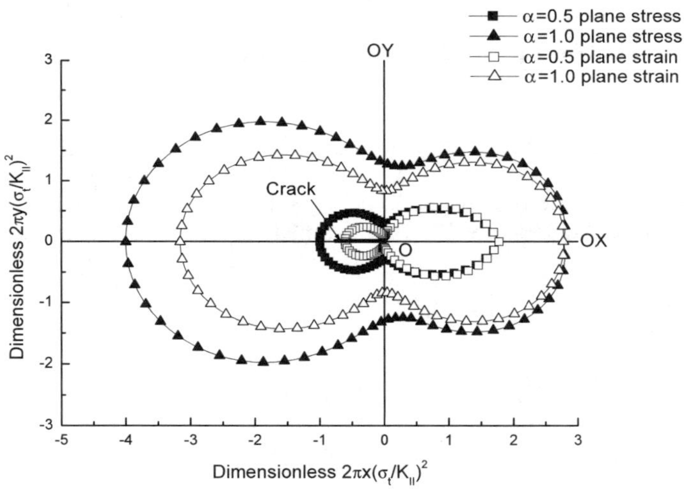

Fig. 13.10 Effect of SD on mode-II crack-tip plastic zone ($b = 0.5, \nu = 1/3$)

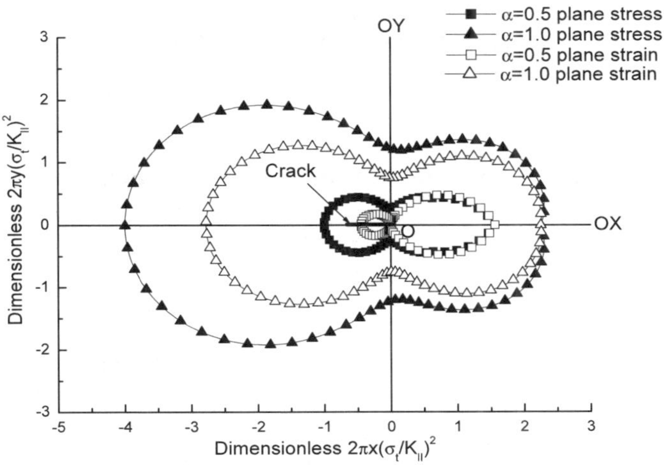

Fig. 13.11 Effect of SD on mode-II crack-tip plastic zone ($b = 1.0, \nu = 1/3$)

13.6 Plastic Zone for Mode-III Crack Tip

For mode-III (plane-stress/strain)
Substituting Eq. (13-14) into Eq. (13-3) leads to

$$r_p = \frac{1}{2\pi}\left(\frac{K_{III}}{\sigma_t}\right)^2\left(1+\frac{\alpha}{1+b}\right)^2, \text{ when } \quad \sigma_1+\alpha\sigma_3 \geq 0$$
$$r_p = \frac{1}{2\pi}\left(\frac{K_{III}}{\sigma_t}\right)^2\left(\alpha+\frac{1}{1+b}\right)^2, \text{when } \quad \sigma_1+\alpha\sigma_3 \leq 0 \tag{13-21}$$

13.7 Shape and Size of Plastic Zone for Non-Conventional Materials

The yield surface will be non-convex if the material parameter $b < 0$ or $b > 1$. The non-convex material may be called the non-conventional materials. The shape and size of plastic zone for non-conventional materials (b=–0.1, –0,25, –0.5) are also plotted. Figures 13.12 and 13.13 are for mode-I plane-stress and plane-strain case respectively, while Figs. 13.14 and 13.15 for mode-II plane-stress and plane-strain case respectively.

13.7.1 Mode-I crack-tip plastic zone for non-conventional materials (plane stress)

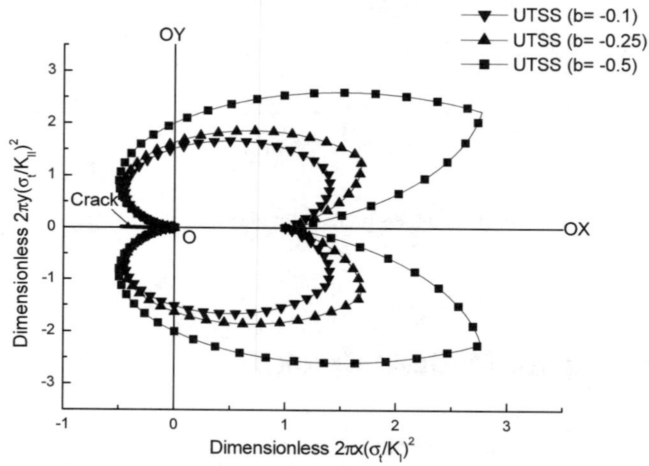

Fig. 13.12 Mode-I crack-tip plastic zone for unconventional materials in plane-stress state ($\alpha = 1, \nu = 1/3$)

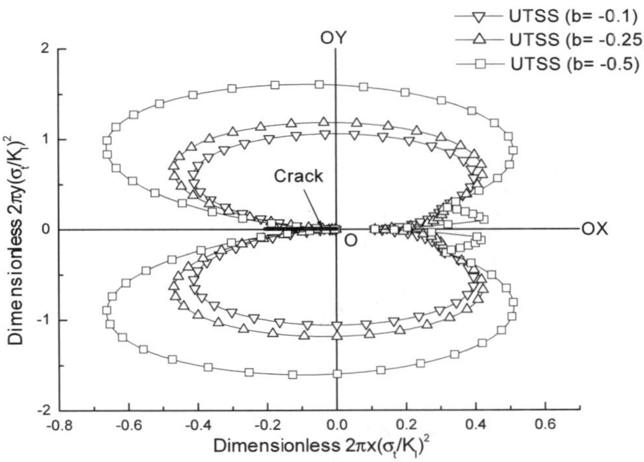

Fig. 13.13 Mode-I crack-tip plastic zone for unconventional materials in plane-strain state ($\alpha = 1, \nu = 1/3$)

13.7.2 Mode-I crack-tip plastic zone for non-conventional materials (plane stress)

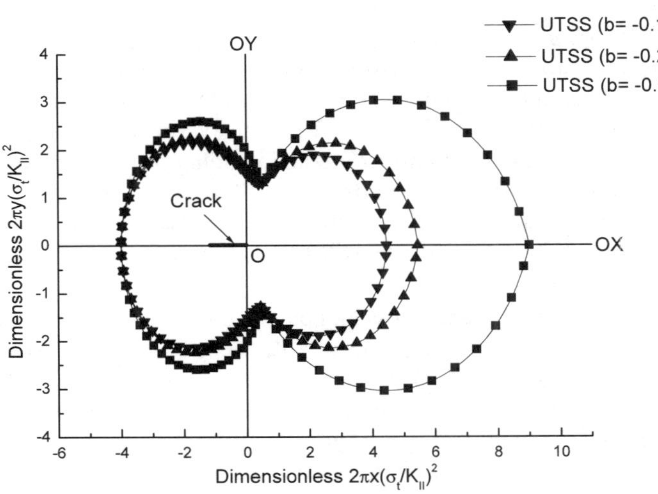

Fig. 13.14 Mode-II crack-tip plastic zone for unconventional materials in plane-stress states ($\alpha = 1, \nu = 1/3$)

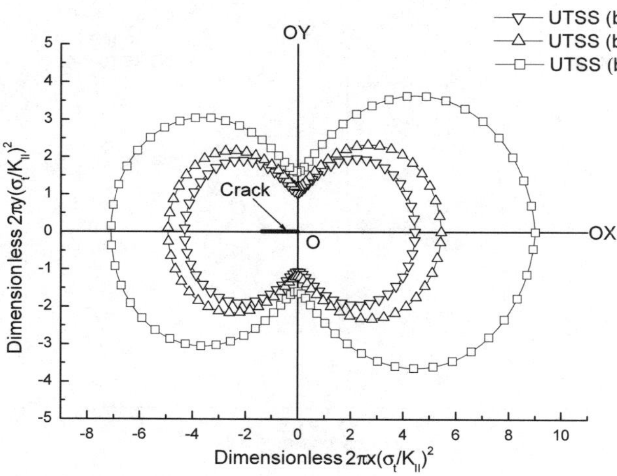

Fig. 13.15 Mode-II crack-tip plastic zone for unconventional materials in plane-strain state ($\alpha = 1.0, \nu = 1/3$)

13.8 Effect of '*b*' Value

Parameter '*b*' represents the influence of the intermediate principal shear stress. For the same material, the extents of the influence are not the same under different crack-tip modes. It reflects in the resulting shape and size of the crack-tip plastic zone. On the other hand, various values of '*b*' also correspond to a spectrum of strength criteria, including the well-known Tresca and Huber-von Mises. In other words, each '*b*' value represents a strength criterion, which is only valid for a group of materials. What follows are observations of results obtained from different criteria through its equivalent value of '*b*'.

- For mode-I: In both plane-stress and plane-strain cases, a general trend is observed that the larger the value of '*b*', the smaller is the size of the plastic zone. In other words, the greater the influence of the intermediate shears, the smaller the size of the plastic zone. It can be seen from Figs. 13.2 and 13.3 that the Tresca criterion (equivalent to $b = 0$) leads to the largest zone, while Yu's earlier full-twin-shear criterion (equivalent to $b = 1$) yields the smallest zone. The size of plastic zone by the Huber-von

Mises criterion is somewhere in-between the Tresca and the Yu's twin-shear criterion, and is almost the same as that by UST with $b = 0.5$.

- For mode-II: Similar trend as in mode-I is observed. The larger the value of 'b', the smaller is the size of the plastic zone. Figure 13.4 shows the plane-stress case, while Fig. 13.5 shows the plane-strain case. In both cases, Tresca criterion (equivalent to $b = 0$) leads to the largest zone, while Yu's full-twin-shear criterion (equivalent to $b=1$) yields the smallest zone. The Huber-von Mises criterion yields zone of intermediate sizes. Their shapes look similar.
- For mode-III: Similar trend as in mode-I and mode-II is observed, except that the plastic zones are all of circular shape.
- For unconventional materials ($b < 0$): Since the same mathematical expressions are used, the same trends are observed. The smaller the value of 'b' (more negative), the larger is the size of the plastic zone. Figures. 13.12 and 13.13 show the results for mode-I plane-stress and plane-strain respectively, while Figs 13.14 and 13.15 show the results for mode-II plane-stress and plane-strain respectively.

13.9 Influence of SD Effect

Parameter 'α' is the ratio of the uniaxial tensile strength to the uniaxial compressive strength. The ratio is a measure of the effect of strength differentials (SD). Definitely, it influences the sizes of the crack-tip plastic zone. Trends are observed in different crack modes but no common trend is noted.

- In mode-I: The weaker-in-tension materials ($\alpha = 0.5$) always result in larger plastic zone (if not less) than the equal-strength materials ($\alpha = 1$), in both plane-stress and plane-strain cases. The sizes in plane-stress cases are usually bigger than those in plane-strain cases. It can be seen in Figs. 13.4 13.5 and 13.6 for $b = 0$, 0.5 and 1.0 respectively.
- In mode-II: Contrary to the trend in Mode-I, the weaker-in-tension materials ($\alpha = 0.5$) always result in smaller plastic zone than the equal-strength materials ($\alpha = 1$), in both plane-stress and plane-strain cases. However, the sizes in plane-stress cases maintain the trend being bigger than those in plane-strain cases. It can be seen in Figs. 13.9, 13.10 and 13.11 for $b = 0$, 0.5 and 1.0 respectively.

- In mode-III: The plastic zones are all of circular shape. The weaker-in-tension materials ($\alpha = 0.5$) always result in smaller plastic zone than the equal-strength materials ($\alpha = 1$), in both plane-stress and plane-strain cases. Again, the sizes in plane-stress cases are usually bigger than those in plane-strain cases.

13.10 Influence of Poisson's Ratio

Results for plane-stress cases are independent of the Poisson's ratio. For Mode-I and -II plane-strain cases, the influence of Poisson's ratio is similar to the trend observed in the variations of 'b' values. The larger the Poisson's ratio, the smaller is the size of the plastic zone. Figure 13.16 shows the effect of Poisson's ratio on mode-I crack-tip plastic zone, while Fig. 13.17 shows the Mode-II.

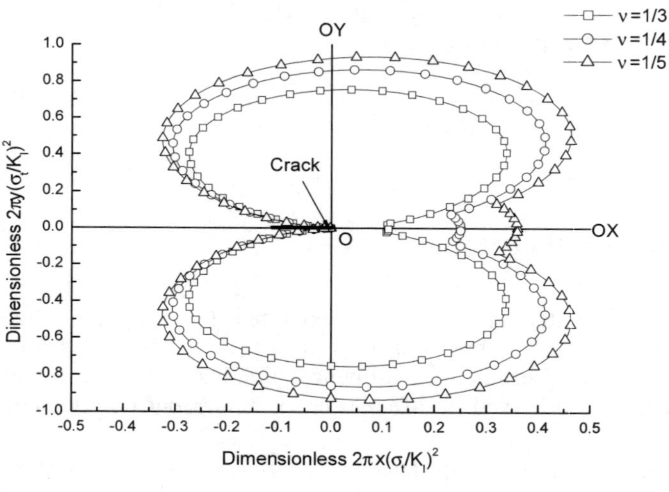

Fig. 13.16 Effect of Poisson's ratio on mode-I crack-tip plastic zone in plane-strain state ($\alpha = 1.0, b = 1.0$)

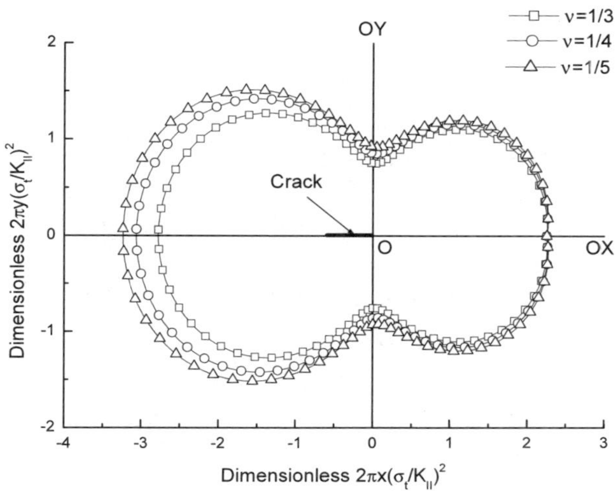

Fig. 13.17 Effect of Poisson's ratio on mode-II crack-tip plastic zone in plane-strain state ($\alpha = 1.0, b = 1.0$)

Summary

This chapter presents the estimations of crack-tip plastic zone under small-scale yielding (SSY) using the unified strength theory (UST). The UST unifies all prevailing strength criteria through a variable parameter 'b'. It provides unified solutions for shapes and sizes of mode-I, -II, -III crack-tip plastic zone in the K-dominant region. The minimum and maximum crack-tip plastic zones can be deduced through the conventionally admissible lower- and upper-bound values of 'b' in the UST. Adopting a proper value of the material parameter 'b' can lead to accurate solutions. Other specific values of 'b' can yield solutions equivalent to those through the Tresca criterion, Huber-von Mises criterion or its varieties. In addition, stretching beyond the normally admissible range for the material parameter 'b' (<0 or >1) can generate a new series of strength criteria for unconventional materials, which strength envelope may not necessarily be convex. Estimations of the crack-tip plastic zones using different values of 'b' are illustrated and compared. The influences on the resulting plastic zone due to strength-differential (SD) effect and Poisson's ratio are also investigated. Embracing the SD effect broadens the applicability to most materials, which have unequal tensile and compressive strength. It demonstrates the robustness of present method in the estimation of shape and size of the crack-tip plastic zone under SSY. It is seen that:

(1) The Yu's unified strength theory is applied to the study of the shape and size of plastic zones of mode I, II, III cracks, and a complete solution of unified closed forms is obtained. The Tresca criterion and the Huber-von Mises criterion are special case and linear approximation of the Yu's theory.
(2) Except for mode III cracks, the boundary curves of mode I, II crack tip plastic zones are composed of two parts, which is resulted from the piece-wise linearity of the Yu's criterion.
(3) Generally speaking, the plastic zone obtained by the Yu's theory is smaller than that by the Huber-von Mises and much smaller than that by the Tresca criterion, showing that a bigger τ_s/σ_s will lead to a smaller plastic zone and weaker ability to resist fracture.
(4) Different values of b in Yu's theory will produce a family of different yield criterion which will help obtain a group of crack tip plastic zones applicable to various materials. The value of σ_t also exerts certain influence on the tip plastic zone, but only affects the plane strain condition (except for mode III cracks).

The results were obtained by analytical solution. The unified strength theory and unified elasto-plastic constitutive model can be also implemented into commercial finite element codes and other FEM codes. The effect of failure criterion on numerical calculations of finite element method (FEM) can be also observed. These problems will be summarized at another book entailed "Computational Plasticity based on the Unified Strength theory"

Problems

Problem 13.1.
The solution of crack tip problem can be also studied by using the slip field theory. Figure P16.1 shows a slip field with the Tresca yield criterion (single-shear theory) for Mode I under plane strain. Can you given a new solution by using the twin-shear slip field method described in Chapter 8.

Problem 13.2.
The solution of crack tip problem can be also studied by using the slip field theory. Figure P16.1 shows a slip field with the Tresca yield criterion for Mode I under plane strain condition. Can you given a new solution by using

the unified slip field theory described in Chapter 9.

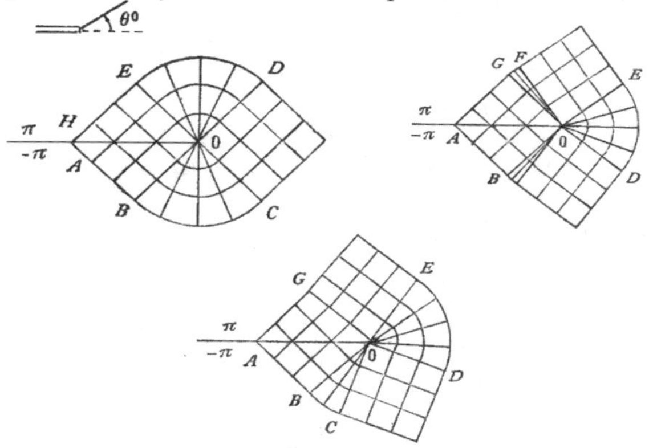

Fig. P16.1 Mode I   Fig. P16.2 Mode II   Fig. P16.3 Mixed Mode III

Problem 13.3.
The solution of crack tip problem can be also studied by using the slip field theory. Figure P13.2 shows a slip field with the Tresca yield criterion (single-shear theory) for Mode II under plane strain condition. Can you given a new solution by using the twin-shear slip field method described in Chapter 8.

Problem 13.4.
The solution of crack tip problem can be also studied by using the slip field theory. Figure P13.2 shows a slip field with the Tresca yield criterion (single-shear theory) for Mode II under plane strain condition. Can you given a new solution by using the unified slip field theory described in Chapter 9.

Problem 13.5.
The solution of crack tip problem can be also studied by using the slip field theory. Figure P13.3 shows a kind of slip field with the Tresca yield criterion (single-shear theory) for Mixed Mode I and II under plane strain. Can you given a new solution by using the unified slip field method described in Chapter 9.

Problem 13.6.
The solution of crack tip problem can be also studied by using the slip field theory. Figure P13.3 shows a kind of slip field with the Tresca yield criterion (single-shear theory) for Mixed Mode I and II under plane strain. Can you given a new solution by using the unified slip field method described in Chapter 9.

References and Bibliography

Barenblatt GI (1962) The mathematical theory of equilibrium cracks in brittle fracture. *Advances in Applied Mechanics*; 7: 55–129.

Broek D (1982) *Elementary Engineering Fracture Mechanics* (3rd ed.) Martinus Nijihoff Publishers, The Hague.

Dugdale DS (1960) Yielding of steel sheets containing slits. *Journal of the Mechanics and Physics of Solids*; 8: 100–108.

Fan SC, Yu MH, Yang SY (2001) On the unification of yield criteria. *Journal of Appllied Mechanics*, 68: 341–343.

Flugge S (1958). Encyclopedia of Physics. New York: Springer-Verlag.

Hult JAH, McClintock FA (1957) Elastic-plastic stress and strain distributions around sharp notches under repeated shear, In: Proceedings of the Ninth International Congress for Applied Mechanics, University of Brussels, 51–58.

Irwin GR (1960) Plastic zone near a crack and fracture toughness. In: Proceedings of the 7th Sagamore Ordnance Materials Conference, New York: Syracuse University Press, 1960: IV-63-78

Irwin GR (1968) Linear fracture mechanics, fracture transition and fracture control. *Engineering Fracture Mechanics*; 1: 241-257.

Kanninen MF, Popelar CH (1985) *Advanced Fracture Mechanics*. Oxford University Press, New York, 563 pages.

Qiang HF, Xu YH, Zhu JH (1998) Unified solutions of mixture type crack tip plastic zone under small scale yielding. In: *Strength Theory: Applications, Developments and prospects for the 21st Century*, MH Yu and SC Fan (eds), Science Press, Beijing, New York, 823–829.

Qiang HF and Lu N (1999) Unified solutions of crack tip plastic zone under small scale yielding, *Chin. J. Mech. Engrg.*, 35(1), 34–38. (in Chinese)

Winstone MR (1984) Influence of prestress on the yield surface of the cast nickel super alloy Mar-Mooz at elevated temperature. Mechanical Behavior of Materials, Pergamon Press, Oxford 1984, Vol. 1, 199–205

Yang Wei. Macro-Micro Scope Fracture Mechanics. Beijing: National Defence Industry Press, 1995.(In Chinese)

Yu MH (1983) Twin-shear stress yield criterion. *Int. Journal of Mechanical Science*, 25(1): 71–74.

Yu MH (2002) Advances in strength theories for materials under complex stress state in the 20th century. *Applied Mechanics Reviews,* 55(3): 169–218.

Yu MH (1983) Twin-shear stress strength criterion. *International Journal of Mechanical Science,* 25(1): 71–74.

Yu MH, He LN, Song LY (985) Twin-shear stress theory and its generalization. *Scientia Sinica* (Science in China) Series A, 28: 1174–1183.

Yu MH, He LN (1991) A new model and theory on yield and failure of materials under complex stress state. In: Jono M, Inoue T. *Mechanical Behavior of Materials*-VI. Pergamon Press. Oxford, 841–846.

Yu Mao-hong, New System of Strength Theory. Xian: Xian Jiaotong University Press, 1992. (In Chinese)

Yu MH (1998) *Twin-shear Theory and Its Applications*. Science Press, Beijing (in Chinese)

Yu Mao-Hong, Advances in strength theories for materials under complex stress state in the 20th Century. *Applied Mechanics Reviews,* 55 (2002) 169-218.

Yu Mao-hong. (2004) *Unified Strength Theory and Its Applications*. Springer, Berlin.

14 Unified Fracture Criterion for Mixed Mode Crack Initiation and Fatigue Crack Growth

14.1 Introduction

Mixed-mode fracture and fatigue crack growth are one of the most commonly used failure forms of material in analyzing practical engineering problems. The research on the mixed fracture criterion and fatigue crack growth is significant in fracture mechanics and engineering. Many research works in this area have been conducted and some criteria for predicating the direction of the crack initiation angle and the critical fracture load of materials have been proposed by energy principles and the stress approach (Khan et al. 2000).

Among the criteria of energy principle type, G-criterion (Hussian et al. 1974), S-criterion (Sih 1974) and T-criterion (Theocaris and Andrianopoulos 1982a, 1982b) are original and more important ones. Other fracture criteria are the maximum tangential stress criterion (Erdogan and Sih 1963), maximum shear stress criterion, maximum stress triaxiality criterion (Kong, Schluter and Dahl 1995), maximum octahedral shear stress criterion etc. The plastic zone radius criterion for crack initiation angle was proposed recently by Golos and Wasiluk (2000), Wasiluk and Golos (2000), Kahn and Khraisheh (2004) and Bian and Kim (2004).

S-criterion states that the crack initiation takes place along the direction with the minimum strain energy density around a constant radius around crack tip and a crack extension starts in the initiation direction when strain energy density reaches a critical value. The region within this constant radius is assumed and the linear elastic fracture mechanics does not hold in this circular region. In fact, there is not any evidence supporting the assumption that this area is circular region with constant radius, which is an intuitively fixed value only. This region is firstly called core region by Theocaris (Theocaris and Andrianopoulos 1982a, 1982b) using the caustic method, and the various fracture criteria maybe applied with high accuracy defined by the radius of the initial curve of the respective caustic created by the different loading modes. Then Theocaris et al. modified S-criterion in two aspects, firstly, total strain energy is separated into two components: a dilatational component that mainly causes cavity nucleation and development around a crack tip and distortional component that changes the shape of an element; secondly, the Huber-von Mises elastic-plastic boundary as the core region is assumed by using a variable radius for the core region, therefore T-criterion was defined that the crack initiation occurs at the direction with maximum dilatational strain energy density along the contour of constant distortional strain energy. Theocaris et al. also performed experiments on polycarbonate (PCBC) specimen to support their criterion (Theocaris et al. 1982c). The investigation mentioned above showed that the introduction (including both size and shape) of core region is very important for definition of various fracture criteria. However, the studies were performed only by using Huber-von Mises yield criterion, because it coincided with distortional energy part in form. Little attention has been paid to investigate the influence of yield criterion on the core region of crack tip.

In fact, the Tresca criterion is the lower bound of yield criteria and the twin-shear stress criterion (or the maximum principal deviatoric stress criterion) is the upper bound of yield criteria for stable and isotropic ductile material (Yu 1983). The Huber-von Mises yield criterion is a special case mediated between the lower bound and upper bound.

A linear Unified strength theory (UST) was proposed by Yu in 1991 and 1992 (see: Yu 2002, 2004). It is obtained by introducing twin-shear element naturally. It has two simple mathematical formulae, piecewise linear yield surfaces and physical significance. Varying the parameter b in the UST derives a family of convex criteria.

Based on the UST, a generalized T-criterion for Strength Difference (SD) or non-SD materials is obtained by Qiang, Yu et al. in 2003 and 2004. A new closed form of plastic core region model and its derivatives are obtained by Qiang in 2004. The Tresca-core region is lower bound, and the twin-shear-core region is upper bound, the Huber-von Mises-core region mediated between these two bounds. They are all the special case of the plastic core region responses of the generalized T-criterion. A unified fracture criterion based directly on the unified strength theory was obtained by Yu, Fan, Che et al. in 2003, 2004.

In this chapter, a variable radius for the plastic core region based on the UST elastic-plastic boundary is introduced and incorporated in the formulation of the T-criterion, the lower bound and upper bound plastic core region responses of T-criterion is derived. Comparisons of the three particular solutions with respect to the Tresca-core region, the Huber-von Mises-core region and the TS-core region are made mutually. a generalized T-criterion is obtained, called the U_T-criterion.

At last, a unified fracture criterion for mixed mode crack is described in detail. It is introduced directly from the formulation of the unified strength theory (Fan 2003, Yu, Fan and Che 2003, 2004). The relations of crack initial angle θ with μ and unified strength parameter b for mode II crack ($\theta \sim \mu \sim b$ curves), cracked initial angle with various fracture criteria, unified fracture criterion of mixed I-II mode for various α, unified fracture criterion of mixed I-III mode for various α , and the unified fracture criterion of mixed II-III mode for various α are obtained by using the numerical calculation method. A series of relations are illustrated in Figs. 14.17 to 14.25. The unified fracture criterion may be also used for the study of fatigue crack growth.

14.2 Main Idea of T-Criterion

Consider an elastic-perfectly plastic plate under plane stress condition containing an internal crack inclined by an angle β to the direction of the uniaxial loading, as show in Fig. 14.1. The stresses around the crack-tip are:

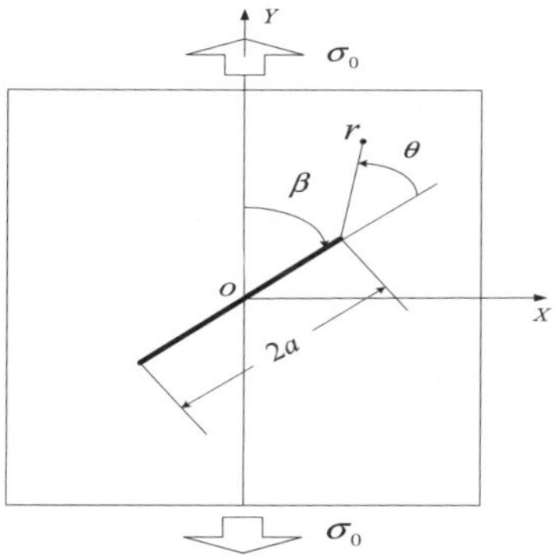

Fig. 14.1 Slant crack under remote uniaxial load

$$\sigma_x = \frac{K_I}{\sqrt{2\pi r}}\left[\cos\frac{\theta}{2}\left(1-\sin\frac{\theta}{2}\sin\frac{3\theta}{2}\right)-\mu\sin\frac{\theta}{2}\left(2+\cos\frac{\theta}{2}\cos\frac{3\theta}{2}\right)\right]=\frac{K_I}{\sqrt{2\pi r}}f_x(\theta)$$

$$\sigma_y = \frac{K_I}{\sqrt{2\pi r}}\left[\cos\frac{\theta}{2}\left(1+\sin\frac{\theta}{2}\sin\frac{3\theta}{2}\right)+\mu\sin\frac{\theta}{2}\cos\frac{\theta}{2}\cos\frac{3\theta}{2}\right]=\frac{K_I}{\sqrt{2\pi r}}f_y(\theta) \quad (14\text{-}1)$$

$$\tau_{xy} = \frac{K_I}{\sqrt{2\pi r}}\left[\cos\frac{\theta}{2}\sin\frac{\theta}{2}\cos\frac{3\theta}{2}+\mu\cos\frac{\theta}{2}\left(1-\sin\frac{\theta}{2}\sin\frac{3\theta}{2}\right)\right]=\frac{K_I}{\sqrt{2\pi r}}f_{xy}(\theta)$$

where σ_x, σ_y. and τ_{xy} are the normal stresses and shear stress in *x*-*y* plane respectively; $K_I=\sigma_0\sqrt{\pi a}\sin^2\beta$, $K_{II}=\sigma_0\sqrt{\pi a}\sin\beta\cos\beta$ and $\mu = K_{II}/K_I$ are stress-intensity factors for mode-I, -II and its notation respectively, σ_0 is uniaxial loading of remote field, $2a$ is the crack length; $f_x(\theta)$, $f_y(\theta)$ and $f_{xy}(\theta)$ are defined as above. Based on T-criterion, the dilatational and distortional parts of the strain energy density are given respectively by:

$$T_V=\frac{1-2\nu}{6E}(\sigma_x+\sigma_y)^2 \quad (14\text{-}2a)$$

$$T_D=\frac{1+\nu}{3E}(\sigma_x^2+\sigma_y^2-\sigma_x\sigma_y+3\tau_{xy}^2) \quad (14\text{-}2b)$$

where E and ν are the modulus of elasticity and Poisson's ratio of the material. Using the notation of equation (1), we obtain from Eqs. (14-2a) and (14-2b) respectively

$$T_V = \frac{(1-2\nu)K_I^2}{12\pi Er}(f_x + f_y)^2 \tag{14-3a}$$

$$T_D = \frac{(1+\nu)K_I^2}{6\pi Er}(f_x^2 + f_y^2 - f_x f_y + 3f_{xy}^2) \tag{14-3b}$$

Since the distortional strain energy is constant along the Huber-von Mises elastic plastic boundary, $T_{D,0}$ can be considered as a material constant, i.e., $T_{D,0} = \frac{(1+\nu)\sigma_s^2}{3E}$. Combining Eqs. (14-3a) and (14-3b), we get boundary of the core region and the dilatational strain energy

$$r = \frac{K_I^2}{2\pi\sigma_s^2}(f_x^2 + f_y^2 - f_x f_y + 3f_{xy}^2) \tag{14-4}$$

$$T_V = \frac{(1-2\nu)\sigma_s^2}{6E}\frac{(f_x + f_y)^2}{f_x^2 + f_y^2 - f_x f_y + 3f_{xy}^2} \tag{14-5}$$

So the T-criterion is defined as

$$T_V\Big|_{\theta=\theta_0} \geq T_{V,cr} \cap T_D = T_{D,0} \tag{14-6}$$

$$\frac{\partial T_V}{\partial\theta}\Big|_{\theta=\theta_0} = 0 \cap \frac{\partial^2 T_V}{\partial\theta^2} < 0 \tag{14-7}$$

where Eq. (14-8) indicates the one has to search for the maximum value of dilatational strain energy along the Huber-von Mises core region boundary for crack propagation, Eq. (14-7) implies the propagation direction of a crack subsequently.

14.3 A Generalization for T-Criterion Using UST

Now the plate is assumed to be made of a rigid perfectly plastic material that obeys the UST. Figure 14.2 shows the limit loci of the UST expressed by principal stresses σ_1 and σ_2 respectively.

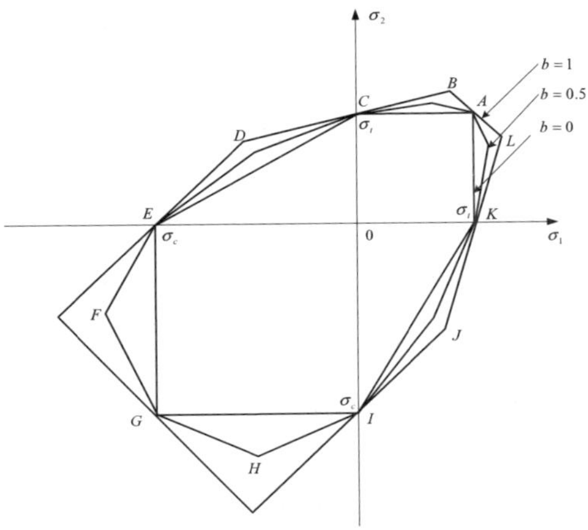

Fig. 14.2 UST in plane stress case

The UST has obviously piecewise linear forms

$$a_i\sigma_1 + b_i\sigma_2 = \sigma_t \quad (i=1\sim12) \tag{14-8}$$

constant a_i and b_i in equation (14-8) for the twelve lines L_i (i=1~12) of AB, BC, CD, DE, EF, FG, GH, HI, IJ, JK, KL and LM in Fig.14.2 are listed in Table 14.1, where $\alpha = \sigma_t/\sigma_c$ is a material uniaxial tension-compression strength ratio; b is a weighting coefficient that reflects the influence of intermediate principal shear stress τ_{12} or τ_{23} on the material strength. When b varies from 0 to 1, a family of convex yield criteria which are suitable for different kinds of materials are deduced. In particular, it become the Tresca criterion when α=1, b=0 and the twin-shear stress yield criterion when α=1, b=1. The Huber-von Mises criterion can be approximated by the UST by using α=1, b=0.5.

Table 14.1 Constants a_i and b_i in UST

	AB (i = 1)	BC (i = 2)	CD (i = 3)	DE (i = 4)	EF (i = 5)	FG (i = 6)
a_i	$b/(1+b)$	$-\alpha b/(1+b)$	$-\alpha/(1+b)$	$-\alpha$	$-\alpha$	$-\alpha/(1+b)$
b_i	$b/(1+b)$	1	1	$b/(1+b)$	$b/(1+b)$	$-\alpha b/(1+b)$
	GH (i = 7)	HI (i = 8)	IJ (i = 9)	JK(i=10)	KL(i=11)	LA (i=12)
a_i	$-\alpha b/(1+b)$	$b/(1+b)$	$1/(1+b)$	1	1	$1/(1+b)$
b_i	$-\alpha/(1+b)$	$-\alpha$	$-\alpha$	$-\alpha/(1+b)$	$-\alpha b/(1+b)$	$b/(1+b)$

For 2-dimensional plane probl ems, the principal stresses are:

$$\left.\begin{aligned}\left.\begin{matrix}\sigma_1\\ \sigma_2\end{matrix}\right\} &= \frac{\sigma_x+\sigma_y}{2} \pm \sqrt{\left(\frac{\sigma_x-\sigma_y}{2}\right)^2+\tau_{xy}^2}\\ \sigma_3 &= \begin{cases}0, & \text{plane stress}\\ \nu(\sigma_1+\sigma_2) & \text{plane strain}\end{cases}\end{aligned}\right\} \tag{14-9}$$

Herein we only discuss plane stress case, for the plane strain case similar procedure can be used. By substituting Eq. (14-1) into Eq. (14-9) and arranging the resulting stresses in the order such that $\sigma_1 \geq \sigma_2 \geq \sigma_3$, we have

$$\begin{aligned}\left.\begin{matrix}\sigma_1\\ \sigma_2\end{matrix}\right\} &= \begin{cases}\dfrac{K_I}{\sqrt{2\pi r}} f_1(\theta)\\ \dfrac{K_I}{\sqrt{2\pi r}} f_2(\theta)\end{cases}\\ \sigma_3 &= 0\end{aligned} \tag{14-10}$$

where

$$\left.\begin{matrix}f_1(\theta)\\ f_2(\theta)\end{matrix}\right\} = \cos\frac{\theta}{2} - \mu\sin\frac{\theta}{2} \pm \frac{1}{2}\sqrt{\sin^2\theta + 2\mu\sin 2\theta + \mu^2(4-3\sin^2\theta)}$$

Since the distortional strain energy is constant along the core region boundary controlled by UST, substituting Eq. (14-10) into Eq. (14-8), a variable radius for core region locating on the segments L_i (i=1~12) are obtained as follows:

$$r = \frac{K_I^2}{2\pi\sigma_t^2}(a_i f_1 + b_i f_2)^2 \qquad (i=1\sim 12) \tag{14-11}$$

Substituting Eq. (14-11) into Eqs. (14-3a) and (14-3b), and combining Eq. (14-10), a generalized T-criterion is introduced by Qiang as follows

$$T_V = \frac{(1-2\nu)\sigma_t^2}{6E}\left(\frac{f_1+f_2}{a_i f_1 + b_i f_2}\right)^2 \qquad (i=1\sim 12) \tag{14-12}$$

$$T_D = \frac{(1+\nu)\sigma_t^2}{3E}\frac{f_1^2+f_2^2-f_1 f_2}{(a_i f_1 + b_i f_2)^2} \qquad (i=1\sim 12) \tag{14-13}$$

The generalized T-criterion is defined as

$$T_V\Big|_{\theta=\theta_0} \geq T_{V,cr} \cap T_D = T_{D,b} \tag{14-14}$$

$$\frac{\partial T_V}{\partial \theta}\Big|_{\theta=\theta_0} = 0 \cap \frac{\partial^2 T_V}{\partial \theta^2} < 0 \tag{14-15}$$

where Eq. (14-14) states the one has to search for the maximum value of dilatational strain energy along the unified Twin Shear (TS)-core region boundary for crack propagation, this core region plastic potential energy is determined by Eq. (14-13). Eq. (14-15) shows the propagation direction of a crack subsequently. When b varies from b=0 to b=1, a family of convex core region boundary which are suitable for different kinds of materials are deduced, i.e., strength differential effect for materials by the parameter α.

In particular, if the material is equal tension-compression, or $\alpha=1$, it becomes Tresca-core region boundary when b=0 and the twin-shear criterion-core region boundary when b=1. The Huber-von Mises core region boundary can be approximated by the UST when b=0.5. Figures 14.3 and 14.4 show the core region boundary around the crack-tip with an internal crack inclined by an different angle β at $\pi/4$ and $\pi/6$, respectively. Furthermore, the core region boundary is also obtained for non-SD materials, it is discussed in section 14.4. The upper bound and the lower bound of core region boundary are obtained. Obviously the shape and size of Huber-von Mises core region are located in the middle between the Tresca-core region (upper bound) and the TS- core region (lower bound).

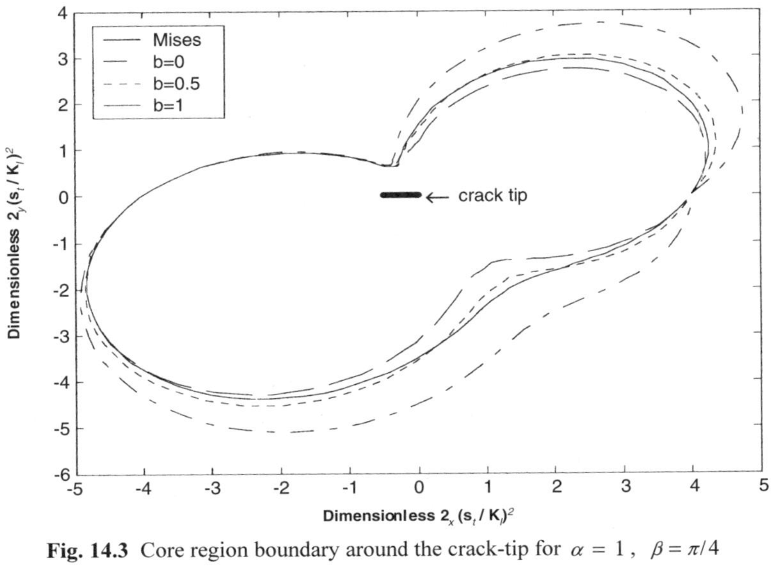

Fig. 14.3 Core region boundary around the crack-tip for $\alpha = 1$, $\beta = \pi/4$

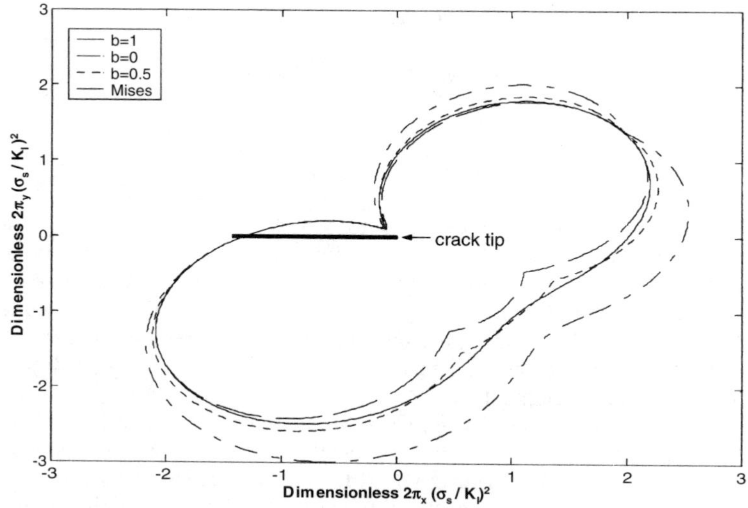

Fig. 14.4 Core region boundary around the inclined crack-tip ($\alpha = 1, \beta = \pi/3$)

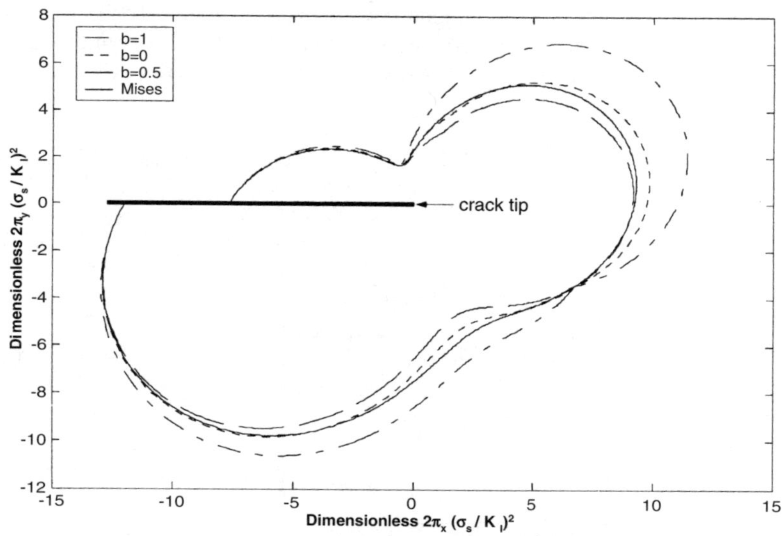

Fig. 14.5 Core region boundary around the inclined crack-tip ($\alpha = 0.8, \beta = \pi/6$)

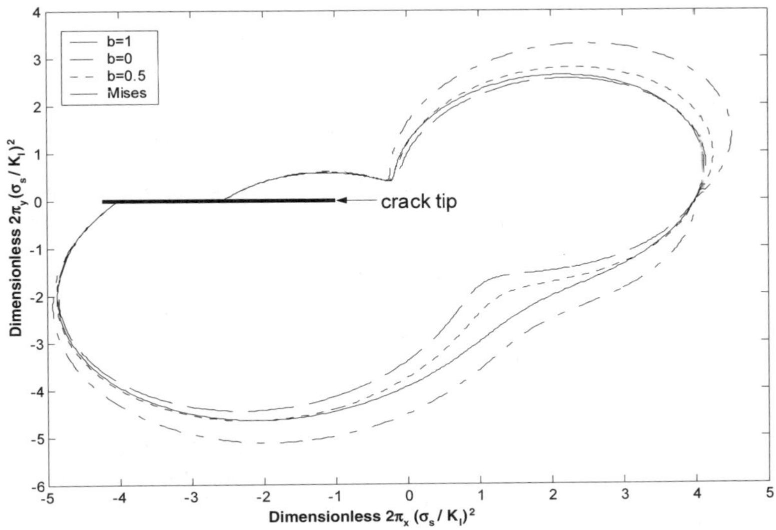

Fig. 14.6 Core region boundary around the inclined crack-tip ($\alpha=0.8$, $\beta=\pi/4$)

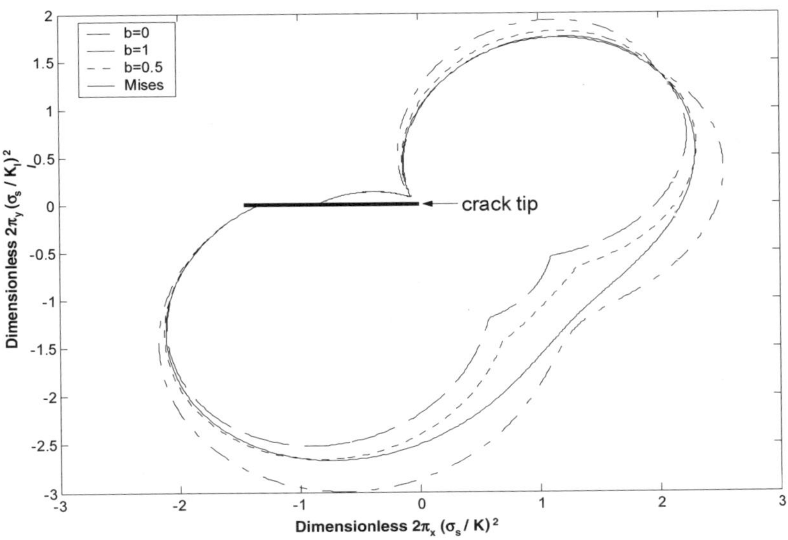

Fig. 14.7 Core region boundary around the inclined crack-tip ($\alpha=0.8$, $\beta=\pi/3$)

14.4 Significance of Parameters b, α and υ

14.4.1 Effect of the unified strength theory parameter 'b'

Parameter b is a material parameter, which represents the influence of the intermediate principal shear stress. For the same material, the extents of the influence are not the same under different crack-tip modes. It reflects in the resulting shape and size of the crack-tip plastic zone. On the other hand, various values of 'b' also correspond to a spectrum of strength criteria, including the well-known Tresca and Huber-von Mises. In other words, each 'b' value represents a strength criterion which is only valid for a group of materials. What follows are observations of results obtained from different criteria through its equivalent value of 'b'.

- For mixed mode: In both plane-stress and plane-strain cases, a general trend is observed that the larger the value of 'b', the smaller is the size of the plastic zone. In other words, the greater the influence of the intermediate shears, the smaller the size of the plastic zone. It can be seen from Figs. 13.3-13.9 that the Tresca criterion (equivalent to $b=0$) leads to the largest zone, while Yu's earlier full-twin-shear criterion (equivalent to $b=1$) yields the smallest zone. The size of plastic zone by Mises criterion is somewhere in-between the Tresca and the Yu's full-twin-shear, and is almost the same as that by UTSS with $b=0.5$.

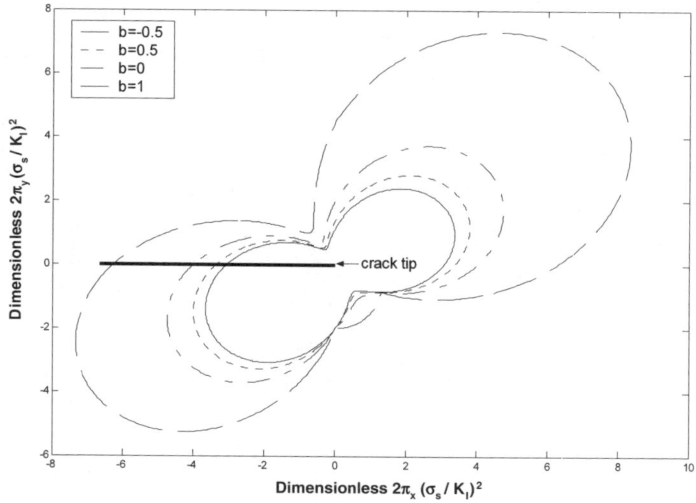

Fig. 14.8 Mixed mode crack-tip plastic zone by different criteria in plane-strain state ($\alpha=1, \nu=1/4, \beta=\pi/4$)

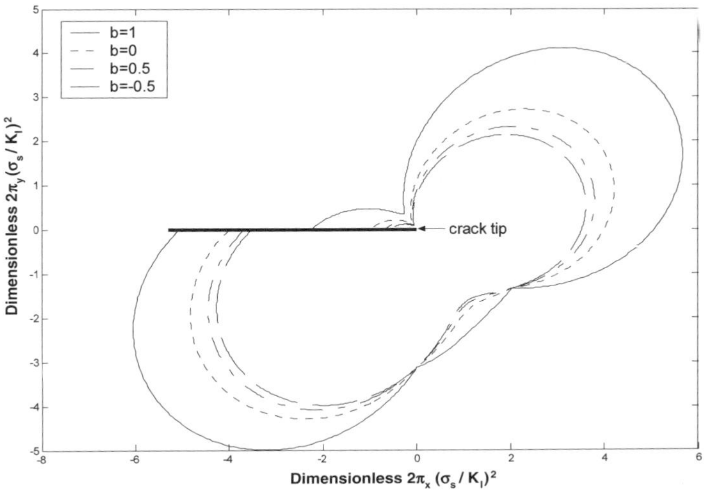

Fig. 14.9 Mixed mode crack-tip plastic zone by different criteria in plane-strain state ($\alpha = 0.5, \nu = 1/4, \beta = \pi/4$)

- For unconventional materials (*b<0*): Since the same mathematical expressions are used, the same trends are observed. The smaller the value of '*b*' (more negative), the larger is the size of the plastic zone. Figures 14.8 and 14.9 show the results for mixed mode SD or non SD effects respectively.

14.4.2 Influence of SD effect

Parameter 'α' is the ratio of the uniaxial tensile strength to the uniaxial compressive strength. The ratio is a measure of the effect of strength differentials (SD). Definitely, it influences the sizes of the crack-tip plastic zone. Trends are observed in different crack modes but no common trend is noted. With mixed mode in plane stress state: The weaker-in-tension materials (α=*0.5*) always result in smaller plastic zone (if not less) than the equal-strength materials (α=*1*). It can be seen in Figs. 13.4-13.7 and 14.10 for *b=0*, *0.5* and *1.0* respectively. On the other hand, in plane strain state: Contrary to the trend in mixed mode, the weaker-in-tension materials (α=*0.5*) always result in larger plastic zone than the equal-strength materials (α=*1*), in both plane-stress and plane-strain cases. It can be seen in Figs 14.3-14.7 and 14.11 for *b*=0, 0.5 and 1.0 respectively.

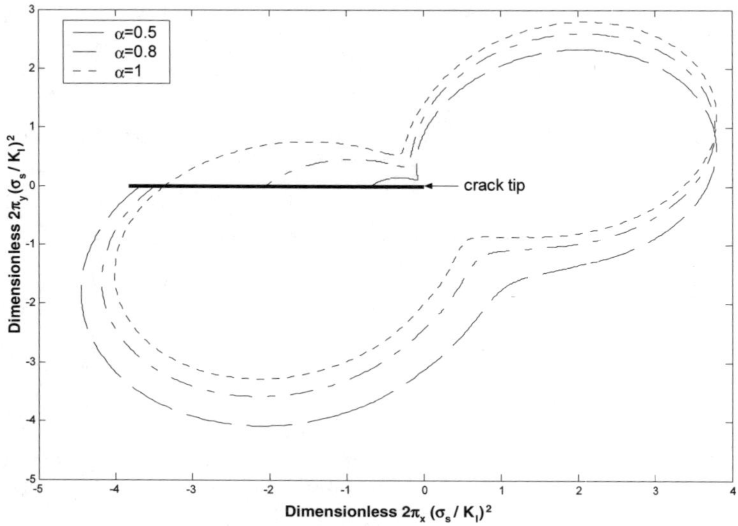

Fig. 14.10 Effect of SD on mixed mode crack-tip plastic zone in plane stress state ($b=0.5, \nu=1/4, \beta=\pi/4$)

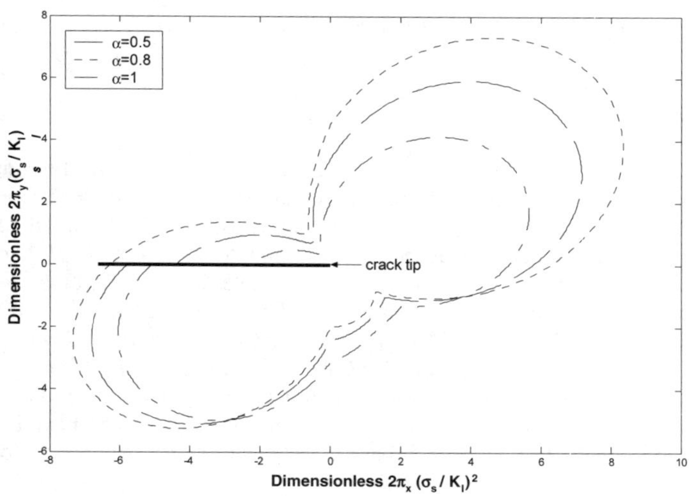

Fig. 14.11 Effect of SD on mixed mode crack-tip plastic zone in plane strain state ($b=0.5, \nu=1/4, \beta=\pi/3$)

14.4.3
Influence of Poisson's ratio

Results for plane-stress cases are independent of the Poisson's ratio. So is for mixed mode crack plane-strain cases, the influence of Poisson's ratio is similar to the trend observed in the variations of 'b' values. The larger the Poisson's ratio, the smaller is the size of the plastic core zone. Figures 14.12 and 14.13 shows the results for SD ($\alpha = 0.5$) or non SD ($\alpha = 1.0$) effect materials with the same inclined angle at $\beta = \pi / 4$.

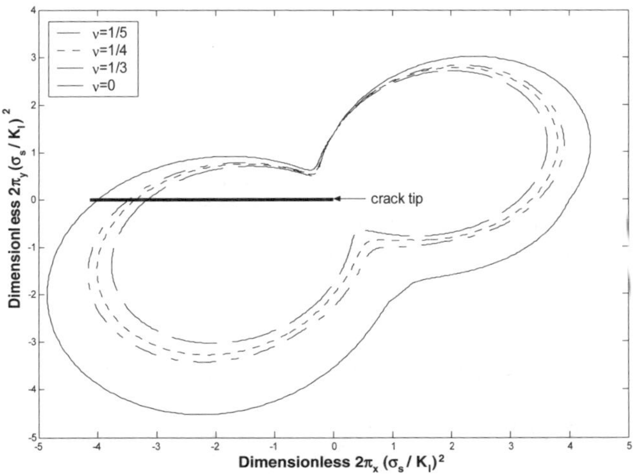

Fig. 14.12 Effect of Poisson's ratio on mixed mode crack-tip plastic zone in plane-strain state (($\alpha = 1.0, b = 0.5, \ \beta = \pi/4$))

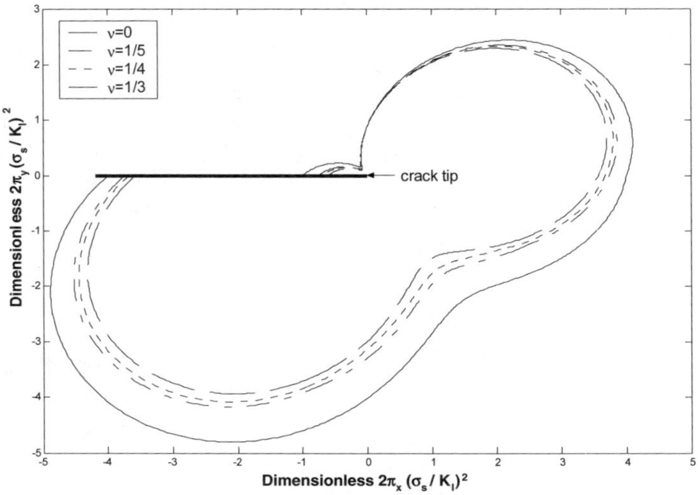

Fig. 14.13 Effect of Poisson's ratio on mixed mode crack-tip plastic zone in plane-strain state ($(\alpha = 0.5, b = 0.5, \beta = \pi/4)$)

14.5 Crack Initiation Angle of The generalized T-criterion

The predication of generalized T-criterion for the crack initiation angle θ_0 versus crack inclination angle β under tensile loading is shown in Fig. 14.14, along with the corresponding predications of S-criterion and G-criterion. The upper bound and lower bound of variation for T-criterion are obtained, the upper bound of variation for crack initiation angle θ_0 versus crack inclination angle β is the case when *b*=0 under Tresca-core region, and the lower bound of variation for crack initiation angle θ_0 versus crack inclination angle β is the case when *b*=1 under TS-core region, similar results can be drawn using S-criterion. The variation law is identical to those of using T-criterion when *b*=0.5 (approximate to Huber-von Mises core region). So we now name the generalized T-criterion as U_T-criterion.

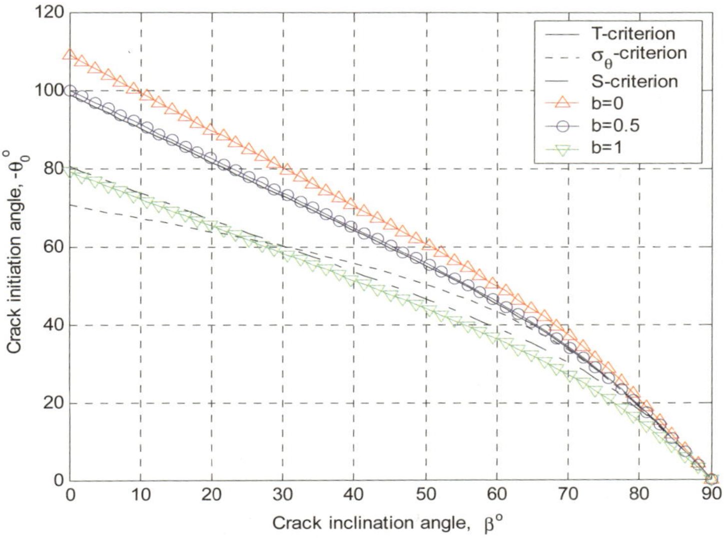

Fig. 14.14 Angle of initial crack growth vs crack inclination using different fracture criterion, tensile loading

14.6 Application of the Unified Strength Theory in Establishing the Mixed Fracture Criterion

The complex stress states are common characteristics between the mixed mode cracks and the unified strength theory. A unified fracture criterion was established by using the unified strength theory by Yu, Fan, and Che et al. in 2003 and 2004. In the unified fracture function, the difference between tension and pressure, the effect of the intermediate principal stress σ_2 and the effect of the intermediate principal shear-stress τ_{12} (or τ_{23}) are taken into account. With the variation of $\alpha = \sigma_t / \sigma_c$ and failure parameter b, a series of mixed mode crack criteria are formed, and can be applied for many materials. The J-integral is calculated through FEM. According to the relation between the result and the stress intensity factor, $K_{\mathrm{I}c}$ can be obtained. Then it is used in the unified fracture function. The unified fracture criterion is compared with those of other mixed mode criterions.

The unified fracture criterion can be generalized to analyzing the initial positions and directions of crack and the growing of crack in fracture mechanics and fatigue problems.

The stress field and the displacement field along the crack front can be expressed by polar coordinates (θ, r) of $X-Y$ plane as follows:

$$\sigma_x = \frac{K_{\mathrm{I}}}{\sqrt{2\pi r}} \cos\frac{\theta}{2}\left(1 - \sin\frac{\theta}{2}\sin\frac{3\theta}{2}\right) - \frac{K_{\mathrm{II}}}{\sqrt{2\pi r}} \sin\frac{\theta}{2}\left(2 + \cos\frac{\theta}{2}\cos\frac{3\theta}{2}\right)$$

$$\sigma_y = \frac{K_{\mathrm{I}}}{\sqrt{2\pi r}}\cos\frac{\theta}{2}\left(1+\sin\frac{\theta}{2}\sin\frac{3\theta}{2}\right)+\frac{K_{\mathrm{II}}}{\sqrt{2\pi r}}\sin\frac{\theta}{2}\cos\frac{\theta}{2}\cos\frac{3\theta}{2}$$
$$\tau_{xy} = \frac{K_{\mathrm{I}}}{\sqrt{2\pi r}}\cos\frac{\theta}{2}\sin\frac{\theta}{2}\cos\frac{3\theta}{2}+\frac{K_{\mathrm{II}}}{\sqrt{2\pi r}}\cos\frac{\theta}{2}\left(1-\sin\frac{\theta}{2}\sin\frac{3\theta}{2}\right)$$
$$\sigma_z = 2\mu\frac{K_{\mathrm{I}}}{\sqrt{2\pi r}}\cos\frac{\theta}{2}-2\mu\frac{K_{\mathrm{II}}}{\sqrt{2\pi r}}\sin\frac{\theta}{2}$$
$$\tau_{xz} = -\frac{K_{\mathrm{III}}}{\sqrt{2\pi r}}\sin\frac{\theta}{2}$$
$$\tau_{yz} = \frac{K_{\mathrm{III}}}{\sqrt{2\pi r}}\cos\frac{\theta}{2} \tag{14-16}$$

The three-dimensional polar coordinates system is shown in Fig.14.15.

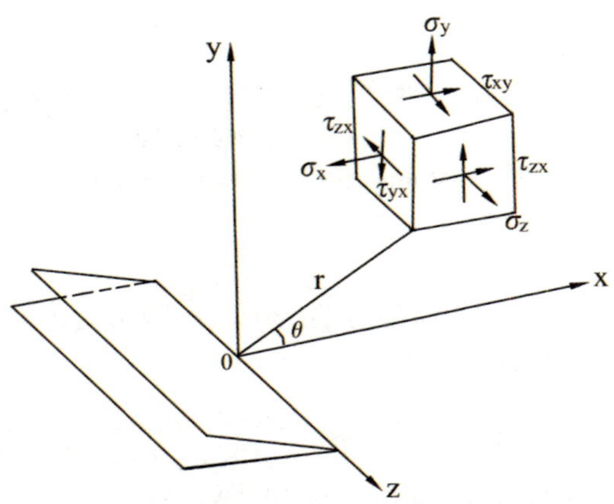

Fig. 14.15 Three-dimensional polar coordinate system

According to the stress state theory, if the six stress components of a point have been given, the stress invariant is:

$$I_1 = \frac{2}{\sqrt{2\pi r}}(1+\mu)\left(K_{\mathrm{I}}\cos\frac{\theta}{2}-K_{\mathrm{II}}\sin\frac{\theta}{2}\right)$$
$$I_2 = A_1K_{\mathrm{I}}^2+A_2K_{\mathrm{I}}K_{\mathrm{II}}+A_3K_{\mathrm{II}}^2+A_4K_{\mathrm{III}}^2 \tag{14-17}$$
$$I_3 = B_1K_{\mathrm{I}}^3+B_2K_{\mathrm{I}}^2K_{\mathrm{II}}+B_3K_{\mathrm{I}}K_{\mathrm{II}}^2+B_4K_{\mathrm{I}}K_{\mathrm{III}}^2+B_5K_{\mathrm{II}}K_{\mathrm{III}}^2+B_6K_{\mathrm{II}}^3$$

where

$$A_1 = \frac{1}{2\pi r}\left(\cos^4\frac{\theta}{2}+4\mu\cos^2\frac{\theta}{2}\right)$$

$$A_2 = -\frac{1}{\pi r}\sin\theta\left(\cos^2\frac{\theta}{2} + 2\mu\right)$$

$$A_3 = \frac{1}{2\pi r}\left(3\sin^2\frac{\theta}{2}\cos^2\frac{\theta}{2} - \cos^2\frac{\theta}{2} + 4\mu\sin^2\frac{\theta}{2}\right)$$

$$A_4 = -\frac{1}{2\pi r}$$

$$B_1 = 2\mu\frac{1}{\sqrt{2\pi r}^3}\cos^5\frac{\theta}{2}$$

$$B_2 = 2\mu\frac{1}{\sqrt{2\pi r}^3}\left(-5\cos^4\frac{\theta}{2}\sin\frac{\theta}{2}\right)$$

$$B_3 = 2\mu\frac{1}{\sqrt{2\pi r}^3}\left(6\sin^2\frac{\theta}{2} - \cos^2\frac{\theta}{2}\right)\cos^3\frac{\theta}{2}$$

$$B_4 = \frac{1}{\sqrt{2\pi r}^3}\left(-\cos^3\frac{\theta}{2}\right)$$

$$B_5 = \frac{1}{\sqrt{2\pi r}^3}\cos^2\frac{\theta}{2}\sin\frac{\theta}{2}$$

$$B_6 = \frac{\mu}{\sqrt{2\pi r}^3}\frac{1}{2}\cos\frac{\theta}{2}\sin\theta(3\cos\theta - 1) \tag{14-18}$$

The three principal stresses $\sigma_1, \sigma_2, \sigma_3$ can be obtained from the characteristic value of the stress matrix:

$$\sigma_1 = \frac{I_1}{3} + R\cos\frac{\varphi}{3}$$

$$\sigma_2 = \frac{I_1}{3} + R\cos\left(\frac{\varphi}{3} - \frac{2\pi}{3}\right)$$

$$\sigma_3 = \frac{I_1}{3} + R\cos\left(\frac{\varphi}{3} + \frac{2\pi}{3}\right) \tag{14-19}$$

where $R = \frac{2}{3}(I_1^2 - 3I_2)^{\frac{1}{2}}$, $\cos\varphi = \dfrac{2I_1^3 - 9I_1I_2 + 27I_3}{2(I_1^2 - 3I_2)^{\frac{3}{2}}}$

$\varphi/3$ corresponding to the stress angle of π -plane; $\varphi/3 = 0° \sim 60°$, and $\sigma_1 \geq \sigma_2 \geq \sigma_3$.

According to the formula of the unified strength theory and Eq. (14-17) and Eq. (14-19), the unified fracture criterion for SD materials can be established as follows:

14.7 Unified Fracture Criterion

The mathematical expression of the unified strength theory is

$$F=\sigma_1-\frac{\alpha}{1+b}(b\sigma_2+\sigma_3)=\sigma_t=\alpha\sigma_c \text{ when } \sigma_2\le\frac{\sigma_1+\alpha\sigma_3}{1+\alpha} \tag{14-20a}$$

$$F'=\frac{1}{1+b}(\sigma_1+b\sigma_2)-\alpha\sigma_3=\sigma_t=\alpha\sigma_c \text{ when } \sigma_2\ge\frac{\sigma_1+\alpha\sigma_3}{1+\alpha} \tag{14-20b}$$

Substituting Eq.(14-17) and Eq.(14-19) into Eq.(14-20), a new unified fracture function in terms of three dimensional stress is obtained (Yu, Fan and Che et al. 2003, 2004). It can be expressed as follows:

$$F_{Fracture}=\sqrt{r}F=\frac{2}{3\sqrt{2\pi}}(1+\mu)(1-\alpha)\left(K_{\mathrm{I}}\cos\frac{\theta}{2}-K_{\mathrm{II}}\sin\frac{\theta}{2}\right)+R\left[\cos\frac{\varphi}{3}+\alpha\frac{b\cos\left(\frac{\varphi}{3}+\frac{\pi}{3}\right)+\cos\left(\frac{\varphi}{3}-\frac{\pi}{3}\right)}{1+b}\right]\sqrt{r} \tag{14-21a}$$

$$F'_{Fracture}=\sqrt{r}F'=\frac{2}{3\sqrt{2\pi}}(1+\mu)(1-\alpha)\left(K_{\mathrm{I}}\cos\frac{\theta}{2}-K_{\mathrm{II}}\sin\frac{\theta}{2}\right)+R\left[\frac{\cos\frac{\varphi}{3}-b\cos\left(\frac{\varphi}{3}+\frac{\pi}{3}\right)}{1+b}+\alpha\cos\left(\frac{\varphi}{3}-\frac{\pi}{3}\right)\right]\sqrt{r} \tag{14-21b}$$

The unified fracture criterion is based on the two hypotheses:

(1) Crack will propagate along the direction of minimum value $F_{Fracture}(F'_{Fracture})$, e.g.

$$\frac{\mathrm{d}F_{Fracture}}{\mathrm{d}\theta}=0 \text{ and } \frac{\mathrm{d}^2F_{Fracture}}{\mathrm{d}\theta^2}\ge0, \text{ when } \theta=\theta_c \tag{14-22}$$

(2) Crack will start to grow when $F_{Fracture}(F'_{Fracture})$ reaches to the critical value F_c, e.g.

$$F_{Fracture}(F'_{Fracture})=F_c, \text{ when } \theta=\theta_c \tag{14-23}$$

Theoretically, according to the formula (14-22), a function $f(\theta_c)=0$ can be obtained. It is difficult, however, to get the solution because of the complexity of the analytic solution. So the problem will be solved through numerical calculation. The process is as follows:

(a) Given a series of the values of θ_i, and according to Eqs. (14-17) ~ (14-23), as for any θ_i, the unified fracture function $F_{Fracture}$ and $F'_{Fracture}$ have a definite value.

(b) Then, the $F_{Fracture}$ - θ curve will be obtained.

(c) From the curves, it can be concluded that when $F_{Fracture}(F'_{Fracture})$ reaches the minimum value, θ will be the initial cracked angle θ_c and $F_{Fracture}(F'_{Fracture})$ will reach the critical F_c, as shown in Fig.14.16.

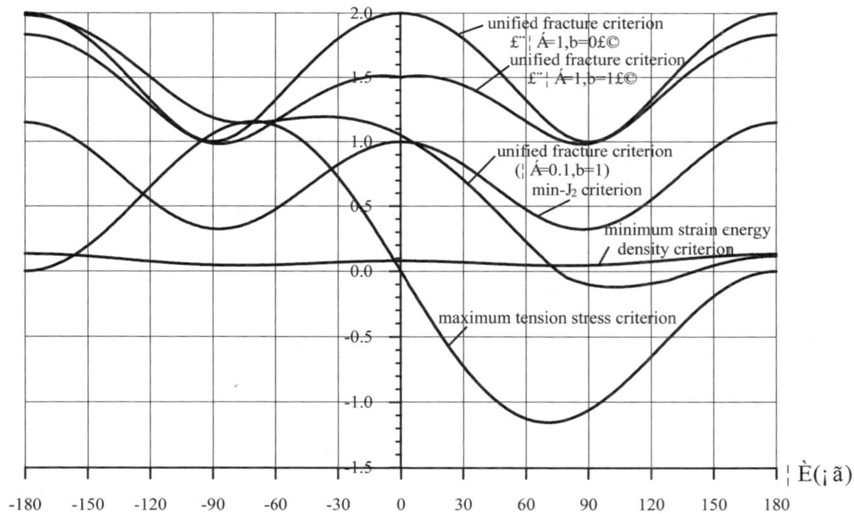

Fig. 14.16 Curves of several fracture criterions

The results in several cases are shown in Figs. 14.17～14.25. The experimental data in Fig.14.17 are given by Zhao and Xu (2002).

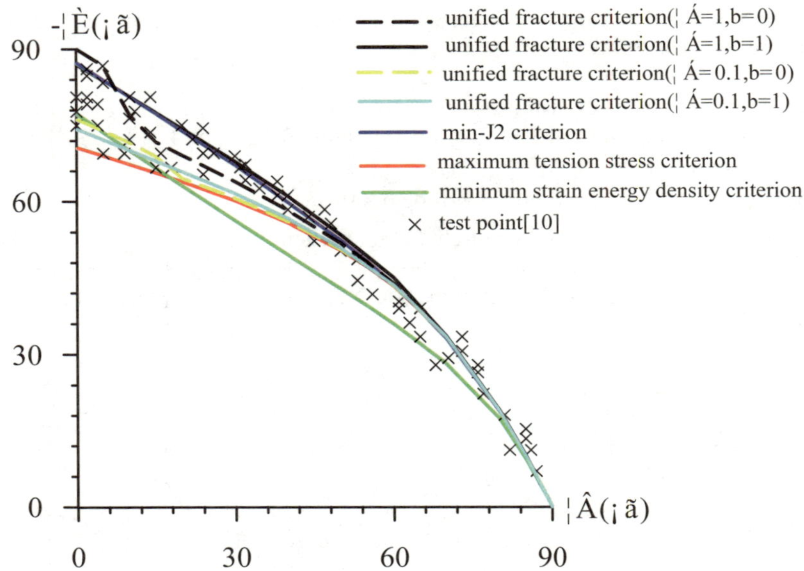

Fig. 14.17 Comparison of cracked angle with various fracture criteria ($\mu = 0.167$)

The relations of crack initial angle θ with μ and unified strength parameter b for mode Ⅱ crack are obtained by Fan (2003) and Yu, Fan, Che et al.(2003, 2004). The relation curves of $\theta \sim \mu \sim b$ are illustrated in Fig.14.18 (a), (b), (c) and (d).

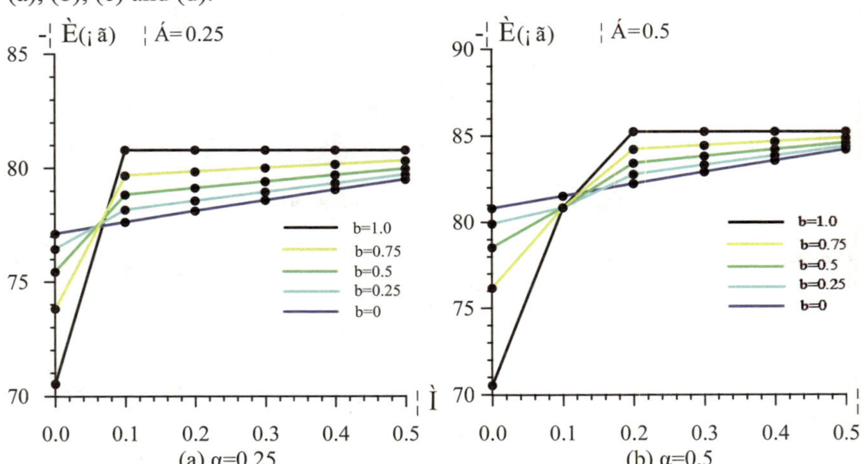

Fig. 14.18 Continued

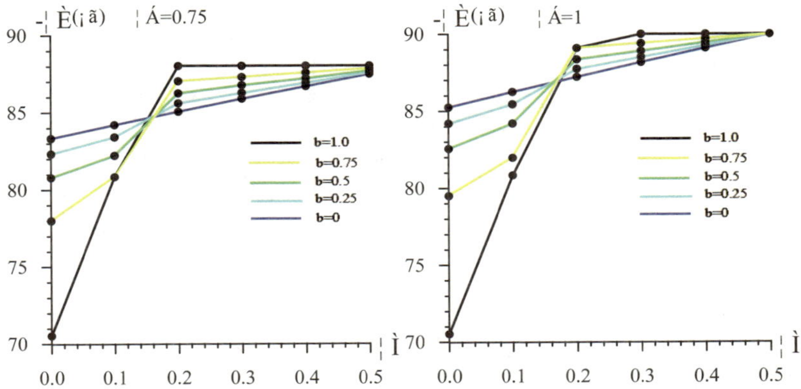

(c) α=0.75 (d) α=1

Fig. 14.18 Crack initial angle θ ~ μ ~ b curves (mode II crack)

14.8 Unified Fracture Criterion of Mixed Mode I-III

Unified fracture criterion of mixed I-III mode for $\alpha=1$, $\alpha=0.5$ and $\alpha=0.3$ materials are illustrated as in Fig.14.19 to Fig.14.21. The test data in Fig. 14.20 are obtained by Zhang , Zhu, He et al. (1993).

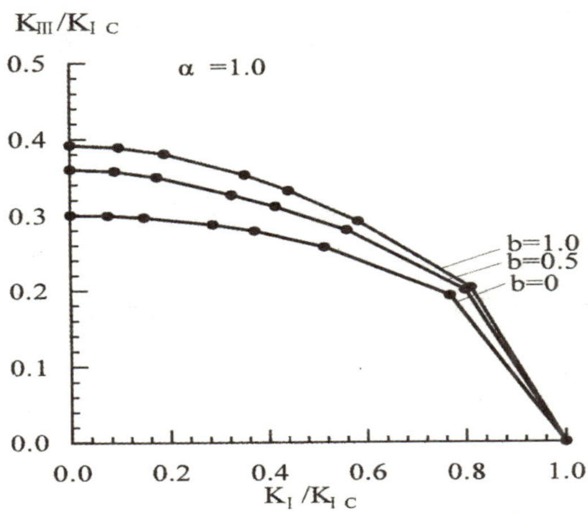

Fig. 14.19 Unified fracture criterion of mixed I-III mode for $\alpha=1$ materials ($\mu=0.2$)

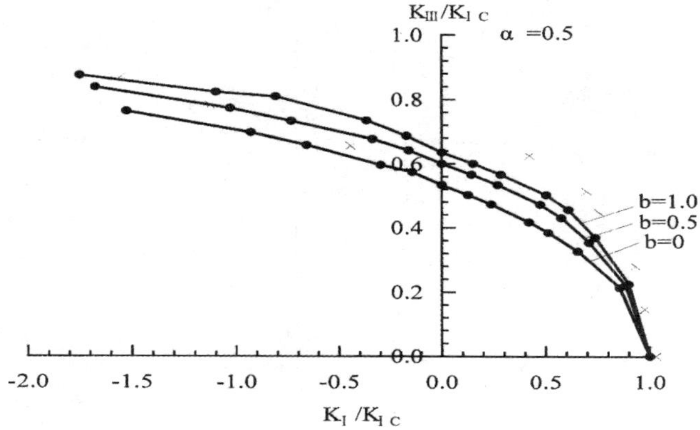

Fig. 14.20 Unified fracture criterion of mixed I-III mode for α=0.5 materials ($\mu = 0.2$)

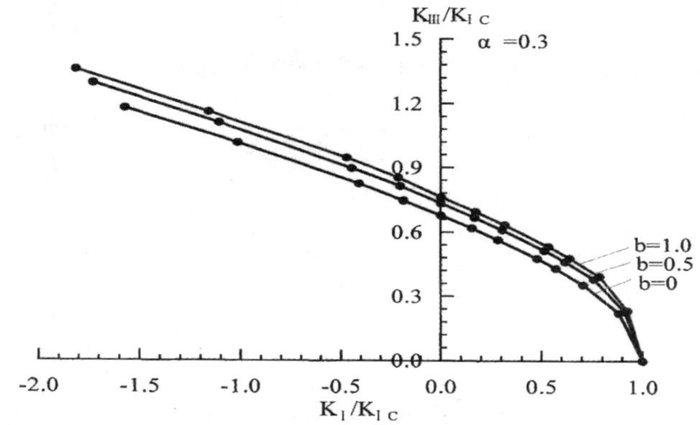

Fig. 14.21 Unified fracture criterion of mixed I-III mode for $\alpha = 0.3$ materials (μ=0.2)

14.9 Unified Fracture Criterion of Mixed Mode II-III

Unified fracture criterion of mixed II-III mode for $\alpha = 1$ and $\alpha = 0.5$ materials are illustrated as in Fig. 14.22 and Fig. 14.23. Relation of $K_{III}/K_{IIc} \sim K_{II}/K_{IIc}$ and $K_{III}/K_{Ic} \sim K_{II}/K_{Ic}$ are given.

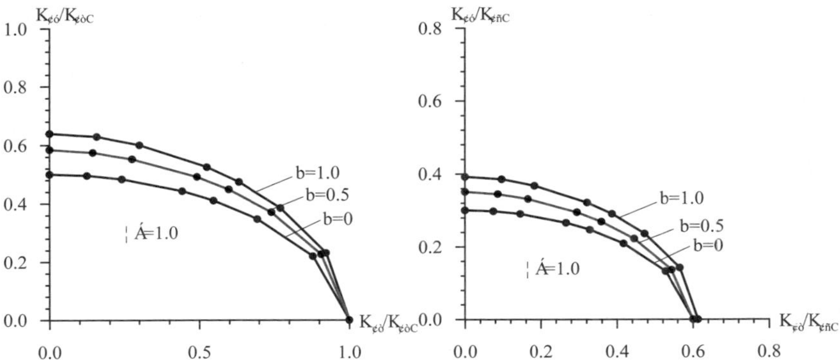

(a)$K_{III}/K_{II} \sim K_{II}/K_{II}$ (b) $K_{III}/K_{I} \sim K_{II}/K_{I}$

Fig. 14.22 Unified fracture criterion of mixed II-III mode for α =1 materials (μ = 0.2)

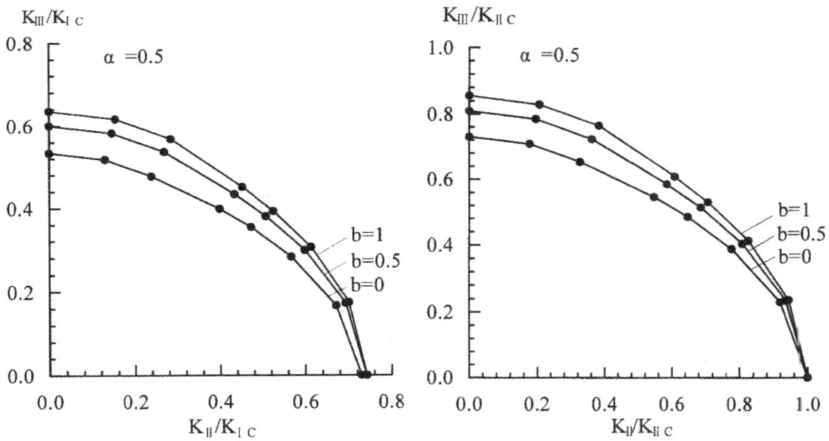

(a) $K_{III}/K_{IC} \sim K_{II}/K_{IC}$ (b) $K_{III}/K_{IIC} \sim K_{II}/K_{IIC}$

Fig. 14.23 Unified fracture criterion of mixed II-III mode for α =0.5 materials (μ = 0.2)

Summary

In this chapter, a variable radius for the plastic core region based on the Unified Strength Theory (UST) elastic-plastic boundary is introduced and incorporated in the formulation of the earlier T-criterion, a new closed form of plastic core region model and its derivatives are obtained, while the lower bound and upper bound plastic core region responses of T-criterion is derived, and variation for the angle of initial crack growth versus crack inclination under different loading condition is obtained also. Comparisons of the three particular solutions with respect to the Tresca-core region, the Huber-von Mises-core region and the Twin-Shear (TS)-core region are made. A U-fracture criterion is also proposed, it is not only for ductile material but also for brittle material.

The unified strength theory is applied to the fracture mechanics. A unified fracture criterion for mixed mode of crack is described. The unified fracture criterion of mixed I-II mode for various α , unified fracture criterion of mixed I-III mode for various α , and the unified fracture criterion of mixed II-III mode for various α are obtained through numerical calculation. A series of relations are illustrated in Figs.14.17 to 14.25. The results show:

(1) In the unified fracture criterion, the SD effect (strength difference between tensile strength and compressive strength of materials), the hydrostatic stress effect, the normal stress effect, the effect of the intermediate principal stress and the effect of intermediate principal shear-stress are taking into account, the unified fracture criterion can be adapted for more materials.

(2) The critical angle and the fracture criterion is consistent with the testing data.

(3) If the material has a good toughness, the plastic zone is so large that the elastic solution will disappear completely. On the tip of the crack, the plastic zones play an important role.

The unified fracture criterion may be used in J-integral for nonlinear cracks.

The concept of the unified fracture criterion may be used in the study of the mixed mode fatigue crack growth (Qian and Fatem 1996; Gao, Alagok, Brown and Miller 1985)

Problems

Problem14.1.
Plotting the plastic zone at crack tip of mode I for $\alpha = 1$ material

Problem14.2.
Plotting the plastic zone at crack tip of mode I for $\alpha = 0.8$ material

Problem14.3.
Plotting the plastic zone at crack tip of mode I for $\alpha = 0.6$ material

Problem14.4.
Plotting the plastic zone at crack tip of mode I for $\alpha = 0.2$ material

Problem14.5.
Plotting the plastic zone at crack tip of mixed I-II mode for $\alpha = 1$ material

Problem14.6.
Plotting the plastic zone at crack tip of mixed I-II mode for $\alpha = 0.8$ material

Problem14.7.
Plotting the plastic zone at crack tip of mixed I-II mode for $\alpha = 0.5$ material

Problem14.8.
Plotting the plastic zone at crack tip of mixed I-II mode for $\alpha = 0.2$ material

Problem14.9.
Plotting the relation curve of mixed I-II mode for $\alpha=1$ material by using the unified fracture criterion.

Problem14.10.
Plotting the relation curve of mixed I-II mode for $\alpha=0.8$ material by using the unified fracture criterion.

Problem14.11.
Plotting the relation curve of mixed I-II mode for $\alpha=0.5$ material by using the unified fracture criterion.

Problem14.12.
Plotting the relation curve of mixed I-II mode for $\alpha=0.2$ material by using the unified fracture criterion.

References and Bibliography

Bian LC and Kim KS (2004) The minimum plastic zone radius criterion for crack initiation direction applied to surface cracks and through-cracks under mixed mode loading. Int. J. of Fatigue, 26: 1169–1178.

Erdogan F and Sih GC (1963) On the crack extension in plates under plane loading and transverse shear. *J. Basic Engrg*, ASME **85**, 519–527

Gao H, Alagok N, Brown MW and Miller KJ (1985) Growth of fatigure cracks under combined mode I and mode II loads. In: Miller KJ and Brown MW eds., *Multiaxial Fatigue,* ASTM STP 853 ASTM, Philadelphia, 184–202.

Golos K and Wasiluk B (2000) Role of plastic zone in crack growth direction criterion under mixed mode loading. *Int. J. of Fracture*, 102: 341353.

Griffith AA (1921) The phenomena of rupture and flow in solids. *Phil. Trans. Roy. Soc.*, A221,163

Hill R (1950) The Mathematical Theory of Plasticity, Clarendon Press, Oxford.

Hussian MA, Pu SL and Underwood J (1974) Strain energy release rate for a crack under combined mode I and mode II, Fracture Analysis ASTM STP 560, 2–28.

Khan SMA and Khraisheh MK (2000) Analysis of mixed mode crack initiation angles under various loading conditions. *Engineering Fracture Mechanics*, 67, 397–419.

Khan SMA and Khraisheh MK (2004) A new criterion for mixed mode fracture initiation based on the crack tip plasticity. Int. J. of Plasticity, 20: 55–84.

Knauss WG and Rosakis AJ eds. (1990) Non-linear fracture : recent advances. Kluwer Academic Publishers, Boston.

Kong XM, Schluter N and DahlW (1995) Effect of triaxial stress on mixed-model fracture. *Engineering Fracture Mech.,* 52(2),379–388.

Qian J and Fatem A (1996) Mixed mode fatigue crack growth: A literature survey. *Engineering Fracture Mech.,*55(6), 969–990.

Qiang HF, Han X and Liu GR (2003) A unified criterion for mixed mode crack initation. Presented at the 5th Euromech Solid Mechanics Conference, Elias C. Aifantis Editor, Greece.

Qiang HF, Yu MH and Yang YC (2004) Unified fracture criterion: A generalization of the T-criterion and the unified strength theory. Presented at the International Symposium on Developments in Plasticity and Fracture Centenary of M.T. HUBER Criterion. August 12–14, 2004, Cracow, Poland.

Rice JR (1968) A path independent integral and the approximate analysis of strain concentrations by notches and cracks, *J. Appl. Mech.,* **35**, 379–386.

Rice JR and Rosengren GF (1968) Plane strain deformation near a crack tip in a power-law hardening material, *JMPS* **16**, 1–12.

Sih GC (1973) Some basic problems in fracture mechanics and new concepts. *Engrg Fract. Mech.* **5**, 365–377.

Sih GC (1974a) Strain-energy-density factor applied to mixed mode crack problems, *International Journal of Fracture* 10, 305–321.
Sih GC and MacDonald, B.(1974b). *Engng Fract.* Mech.,361–386
Song L, Yu MH (2000) Research on the three-dimensional generalized twin-shear fracture criterion. *Chinese J. of Rock Mechanics and Engineering*, **19**(6), 692–696.
Tian CH, Ren MF and Chen HR (2001) Compound crack's brittle collapse principal strain factor criterion. *Journal of Dalian University of Technology,* **19(2),** 140
Theocaris PS and Andrianopoulos NP (1982a) The T-criterion applied to ductile fracture, *International Journal of Fracture*, 20, R125-R130.
Theocaris PS and Andrianopoulos NP (1982b) The Mises elastic-plastic boundary as the core region in fracture criteria, *Engineering Fracture Mechanics*, 16, 425–432.
Theocaris PS, Kardomateas GA and Andrianopoulos NP (1982c) Experimental study of the T-criterion in ductile fracture, *Engineering Fracture Mechanics*, 17, 439–447.
Theocaris PS (1981) The caustic as a means to define the core region in brittle fracture. *Engineering Fracture Mechanics*, 14, 353–362.
Wasiluk B and Golos K (2000) Prediction of crack growth direction under plane stress for mixed mode I and II loading. *Fatigue and Fracture of Engineering Materials and Structures,* 23:381–386
Yehia NAB (1985) On the use of the T-criterion in fracture mechanics. *Engineering Fracture Mechanics* 22: 189-199
Yu MH (1983) Twin-shear stress strength criterion. *In.t J. of Mechanical Science*, 25, 71–74.
Yu Mao-hong and He Li-nan, A new model and theory on yield and failure of material under complex stress state. In: *Mechanical Behavior of Materials- VI*, Inoue et al. eds. Pergamon Press, Oxford, 1991,Vol.**3**, 841–846
Yu MH (2002) Advances in strength theories for materials under complex stress state in the 20th century. *Applied Mechanics Reviews*, 55, 169–218.
Yu MH, Fan W and Che AL, Yoshimine M and Iwatate T (2003) Application of the Unified Strength theory in analyzing fracture strength. Presented at the Fifth Int. Conf. on Fracture and Strength of Solids and the Second Int. Conf. On Physics & Chemistry of Fracture and Failure Prevention, , October 20–22, 2003, Sendai, Japan.
Yu Mao-Hong. (2004) *Unified Strength Theory and Its Applications*. Berlin: Springer.
Yu MH, Fan W and Che AL, Yoshimine M and Iwatate T (2004) Application of the Unified Strength theory in analyzing fracture strength. *Key Engineering Materials,* Vols. 261–263: 111–116.
Zhang GD (1988) Research on the critical curve curface of the concrete's three-dimensional fracture. *Journal of Hehai University,* **1,** 90−96.
Zhang LB, Zhu WX, He M and Xu DY (1993) Experimental study of critical fracture parameter for concrete. *Proc. 5th Symposium on Fracture and Strength of Rock and Concrete*, National Defense University Press, 78–84.
Zhao YH and Xu SL (2002) Minimum J_2 criterion of I ~ II mixed crack. *Engineering Mechanics*, **19**(4), 94–98.
Zhu WX, Shou CH and Zhang LB (1993) Analysis on fracture mechanics of the minor structure, *Proc. 5th Symposium on Fracture and Strength of Rock and Concrete*, National Defense University Press, 273–277.

15 Limit Load and Shakedown Load of Pressure Vessel

15.1 Introduction

The elastic limit, plastic limit and shakedown analysis are important in analysis and design of engineering structures. This chapter deals with the limit analysis and shakedown limit analysis of pressure vessel.

Limit analysis and design of structures is highly developed. The basic techniques were given in several textbooks and papers (Symonds and Neal 1951; Neal 1956; Hodge 1959, 1963; Baker and Heyman 1969; Heyman 1971; Save and Massonnet 1972; Horne 1979; Mrazik et al. 1987)

Exploitation of the strength reserve of the load-bearing capacity allows for the design, in many cases of structures with increased admissible loads or decreased cross-sections and the weight of structure. This results in a reduction in material consumption and cost.

The savings are achieved in three ways: first, the most highly stressed cross-sections are better exploited by transferring part of the load to those that are under-stressed in the elastic state. Second, the number of fully exploited cross-sections is increased by the redistribution of the internal forces throughout the statically indeterminate structure. Third, the limit-bearing capacity of structures may be increased by using an advanced strength theory or yield criterion.

The Tresca yield criterion, the Huber-von Mises yield criterion and the maximum principal stress criterion are usually used to perform the limit analysis of metallic structures. The Mohr-Coulomb criterion is usually used to perform the limit analysis for geomaterials and geotechnical engineering. Each adapts only for one kind of metallic material or geomaterials. Moreover, both the maximum principal stress criterion and the single-shear theory (Tresca and Mohr-Coulomb criteria) only consider the effect of one or two principal stresses. In addtion, the Huber-von Mises criterion is not convenient to use in analytical solutions because of its nonlinear formula.

In 1991, a new linear unified strength theory was proposed (Yu and He 1991; Yu 1992). The yield loci of the unified strength theory cover all regions of convex yield criteria. The Tresca yield criterion, the Huber-von Mises yield criterion, the twin-shear yield criterion, the Mohr-Coulomb theory, the generalized twin-shear failure criterion and a series of new linear yield criteria are special cases or linear approximations of the unified strength theory. It provides us with a new available unified strength theory and approach to study the load-carrying capacities and the shakedown load of structures. Systematic results can be obtained by using the unified strength theory.

The limit analyses of thick-walled cylinder, gun barrels and wellbore by using the twin-shear yield criterion and the unified strength theory were given by Li and Zhang (1998), Liu, Ni and Yan (1998), Ni, Liu et al. (1998), and Zhao et al (1999). The plastic limit analyses based on the unified strength theory for thick-walled cylinder under the combined action of both an inner pressure and axial force were obtained by Feng, Zhang and Han (2004).

The unified strength theory is also used to analyse the shakedown load of thick-walled cylinder (Xu and Yu 2004, 2005). The effects of SD (strength difference in tension and compression), the intermediate principal stress and the effect of yield criterion on the shakedown load of the thick-walled cylinder can be evaluated. It is seen that the applications of the unified strength theory to plastic analysis and shakedown analysis of structure are very effective.

In this chapter, the unified strength theory is used to obtain the unified solutions of plastic limit of pressure vessel and shakedown limit of pressure vessel. These results can be suitable for a wide range of materials and engineering.

15.2 Theorems of Limit Analysis of Structures

The theorems of limit analysis were first presented by Gvozdew in 1938 and independently proved by Hill in 1951 for rigid perfectly plastic materials and by Drucker et al. in 1951 for elastic perfectly plastic materials. The general forms of the theorems of limit analysis are described as follows.

15.2.1 Lower-Bound Theorem

If an equilibrium distribution of stress can be found that balances the applied load and is everywhere below yield or at yield, the structure will not collapse or will just be at the point of collapse. This gives a lower bound on the limit load and is called the lower bound theorem. The maximum lower bound is the limit load.

We define a statically admissible stress field as one that is in internal equilibrium, is in equilibrium with the external load λ_p and nowhere exceeds the yield limit. The multiplier λ is used and structure acts under the load λ_p as λ is slowly increased from zero. The multiplier λ corresponding to such a statically admissible stress field is called a statically admissible multiplier. The lower bound theorem can be stated as follows: the limit load factor λ^0 is the largest statically admissible multiplier λ^-, i. e.

$$\lambda^- \leq \lambda^0 \tag{15-1}$$

15.2.2
Upper-Bound Theorem

The structure will collapse if there is any compatible pattern of plastic deformation for which the rate at which the external forces work is equal to or exceeds the rate of internal dissipation. This gives the upper bound on the limit or collapse load and is called the upper bound theorem. The minimum upper bound is the limit load.

The upper bound theorem can be stated as follows: the limit load factor λ^0 is the smallest kinematically admissible multiplier λ^+, i. e.

$$\lambda^+ \geq \lambda^0 \tag{15-2}$$

The above theorems furnish the limit load with upper and lower bounds. They can be summarized by the relation

$$\lambda^- \leq \lambda^\circ \leq \lambda^+ \tag{15-3}$$

15.3
Unified Solution of Limit Pressures for Thin-Walled Pressure Vessel

Thin-walled vessels and thick-walled cylinders are used widely in industry, for instance, as pressure vessels, pipes and gun tubes. In many applications the wall thickness of the cylinder is constant, and the cylinder is subjected to a uniform internal pressure p. The deformations of the cylinder are symmetric with respect to the axis of symmetry of the cylinder under such conditions.

Furthermore, the deformations at a cross section sufficiently far removed from the junction of the cylinder and its end caps are practically independent of the axial coordinate z. In particular, if the cylinder is open (no end caps) and unconstrained, it undergoes axisymmetric deformations from pressure p, which are independent of z. If the deformation of a cylinder is constrained by end caps, the displacements and stresses at cylinder cross sections near the end cap junctions differ from those at sections far away from the end cap junctions. In this chapter, we consider the stresses and strength at sections far away from the end caps. The study of stresses and strength near the junction of the end caps and the cylinder lies outside of the scope of this book. This problem often is treated by the finite element method (FEM).

Since only axially symmetrical loads and constraints are permitted, the solution is axisymmetrical, that is, a function of radial coordinate r only. In the case of a thin-walled cylinder, the difference between stresses at the inner wall and the outer wall is small when the thickness t is much less than the diameter of vessel (many vessels have a thickness-to-diameter ratio less than 1/20). The stresses may be independent of the radial coordinate r.

The elastic limit and plastic limit of thin-walled vessels and thick-walled cylinders will be studied using the unified strength theory.

Consider the stresses in a simple pressure vessel where fluid under pressure acts on the inner surface of the vessel wall. The pressures cause a circumferential stress (or hoop stress) σ_1 and longitudinal stress σ_2. Relationships between pressure p, the dimensions of the vessel and stresses will now be established partly because the stress formulas are not only well known in any course of engineering interest, but also to develop some principles that have wider application. As can be found in any textbook on mechanics of materials, three principal stresses of element are

$$\sigma_1 = \frac{pD}{2t}, \qquad \sigma_2 = \frac{pD}{4t}, \qquad \sigma_3 = 0 \tag{15-4}$$

The unified strength theory is

$$F = \sigma_1 - \frac{\alpha}{1+b}(b\sigma_2 + \sigma_3) = \sigma_t, \qquad \text{when} \quad \sigma_2 \le \frac{\sigma_1 + \alpha\sigma_3}{1+\alpha} \tag{15-5a}$$

$$F' = \frac{1}{1+b}(\sigma_1 + b\sigma_2) - \alpha\sigma_3 = \sigma_t, \qquad \text{when} \quad \sigma_2 \le \frac{\sigma_1 + \alpha\sigma_3}{1+\alpha} \tag{15-5b}$$

where σ_1, σ_2 and σ_3 are the principal stresses and they satisfy $\sigma_1 \geq \sigma_2 \geq \sigma_3$. $\alpha = \sigma_t/\sigma_c$ is tensile-compressive strength ratio of the material. The coefficient b ($0 \leq b \leq 1$) is a material parameter that reflects the influence of intermediate principal shear stress as well as the influence of intermediate principal stress on plastic behavior of the material, which is often called the yield criterion coefficient.

The unified strength theory can also be expressed in terms of material parameter m as follows:

$$\sigma_1 - \frac{1}{m(1+b)}(b\sigma_2 + \sigma_3) = \sigma_t \qquad \left(\sigma_2 \le \frac{m\sigma_1 + \sigma_3}{m+1}\right) \tag{15-6a}$$

$$\frac{1}{1+b}(\sigma_1 + b\sigma_2) - \frac{\sigma_3}{m} = \sigma_t \qquad \left(\sigma_2 \ge \frac{m\sigma_1 + \sigma_3}{m+1}\right) \tag{15-6b}$$

where $m = \sigma_c / \sigma_t$ is compressive-tensile strength ratio of the material. The ratio m is an index of the material strength difference effect and $m \ge 1$ in general.

When b varies from 0 to 1, the unified strength theory can be simplified to the Tresca criterion ($\alpha = m = 1$ and b=0), the linear approximation of Mises criterion ($\alpha = m = 1$ and b=0.5), the twin shear criterion ($\alpha = m = 1$ and b=1), the Mohr-Coulomb criterion ($\alpha \neq m \neq 1$ or $m > 1$ and b=0), the generalized twin shear criterion ($\alpha \neq m \neq 1$, i.e. $\alpha < 1$ or $m > 1$ and b=1), and a series of new criteria (other values of m and b). Hence, the unified strength theory can be applied to more than one kind of material. The yield loci of the unified strength theory in the deviatoric plane for non-SD and SD materials can be seen in Chapter 4.

The stress state of thin-walled vessels satisfies the condition of the unified strength theory

$$\sigma_2 = \frac{1}{2}(\sigma_1 + \sigma_3) \le \frac{\sigma_1 + \alpha\sigma_3}{1+\alpha}$$

We have to use the first formula Eq. (15-5a). Substituting Eq. (15-4) into the first equation of the unified strength theory Eq. (15-5a), we can get the following expression for the yield condition of thin-walled vessel:

$$F = \sigma_1 - \frac{\alpha}{1+b}(b\sigma_2 + \sigma_3) = \frac{pD}{2t} - \frac{\alpha b}{1+b}\frac{pD}{4t} = \sigma_t \tag{15-7}$$

The limit pressure can be obtained as follows

$$p_e = \frac{1+b}{2+2b-\alpha b}\frac{4t}{D}\sigma_t \tag{15-8}$$

If the material has an allowable tensile stress of $[\sigma] = \sigma_t / n$, the allowable limit pressure is

$$[p] = \frac{1+b}{2+2b-\alpha b}\frac{4t}{D}[\sigma] \qquad \text{15-9)}$$

If the internal pressure p and allowable stress $[\sigma]$ are given, the wall thickness is required by the following condition

$$t \ge \frac{2+2b+\alpha b}{1+b}\frac{pD}{4[\sigma]} \tag{15-10}$$

The relationships between limit pressure and wall thickness and the parameter of strength theory b in the unified strength theory are illustrated in Figs. 15.1 and 15.2.

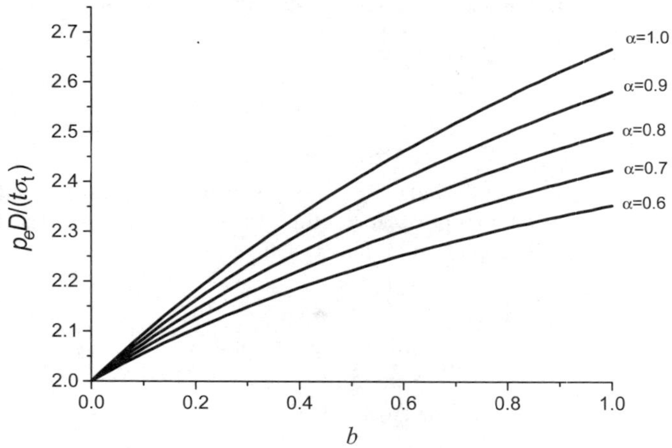

Fig. 15.1 Relation of limit pressure to parameter *b*

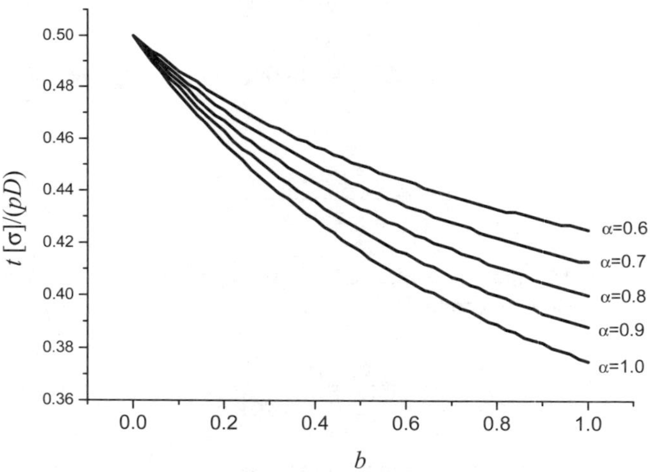

Fig. 15.2 Relation of wall thickness to parameter *b*

15.4 Unified Solution of Elastic Limit Pressure for Thick-Walled Cylinders

Thick-walled cylinders are used widely in engineering. It is important to know the elastic and plastic limit internal pressures of hollow cylinders. A considerable amount of work has been done on the problem of elasto–plastic analysis in a thick-walled cylinder under internal pressure. Various solutions have been obtained by

Turner (1909), Nadai (1931), Manning (1945), Hill et al. (1947), Allen and Sopwith (1951), Crossland and Bones (1958). The subject was thoroughly discussed in the books by Hill (1950), Prager and Hodge (1951), Johnson and Mellor (1962), Mendelson (1968) and Chakrabarty (1987).

Conventionally, the Tresca yield criterion and the Huber-von Mises yield criterion were adopted to derive the elastic and plastic pressure. Some solutions used the Tresca yield criterion. Other used the Huber-von Mises yield criterion. Complete incompressibility is assumed in both the elastic and plastic regions. However, the Tresca yield criterion ignores the effect of the intermediate principal stress on yield. It leads to conservative predictions of limit pressures. The Huber-von Mises yield criterion is not convenient to use due to its nonlinear mathematical expressions.

The twin-shear yield criterion proposed by Yu in 1961 (Yu 1961a, 1983) has been used to study the limit pressure of thick-walled cylinders by Li (1988) and Huang and Zeng (1989). The generalized twin-shear strength theory (Yu et al.1985) was also used to obtain the limit pressure of thick-walled cylinders and hollow spherical shells by Ni et al. (1998) and Zhao (1999). It was applied to gun barrels by Liu et al. (1998). The elastic limit pressure, plastic limit pressure and autofrettage pressure in autofretted gun barrel were studied by Liu and Ni (1998). Modern gun barrels are made of high-strength steel having different strengths in tension and compression. So, their solutions are more difficult.

The unified yield criterion (Yu and He 1991b, Yu et al.1992) was used to derive the limit pressure for thick-walled tubes with different end conditions, e.g., the open-end condition, the closed-end condition and the plane strain condition (Wang and Fan 1998). The effects of yield criteria on elastic and plastic limit pressure for thick-walled tubes using the unified yield criterion were illustrated and discussed. These results can be used for those materials with identical yield stress in tension and compression.

For pressure-sensitive materials, the generalized failure criterion considering the effect of strength differences in tension and compression (SD effect) has to be used. The unified strength theory takes all the stress components into account and satisfies the basic characteristics of materials under complex stress states as summarized in Chap. 4. It is suitable for use in most cases. The unified solutions of elastic and plastic limit pressures for thick-walled cylinders are studied below.

Let us consider a thick-walled cylinder under an internal pressure p and a longitudinal force P (Fig.12-3). The inner and outer radii of the cylinder are r_a and r_b, respectively. The cylinder is assumed so large that planar transverse sections remain plane during the expansion. This means that the longitudinal strain ε_z is independent of the radius to the element.

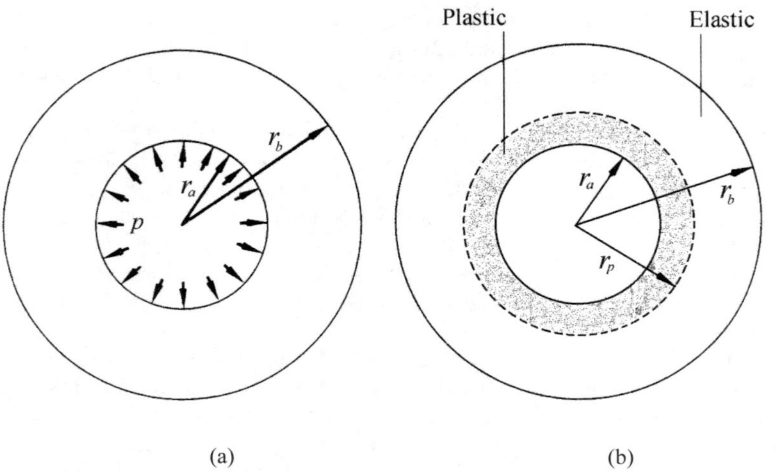

Fig. 15.3 Thick-walled cylinder

The stresses and strains sufficiently far away from the ends do not vary along the length of the cylinder, and the equation of equilibrium is

$$\frac{d\sigma_r}{dr} = \frac{\sigma_\theta - \sigma_r}{r} \tag{15-12}$$

The z-axis of the cylindrical coordinates (r, θ, z) is taken along the axis of the tube. The longitudinal stress in the elastic state may be written from Hooke's law as

$$\sigma_z = \nu(\sigma_r + \sigma_\theta) \tag{15-13}$$

where E is Young's modulus, and ν is Poisson's ratio. The radial strain ε_r and the circumferential strain ε_θ are

$$\begin{aligned} \varepsilon_r &= \frac{1+\nu}{E}\left[(1-\nu)\sigma_r - \nu\sigma_\theta\right] \\ \varepsilon_\theta &= \frac{1+\nu}{E}\left[(1-\nu)\sigma_\theta - \nu\sigma_r\right] \end{aligned} \tag{15-14}$$

The compatibility equation is

$$\frac{d}{dr}(\sigma_r + \sigma_\theta) = 0$$

It follows that $\sigma_r + \sigma_\theta$ and σ_z have constant values at each stage of the elastic expansion. Integrating Eq. (15-12) and using the boundary conditions $\sigma_r = 0$ at $r = r_b$, and $\sigma_r = -p$ at $r = r_a$, the stresses are given as follows:

$$\sigma_r = -p\left(\frac{r_b^2}{r^2}-1\right)\left(\frac{r_b^2}{r_a^2}-1\right)^{-1}, \quad \sigma_\theta = p\left(\frac{r_b^2}{r^2}+1\right)\left(\frac{r_b^2}{r_a^2}-1\right)^{-1} \tag{15-15}$$

This is the famous Lame's solution. If the resulting longitudinal load is denoted by P, the axial stress σ_z is $P/\pi(r_b^2 - r_a^2)$, since this stress is constant over the cross section. In particular, $P = 0$ for the open-end condition and $P = \pi r_a^2 p$ for the closed-end condition. The plane strain condition ($\varepsilon_z = 0$), sometimes considered for its simplicity, gives σ_z directly from Eqs. (15-13) and (15-15). Hence

$$\sigma_z = \frac{p}{K^2-1} \quad \text{closed end}$$

$$\sigma_z = 0 \quad \text{open end}$$

$$\sigma_z = \frac{2\nu p}{K^2-1} \quad \text{plane strain} \tag{15-16}$$

The axial strain is obtained from Eqs. (15-14) and (15-16) as

$$\varepsilon_z = \frac{(1-2\nu)p}{(K^2-1)E} \quad \text{closed end}$$

$$\varepsilon_z = \frac{-2\nu p}{(K^2-1)E} \quad \text{open end}$$

$$\varepsilon_z = 0 \quad \text{plane strain} \tag{15-17}$$

In all the three cases, σ_z is the intermediate principal stress. For the closed-end condition, σ_z is exactly the mean of the other two principal stresses. If a material is assumed to be incompressible in both the elastic and plastic range, σ_z for the plane strain condition is identical to the closed-end condition. It can be seen that $\sigma_1 = \sigma_\theta$, $\sigma_2 = \sigma_z$, $\sigma_3 = \sigma_r$, and

$$\sigma_2 = \frac{1}{2}(\sigma_1 + \sigma_3) \le \frac{\sigma_1 + \alpha\sigma_3}{1+\alpha} \tag{15-18}$$

So, we need choose the first equation of the unified strength theory.

$$\sigma_1 - \frac{\alpha}{1+b}(b\sigma_2 + \sigma_3) = \sigma_t \tag{15-19}$$

Substituting Eq. (15-18) into the above equation, we have the unified strength theory in the case of a thick-walled cylinder with closed-end and plane strain condition as follows:

$$\frac{2+(2-\alpha)\alpha}{2(1+b)}\sigma_\theta - \frac{\alpha(2+b)}{2(1+b)}\sigma_r = \sigma_t \tag{15-20}$$

or

$$\sigma_\theta - \frac{\alpha}{1+b}\sigma_r = \sigma_t \quad \text{(open end)} \tag{15-21}$$

Substituting σ_θ and σ_r (Eq.12-15) into above equations, we have

$$[2+(2-\alpha)b]\left(\frac{p}{K^2-1}\right)\left(\frac{r_b^2}{r^2}+1\right)+\alpha(2+b)\left(\frac{p}{K^2-1}\right)\left(\frac{r_b^2}{r^2}-1\right)=2(1+b)\sigma_t \tag{15-22}$$

This is suitable for closed-end and plane strain condition for incompressible materials.

Hence, the following elastic limit pressure in terms of the unified strength theory can be obtained:

$$p_e = \frac{(1+b)(K^2-1)\sigma_t}{K^2(1+b+\alpha)+(1+b)(1-\alpha)} \quad \text{closed end} \tag{15-23}$$

$$p_e = \frac{(1+b)(K^2-1)\sigma_t}{(1+b)(K^2+1)+\alpha(K^2-1)} \quad \text{open end} \tag{15-24}$$

$$p_e = \frac{(1+b)(K^2-1)\sigma_t}{K^2(1+b+\alpha)+(1+b)(1-\alpha)} \quad \text{plane strain} \tag{15-25}$$

The elastic limit pressure in the plane strain state for an incompressible material is the same as the closed-end condition.

If material is assumed to be compressible, the values of elastic limit pressures p_e for the three end conditions will differ marginally from one another for usual values of Poisson's ratio ν, where the lowest elastic limit pressure corresponds to open ends.

These unified solutions are general solutions adopted for different materials. The limit pressure for closed-ends in terms of the Mohr-Coulomb strength theory (single-shear theory) was discussed. The twin-shear strength theory was used to obtain the limit pressure of thick-walled cylinders by Liu et al. (1994) and Ni et al. (1998). It can be also introduced from the unified solution when b=0 and b=1.

$$p_e = \frac{K^2 - 1}{(1+\alpha)K^2 + (1-\alpha)}\sigma_t \qquad \text{(Mohr–Coulomb strength theory)} \quad (15\text{-}26)$$

$$p_e = \frac{2(K^2 - 1)}{(2+\alpha)K^2 + 2(1-\alpha)}\sigma_t \qquad \text{(twin-shear strength theory)} \quad (15\text{-}27)$$

If the yield stresses in tension and compression are equal, i.e., $\alpha = 1$, or $\sigma_t = \sigma_c = \sigma_y$. Eqs. (15-23)– (15-25) are simplified to

$$p_e = \frac{(1+b)(K^2 - 1)}{K^2(2+b)}\sigma_y \qquad \text{closed end} \quad (15\text{-}28)$$

$$p_e = \frac{(1+b)(K^2 - 1)}{K^2(2+b)+b}\sigma_y \qquad \text{open end} \quad (15\text{-}29)$$

$$p_e = \frac{(1+b)(K^2 - 1)}{K^2(2+b)+b(1-2\nu)}\sigma_y \qquad \text{plane strain} \quad (15\text{-}30)$$

These results are identical with the solutions from Yu's unified yield criterion obtained by Wang and Fan in 1998. If $\alpha = 1$, $b = 1$, the unified strength theory become the twin-shear yield criterion (Yu 1961a); for $\alpha = 1$, b=1/2, the unified strength theory becomes the linear Huber-von Mises yield criterion.

The elastic limit pressure for a Tresca material at closed-end, open-end and plane strain conditions can be obtained from Eqs. (15-28)–(15-30) when $\alpha = 1,\ b = 0$. They are identical:

$$p_e = \frac{K^2 - 1}{2K^2}\sigma_y \qquad (15\text{-}31)$$

The elastic limit pressure for the Huber-von Mises material may be approximately obtained from the unified solution (when $\alpha = 1,\ b = 1/2$).

$$p_e = \frac{3(K^2-1)}{5K^2}\sigma_y \qquad \text{closed end} \qquad (15\text{-}32)$$

$$p_e = \frac{3(K^2-1)}{5K^2+1}\sigma_y \qquad \text{open end} \qquad (15\text{-}33)$$

$$p_e = \frac{3(K^2-1)}{5K^2+(1-2\nu)}\sigma_y \qquad \text{plane strain} \qquad (15\text{-}34)$$

The classical solutions for Huber-von Mises materials are

$$p_e = \frac{K^2-1}{\sqrt{3K^2}}\sigma_y \qquad \text{closed end} \qquad (15\text{-}35)$$

$$p_e = \frac{K^2-1}{\sqrt{3K^4+1}}\sigma_y \qquad \text{open end} \qquad (15\text{-}36)$$

$$p_e = \frac{K^2-1}{\sqrt{3K^4+(1-2\nu)^2}}\sigma_y \qquad \text{plane strain} \qquad (15\text{-}37)$$

Comparing these results, we can see the difference between the Huber-von Mises material and the unified yield criterion material with b=1/2 (linear Huber-von Mises material) is very small. The difference is less than 0.38%.

The elastic limit pressure in view of the twin-shear yield criterion can be obtained from the unified solution from $a = 1,\ b = 1$.

$$p_e = \frac{2(K^2-1)}{3K^2}\sigma_y \qquad \text{closed end} \qquad (15\text{-}38)$$

$$p_e = \frac{2(K^2-1)}{3K^2+1}\sigma_y \qquad \text{open end} \qquad (15\text{-}39)$$

$$p_e = \frac{2(K^2-1)}{3K^2+(1-2\nu)\sigma_y} \qquad \text{plane strain} \qquad (15\text{-}40)$$

The difference between the solution for the Tresca material and the solution for the twin-shear material is 33.4%.

Equation (15-39) was given by Li (1988) and Huang and Zheng (1989). It is the same as the solution by using the unified yield criterion.

It is interesting to note that all the previous solutions can be deduced from the unified solution in terms of the unified strength theory. The various unified solutions are illustrated in Figs. 15.4 to 15.6.

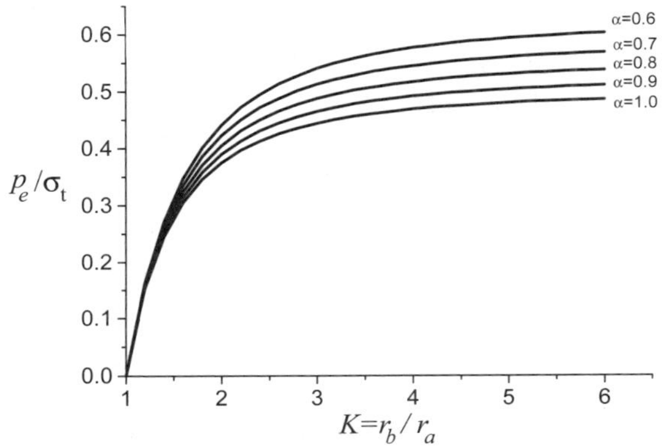

Fig. 15.4 Relation of elastic pressure to $K=r_b/r_a$ (b=0.0)

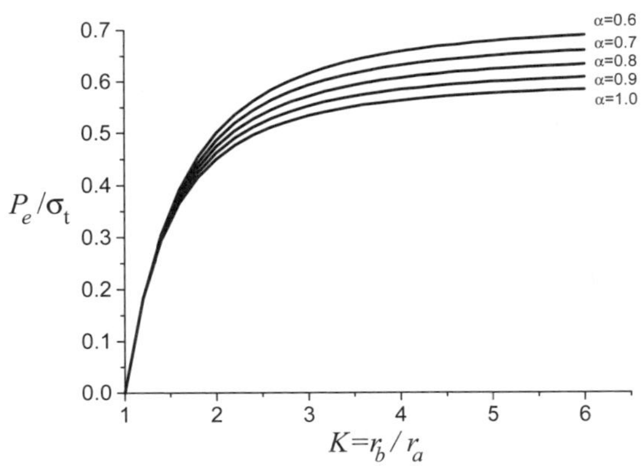

Fig. 15.5 Relation of elastic pressure to $K=r_b/r_a$(b=0.5)

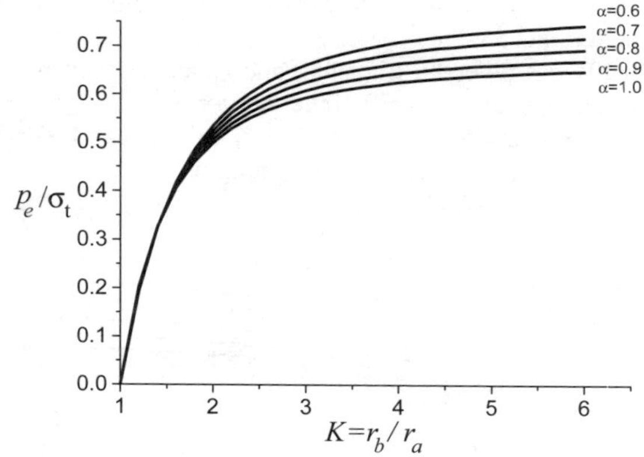

Fig. 15.6 Relation of elastic pressure to $K=r_b/r_a$ (b=1.0)

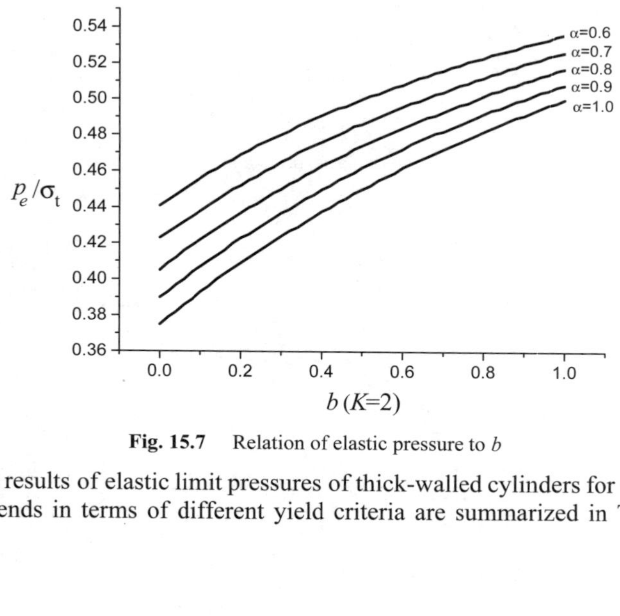

Fig. 15.7 Relation of elastic pressure to b

Various results of elastic limit pressures of thick-walled cylinders for closed ends and open ends in terms of different yield criteria are summarized in Tables 15.1 and 15.2.

Table 15.1 Summary of elastic limit pressures for closed-end conditions

	Materials	Elastic limit pressures	Failure criterion used
1	SD material $\alpha \neq 1$	$p_e = \dfrac{(1+b)(K^2-1)\sigma_t}{K^2(1+b+\alpha)+(1+b)(1-\alpha)}$	Unified strength theory
2	SD material $\alpha \neq 1$	$p_e = \dfrac{K^2-1}{(1+\alpha)K^2+(1-\alpha)}\sigma_t$	Unified strength theory b=0, Mohr–Coulomb
3	SD material $\alpha \neq 1$	$p_e = \dfrac{2(K^2-1)}{K^2(2+\alpha)+2(1-\alpha)}\sigma_t$	Unified strength theory b=1, twin-shear theory
4	$\alpha = 1$ materials	$p_e = \dfrac{(1+b)(K^2-1)}{K^2(2+b)}\sigma_y$	Unified yield criterion
5	$\alpha = 1$ materials	$p_e = \dfrac{K^2-1}{2K^2}\sigma_y$	Unified yield criterion b=0, Tresca criterion
6	$\alpha = 1$ materials	$p_e = \dfrac{K^2-1}{\sqrt{3}K^2}\sigma_y$	von Mises yield criterion
7	$\alpha = 1$ materials	$p_e = \dfrac{3(K^2-1)}{5K^2}\sigma_y$	Unified yield criterion b=1/2
8	$\alpha = 1$ materials	$p_e = \dfrac{2(K^2-1)}{3K^2}\sigma_y$	Unified yield criterion b=1, twin-shear criterion

Table 15.2 Summary of elastic limit pressures for open-end conditions

	Materials	Elastic limit pressures	Failure criterion used
1	SD material $\alpha \neq 1$	$\dfrac{(1+b)(K^2-1)\sigma_t}{(1+b)(K^2+1)+\alpha(K^2-1)}$	Unified strength theory
2	SD material $\alpha \neq 1$	$p_e = \dfrac{K^2-1}{(1+\alpha)K^2+(1-\alpha)}\sigma_t$	Unified strength theory $b=0$, Mohr–Coulomb
3	SD material $\alpha \neq 1$	$p_e = \dfrac{2(K^2-1)}{K^2(2+\alpha)+2(1-\alpha)}\sigma_t$	Unified strength theory b=1, twin-shear heory
4	$\alpha = 1$ materials	$p_e = \dfrac{(1+b)(K^2-1)}{K^2(2+b)+b}\sigma_y$	Unified yield criterion
5	$\alpha = 1$ materials	$p_e = \dfrac{K^2-1}{2K^2}\sigma_y$	Unified yield criterion $b=0$, Tresca criterion
6	$\alpha = 1$ materials	$p_e = \dfrac{K^2-1}{\sqrt{3K^4+1}}\sigma_y$	von Mises yield criterion
7	$\alpha = 1$ materials	$p_e = \dfrac{3(K^2-1)}{5K^2+1}\sigma_y$	Unified yield criterion when $b=1/2$
8	$\alpha = 1$ materials	$p_e = \dfrac{2(K^2-1)}{3K^2+1}\sigma_y$	Unified yield criterion $b=1$, twin-shear criterion

When a uniform pressure p is applied externally to a thick-walled cylinder of wall ratio r_b/r_a, the elastic stress distribution of σ_r and σ_θ is obtained from Eq. (15-15) by interchanging r_a and r_b. In this case, both stresses are negative, where σ_θ is more compressive than σ_r.

15.5 Unified Solution of Plastic Limit Pressure for Thick-Walled Cylinder

15.5.1 Stress Distribution

As the internal pressure gradually increases beyond elastic limit p_e, a plastic zone will begin at the inner surface and spread outward from the inner radius to the radius of an elastic-plastic boundary, then toward to the outer surface. The elastic-plastic boundary at any stage has radius r_c. In the elastic region, ($r_c \le r \le r_b$), the radial and circumferential stresses are obtained from Lame's equations using the boundary condition σ_r=0 at r=r_b and the fact that the material at r=r_c is stressed to the yield point. The pressure reaches its maximum value when the plastic zone reaches the outer surface of the thick-walled tube.

The elastic part of the elastic-plastic thick-walled tube may be considered as a new tube with inner radius r_c and outer radius r_b, with an internal pressure p_e. The stress distribution in the elastic region for an incompressible material is easily shown to be

$$\sigma_\theta = \frac{p_e r_c^2}{r_b^2 - r_c^2}\left(1 + \frac{r_b^2}{r^2}\right) \tag{15-41}$$

$$\sigma_r = \frac{p_e r_c^2}{r_b^2 - r_c^2}\left(1 - \frac{r_b^2}{r^2}\right) \tag{15-42}$$

$$\sigma = \frac{1}{2}(\sigma_\theta + \sigma_r) \tag{15-43}$$

where

$$p_e = \frac{2(1+b)(r_b^2 - r_c^2)}{(2+2b-\alpha b)(r_b^2 + r_c^2) + \alpha(2+b)(r_b^2 - r_c^2)}\sigma_t \tag{15-44}$$

15.5.2 Plastic Zone in the Elasto–Plastic Range

In the plastic zone, the material is assumed perfectly elastic–plastic, so the stress state satisfies Eq. (15-5) or Eq. (15-5′) when the unified strength theory is adopted as the yield criterion. According to the stress state condition Eq. (10-5), the first equation of the unified strength theory Eq. (15-5) must be used. Hence, we have

$$\frac{2+(2-\alpha)b}{2(1+b)}\sigma_\theta - \frac{\alpha(2+b)}{2(1+b)}\sigma_r = \sigma_t \tag{15-45}$$

Substituting into the equilibrium equation Eq. (15-12) gives

$$\frac{d\sigma_r}{dr} + \frac{2(1+b)(1-\alpha)}{2+(2-\alpha)b}\frac{\sigma_r}{r} - \frac{2(1+b)}{2+(2-\alpha)b}\frac{\sigma_t}{r} = 0 \tag{15-46}$$

The general solution of this differential equation is

$$\sigma_r = \frac{c}{r\left[\dfrac{2(1+b)(1-\alpha)}{2+(2-\alpha)b}\right]} + \frac{\sigma_t}{1-\alpha} \tag{15-47}$$

The integral constant can be determined by the boundary condition $r=r_a$, $\sigma_r=-p$.

$$-p = \frac{c}{r_a\left[\dfrac{2(1+b)(1-\alpha)}{2+(2-\alpha)b}\right]} + \frac{\sigma_t}{1-\alpha}$$

Therefore

$$c = \left(-p - \frac{\sigma_t}{1-\alpha}\right)A^{\frac{2(1+b)(1-\alpha)}{2+(2-\alpha)b}} \tag{15-48}$$

The stress distribution in the plastic region $(r_a \le r \le r_c)$ is

$$\sigma_r = -\left(p + \frac{\sigma_t}{1-\alpha}\right)\left(\frac{r_a}{r}\right)^{\frac{2(1+b)(1-\alpha)}{2+(2-\alpha)b}} + \frac{\sigma_t}{1-\alpha} \tag{15-49}$$

$$\sigma_\theta = \frac{2(1+b)\sigma_t}{2+(2-\alpha)b} - \frac{\alpha(2+b)}{2+(2-\alpha)b}\left[\left(p+\frac{\sigma_t}{1-\alpha}\right)\left(\frac{r_a}{r}\right)^{\frac{2(1+b)(1-\alpha)}{2+(2-\alpha)b}} + \frac{\sigma_t}{1-\alpha}\right] \tag{15-50}$$

$$\sigma_z = \frac{1}{2}(\sigma_r + \sigma_\theta) \tag{15-51}$$

Equations (15-49)–(15-51) give the stresses of thick-walled cylinder at the plastic region. Note that no stress–strain relation was needed to obtain these stresses; the problem is therefore statically determinate.

15.5.3 Plastic Zone Radius in the Elasto–Plastic Range

The boundary pressure equation (15-44) of the elastic zone gives the pressure required to cause the plastic zone to reach a radius r_c, or alternatively, for a given internal pressure p. Equation (15-44) could be solved for the plastic zone radius r_c.

When plastic zone radius r_c increases from $r_c=r_a$ to $r_c=r_b$, the pressure steadily increases with the plastic zone radius r_c.

The stress continuity of radial stress σ_r across $r=r_c$ requires that

$$\sigma_{r=r_c}\ \text{(elastic zone)} = \sigma_{r=r_c}\ \text{(plastic zone)}$$

Substituting the radial stress equation (15-42) in the elastic zone and the radial stress equation (15-47) in the plastic zone into the above continuous condition, the relation of pressure p with plastic zone radius is obtained as follows:

$$p = \left(\frac{r_c}{r_a}\right)^{\frac{2(1+b)(1-\alpha)}{2+(2-\alpha)b}}\left[\frac{2(1+b)(r_b^2-r_c^2)}{(2+2b-\alpha b)(r_b^2+r_c^2)+\alpha(2+b)(r_b^2-r_c^2)} + \frac{1}{1-\alpha}\right]\sigma_t - \frac{1}{1-\alpha}\sigma_t \tag{15-52}$$

If ratio of the external radius r_b to internal radius r_a is $K = r_b/r_a = 2$. An example of the relation of pressure versus the plastic zone radius is illustrated in Fig. 15.8.

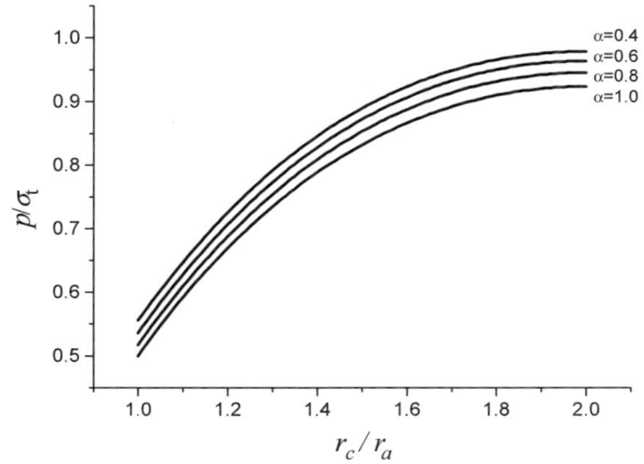

Fig. 15.8 Plastic zone radius versus applied pressure for different α (K=2, b=1.0)

15.5.4 Plastic Limit Pressure

1. Plastic Limit Pressure for SD Materials

When r_c becomes equal to r_b, the thick-walled tube is completely plastic. The plastic limit pressure for a thick-walled cylinder is, therefore, obtained as

$$p_p = \frac{\sigma_t}{1-\alpha}\left\{K^{\frac{2(1+b)(1-\alpha)}{2+2b-\alpha b}} - 1\right\} \tag{15-53}$$

This solution for thick-walled cylinders for closed-end or plane strain with incompressible materials is a general solution adopted for most materials. It may be referred as the unified solution of plastic limit pressure for thick-walled cylinders.

When b=0, the plastic limit pressure in terms of the Mohr-Coulomb theory is deduced from the unified solution

$$p_p = \frac{\sigma_t}{1-\alpha}[K^{(1-\alpha)} - 1] \tag{15-54}$$

When b=1, the unified solution became the plastic limit pressure in terms of the twin-shear strength theory as follows

$$p_p = \frac{\sigma_t}{1-\alpha}\left[K^{\frac{4(1-\alpha)}{4-\alpha}} - 1\right] \tag{15-55}$$

2. Plastic Limit Pressure for $\alpha = 1$ Materials

The unified solution that obeys the unified yield criterion can be obtained from the unified solution in terms of the unified strength theory by taking the limit $\alpha \to 1$. The plastic limit pressure of a thick-walled cylinder using the unified yield criterion is expressed as follows:

$$p_P = \frac{2(1+b)}{2+b} \ln K \tag{15-56}$$

If b=0, the limit pressure in terms of the Tresca yield criterion is obtained as

$$p_p = \sigma_t \ln K \tag{15-57}$$

This special case of the unified solution is identical to the classical solution.

If b=1/2, the plastic limit pressure in terms of the linear Huber-von Mises yield criterion is obtained as follows:

$$p_P = \frac{6}{5} \sigma_t \ln K \tag{15-58}$$

If b=1, the plastic limit pressure in terms of the twin-shear yield criterion is obtained as

$$p_p = \frac{4}{3} \sigma_t \ln K \tag{15-59}$$

Equation (15-59) is identical with the plastic limit pressure given by using the twin-shear strength theory.

A series of solutions can be deduced from the unified solution in terms of the unified strength theory. The relation of the plastic limit pressure to the strength theory parameter b and the thickness of a cylinder are shown in Figs. 15.9 and 15.10. The marked effect of failure criterion on structural analysis can be observed.

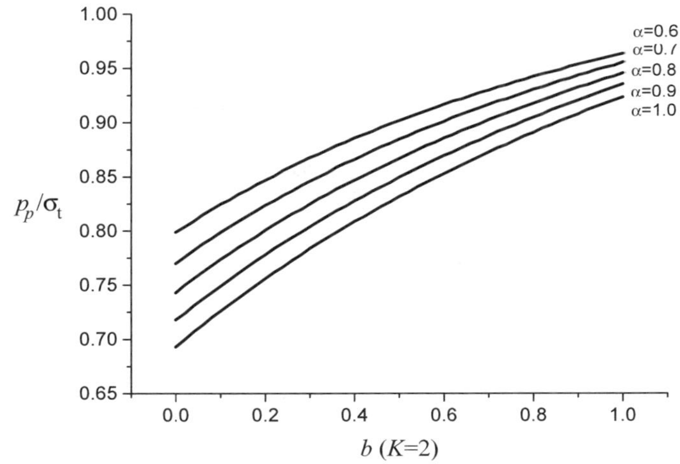

Fig. 15.9 Relation of plastic limit pressure to the parameter b

Figures 15.10–10.12 and Tables 15.1–15.2 show the elastic and plastic limit pressures in terms of different yield criteria. It can be seen that the elastic limit pressure in terms of the unified strength theory increases monotonically when b varies from 0 to 1 for all the three end conditions. The elastic limit pressure in terms of the Tresca criterion equals those obtained by using the unified strength theory when b=0 and α =1. The elastic limit pressure in terms of the Huber-von Mises criterion equals those obtained using the unified strength theory when $b \cong 0.4$. Therefore, it can be concluded that the Huber-von Mises and the Tresca criteria are encompassed in the unified strength theory with regard to the elastic limit pressure.

The maximum elastic limit pressure in terms of the unified strength theory is obtained when b=1. It is 33.4% and 15.5% higher than those obtained using the Tresca yield criterion and the Huber-von Mises yield criterion, respectively. It was also found that the higher values obtained from the unified strength theory were insensitive to the variations in the inner-to-outer-radius ratio in different end conditions. In the case of plastic limit pressure, almost the same conclusions can be drawn.

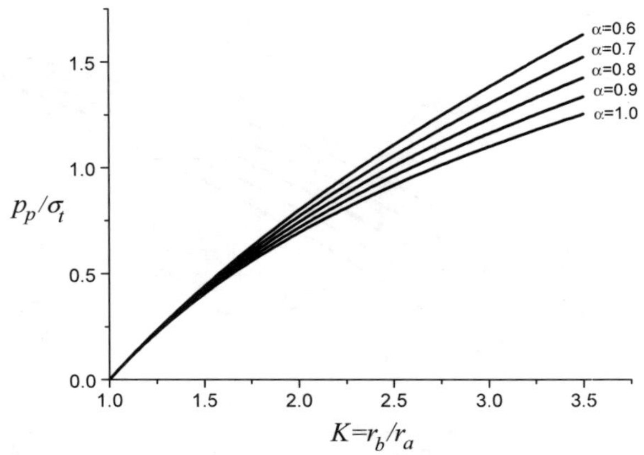

Fig. 15.10 Relation of plastic limit pressure to the thickness of tube (b=0.0)

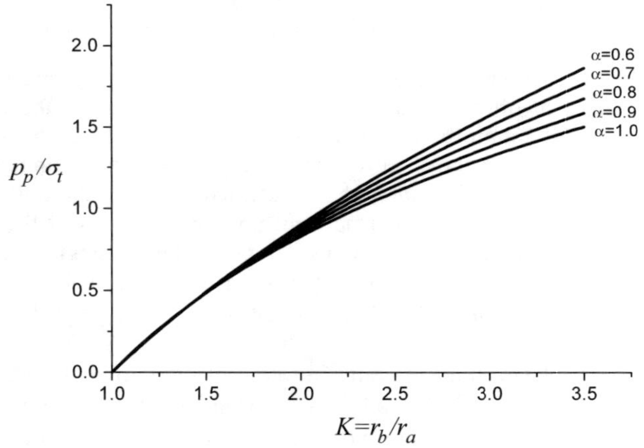

Fig. 15.11 Relation of plastic limit pressure to the thickness of tube (b=0.5)

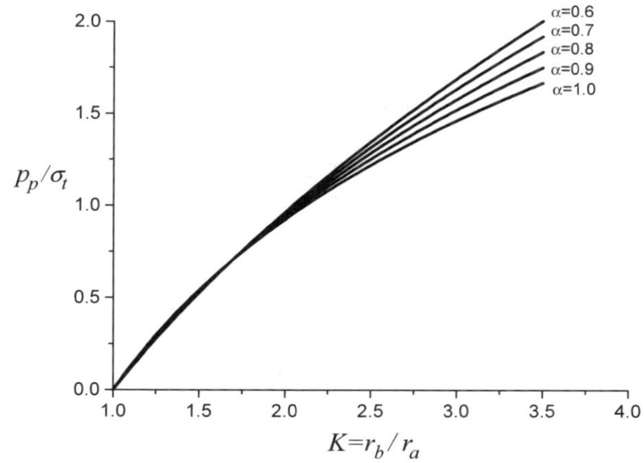

Fig. 15.12 Relation of plastic limit pressure to the thickness of tube (b=1.0)

The elastic limit pressure and plastic limit pressure, which are the two important parameters in the design of cylinders, were derived using the unified strength theory. It was found that the ratio of elastic-plastic limit pressures derived from different criteria could differ from one to another by as much as 33.4%. If the unified strength theory is used instead of the Tresca or the Huber-von Mises criterion in design, it could lead to substantial savings in the amount of material required.

15.6 Statical Shakedown Theorem (Melan theorem)

Many engineering structures or components are subjected to mechanical or other loads varying with time, and in many cases only the domain within which these loads change can be estimated, while the detailed loading path is unknown or very complex due to various reasons. The shakedown is a necessary condition for safety of such kinds of structures. The shakedown theory has constituted a well-established branch of plasticity theory.

Shakedown theory of structures provides a powerful tool to solve such kind of problem. A structure in a nonshakedown or inadaptation condition under varying loads may fail by one of two failure modes, namely alternating plasticity or incremental plastic collapse (see Kachanov 1971, Martin 1975). The structure will be shakedown if neither of the failure modes occurs.

The methods of shakedown analysis came into existence in the 1930s. Melan (1936) and Koiter (1956) proved the two crucial shakedown theorems, namely the static shakedown theorem (the Melan's or the lower bound shakedown theorem) and the dynamic shakedown theorem (the Koiter's or the upper bound shakedown theorem), which constitute the backbone in shakedown theory of elastoplastic structures. Accordingly, the numerous methods of shakedown analysis developed thereafter can be divided into two classes, i.e. the static and the dynamic shakedown analysis methods.

In recent years, the shakedown analysis of elasto-plastic structure has increasingly gained importance in engineering due to the requirements of modern technologies such as nuclear power plants, the chemical industry, aeronautical and astronautical technologies as well as the electrical and electronical elements. The shakedown theory has been applied with success in a number of engineering problems such as construction of nuclear reactors, highways and railways, and employed as one of the tools of structural design and safety assessment in some design standards, rules and regulations (see König and Maier 1987; Feng and Liu 1997, Weichert and Maier 2000). The shakedown theory has rapidly been developed (see Cocchetti and Maier 1998; Maier 2001).

We denote σ^*_{ij} , ε^*_{ij} the instantaneous values of the stresses and strains in the corresponding perfectly elastic body, and denote $\sigma_{ij}, \varepsilon_{ij}$ the instantaneous values of the stresses and strains in the actual elasto-plastic state. Let σ^0_{ij} , ε_{ij}^0 be the residual stresses and strains in the body, given by the difference

$$\sigma_{ij}^0 = \sigma_{ij} - \sigma_{ij}^* , \tag{15-60}$$

$$\varepsilon_{ij}^0 = \varepsilon_{ij} - \varepsilon_{ij}^* , \tag{15-60}$$

and let ε_{ij}^{0e} be the elastic strains corresponding to the residual stresses.

All the above stresses and strains are slowly-changing functions of time because the loads are variable. Note also that the strains ε^*_{ij} , ε_{ij} are kinematieally possible, i.e. they satisfy the compatibility conditions, and the corresponding displacements satisfy prescribed kinematic boundary conditions.

Melan shakedown theorem

Melan shakedown theorem can be described as follows (see Kachanov 1971, Martin 1975):

Shakedown cannot occur if there does not exist any time-independent field of residual stresses $\bar{\sigma}_{ij}$ such that the sum ($\bar{\sigma}_{ij} + \sigma_{ij}^e$) is admissible (necessary condition).

This means that this stress field ($\bar{\sigma}_{ij} + \sigma_{ij}^e$) is safe if no arbitrary load-variation in the prescribed limits causes the yield surface $f(\sigma_{ij}^e + \bar{\sigma}_{ij})$ to be reached, i.e.

$$f(\sigma_{ij}^e + \bar{\sigma}_{ij}) < 0 \tag{15-60}$$

The necessary condition is obvious: if there is no distribution of residual stresses for which $f(\sigma^e_{ij}+\overline{\sigma}_{ij})<0$, then by definition shakedown cannot occur.

Inverse, shakedown occurs if it is possible to find a field of fictitious residual stresses $\overline{\sigma}_{ij}$, independent of time, such that for any variations of loads within the prescribed limits the sum of this field with the stress field σ^e_{ij} in a perfectly elastic body is safe (sufficient condition).

The residual stresses field is expediently chosen such that the region of admissible load variation is greatest. The Melan's theorem given a low bound of limit load.

As an example, the shakedown load of pressure vessel based on the unified strength theory is studied in the next section.

15.7 Unified Solution of Shakedown Pressure for Thick-Walled Cylinder

Correct prediction of the load-bearing capacity of structures is a crucial task in analysis and design of engineering structures. The plastic limit load of structures determined by the limit analysis or the slip-line analysis is often used as an index of the load-bearing capacity of the structure subject to a monotonic loading. When the loading is a repeated loading, however, the structures fail at the load lower than the plastic limit load. This is because the failure of structures subject to the repeated loading is by the gradual deterioration due to the alternating plasticity or the incremental plasticity, not by the sudden collapse like that of structures subject to the monotonic loading. If the load does not exceed the critical value, the structure subject to the repeated loading may behave plastically at first and then elastically, and thus no further plastic deformation takes place in the structure. When this happens we say the structure shakes down to the repeated loading. If the load exceed the critical value, the structure does not shakedown, and fails due to the alternating plasticity or the incremental plasticity. This critical load level, below which the structure shakes down and above which it fails, is called shakedown load. Hence, the shakedown load is an important parameter, and we often regard it as the load-bearing capacity of the structure subject to the repeated loading.

Long thick-walled cylinders are commonly used as gun barrels and pressure vessels in engineering. They are often subjected to the repeated internal pressure. Hence, it is necessary to conduct the shakedown analysis in order to determine the shakedown load of the cylinder. The shakedown problem of cylinder is readily solvable in context of the classical plasticity, and the analytical solution can be found in some literatures. However, it should be noted that the above solution is obtained based on the Tresca yield criterion, and the analytical solution based on the Huber-von Mises criterion is not readily obtained in most cases due to the nonlinear expression of the criterion. The Tresca yield criterion only involves the two principal stresses, i.e. the maximum principal stress and the minimum principal stress, and ignores the compressive-tensile strength difference effect of materials (denoted by SD effect for short hereafter). Thus, this classical solution can only be applied to the cylinder made of materials without both the SD and the intermediate principal stress effect. In practice, however, the compressive strength is observed not to be identical to the tensile strength and the intermediate principal stress effect to have significant influences for some widely used metals such as high-strength steels, high-strength aluminum alloys and special steels etc.(Chait 1972; Drucker 1973; Casey and Sullivan 1985). These properties may invalidate the result by the classical plasticity. Hence, it is of great importance to develop a new approach in which the SD effect and intermediate principal stress effect are considered.

In view of this fact, an elastoplastic model incorporating the unified strength theory is suggested for shakedown analysis of thick-walled cylinder. A closed-form unified solution of shakedown load for cylinders was given by Xu and Yu in 2004. This solution involves the two parameters of the unified strength theory m and b, they can reflect both the intermediate principal stress and the SD effect in a quantitative manner. By changing the values of m and b, the solution is applicable to many kinds of cylinders made of materials with the SD effect and the intermediate principal stress effect. The effects of SD and intermediate principal stress on the shakedown load of the thick-walled cylinder can be evaluated.

Consider a plane strain thick-walled cylinder under uniform internal pressure p, whose internal and external radii are r_i and r_e, respectively. For sake of simplicity, we assume that the material is impressible and elastic-perfectly plastic and its Bauschinger effect is too small to be regarded. If the pressure p is moderate, the thick-walled cylinder is in elastic state. Then, the stress field of the cylinder is given by the Lame Solutions as follows

$$\sigma_r = \frac{r_i^2 p}{r_e^2 - r_i^2}\left(1 - \frac{r_e^2}{r^2}\right) \tag{15-61a}$$

$$\sigma_\theta = \frac{r_i^2 p}{r_e^2 - r_i^2}\left(1 + \frac{r_e^2}{r^2}\right) \tag{15-61b}$$

$$\sigma_z = \frac{r_i^2 p}{r_e^2 - r_i^2} \tag{15-61c}$$

Obviously, σ_θ is the major principal stress, σ_z the intermediate principal stress and σ_r the minor principal stress, and they satisfy

$$\sigma_z \le \frac{m\sigma_\theta + \sigma_r}{m+1}.$$

Hence, the yield condition satisfied by σ_θ, σ_z and σ_r is expressed by

$$\sigma_\theta - \frac{1}{m(1+b)}(b\sigma_z + \sigma_r) = \sigma_t \tag{15-62}$$

It is clear that the magnitude of σ_t takes the maximum value on the internal wall of the cylinder. Hence, yielding will start from the internal wall of the cylinder when the internal pressure attains

$$p_e = \frac{m(1+b)(r_e^2 - r_i^2)}{(m+1+mb)r_e^2 + (m-1)(1+b)r_i^2}\sigma_t \tag{15-63}$$

where p_e is the elastic limit pressure of the cylinder.

When the internal pressure exceeds p_e, a plastic zone spreads out from the inner radius. If the plastic zone reaches the radius r_p, the cylinder will be divided into two parts: the plastic zone ($r_i \le r \le r_p$) and the elastic zone ($r_p \le r \le r_e$). Using the Lame solution, the boundary condition $\sigma_r = 0$ at $r = r_e$, and the fact that the material at $r = r_p$ is stressed to the yield point, we obtain the stress components in the elastic zone as

$$\sigma_r = \frac{r_p^2 p_p}{r_e^2 - r_p^2}\left(1 - \frac{r_e^2}{r^2}\right) \tag{15-64a}$$

$$\sigma_\theta = \frac{r_p^2 p_p}{r_e^2 - r_p^2}\left(1 + \frac{r_e^2}{r^2}\right) \tag{15-64b}$$

$$\sigma_z = \frac{r_p^2 p_p}{r_e^2 - r_p^2} \tag{15-64c}$$

where P_p is the associated radial pressure acting on the elastoplastic interface under the internal pressure p. It equals

$$p_p = \frac{m(1+b)(r_e^2 - r_p^2)}{(m+1+mb)r_e^2 + (m-1)(1+b)r_p^2}\sigma_t$$

According to the equilibrium equation

$$\frac{d\sigma_r}{dr} + \frac{\sigma_r - \sigma_\theta}{r} = 0\ , \tag{15-65}$$

yield condition, the boundary condition $\sigma_r = p$ at $r = r_i$ and the impressible condition of materials, we obtain the stress components in the plastic zone

$$\sigma_r = -\left(p + \frac{m\sigma_t}{m-1}\right)\left(\frac{r_i}{r}\right)^{\frac{2(m-1)(1+b)}{2m+2mb-b}} + \frac{m}{m-1}\sigma_t \tag{15-66a}$$

$$\sigma_\theta = -\frac{(2+b)}{2m+2mb-b}\left(p+\frac{m\sigma_t}{m-1}\right)\left(\frac{r_i}{r}\right)^{\frac{2(1+b)(m-1)}{2m+2mb-b}} + \frac{m}{m-1}\sigma_t \quad (15\text{-}66b)$$

$$\sigma_z = -\frac{1+m+mb}{2m+2mb-b}\left(p+\frac{m\sigma_t}{m-1}\right)\left(\frac{r_i}{r}\right)^{\frac{2(1+b)(m-1)}{2m+2mb-b}} + \frac{m}{m-1}\sigma_t \quad (12\text{-}66c)$$

Using the continuity of σ_r across $r = r_p$, we have

$$p = \frac{m\sigma_t}{m-1}\left[\frac{(2m+2mb-b)r_e^2}{(m+1+mb)r_e^2+(m-1)(1+b)r_p^2}\left(r_p/r_i\right)^{\frac{2(m-1)(1+b)}{2m+2mb-b}} - 1\right] \quad (15\text{-}67)$$

This provides the relationship between the internal pressure p and the radius of the plastic zone r_p.

If we increase the pressure p, the plastic zone will further expand and the elastic-plastic interface will move gradually to the external wall of the cylinder. Setting $r_p = r_e$ in the above expression, the internal pressure becomes

$$p_s = \frac{m\sigma_t}{m-1}\left[\left(r_e/r_i\right)^{\frac{2(m-1)(1+b)}{2m+2mb-b}} - 1\right] \quad (15\text{-}68)$$

which is the plastic limit pressure of the cylinder.

When $p_e < p < p_s$, the cylinder is rendered partially plastic. In that case, if the cylinder is unloaded, it will be left with residual stress. When p is not large, the unloading process is purely elastic, and the residual stress can be obtained by superposition of the elastic unloading stress and the elastic-plastic loading stress. The following expressions are the residual stresses in the zone adjacent to the internal wall of the cylinder ($r_i \le r \le r_p$)

$$\sigma_r^r = -\left(p+\frac{m\sigma_t}{m-1}\right)\left(\frac{r_i}{r}\right)^{\frac{2(m-1)(1+b)}{2m+2mb-b}} + \frac{m}{m-1}\sigma_t - \frac{r_i^2 p}{r_e^2-r_i^2}\left(1-\frac{r_e^2}{r^2}\right) \quad (12\text{-}69a)$$

$$\sigma_\theta^r = -\frac{(2+b)}{2m+2mb-b}\left(p+\frac{m\sigma_t}{m-1}\right)\left(\frac{r_i}{r}\right)^{\frac{2(m-1)(1+b)}{2m+2mb-b}} + \frac{m}{m-1}\sigma_t - \frac{r_i^2 p}{r_e^2-r_i^2}\left(1+\frac{r_e^2}{r^2}\right) \quad (15\text{-}69b)$$

$$\sigma_z^r = -\frac{1+m+mb}{2m+2mb-b}\left(p+\frac{m\sigma_t}{m-1}\right)\left(\frac{r_i}{r}\right)^{\frac{2(m-1)(1+b)}{2m+2mb-b}} + \frac{m}{m-1}\sigma_t - \frac{r_i^2 p}{r_e^2-r_i^2} \quad (15\text{-}69c)$$

Setting $r = r_i$ in above expressions, we obtain the residual stresses on internal wall of the cylinder as follows

$$\sigma_r^r = 0 \tag{15-70a}$$

$$\sigma_\theta^r = -\left[\frac{(2+2b)}{2m+2mb-b} + \frac{r_e^2 + r_i^2}{r_e^2 - r_i^2}\right] p + \frac{(2+2b)m}{(2m+2mb-b)}\sigma_t \tag{15-70b}$$

$$\sigma_z^r = -\frac{1}{2}\cdot\left[\frac{(2+2b)}{2m+2mb-b} + \frac{r_e^2 + r_i^2}{r_e^2 - r_i^2}\right] p + \frac{(1+b)m}{(2m+2mb-b)}\sigma_t \tag{15-70c}$$

Obviously, σ_r^r, σ_z^r and σ_θ^r on the internal wall are in turn the major principal stress, the intermediate principal stress and the minor principal stress, and $\sigma_z^r \geq \dfrac{m\sigma_r^r + \sigma_\theta^r}{m+1}$. Hence, the yield condition is satisfied by σ_r^r, σ_z^r and σ_θ^r on the internal wall is

$$\frac{1}{1+b}(\sigma_r + b\sigma_z) - \frac{\sigma_\theta}{m} = \sigma_t \tag{15-71}$$

From Eqs. (15-83) and (15-84), we see that the internal wall of the cylinder yields when the internal pressure reaches

$$p_{\max} = \frac{2m(m+1)(1+b)(b+2)/(2m+2mb-b)/(2-mb+2b)}{(2+b)/(2m+2mb-b)+(r_e^2+r_i^2)/(r_e^2-r_i^2)}\sigma_t \tag{15-72}$$

Evidently, if $p < p_{\max}$ (p should be less than the plastic limit pressure p_s here), a secondary yielding does not take place at the internal wall of the unloaded cylinder. Further, we can demonstrate that the residual stress induced by the cycle of loading-unloading will not produce any new plastic deformation in the whole section of the cylinder when this condition is satisfied. Hence, the shakedown condition for a thick-walled cylinder under repeated loading and unloading is that the internal pressure p is less than the critical value $p_{\text{shakedown}}$ or p_{plastic}.

$$p_{\max,shakedown} = \min\left\{\frac{2m(m+1)(1+b)(2+b)/(2m+2mb-b)/(2-mb+2b)}{(2+b)/(2m+2mb-b)+(r_e^2+r_i^2)/(r_e^2-r_i^2)}\sigma_t\right\} \tag{15-73a}$$

$$p_{\max,plastic} = \min\left\{\frac{m\sigma_t}{m-1}\left[(r_e/r_i)^{\frac{2(m-1)(1+b)}{2m+2mb-b}} - 1\right]\right\} \tag{15-73b}$$

which is often called the shakedown load of the thick-walled cylinder. Setting m=1 and b=0 in the above expression, we obtain the shakedown load of the thick-walled cylinder in the form

$$p_{max} = \min\left\{\sigma_t\,(1 - r_i^2/r_e^2),\ \sigma_t \ln(r_e/r_i)\right\} \tag{15-74}$$

This is in agreement with the expression of the shakedown load of cylinder from the classical plasticity.

Clearly, the shakedown load given by the Eq. (15-74) is correlated with the compressive-tensile strength ratio m and the yield criterion coefficient b. Hence, the present approach has the capability to reflect the effects of SD and intermediate principal stress on the shakedown load of the cylinder in a quantitative manner. This is not the case for the classical solution.

15.8 Effects of Yield Function on the Plastic Limit Pressure and Shakedown Pressure of Thick-Walled Cylinders

Various results for plastic limit pressures of thick-walled cylinders for closed ends in terms of different yield criteria are summarized in Table 15.3. It is seen that the effect of yield function is obvious.

In order to investigate the effects of SD and intermediate principal stress on the shakedown load of thick-walled cylinder, analytical results are calculated from the derived closed-form solution. They are shown in Figs. 15.13 and 15.14, in which the abscissas denote the wall ratio of the cylinder r_e/r_i and the ordinates the shakedown load p_{max}/σ_t.

Figure 15.13 shows the effect of intermediate principal stress of materials on the shakedown load. The results in Fig.15.13 (m=1) are for non-SD materials.

Table 15.3. Summary of plastic limit pressures for closed-end conditions

	Materials	Plastic limit pressure	Failure criterion used
1	SD material $\alpha \neq 1$	$p_p = \frac{\sigma_t}{1-\alpha}\left(K^{\frac{2(1+b)(1-\alpha)}{2+2b-\alpha b}} - 1\right)$	Unified strength theory
2	SD material $\alpha \neq 1$	$p_p = \frac{\sigma_t}{1-\alpha}\left(K^{(1-\alpha)} - 1\right)$	Unified strength theory b=0, Mohr–Coulomb
3	SD material $\alpha \neq 1$	$p_p = \frac{\sigma_t}{1-\alpha}\left(K^{\frac{4(1-\alpha)}{4-\alpha}} - 1\right)$	Unified strength theory b=1, twin-shear theory
4	$\alpha = 1$ materials	$p_p = \frac{2(1+b)}{2+b}\ln K$	Unified yield criterion, $\alpha = 1$
5	$\alpha = 1$ materials	$p_p = \sigma_t \ln K$	Tresca yield criterion $\alpha = 1,\ b = 0$
6	$\alpha = 1$ materials	$p_p = \frac{6}{5}\sigma_t \ln K$	Unified yield criterion when $b = 1/2$
7	$\alpha = 1$ materials	$p_p = \frac{4}{3}\sigma_t \ln K$	Twin-shear yield criterion $\alpha = 1,\ b = 1$

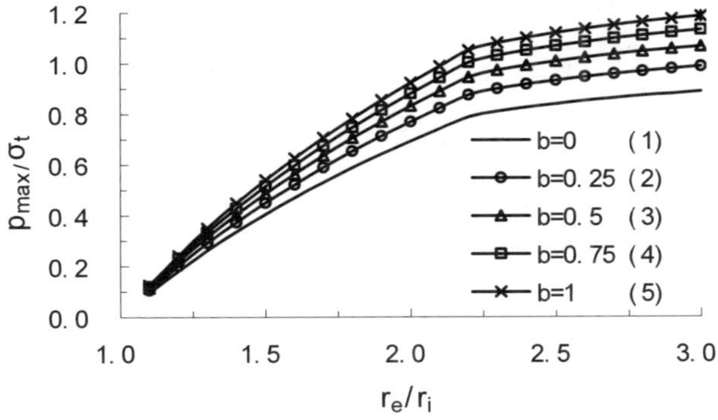

Fig. 15.13 The shakedown load for different values of parameter b (m=1)

The curve (1) at the bottom in Fig. 15.13 (b=0 and m=1) is suitable for materials without both the SD and the intermediate principal stress effect, which is just the result of the classical solution. The special case of the present solution at m=1 and b=0.5 is shown by curve (3) at the middle in Fig. 15.13; it is a good approximation to the result of the Huber-von-Mises criterion. The curve (5) at the top in Fig. 15.13 (b=1 and m=1) is the result of the twin-shear stress yield criterion.

Figure 15.14 (m=1.1) and Fig.15.15 (m=1.2) are the shakedown load for materials with SD effect. We can see from these figures that the shakedown load is correlated with the yield criterion parameter b, i.e. with the effect of intermediate principal stress of materials, and the larger the coefficient b, the larger the shakedown load p_{max}. Consequently, for a given compressive-tensile strength ratio m, the case of b=0 (at the bottom in Fig.15.14 and 15.15, which corresponds to the Tresca criterion or the Mohr-Coulomb criterion) gives the smallest value of p_{max}, while that of b=1 (at the top in Fig.15.14 and 15.15, which corresponds to the twin-shear stress yield criterion or the generalized twin shear criterion) gives the largest value. Therefore, the shakedown load of the cylinder is underestimated when the effect of the intermediate principal stress of materials is neglected.

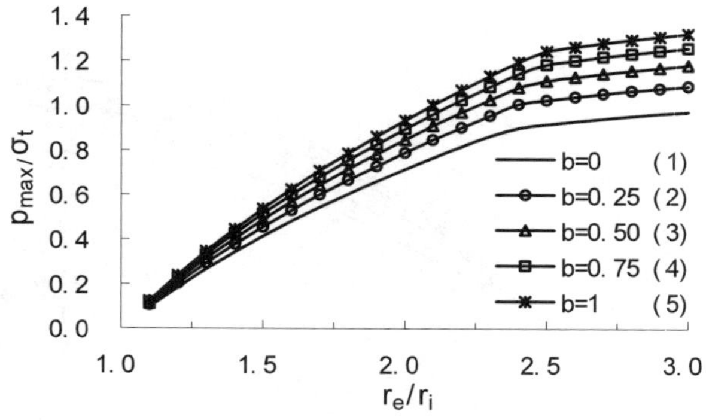

Fig. 15.14 The shakedown load for different values of parameter b (m=1.1)

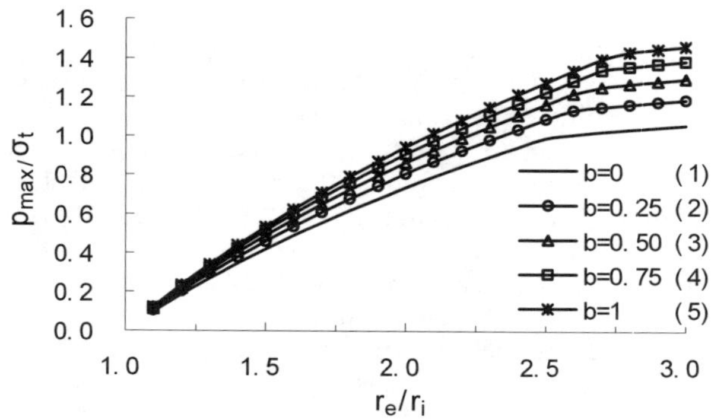

Fig. 15.15 Shakedown load for different values of parameter b (m=1.2)

Figure 15.16 shows the effect of SD of materials on the shakedown load of the cylinder. The results with respect to b=0 are shown in Fig. 15.16, which can also be obtained with the Mohr-Coulomb criterion, and are suitable for materials without the intermediate principal stress effect. The curve (1) in Fig. 15.16 (b=0 and m=1) is the result of the classical solution.

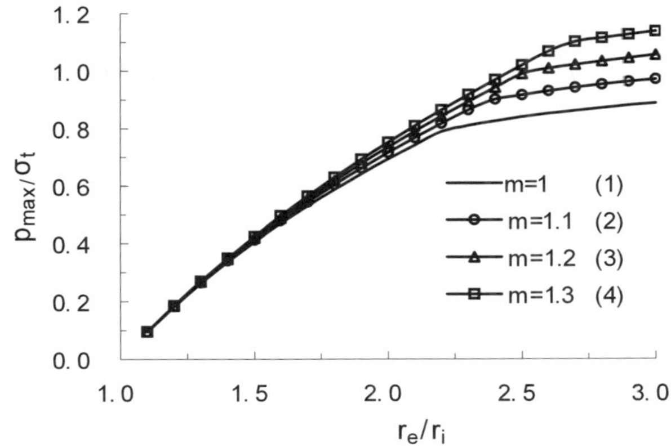

Fig. 15.16 Shakedown load for different values of parameter *m* (b =0)

Figure 15.17 (b=0.5) and Fig.15.18 (b=1, i.e. the generalized twin shear criterion) are suitable for materials with the intermediate principal stress effect.

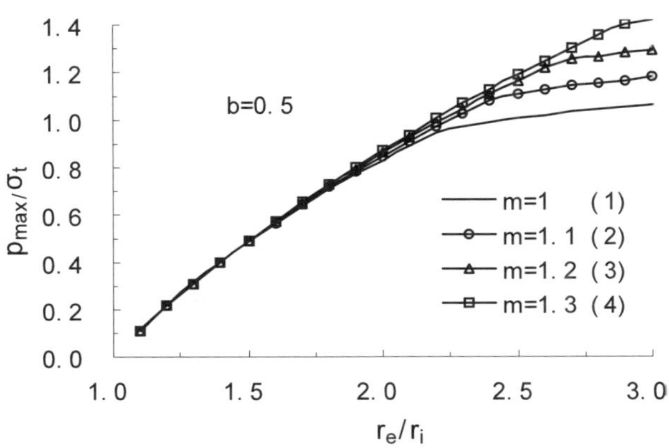

Fig. 15.17 Shakedown load for different values of parameter *m* (b =0.5)

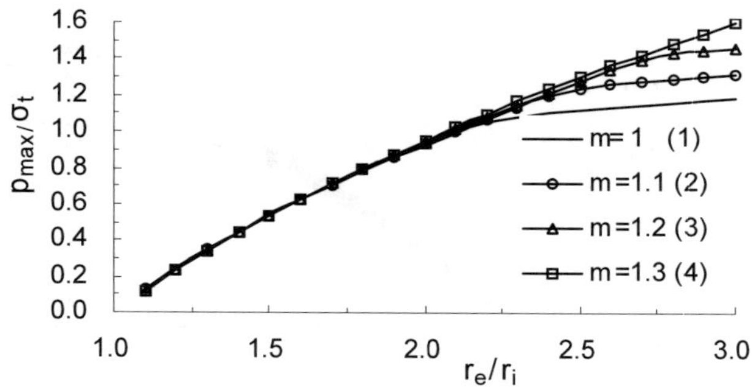

Fig. 15.18 The shakedown load for different values of parameter *m* (*b*=1)

It is seen from these figures that the shakedown load is dependent on the compressive-tensile strength ratio *m*, i.e., dependent on the SD effect of the materials, and the shakedown load will increase as the parameter *m* increases. Consequently, the shakedown load of the cylinder is underestimated when the SD effect of materials is disregarded. It is also seen that the SD effect of materials on the shakedown load of the cylinder is insignificant when the wall ratio is small, whereas it is prominent when the wall ratio is big. Therefore, the SD effect of materials should be taken into account in shakedown analysis of the cylinder especially for the bigger wall ratio of the cylinder.

15.11 Connection between Shakedown Theorem and Limit Load Theorem

Remark on the connection between shakedown theorems and limit load theorems, Koiter has drawn attention to the fact that the limit load theorems are a consequence of the shakedown theorems if it is assumed that the prescribed limits of load variation coincide.

On the basis of the unified strength theory, a shakedown analysis of a thick-walled cylinder under internal pressure was carried out, and the unified analytical solution of shakedown load for a cylinder was derived by Xu and Yu (2004, 2005). This solution not only includes the existing classical solution as its special case but gives a series of new results.

It is noted that this solution consist of two parts: the limit pressure and shakedown pressure

$$p_{\max,shakedown} = \min\left\{\frac{2m(m+1)(1+b)(2+b)/(2m+2mb-b)/(2-mb+2b)}{(2+b)/(2m+2mb-b)+(r_e^2+r_i^2)/(r_e^2-r_i^2)}\right\}\sigma_t \qquad (15\text{-}75a)$$

$$p_{\max, plastic} = \min \left\{ \frac{m\,\sigma_t}{m-1} \left[\left(r_e / r_i \right)^{\frac{2(m-1)(1+b)}{2m+2mb-b}} - 1 \right] \right\} \quad (15\text{-}75b)$$

Two curves will be crossed when the limit pressure equals the shakedown pressure, i.e.

$$\frac{m\sigma_t}{m-1}\left[\left(r_e / r_i \right)^{\frac{2(m-1)(1+b)}{2m+2mb-b}} - 1 \right] =$$

$$= \frac{2m(m+1)(1+b)(2+b)/(2m+2mb-b)/(2-mb+2b)}{(2+b)/(2m+2mb-b)+(r_e^2+r_i^2)/(r_e^2-r_i^2)}\sigma_t$$

The solution involves the two parameters m and b originally appeared in the unified strength theory, and can reflect both the SD and the effect of intermediate principal stress. With the varying of m and b, the present solution gives a series of values of shakedown load that are suitable for cylinders made of material with the SD or non-SD effect.

In order to investigate the effects of SD and intermediate principal stress on the shakedown load, the graphical alternatives to the analytical solution are given. They show that both the SD and the intermediate principal stress have influences on the shakedown load, and the more pronounced the two effects, the larger the shakedown load. Hence, for the cylinder made of material with the SD or non-SD effect, the classical solution underestimates the shakedown load, and it is thereby of great significance for the shakedown analysis to take the two effects into consideration.

Although the present approach can simulate materials with the effects of SD and intermediate principal stress, it should be noted that other important properties such as the Bauschinger effect, the strain-hardening effect have not been dealt with in the present solution(see: Feng XQ and Yu SW 1995).

Summary

Theoream of limit analysis and theorem of shakedown are used for the unified solutions of limit load and shakedown load of pressure vessel. The unified strength theory is used to obtain the unified solutions of plastic limit and shakedown analysis of pressure vessel. These results can be suitable for a wide range of materials and engineering.

In these solutions, the SD effects (effect of compressive-tensile strength difference) and effect of intermediate principal stress on the plastic limit loads and shakedown loads of thick-walled cylinder under uniform internal pressure is presented. The unified analytical solutions of limit loads and shakedown loads for cylinder are derived. With changing of the two parameters α and b (or m and b), a series of values of limit loads and shakedown loads can be obtained from the current solution, which include both the result from the classical plasticity and a series of new results. These solutions are suitable for materials with the SD effect and the intermediate principal stress effect.

Finally, the graphical alternatives to the analytical solution are presented to examine the effects of strength difference and intermediate principal stress on the limit loads and the shakedown loads. They show that the limit loads and the shakedown loads are dependent on both the strength difference in tension and compression of material and the effect of intermediate principal stress. The limit loads and shakedown loads are underestimated if these two effects are neglected. The unified strength theory gives us a basic theory for using in strength design of engineering structures. It also gives a method to increased the admissible loads or decreased cross-sections and the weight of structure. This results in a reduction in material consumption, energy and a reduction in environmental pollution and the cost of structures.

Problems

Problem 15.1

Compare the solutions of limit analysis and shakedown analysis.

Problem 15.2

Determine the shakedown load of a pressure cylinder by using the Tresca yield criterion.

Problem 15.3

Determine the shakedown load of a pressure cylinder by using the Mohr-Coulomb strength theory

Problem 15.4

Determine the shakedown load of a pressure cylinder by using the twin-shear yield criterion

Problem 15.5

Determine the shakedown load of a pressure cylinder by using the twin-shear strength theory.

Problem 15.6

Determine the limit load of a cylinder under tension and inter- pressure by using the twin-shear strength theory

Problem 15.7

Determine the shakedown load of a cylinder under tension and inter-pressure by using the twin-shear strength theory

References and Bibliography

American Society of Mechanical Engineers (1995) Cases of ASME Boiler and Pressure Vessel Code, Case N-47. ASME, New York.

Baker JF and Heyman J (1969) *Plastic Design of Frames: Fundamatials.* Cambridge University Press. London.

Carvelli V, Cen Z, Liu,Y and Mater G (1999) Shakedown analysis of defective pressure vessels by a kinematic approach. *Arch. Appl. Mech.*, 69: 751–764.

Casey J and Sullivan TD (1985) Pressure dependency, strength-differential effect and plastic volume expansion in metals, *International Journal of Plasticity*, **1**, pp. 39–61.

Chait R (1972) Factors influencing the strength differential of high strength steels, *Metallurgical Transactions*, **3**, pp. 365–371.

Chakrbarty, J., 1987, *Theory of Plasticity*, McGraw-Hill Book Co., Inc., New York, N.Y.

Chen WF (1975) *Limit Analysis and Soil Plasticity*. Elsevier, Amsterdam.

Cocchetti G and Maier G (1998) Static shakedown theorems in piecewise linearized poroplasticity. *Archive of Applied Mechanics*, 68: 651–661.

Crossland B and Bones JA (1958) Behaviour of thick-walled steel cylinders subjected to internal pressure. *Proc. Inst. Mech. Engrg.* **172**, 777.

Drucker DC and Prager W (1952) Soil mechanics and plastic analysis for limit design. *Quart. Appl. Math.* **10**, 157–165.

Drucker DC, Prager W and Greenberg HJ (1952) Extended limit design theorems for continuous media. *Quart. Appl. Math.* **9**, 381–389.

Drucker DC (1973) Plasticity theory, strength differential (SD) phenomenon and volume expansion in metals and plastics, *Metallurgical Transactions*, **4**, pp. 667–673.

Fan SC, Yu MH and Yang SY (2001) On the unification of yield criteria, *Journal of Applied Mechanics*, **68**, pp. 341–343.

Feng JJ, Zhang JY, Zhang P and Han JF (2004). Plastic limit load analysis of thick-walled tube based on twin-shear unified strength theory. *Acta Mechanica Solid Sinica*, 25(2): 208–212 (in Chinese).

Feng X and Gross D (1999) A global/local shakedown analysis method of elastoplastic cracked structures.

Feng X and Gross D (1999) A global/local shakedown analysis method of elastoplastic cracked structures. *Engineering Fracture Mechanics*, 63:179-192.

Feng XQ and Liu XS (1997) On shakedown of three-dimensional elastoplastic strain-hardening structures. *Int. J. Plasticity*, 12: 1241–1256.

Feng XQ and Yu SW (1995) Damage and shakedown analysis of dtructures with strain-hardening. Int. J. Plasticity, 11:237–249.

Guowei M., Iwasaki S. , Miyamoto Y. and Deto H., Plastic Limit Analysis of Circular Plates with Respect to the Unified Yield Criterion, *Int. J. of Mechanical Science*,1998, 40(10): 963–976.

Gvozdew AA (1938) The determination of the value of the collapse load for statically indeterminate systems undergoing plastic deformation. *Proceedings of the conference on lastic deformations,* Akademiia Nauk SSSR, Moscow, pp 19–33. (Translated into English by Haythornthwaite RM (1960). *Int. J. Mech. Sci.* 1960, **1**, 322–355).

Hamilton R, Boyle JT, Shi and Mackenzie D (1996) A simple upper-bound method for calculating approximate shakedown loads. *J. Pressure Vessel Technol.* ASME, 120: 195–199.

Heyman J (1971) *Plastic Design of Frames: Applications.* Cambridge University Press, London.

Hill R (1951) On the state of stress in a plastic-rigid body at the yield point. *Phil. Mag.* **42**, 868–875.

Hodge PG Jr (1959) *Plastic Analysis of Structures*. McGraw-Hill, New York.

Horne MR (1979) *Plastic Theory of Structures*. Pergmon Press, Oxford.

Huang WB and Zeng GP (1989) Solving some plastic problems by using the Twin shear stress criterion. *Acta Mechanica Sinica* **21**(2), 249–256 (in Chinese, English abstract).

Ivey, H.J., 1961, Plastic stress-strain relations and yield surfaces for aluminium alloys, *J. Mech. Eng. Sci.*, **3**(1): 15–31.
Johnson, W. and Mellor, P.B., 1973, *Engineering Plasticity*, Van Nostrand Reinhold, London.
Kachanov, L.M., 1971, *Foundations of theTheory of Plasticity*, Northhollod Publication Co., Amsterdam, The Netherlands.
Kamenjarzh LA and Weichert D (1992) On kinematic upper bounds of the safety factor in shakedown theory. *Int. J. Plast.*, 8: 827–837.
Kameniarzh JA (1996) *Limit Analysis of Solids and Strnctures*. CRC Press, Boca Baton. FL.
Koiter WE (1956) A New General Theorem on Shakedown of Elastic-Plastic Struclures. *Proc. K. Ned. Akad. Wet.*, B59: 24–34,
Koiter WT (1960) General Theorems for Elastic Plastic Solids. *Progress in Solid Mechanics,* Vol. 1, J, N. Sneddon and R. Hill eds. North-Holland. Amsterdam, 165−221.
König JA (1987) *Shakedown of Elastic-Plastic Structures*. Elsiver, Amsterdam.
König JA and Maier G (1987) Shakedown anlysis of elastic-plastic structures: A review of recent developments. *Nucl. Eng. Design*, 66: 81–95.
Li JC and Zhang YQ (1998) Limit analysis of a wellbore based on the twin-shear strength theory. In: *Strength Theory: Applications, Developments and Prospects for the 21st Century.* Yu MH and Fan SC eds. Science Press, New York, Beijing, 1103-1108.
Li YM (1998) Elastoplastic limit analyse with a new yield criterion (twin-shear yield criterion). *J. Mech. Strength* **10**(3): 70–74 (in Chinese).
Liu XQ, Ni XH, Yan S et al (1998) Application of the twin shear strength theory in strength-calculation of gun barrels. In: *Strength Theory: Applications, Developments and Prospects for the 21st Century.* Yu MH and Fan SC eds. Science Press, New York, Beijing, 1039–1042.
Ma GW, Iwasaki S, Miyamoto Y and Deto H (1999c) Dynamic plastic behavior of circular plates using the Unified yield criterion, *Int. J. Solids & Structure,* 36(**3**):3257–3275.
Ma GW, Hao H and Miyamoto Y (2001) Limit angular velocity of rotating disc with unified yield criterion. *Int. J. Mech. Sci.* **43**, 1137–1153.
Maier G, Carvelli V and Coccbetti G (2000) On direct methnds for shakedown and limit analysis. *Eur. J. Mech. A/Solids* (Special lssue), 19, S79–S100.
Maier G (2001) On some issues in shakedown analysis. *J. of Appl. Mechanics*. 68:799–808.
Martin J.B (1975) *Plasticity: Fundamentals and General Results*. The MIT Press.
Mendelson A (1968) *Plasticity: Theory and Application*. MaCmillan, New York.
Mrazik A, Skaloud M and Tochacek M (1987) *Plastic Design of Steel Structures*. Ellis Horwood, New York.
Mroz Z, Weichert D and Dorosz S eds. (1995) *Inelastic Behavior of Structures Under Variable Loads*, Kluwer, Dordrecht. Fbe Netherlands.
Naghdi, P.M., Essenburg, F. and Koff, W., 1958, An experimental study of initial and subsequent yield surfaces in plasticity, *ASME, J. Appl. Mech.*, **25**(2): 201–209.
Neal BG (1956) *The Plastic Methods of Structural Analysis*. Wiley, New York (2nd edn. 1963).
Ni XH, Liu XQ, Liu YT and Wang XS (1998) Calculation of stable loads of strength-differential thick cylinders and spheres by the twin shear strength theory. *Strength Theory: Applications, Developments and Prospects for the 21st Century.* Yu MH and Fan SC eds. Science Press, New York, Beijing, 1043–1046.
Ponter ARS and Carter KF (1997) Shakedown state simulation techniques based on linear elastic solutions. *Contput. Methods Appl. Mech. Eng*., 140: 259–279.
Richmond, O. and Spitzig, W.A., 1980, Pressure dependence and dilatancy of plastic flow, *Theoretical and applied Mechanics, 15th, ICTAM,* 377–386.
Save MA and Massonnet CE (1972) *Plastic Analysis and Design of Plates, Shells and Disks.* North-Holland, Amsterdam.
Symonds PS and Neal BG (1951) Recent progress in the plastic methods of structural analysis. *J. Franklin Inst.* **252**, 383–407, 469–492.
Wang F (1998) Non-linear Analysis of RC Plate and Shell Using Unified Strength Theory", Ph.D. Thesis of Nanyang Technological University, Singapore.

Wang F and Fan SC (1998) Limit pressures of thick-walled tubes using different yield criteria. *Strength Theory: Applications, Developments and Prospects for the 21st Century.* Yu MH and Fan SC eds. Science Press, New York, Beijing, 1047–1052.

Wang YB, Yu MH, Li JC et al,(2002a), Unified plastic limit analysis of metal circular plates subjected to border uniformly distributed loading. *J. Mechanical Strength* **24**: 305–307 (in Chinese, English abstract).

Wang YB, Yu MH, Wei XYet al. (2002b) Unified plastic limit analyses of circular plates under uniform annulad. *Engineering Mechanics* **19**: 84–88 (in Chinese, English abstract).

Wei XY (2001) *Investigation of Long Rod Penetrating Target.* PhD. thesis, Xi'an Jiaotong Uni., Xi'an, China (in Chinese, English abstract).

Wei XY and Yu MH (2001) Unified plastic limit of clamped circular plate with strength differential effect in tension and compression. *Chinese Quart. Mechanics* **22**: 78–83 (in Chinese, English abstract).

Wei XY and Yu MH (2002b) Unified solutions for plastic limit of annular plate. *J. of Mechanical Strength* **24**: 140–143 (in Chinese, English abstract).

Weichert D and Maier G eds. (2000), *Nonelastic Analysis of Structures Under Variable Repeated Loads.* Kluwer, Dordrecht, The Netherlands.

Xu SQ and Yu MH (2004) Unified analytical solution to shakedown problem of thick-walled cylinder. *Chinese J of Mechanical Engineering,* 40(9):23–27 (in Chinese).

Xu SQ and Yu MH (2005) Shakedown analysis of thick cylinder based on the unified strength theory. *J of Pressure Vessel and Pipes*, to be published.

Yu MH (1961a) General behaviour of isotropic yield function. *Res. Report of Xi'an Jiaotong University.* Xi'an, China (in Chinese).

Yu MH (1961b) Plastic potential and flow rules associated singular yield criterion. *Res. Report of Xi'an Jiaotong University.* Xi'an, China (in Chinese).

Yu MH (1983) Twin shear stress yield criterion. *Int. J. of Mech. Sci.* **25**, 71–74.

Yu MH and He LN (1983) Non-Schmid effect and twin shear stress criterion of plastic deformation in crystals and polycrystalline metals. *Acta Metallurgica Sinica* **19**(5), B190–196 (in Chinese, English abstr.).

Yu MH, He LN and Song LY (1985) Twin shear stress theory and its generalization. *Scientia Sinica* (*Sciences in China*), English edn. Series A, **28**(11):1174–1183.

Yu Mao-hong and He Li-nan (1991) A new model and theory on yield and failure of materials under complex stress state, *Proceedings, Mechanical Behaviour of Materials-VI*, **3**, pp.851-856

Yu MH (1998) *Twin Shear Theory and Its Application.* Science Press, Beijing (in Chinese).

Yu MH (2002a) Advances in strength theories for materials under complex stress state in the 20th Century. *Applied Mechanics Reviews* ASME, **55**(3),:169–218.

Yu MH (2002b) *Concrete Strength Theory and Its Applications.* Higher Education Press, Beijing (in Chinese).

Yu Mao-Hong (2004) *Unified Strength Theory and Its Applications.* Berlin: Springer,.

Zhao DW (2004) *Mathematical Solutions for Forming Mechanics of Continum.* North-East University Press. Shenyang.

Zhao DW, Xu JZ, Yang H, Li XH and Wang GD (1998) Application of twin shear stress yield criterion in axisymmetric indentation of a semi-infinite medium. In: *Strength Theory: Applications,Developments and Prospects for the 21st Century.* Yu MH and Fan SC eds, Science Press, Beijing and New York, 1079-1084.

Zhao JH, Zhang YQ, Li JC (1999) The solutions of some plastic problems in plane strain by using the unified strength theory and the unified slip field theory. *Chinese J of Mechanical Engineering* **35**(6),61-66 (in Chinese).

Zhuang JH (1998) The analysis of limit inner pressure of thick-walled tube and sphere shell with different strength in tension and compression by theory of generalized twin-shear strength theory. *Mechanics and Practices*, 29(4): 39-41 (in Chinese).

Author index

Subject index